Combinatorics

NATO ADVANCED STUDY INSTITUTES SERIES

Proceedings of the Advanced Study Institute Programme, which aims at the dissemination of advanced knowledge and the formation of contacts among scientists from different countries

The series is published by an international board of publishers in conjunction with NATO Scientific Affairs Division

A	Life Sciences	Plenum Publishing Corporation
B	Physics	London and New York
C	Mathematical and Physical Sciences	D. Reidel Publishing Company Dordrecht and Boston
D	Behavioral and Social Sciences	Sijthoff International Publishing Company Leiden
E	Applied Sciences	Noordhoff International Publishing Leiden

Series C – Mathematical and Physical Sciences

Volume 16 - Combinatorics

Combinatorics

Proceedings of the NATO Advanced Study Institute held at Nijenrode Castle, Breukelen, The Netherlands 8-20 July, 1974

edited by

M. HALL, Jr. and J. H. VAN LINT

D. Reidel Publishing Company

Dordrecht-Holland / Boston-U.S.A.

Mathematical Centre

Amsterdam-Holland

Published in cooperation with NATO Scientific Affairs Division

Library of Congress Cataloging in Publication Data

Advanced Study Institute on Combinatorics, Breukelen, Netherlands, 1974.
Combinatorics.

(NATO advanced study institute series : Series C, mathematical and physical sciences ; v. 16)
Includes bibliographical references.
1. Combinatorial analysis—Congresses. I. Hall, Marshall, 1910–
II. Lint, Jacobus Hendricus van, 1932–
III. North Atlantic Treaty Organization. IV. Title. V. Series.
QA164.A34 1974 511′.6 75–8819
ISBN 90–277–0593–3

Published by D. Reidel Publishing Company
P.O. Box 17, Dordrecht, Holland

Sold and distributed in the U.S.A., Canada, and Mexico
by D. Reidel Publishing Company, Inc.
306 Dartmouth Street, Boston, Mass. 02116, U.S.A.

Printed in the Netherlands by D. Reidel, Dordrecht

CONTENTS

PREFACE

Combinatorics has come of age. It had its beginnings in a number of puzzles which have still not lost their charm. Among these are EULER'S problem of the 36 officers and the KÖNIGSBERG bridge problem, BACHET's problem of the weights, and the Reverend T.P. KIRKMAN'S problem of the schoolgirls. Many of the topics treated in ROUSE BALL'S *Recreational Mathematics* belong to combinatorial theory.

All of this has now changed. The solution of the puzzles has led to a large and sophisticated theory with many complex ramifications. And it seems probable that the four color problem will only be solved in terms of as yet undiscovered deep results in graph theory. Combinatorics and the theory of numbers have much in common. In both theories there are many problems which are easy to state in terms understandable by the layman, but whose solution depends on complicated and abstruse methods. And there are now interconnections between these theories in terms of which each enriches the other.

Combinatorics includes a diversity of topics which do however have interrelations in superficially unexpected ways. The instructional lectures included in these proceedings have been divided into six major areas: 1. *Theory of designs;* 2. *Graph theory;* 3. *Combinatorial group theory;* 4. *Finite geometry;* 5. *Foundations, partitions and combinatorial geometry;* 6. *Coding theory.* They are designed to give an overview of the classical foundations of the subjects treated and also some indication of the present frontiers of research.

Without the generous support of the North Atlantic Treaty Organization, this *Advanced Study Institute on Combinatorics* would not have been possible, and we thank them sincerely. Thanks are also due to the National Science Foundation for the support of some advanced students, in addition to the support of those with their own NSF grants. The IBM Corporation has kindly given us financial support to supplement the NATO grant. The Xerox Corporation has helped with donations of material and equipment.

Finally we must acknowledge the extensive activities of the Mathematical Centre of Amsterdam in making all the arrangements necessary for holding this conference and preparing these proceedings.

M. HALL, Jr.
J.H. VAN LINT

PART 1

THEORY OF DESIGNS

FINITE GEOMETRY

CODING THEORY

INDETERMINATES AND INCIDENCE MATRICES *)

H.J. RYSER

California Institute of Technology, Pasadena, Cal. 91109, USA

1. INTRODUCTION

We let

(1.1) $X = \{x_1,\ldots,x_n\}$

denote a non-empty set of n elements. We call such a set an n-*set*. We let

(1.2) $X_1,\ldots,X_m$

denote m not necessarily distinct subsets of X. We refer to this collection of subsets of X as a *configuration* and remark that configurations occur in great profusion throughout the combinatorial literature.

We now let F denote an arbitrary field. We interconnect F and our configuration by regarding the elements $x_1,\ldots,x_n$ of X as n independent indeterminates with respect to the field F. This simple device immediately imposes an algebraic structure on our original configuration and allows us to carry out various algebraic manipulations within the polynomial ring

(1.3) $F[x_1,\ldots,x_n]$.

We exploit this algebraic structure further by the introduction of an incidence matrix A. We set $a_{ij} = 1$ if $x_j \in X_i$ and we set $a_{ij} = 0$ if $x_j \notin X_i$. In these equations the 1 and the 0 are the identity element and the zero element, respectively, of the field F. The resulting matrix

---

*) This research was supported in part by the Army Research Office-Durham under Grant DA-ARO-D-31-124-72-G171 and the National Science Foundation under Grant GP-36230X.

M. Hall, Jr. and J. H. van Lint (eds.), Combinatorics, 3-17.

$$(1.4) \qquad A = [a_{ij}] \qquad (i=1,\ldots,m;\ j=1,\ldots,n)$$

of size $m \times n$ is the *incidence matrix* for the subsets $X_1,\ldots,X_m$ of X. Row i of A displays the subset X_i and column j of A displays the occurrences of the element x_j among the subsets. Thus A gives us a complete description of the subsets and the occurrences of the elements within the subsets.

We may now write

$$(1.5) \qquad \begin{pmatrix} y_1 \\ \vdots \\ y_m \end{pmatrix} = A \begin{pmatrix} x_1 \\ \vdots \\ x_n \end{pmatrix},$$

where the component y_i of the vector on the left side of (1.5) is precisely the sum of the elements of the subset X_i of X. The equation (1.5) opens the door for important matrix manipulations. For example, equation (1.5) and its transpose allow us to associate with our configuration the quadratic form

$$(1.6) \qquad y_1^2 + \ldots + y_m^2 = (x_1,\ldots,x_n)A^TA \begin{pmatrix} x_1 \\ \vdots \\ x_n \end{pmatrix}.$$

We point out that this quadratic form has been exploited with great success in the study of block designs. (See, for example, [3], [11], [19].)

Indeterminates may be associated with a given configuration in various ways. In what follows we discuss in some detail two such associations. The one deals with n independent indeterminates that represent the elements of the n-set X and the other involves mn independent indeterminates associated with the positions of the incidence matrix A of size $m \times n$. Much of the material that we discuss is in the very early stages of its development. But we anticipate that these topics hold great potential for further study. Our concluding remarks deal with indeterminates and the incidence matrix A of a (v,k,λ)-design.

2. A FUNDAMENTAL MATRIX EQUATION FOR FINITE SETS

We return to a configuration of subsets $X_1,\ldots,X_m$ of an n-set $X = \{x_1,\ldots,x_n\}$, where we regard the elements $x_1,\ldots,x_n$ as n independent indeterminates with respect to a field F. This configuration now has an incidence matrix A of size $m \times n$. We also denote by X the diagonal matrix of order n

(2.1) $$X = \mathrm{diag}[x_1,\ldots,x_n],$$

and we then form the matrix equation

(2.2) $$AXA^T = Y,$$

where A^T denotes the transpose of the matrix A. All of the matrices in (2.2) have elements in the polynomial ring $F[x_1,\ldots,x_n]$.

The fundamental matrix equation (2.2) is a most remarkable one because it contains a vast amount of information in a highly compact form. The matrix Y is a symmetric matrix of order m and we know the structure of this matrix explicitly. Thus the matrix Y has in its (i,j) position the sum of the indeterminates in the set intersection $X_i \cap X_j$, and it follows that the matrix Y gives us an explicit representation for all of these set intersections. In particular, the elements on the main diagonal of Y display the subsets $X_1,\ldots,X_m$ of our original configuration. The matrix equation (2.2) was introduced by RYSER in [20] and [22]. Some other investigations that deal with matrices and set intersections include [9], [10], [14], [21].

The matrix Y involves the indeterminates $x_1,\ldots,x_n$ and we write

(2.3) $$Y = Y(x_1,\ldots,x_n).$$

We may assign the indeterminates in the matrix equation (2.2) arbitrary values of the field F, and each such assignment produces a new matrix equation that must be satisfied by the incidence matrix A of our configuration. Suppose, for example, that we have $m = n$. Then A is a square matrix of order n and suppose further that the matrix A is non-singular. Then if we assign x_i the value e_i in F it follows that the matrix $Y(e_1,\ldots,e_n)$ is congruent to the diagonal matrix

(2.4) $$E = \mathrm{diag}[e_1,\ldots,e_n]$$

with respect to the field F, and this congruence relationship remains valid for arbitrary choices of the e_i in F.

In many problems it is desirable to select F as the field of rational numbers or some extension field of this field. The incidence matrix A is then a (0,1)-*matrix* of size $m \times n$. If we now set $x_1 = \ldots = x_n = 1$, then

(2.2) reduces to the classical equation

$$AA^T = Y(1,\ldots,1). \tag{2.5}$$

In this case the matrix $Y(1,\ldots,1)$ on the right-hand side of (2.5) reveals the cardinalities of the set intersections.

The following theorem appears in [20] and affords a good illustration of the type of result that is motivated by the matrix equation (2.2). The proof of the theorem uses techniques similar to those employed by VAN LINT & RYSER [15] in their study of block designs with repeated blocks.

THEOREM 2.1. *Suppose that* A *is a* (0,1)*-matrix of order* n *and suppose that* A *satisfies the matrix equation*

$$AEA^T = D, \tag{2.6}$$

where A^T *is the transpose of the matrix* A*. Suppose further that the matrices* D *and* E *are real (or complex) diagonal matrices of order* n *and that* D *is non-singular. Then it follows that* A *is a permutation matrix.*

PROOF. The matrix D is non-singular so that we may write

$$AEA^TD^{-1} = EA^TD^{-1}A = I, \tag{2.7}$$

where I is the identity matrix of order n. Hence it follows that

$$A^TD^{-1}A = E^{-1}. \tag{2.8}$$

We now let

$$D = \mathrm{diag}[d_1,\ldots,d_n], \quad E = \mathrm{diag}[e_1,\ldots,e_n] \tag{2.9}$$

and inspect the main diagonal of (2.8). Then we obtain

$$A^T \begin{pmatrix} 1/d_1 \\ \vdots \\ 1/d_n \end{pmatrix} = \begin{pmatrix} 1/e_1 \\ \vdots \\ 1/e_n \end{pmatrix}. \tag{2.10}$$

We note that in (2.10) we have made strong use of the fact that A is a (0,1)-matrix. We now multiply (2.10) by AE and this gives

(2.11) $$A\begin{pmatrix}1\\ \vdots\\ 1\end{pmatrix} = \begin{pmatrix}1\\ \vdots\\ 1\end{pmatrix}.$$

Thus each of the row sums of the matrix A is equal to 1. But A is a non-singular (0,1)-matrix and hence it follows that A is a permutation matrix.□

Thus far our discussion has been motivated entirely by combinatorial considerations. But the matrix equation (2.2) also suggests some algebraic questions that are of considerable interest in their own right. The following theorem illustrates this point [22].

THEOREM 2.2. *Suppose that* Y *is a matrix of order* $n \geq 3$ *and such that every element of* Y *is a linear form in the* n *independent indeterminates* $x_1,\ldots,x_n$ *with respect to a field* F. *We let*

(2.12) $$X = \mathrm{diag}[x_1,\ldots,x_n]$$

and we suppose that the determinant of Y *satisfies*

(2.13) $$\det(Y) = cx_1 \cdots x_n,$$

where $c \neq 0$ *and* $c \in F$. *We suppose further that every element of* Y^{-1} *is a linear form in* $x_1^{-1},\ldots,x_n^{-1}$ *with respect to the field* F. *Then there exist matrices* A *and* B *of order* n *with elements in* F *such that*

(2.14) $$AXB = Y.$$

The proof of theorem 2.2 is available in [22]. Further theorems that imply a factorization of Y of the form (2.14) are also valid. These results deal with compound matrices and do not require Y to be a square matrix [22]. Much earlier investigations by KANTOR [13], FROBENIUS [8], and SCHUR [24] study related problems but with X a matrix of size $m \times n$ and such that the elements of X are mn independent variables over the complex field. A more recent account of this older theory appears in [16].

3. THE FORMAL INCIDENCE MATRIX

We now let

(3.1) $$A = [a_{ij}] \qquad (i=1,\ldots,m;\ j=1,\ldots,n)$$

denote a matrix of size $m \times n$ with elements in a field F. We may still regard A as the *incidence matrix* for our configuration of m subsets of an n-set, where the non-zero elements of A play the role of the identity element of the field in our earlier representation. One of the most remarkable general theorems in combinatorial matrix theory is the following theorem of KÖNIG. (The theorem is also frequently referred to as the König-Egerváry theorem or the Frobenius-König theorem.) Throughout our discussion a *line* of a matrix designates either a row or a column of the matrix.

THEOREM 3.1. *Suppose that* A *is a matrix of size* $m \times n$ *with elements in a field* F. *Then the minimal number of lines in* A *that cover all of the non-zero elements in* A *is equal to the maximal number of non-zero elements in* A *with no two of the non-zero elements on a line.*

An enormous literature centers around this theorem and the related theorems of P. HALL [12], DILWORTH [4], and FORD & FULKERSON [7]. A detailed discussion of these topics is available in the recent book by MIRSKY [17]. KÖNIG's theorem is frequently stated in the terminology of (0,1)-matrices. But the nature of the theorem is such that it holds quite generally for an arbitrary rectangular array in which all of the elements of the array have been partitioned into exactly two components.

At this point we disregard our earlier notation and we let

$$X = [x_{ij}] \qquad (i=1,\dots,m;\ j=1,\dots,n) \tag{3.2}$$

denote the matrix of size $m \times n$, where the elements of X are mn independent indeterminates with respect to the field F. We call the Hadamard product

$$M = A * X = [a_{ij}x_{ij}] \tag{3.3}$$

the *formal incidence matrix* associated with A. The elements of M belong to the polynomial ring

$$F^* = F[x_{11}, x_{12}, \dots, x_{mn}]. \tag{3.4}$$

The formal incidence matrix is useful in various combinatorial investigations [2],[6],[18],[23],[26].

The maximal number of non-zero elements in A with no two of the non-zero elements on a line is called the *term rank* of A. It turns out that this important combinatorial invariant of A is equal to an algebraic invariant of M. *The term rank of* A *is equal to the rank of* M. This observation is due to EDMONDS [6] and may be derived as follows. We note that a submatrix of M of order r has a non-zero determinant if and only if the corresponding submatrix of A has term rank r. This is a consequence of the definition of the formal incidence matrix. But the rank of a matrix is equal to the maximal order of a square submatrix with a non-zero determinant. Hence we obtain the desired conclusion.

We next discuss another basic combinatorial property of A in terms of an algebraic property of M. We now deal with square matrices of order n with elements in a field F. We say that a matrix A of order $n > 1$ is *fully indecomposable* provided that A does not contain a zero submatrix of size $r \times (n-r)$, for some integer r in the interval $1 \leq r \leq n-1$. It follows that the non-zero elements of a fully indecomposable matrix A of order $n > 1$ cannot be covered by n lines that are composed of both rows and columns of A. In case the matrix A is of order $n = 1$ then we say that A is *fully indecomposable* provided that A is not the zero matrix of order 1. We may conclude at once from theorem 3.1 that a fully indecomposable matrix A of order n has term rank n. Hence our earlier observation implies that a fully indecomposable matrix A has $\det(M) \neq 0$. But it is clear that $\det(M) \neq 0$ does not in general imply that the matrix A is fully indecomposable. However, the following theorem in a recent paper by RYSER [23] shows that a fully indecomposable matrix A is characterized by a somewhat deeper algebraic property of det(M).

THEOREM 3.2. *Suppose that* A *is a matrix of order* n *with elements in a field* F *and let* $M = A * X$ *denote the formal incidence matrix associated with* A. *Then the matrix* A *is fully indecomposable if and only if* det(M) *is an irreducible polynomial in the polynomial ring*

$$F^* = F[x_{11}, x_{12}, \dots, x_{nn}]. \tag{3.5}$$

We do not attempt a derivation of theorem 3.2 here. But we remark that it is easy to show that if the polynomial det(M) is irreducable in F^* then the matrix A is fully indecomposable. The proof of the converse proposition is more difficult.

We describe briefly a lemma of some intrinsic interest used in the derivation of the converse. A *diagonal product* of a matrix of order n is a product of n elements of the matrix with no two of the elements on a line.

We now let X_r denote a submatrix of X of order r. We designate the u = r! diagonal products of X_r by

$$(3.6) \qquad y_1, \ldots, y_u.$$

We say that the polynomial f has an *indeterminate pattern* based on X_r provided that

$$(3.7) \qquad f = \sum_{i=1}^{u} a_i y_i,$$

where the coefficients a_i are in F and not all of the a_i are zero. A polynomial with an indeterminate pattern based on X_r is homogeneous and of degree r over F. We note that if det(M) ≠ 0, then det(M) is an example of a polynomial with an indeterminate pattern based on X. Two submatrices B and C of orders r and n-r, respectively, of a matrix A of order n are called *complementary* provided that they are formed from complementary sets of lines of A. We are now ready to state the lemma used in the derivation of theorem 3.2 [23].

LEMMA 3.1. *Suppose that* h *is a polynomial with an indeterminate pattern based on* X *and suppose that in* F^* *we have*

$$(3.8) \qquad h = fg,$$

where f *and* g *are polynomials of positive degrees* r *and* n-r, *respectively. Then it follows that the polynomials* f *and* g *have indeterminate patterns based on* X_r *and* X_{n-r}, *respectively, where* X_r *and* X_{n-r} *are complementary submatrices of* X *of orders* r *and* n-r, *respectively.*

We now let A denote a matrix of order n with elements in a field F and we suppose that A is of term rank n. Then it follows that there exist permutation matrices P and Q of order n such that

$$(3.9) \qquad PAQ = \begin{bmatrix} A_1 & 0 & \cdots & 0 \\ * & A_2 & \cdots & 0 \\ \vdots & \vdots & & \vdots \\ * & * & \cdots & A_r \end{bmatrix},$$

where the matrices $A_1, A_2, \ldots, A_r$ are fully indecomposable. These matrices are called the *fully indecomposable components* of A. A theorem of DULMAGE & MENDELSOHN [1],[5] asserts the following.

THEOREM 3.3. *Suppose that* A *is a matrix of order* n *with elements in a field* F *and suppose that* A *is of term rank* n. *Then the fully indecomposable components of* A *are unique apart from order and row and column permutations within components.*

PROOF. Our proof follows [23] and is based on algebraic properties of the formal incidence matrix $M = A * X$. The matrix A is of term rank n so that we know that $\det(M) \neq 0$. We let $A_1,\ldots,A_r$ and $B_1,\ldots,B_s$ denote two sets of fully indecomposable components of A. Suppose that we apply certain permutations to the rows and the columns of A and also apply the identical permutations to the rows and the columns of M. Then we observe that the zero elements in both of the permuted matrices occupy the identical positions. Thus we may write

$$(3.10) \qquad \det(M) = \pm f_1 \cdots f_r = \pm g_1 \cdots g_s,$$

where each of the polynomials f_i and g_j in (3.10) has an indeterminate pattern based on the appropriate submatrix of X. Each of these polynomials uniquely describes its associated submatrix of X. Moreover, this submatrix of X corresponds in A to a fully indecomposable component of A. Hence by theorem 3.2 we may conclude that the polynomials f_i and g_j are irreducible polynomials in F^*. But F^* is a unique factorization domain and this means that $r = s$ and the f_i and the g_j are the same apart from order and scalar factors. Hence it follows that the fully indecomposable components A_i and B_j are the same apart from order and row and column permutations within components. □

The preceding proof is especially intriguing because the uniqueness of the fully indecomposable components is now a natural consequence of the unique factorization property of the polynomial ring F^*.

Suppose that A is a fully indecomposable matrix of order n with elements in a field F and let $M = A * X$ denote the formal incidence matrix associated with A. Then by theorem 3.2 we may associate with A the irreducible polynomial $\det(M)$ in F^*. This correspondence between fully indecomposable matrices and irreducible polynomials is a most remarkable one because the polynomial $\det(M)$ determines the matrix A uniquely apart from multiplication of A on the left and the right by diagonal matrices [23]. Our derivation of this fact requires the following theorem of SINKHORN & KNOPP [25].

THEOREM 3.4. *Suppose that* A *is a fully indecomposable matrix of order* n *with elements in a field* F *and suppose that all of the non-zero diagonal products of* A *are equal. Then there exists a unique matrix* B *of order* n *with non-zero elements and of rank one such that* $b_{ij} = a_{ij}$ *whenever* $a_{ij} \neq 0$.

The following theorem concerning the correspondence between fully indecomposable matrices and irreducible polynomials is now a fairly easy consequence of theorem 3.4. Details of the proof are available in [23].

THEOREM 3.5. *Suppose that* A *and* B *are fully indecomposable matrices of order* n *with elements in a field* F *and let* $M = A * X$ *and* $N = B * X$ *denote the formal incidence matrices associated with* A *and* B, *respectively. Suppose further that*

(3.11) $$\det(M) = c \det(N) \neq 0,$$

where c *is a scalar in* F. *Then there exist diagonal matrices* D *and* E *with elements in* F *such that*

(3.12) $$DAE = B.$$

4. SYMMETRIC BLOCK DESIGNS

We recall that a (v,k,λ)-*design (symmetric block design)* is a configuration of subsets $X_1,\ldots,X_v$ of a v-set $X = \{x_1,\ldots,x_v\}$ subject to the following postulates:

(4.1) each X_i is a k-subset of X;

(4.2) each $X_i \cap X_j$ for $i \neq j$ is a λ-subset of X;

(4.3) the integers v, k and λ satisfy $0 < \lambda < k < v-1$.

These postulates imply that the incidence matrix A of a (v,k,λ)-design is a (0,1)-matrix of order v that satisfies the matrix equation

(4.4) $$AA^T = (k-\lambda)I + \lambda J,$$

where A^T is the transpose of the matrix A, I is the identity matrix of order v, and J is the matrix of 1's of order v. One may show that the incidence matrix A of a (v,k,λ)-design is normal, namely,

(4.5) $$AA^T = A^TA,$$

and that the parameters v, k and λ satisfy the relationship

(4.6) $$k - \lambda = k^2 - \lambda v.$$

Of special importance are the (v,k,λ)-designs with $\lambda = 1$. These configurations are called *finite projective planes*. One of the main unresolved problems in the study of block designs is the determination of the precise range of values of v, k and λ for which (v,k,λ)-designs exist. Detailed discussions of these topics are available in the books by DEMBOWSKI [3], HALL [11], and RYSER [19].

In what follows we make some observations concerning indeterminates and the incidence matrix A of a (v,k,λ)-design. We look mainly at the matrix equation (2.2). Then we have

(4.7) $$AXA^T = Y,$$

where

(4.8) $$X = \mathrm{diag}[x_1,\ldots,x_v].$$

An appropriate analysis of the matrix equation (4.7) could conceivably yield important breakthroughs concerning the non-existence of (v,k,λ)-designs. One possible attack is an ingenious assignment of values e_i to the indeterminates x_i so that the resulting matrix equation contains a contradiction.

We now prove a new result that illustrates this idea.

THEOREM 4.1. *Suppose that* A *is the incidence matrix of a* (v,k,λ)*-design and suppose that* A *satisfies the matrix equation*

(4.9) $$AEA^T = C,$$

where E *is a diagonal matrix such that all of the diagonal elements of* E *are equal to* ± 1. *Suppose further that all of the off-diagonal elements of* C *are also equal to* ± 1. *Then it follows that* $\lambda = 1$ *and thus* A *is the incidence matrix of a finite projective plane.*

PROOF. The matrix A is the incidence matrix of a (v,k,λ)-design and A satisfies the matrix equation (4.9). We let c_i denote the element in position i of the main diagonal of C and we inspect the main diagonal of (4.9). Then we obtain

$$AE\begin{pmatrix}1\\ \vdots\\ 1\end{pmatrix} = \begin{pmatrix}c_1\\ \vdots\\ c_v\end{pmatrix}. \tag{4.10}$$

We note that in (4.10) we have again made strong use of the fact that A is a (0,1)-matrix. We now multiply (4.10) by its transpose and we thereby obtain

$$AEJE^TA^T = [c_ic_j]. \tag{4.11}$$

We next square the matrix equation (4.9) and by (4.4) and (4.5) we have

$$AE((k-\lambda)I + \lambda J)EA^T = C^2. \tag{4.12}$$

But

$$E^2 = I \tag{4.13}$$

and hence by (4.4) and (4.11) we have

$$(k-\lambda)^2I + \lambda(k-\lambda)J + \lambda[c_ic_j] = C^2. \tag{4.14}$$

The off-diagonal elements of C are equal to ±1 and hence an inspection of the main diagonal of (4.14) implies

$$(k-\lambda)^2 + \lambda(k-\lambda) + \lambda c_i^2 = c_i^2 + (v-1). \tag{4.15}$$

Then by (4.6) we have

$$(\lambda-1)c_i^2 = \frac{k(k-1)}{\lambda} - k(k-\lambda) \geq 0. \tag{4.16}$$

We may rewrite (4.16) in the form

$$\frac{\lambda^2-1}{\lambda} \geq \frac{k(\lambda-1)}{\lambda}. \tag{4.17}$$

Now the assumption $\lambda > 1$ implies

$$\lambda \geq k-1. \tag{4.18}$$

But by (4.3) we have $\lambda \leq k-1$ and hence $\lambda = k-1$. But then $k = v-1$ and this contradicts (4.3). Hence we have $\lambda = 1$ and A is the incidence matrix of a finite projective plane. □

We note that a converse type proposition is immediate. Thus suppose that A is the incidence matrix of a finite projective plane. Then each off-diagonal element of the matrix Y in (4.7) is equal to some indeterminate x_i. Hence it follows that an arbitrary assignment of values ±1 to the indeterminates x_i in (4.7) gives us a matrix equation of the form (4.9) that satisfies all of the requirements of theorem 4.1.

We conclude with a few remarks on the formal incidence matrix

$$M = A * X = [a_{ij}x_{ij}], \tag{4.19}$$

where A is the incidence matrix of a (v,k,λ)-design and

$$X = [x_{ij}] \tag{4.20}$$

is the matrix of v^2 independent indeterminates with respect to the rational field. It is easy to verify that the incidence matrix A of a (v,k,λ)-design is fully indecomposable and hence by theorem 3.2 we have that det(M) is an irreducible polynomial in the polynomial ring of the v^2 independent indeterminates with respect to the rational field.

One is now tempted to study the matrix equations

$$(A * X)A^T = Y(x_{11},x_{12},\ldots,x_{vv}) \tag{4.21}$$

and

$$(A * X)(A * X)^T = Z(x_{11},x_{12},\ldots,x_{vv}). \tag{4.22}$$

These matrix equations afford vastly greater substitution possibilities than does the matrix equation (4.7). For example, suppose that A is the incidence matrix of a finite projective plane and that we replace the matrix X on the left sides of (4.21) and (4.22) by an arbitrary (1,-1)-matrix of order v. Then the resulting matrices on the right sides of (4.21) and (4.22) have the property that all of their off-diagonal elements are also equal to ±1.

REFERENCES

[1] BRUALDI, R.A., *Permanent of the product of doubly stochastic matrices*, Proc. Cambridge Philos. Soc., 62 (1966) 643-648.

[2] BRUALDI, R.A. & H. PERFECT, *Extension of partial diagonals of matrices I*, Monatsh. Math., 75 (1971) 385-397.

[3] DEMBOWSKI, P., *Finite geometries*, Ergebnisse der Mathematik 44, Springer Verlag, Berlin, 1968.

[4] DILWORTH, R.P., *A decomposition theorem for partially ordered sets*, Ann. of Math. (2), 51 (1950) 161-166.

[5] DULMAGE, A.L. & N.S. MENDELSOHN, *Coverings of bipartite graphs*, Canad. J. Math., 10 (1958) 517-534.

[6] EDMONDS, J., *Systems of distinct representatives and linear algebra*, J. Res. Nat. Bur. Standards Ser. B, 71 (1967) 241-245.

[7] FORD Jr., L.R. & D.R. FULKERSON, *Flows in networks*, Princeton University Press, Princeton, 1962.

[8] FROBENIUS, G., *Über die Darstellung der endlichen Gruppen durch lineare Substitutionen*, Sitzungsberichte Berliner Akademie (1897) 994-1015.

[9] GOODMAN, A.W., *Set equations*, Amer. Math. Monthly, 72 (1965) 607-613.

[10] HALL Jr., M., *A problem in partitions*, Bull. Amer. Math. Soc., 47 (1941) 804-807.

[11] HALL Jr., M., *Combinatorial theory*, Blaisdell, Waltham, Mass., 1967.

[12] HALL, P., *On representatives of subsets*, J. London Math. Soc., 10 (1935) 26-30.

[13] KANTOR, S., *Theorie der Äquivalenz von linearen* ∞^λ*-Scharen bilinearer Formen*, Sitzungsberichte Münchener Akademie (1897) 367-381.

[14] KELLY, J.B., *Products of zero-one matrices*, Canad. J. Math., 20 (1968) 298-329.

[15] LINT, J.H. VAN & H.J. RYSER, *Block designs with repeated blocks*, Discrete Math., 3 (1972) 381-396.

[16] MARCUS, M. & F. MAY, *On a theorem of I. Schur concerning matrix transformations*, Arch. Math. (Basel), 11 (1960) 401-404.

[17] MIRSKY, L., *Transversal theory*, Academic Press, New York, 1971.

[18] PERFECT, H., *Symmetrized form of P. Hall's theorem on distinct representatives*, Quart. J. Math. Oxford Ser. 2, 17 (1966) 303-306.

[19] RYSER, H.J., *Combinatorial mathematics*, Carus Math. Monograph No. 14, Math. Assoc. Amer., Wiley, New York, 1963.

[20] RYSER, H.J., *A fundamental matrix equation for finite sets*, Proc. Amer. Math. Soc., 34 (1972) 332-336.

[21] RYSER, H.J., *Intersection properties of finite sets*, J. Combinatorial Theory A, 14 (1973) 79-92.

[22] RYSER, H.J., *Analogs of a theorem of Schur on matrix transformations*, J. Algebra, 25 (1973) 176-184.

[23] RYSER, H.J., *Indeterminates and incidence matrices*, Linear and Multilinear Algebra, 1 (1973) 149-157.

[24] SCHUR, I., *Einige Bemerkungen zur Determinantentheorie*, Sitzungsberichte Berliner Akademie (1925) 454-463.

[25] SINKHORN, R. & P. KNOPP, *Problems involving diagonal products in non-negative matrices*, Trans. Amer. Math. Soc., 136 (1969) 67-75.

[26] TUTTE, W.T., *The factorization of linear graphs*, J. London Math. Soc., 22 (1947) 107-111.

CONSTRUCTIONS AND USES OF PAIRWISE BALANCED DESIGNS *)

R.M. WILSON

Ohio State University, Columbus, Ohio 43210, USA

1. INTRODUCTION

A *pairwise balanced design* (PBD) of index unity is a pair $(X,\mathcal{A})$ where X is a set (of *points*) and $\mathcal{A}$ a class of subsets A of X (called *blocks*) such that any pair of distinct points of X is contained in exactly one of the blocks of $\mathcal{A}$ (and we may also require $|A| \geq 2$ for each $A \in \mathcal{A}$). Such systems are also known as linear spaces. PBD's where all blocks have the same size $|A| = k$ are known as *balanced incomplete block designs* (BIBD's) of index $\lambda = 1$, as $2-(v,k,1)$ designs, and as Steiner systems $S(2,k,v)$. The more general concept, where multiple block sizes are allowed, was introduced by BOSE, SHRIKHANDE & PARKER [4] and H. HANANI [9], and played important roles in their respective work on orthogonal Latin squares and BIBD's.

The purpose of this paper is to present some of the methods for constructing designs and to briefly indicate how PBD's have and can be used in the construction of related combinatorial structures. We also take this opportunity to add various remarks and indicate variations on proofs of known results. Many of the proofs, if not omitted, will be very brief.

A $\mathrm{PBD}[K,v]$ is to be a PBD $(X,\mathcal{A})$ where $|X| = v$ and $|A| \in K$ for every $A \in \mathcal{A}$. Here K is a (finite or infinite) set of positive integers. For the case where K consists of a single positive integer k, we write $B[k,v]$ in place of $\mathrm{PBD}[\{k\},v]$.

We observe that the existence of a $\mathrm{PBD}[K,v]$ (with $v > 0$) implies

(i) $v \equiv 1 \pmod{\alpha(K)}$, and

(ii) $v(v-1) \equiv 0 \pmod{\beta(K)}$,

where $\alpha(K)$ is the greatest common divisor of the integers $\{k-1 : k \in K\}$ and

---

*) This research was supported in part by NSF Grant GP-28943 (O.S.U.R.F. Project No. 3228-A1).

M. Hall, Jr. and J. H. van Lint (eds.), Combinatorics, 19-42. *All Rights Reserved.*

$\beta(K)$ is the greatest common divisor of the integers $\{k(k-1) : k \in K\}$. Here (i) follows from the fact that the blocks containing a given point of a PBD partition the remaining v-1 points; and (ii) follows since the $\binom{v}{2}$ pairs of points are partitioned by the blocks.

The above conditions (i) and (ii) are "asymptotically sufficient" for the existence of a PBD[K,v]. The following theorem is proved in [24] using the methods discussed in this paper.

THEOREM 1.1. *Given* K, *there exists a constant* c_K *such that designs* PBD[K,v] *exist for all* $v \geq c_K$ *which satisfy the congruences* $v \equiv 1 \pmod{\alpha(K)}$ *and* $v(v-1) \equiv 0 \pmod{\beta(K)}$.

For example, PBD[{7,8,9},v] exist for all large integers v; PBD[{5,7},v] exist for all large odd integers v; and B[6,v] exist for all large $v \equiv 1,6,16,$ or 21 (mod 30). Complete characterizations of the set $\mathbb{B}(K)$ of positive integers v for which there exist designs PBD[K,v] are known only for a few sets K (see section 5).

In section 2, we use difference methods and finite fields to construct BIBD's. As far as the author is aware, this provides the only known method of ascertaining for each k the existence of a design B[k,v] for some v > k.

In section 3 we attempt to communicate the flavor of some purely combinatorial techniques for constructing PBD's. We give some very general methods, but do not attempt to survey the tremendous variety of constructions which are known for, say, Steiner triple systems.

The application of pairwise balanced designs in the construction of related combinatorial systems is illustrated in several places. We mention in section 3 how PBD's were used in the construction of resolvable designs. In section 4, we point out the use of PBD's in the construction of certain quasigroups (Latin squares). The results on PBD's and the difference method are used in section 6 to obtain infinite classes of what we call edge-decompositions of complete graphs.

2. CONSTRUCTION OF DESIGNS BY DIFFERENCE METHODS

The "method of differences" introduced by R.C. BOSE [1] has been an effective method for the construction of designs B[k,v], especially for small values of k. We consider below the simplest case of the method.

Let G be an abelian group of order $v = k(k-1)t+1$. By a *simple difference family* D[k,v] in G, we mean a family $A_1, A_2, \ldots, A_t$ of k-subsets of G such that every non-zero group element occurs exactly once in the list of differences

$$(x-y : x,y \in A_i;\ x \neq y;\ i=1,2,\ldots,t).$$

This includes the case of planar difference sets (t = 1).

Examples include the D[3,13] ({1,3,9}, {2,6,5}) in Z_{13} and the D[4,25] ({(0,0), (1,0), (0,1), (2,2)}, {(0,0),(2,0), (0,2), (4,4)}) in $Z_5 \times Z_5$.

The existence of a simple difference family D[k,v] implies the existence of a design B[k,v]. For the design, take X = G and $\mathcal{A} = \{A_i + g : i=1,2,\ldots,t;\ g \in G\}$.

The existence of simple difference families D[3,v] in cyclic groups of order v = 6t+1 was established in 1939 by R. PELTESOHN [16], and a construction of R.C. BOSE [1], also in 1939, shows that D[3,v] always exist in elementary abelian groups of order v = 6t+1.

We conjecture that for each fixed k, simple difference families D[k,v] exist in all but finitely many abelian groups of orders $v = k(k-1)t+1$. The following theorem provides some evidence towards this conjecture.

THEOREM 2.1. *If* $q = k(k-1)t+1$ *is a prime power and*

$$q > e^{k^2+2k \log k},$$

then there exists a simple difference family D[k,q] *in the elementary abelian group of order* q, *and hence a design* B[k,q].

This theorem in a weaker form (with the bound $q > [k(k-1)/2]^{k(k-1)}$) was proved in [21]. While this improvement in the bound is significant, it surely is still very far from best possible. (Note that here e denotes the exponential e = 2.718.... , in distinction to its use in [21].)

The main idea of the proof is to exploit the multiplicative structure of finite fields to find simple difference families in their additive groups. Proposition 2.1 below reduces the problem to finding a single k-subset with a certain property. It remains to establish the existence of such k-subsets for large q, and we take this opportunity to give a proof which is more combinatorial in nature than that of [21] (i.e., we avoid the use of character sums).

Let $q = mf+1$ be a prime power and let $F = GF(q)$ be the field with q elements. The cyclic multiplicative group of F has a unique subgroup C_0 of index m (and order $f = (q-1)/m$). The multiplicative cosets $C_0, C_1, \ldots, C_{m-1}$ of C_0 are the *cyclotomic classes* of index m. They evidently partition $F - \{0\}$.

PROPOSITION 2.1. *Let* $q = k(k-1)t+1$ *be a prime power and put* $m = \frac{1}{2}k(k-1)$. *If there exists a* k*-tuple* $(a_1, a_2, \ldots, a_k)$ *of elements of* $GF(q)$ *such that the* m *differences*

$$(a_j - a_i : 1 \leq i < j \leq k)$$

form a system of representatives for the cyclotomic classes $C_0, C_1, \ldots, C_{m-1}$ *of index* m *in* $GF(q)$*, then there exists a simple difference family* $D[k,q]$ *in the additive group of* $GF(q)$.

PROOF. Let $A = \{a_1, \ldots, a_k\}$. Since $2m$ divides $q-1$, -1 will belong to C_0. Let S be a system of representatives for the cosets of the factor group $C_0/\{1,-1\}$ (i.e., S consists of half of the elements of C_0, one element from each pair $\{x,-x\} \subseteq C_0$).

We claim that $(sA : s \in S)$ is a simple difference family, where $sA = \{sa_1, \ldots, sa_k\}$. To see this, we need only observe that for each i,j $(1 \leq i < j \leq k)$, we find among the differences from $(sA : s \in S)$

$$sa_j - sa_i \quad \text{and} \quad sa_i - sa_j, \quad s \in S,$$

or

$$(\pm s(a_j - a_i) : s \in S) ,$$

which exhaust precisely the cyclotomic class represented by $a_j - a_i$. □

Examples for $k = 4$ include $(0,1,3,24)$ in $GF(37)$ and $(0,1,5,11)$ in $GF(61)$.

For prime powers $q \equiv 1 \pmod{k(k-1)}$, let $N(k,q)$ denote the number of k-tuples $(a_1, \ldots, a_k)$ with the property required in proposition 2.1. We prove that $N(k,q) > 0$, i.e. that such k-tuples exist, for q sufficiently large with respect to k by showing that

$$N(k,q) = \frac{m!}{m^m} q(q-1)\dots(q-k+1) + O(q^{k-\frac{1}{2}}),$$

where $m = \binom{k}{2}$. This is stated in a more exact form as theorem 2.2 below. The corollary 2.1 of this theorem, together with proposition 2.1, will complete the proof of theorem 2.1.

LEMMA 2.1. *Let* $\alpha_1, \alpha_2, \dots, \alpha_n$ *be real numbers,*

$$\bar{\alpha} = \frac{1}{n}(\alpha_1 + \dots + \alpha_n)$$

their mean, and

$$V = \frac{1}{n}\sum_{i=1}^{n} (\alpha_i - \bar{\alpha})^2 = \frac{1}{n}\left(\sum_{i=1}^{n} \alpha_i^2\right) - \bar{\alpha}^2$$

their variance. If we put $N = \alpha_1 + \dots + \alpha_l$ *for some* l, $0 \le l \le n$, *then*

$$|N - l\bar{\alpha}|^2 \le l(n-l)V \le \frac{1}{4}n^2 V.$$

PROOF. It is sufficient to establish the inequality in the case $\bar{\alpha} = 0$. For this case, let $\underline{a}$ denote the vector $(\alpha_1, \dots, \alpha_n)$, $\underline{u} = (1,1,\dots,1)$, and $\underline{w} = (1,\dots,1,0,\dots,0)$ (ones in the first l coordinates, zeros in the last $n-l$). Then the dot product $\langle \underline{a}, \underline{u} \rangle$ is 0 and by the Cauchy-Schwarz inequality,

$$|\langle \underline{a}, \underline{w} \rangle| = |\langle \underline{a}, \underline{w} - \beta\underline{u} \rangle| \le \|\underline{a}\| \cdot \|\underline{w} - \beta\underline{u}\|$$

for any real β. Taking $\beta = (n-l)/n$, we obtain

$$(\alpha_1 + \dots + \alpha_l)^2 \le (\alpha_1^2 + \dots + \alpha_n^2)\,\frac{l(n-l)}{n},$$

which is the desired inequality. □

For the following lemma, we fix a prime power $q = mf+1$ and write $F = GF(q)$, $Z_m = \{0,1,\dots,m-1\}$. C_i, $i \in Z_m$, are the cyclotomic classes of index m. For a set X, X^r denotes the set of all r-tuples $(x_1,\dots,x_r)$ of elements of X, and $X^{(r)}$ denotes the subset of X^r consisting of $(x_1,\dots,x_r)$ with $x_1,\dots,x_r$ distinct. Thus if $|X| = n$, then $|X^r| = n^r$ and $|X^{(r)}| = n^{(r)} = n(n-1)(n-2)\dots(n-r+1)$.

For $i_1, i_2, \dots, i_r \in Z_m$ and distinct $a_1, a_2, \dots, a_r \in F$, let

$$E_{i_1 i_2 \dots i_r}(a_1, a_2, \dots, a_r)$$

denote the number of field elements $x \in F$ such that $x-a_1 \in C_{i_1}$, $x-a_2 \in C_{i_2}, \dots, x-a_r \in C_{i_r}$.

LEMMA 2.2. *The mean value of* $E_{(i)}(a) = E_{i_1 \dots i_r}(a_1, \dots, a_r)$ *(over the* $m^r q^{(r)}$ *choices of* $(i) = (i_1, \dots, i_r) \in Z_m^r$ *and* $(a) = (a_1, \dots, a_r) \in F^{(r)}$*) is* $(q-r)m^{-r}$, *and the variance* V_r *of these* $m^r q^{(r)}$ *quantities is*

$$V_r = \frac{q(q-1)}{m^r q^{(r)}} \left[\frac{q-m-1}{m}\right]^{(r)} + \frac{q-r}{m^r} - \left(\frac{q-r}{m^r}\right)^2 < \frac{q-r}{m^r} .$$

PROOF. It is immediate that for $(a) \in F^{(r)}$,

$$\sum_{(i)} E_{(i)}(a) = q-r ,$$

and then

$$\sum_{(i),(a)} E_{(i)}(a) = q^{(r+1)} .$$

So the mean is as claimed.

With the usual notation $E^{(2)} = E(E-1)$, $[E_{(i)}(a)]^{(2)}$ is the number of pairs $(x,y) \in F^{(2)}$ such that $x-a_j$ and $y-a_j$ both belong to the class C_{i_j} for $j=1,2,\dots,r$. Then for fixed $(a) \in F^{(r)}$,

$$\sum_{(i)} [E_{(i)}(a)]^{(2)}$$

is the number of $(x,y) \in F^{(2)}$ such that $x-a_j$ and $y-a_j$ belong to the same cyclotomic class for $j=1,2,\dots,r$; and

$$\sum_{(i),(a)} [E_{(i)}(a)]^{(2)}$$

counts the number of $(r+2)$-tuples $(a_1, \dots, a_r;\ x,y)$ with $(a) \in F^{(r)}$, $(x,y) \in F^{(2)}$, and such that $x-a_j$ and $y-a_j$ are in the same class. For fixed $(x,y) \in F^{(2)}$, such an $(r+2)$-tuple is obtained by choosing $a_1, \dots, a_r$ as distinct elements of the set $S(x,y)$ of c with $x-c$ and $y-c$ in the same class.

Now x-c and y-c are in the same class if and only if $x-c = b(y-c)$ for some $b \in C_0$. But for each of the $(q-m-1)/m$ elements $b \in C_0$, $b \neq 1$, there is a unique such solution c; that is, $|S(x,y)| = (q-m-1)/m$ (independent of x,y). So

$$\sum_{(i),(a)} [E_{(i)}(a)]^{(2)} = q(q-1)\left[\frac{q-m-1}{m}\right]^{(r)} .$$

The computation of the variance V_r is now straightforward, and the inequality $V_r < m^{-r}(q-r)$ is elementary. □

THEOREM 2.2. *Let* $m = \frac{1}{2}k(k-1)$. *Then for prime powers* $q = 2mt+1$,

$$\left|N(k,q) - \frac{m!}{m^m} q^{(k)}\right| < m^{\frac{1}{2}(k-1)} q^{k-\frac{1}{2}} .$$

PROOF. For $0 \le r \le k$, let M_r denote the set of $(a_1,\ldots,a_r) \in F^{(r)}$ such that the differences $a_j - a_i$, $1 \le i < j \le r$, represent $\binom{r}{2}$ distinct cyclotomic classes and write $M_r = |M_r|$. Thus $M_0 = 1$, $M_1 = q$, $M_2 = q(q-1)$, and $M_k = N(k,q)$. The members of M_{r+1} may be partitioned according to their first r coordinates (which determine a member of M_r) and we evidently have

$$M_{r+1} = \sum_{(a)\in M_r} E'(a) ,$$

where $E'(a_1,\ldots,a_r)$ is the number of x such that $(a_1,\ldots,a_r,x) \in M_{r+1}$. But clearly

$$E'(a_1,\ldots,a_r) = \sum E_{i_1\ldots i_r}(a_1,\ldots,a_r) ,$$

where the sum is extended over the $[m-\binom{r}{2}]^{(r)}$ choices of distinct $i_1,\ldots,i_r$ chosen from the set of $m-\binom{r}{2}$ elements $i \in Z_m$ for which the class C_i is *not* represented by a difference from $(a_1,\ldots,a_r)$.

In summary, M_{r+1} is a sum of $M_r[m-\binom{r}{2}]^{(r)}$ of the quantities $E_{(i)}(a)$. By lemmas 2.1 and 2.2,

$$\left|M_{r+1} - \frac{q-r}{m^r}[m-\tbinom{r}{2}]^{(r)} M_r\right| \le (\tfrac{1}{4}m^r q^{(r)} V_r)^{\frac{1}{2}} < (\tfrac{1}{4}m^r q^{2r+1})^{\frac{1}{2}} .$$

With

$$c_r = \frac{q-r}{m^r}\,[m-\tbinom{r}{2}]^{(r)}, \qquad d_r = \tfrac{1}{2}m^{r/2}\, q^{r+\frac{1}{2}},$$

we have $|M_{r+1}-c_rM_r| < d_r$ for $r=0,1,\ldots,k-1$. Using the triangle inequality inductively, and $M_0 = 1$, we arrive at

$$|M_{r+1}-c_rc_{r-1}\cdots c_1c_0| < d_r+c_rd_{r-1}+c_rc_{r-1}d_{r-2}+\cdots+c_rc_{r-1}\cdots c_1d_0.$$

Now $c_r < q$ and

$$c_{k-1}c_{k-2}\cdots c_1c_0 = \frac{m!}{m^m}\,q^{(k)},$$

so

$$|M_k-\frac{m!}{m^m}\,q^{(k)}| < \tfrac{1}{2}q^{k-\frac{1}{2}}(m^{(k-1)/2}+m^{(k-2)/2}+\cdots+1) < m^{(k-1)/2}q^{k-\frac{1}{2}}. \quad \square$$

COROLLARY 2.1. *If* $q = k(k-1)t+1$ *is a prime power, then* $N(k,q) > 0$ *whenever*

$$q > e^{k^2}\,k^{2k}.$$

PROOF. By theorem 2.2,

$$N(k,q) > \frac{m!}{m^m}\,q^{(k)} - m^{\frac{1}{2}(k-1)}\,q^{k-\frac{1}{2}},$$

$$q^{\frac{1}{2}-k}\,N(k,q) > \frac{m!}{m^m}\,\sqrt{q}\,\left(\frac{q^{(k)}}{q^k}\right) - m^{\frac{1}{2}(k-1)},$$

where $m = k(k-1)/2$. Using the inequality $m!/m^m > e^{-m}$ (which is immediate on noticing that $m^m/m!$ is one of the terms in the power series expansion of e^m) and the (very poor) inequality $q^{(k)}/q^k > e^{-k/2}$ for q satisfying the hypothesis,

$$q^{\frac{1}{2}-k}\,N(k,q) > e^{-k^2/2}\,\sqrt{q} - k^k.$$

The assertion of the corollary is now clear. $\square$

3. CONSTRUCTION OF DESIGNS BY COMPOSITION METHODS

In this section we discuss a class of recursive methods for the construction of PBD's. We make no attempt to be complete, but just enumerate certain principles and illustrations that the author finds of interest.

A concept which has played an important role in the construction of BIBD's and sets of orthogonal Latin squares is that of a *group divisible design* (GDD). We use the term to mean a triple $(X,\mathcal{S},\mathcal{A})$ where (i) X is a set (of *points*), (ii) $\mathcal{S}$ is a class of non-empty subsets of X (called *groups*) which partition X, (iii) $\mathcal{A}$ is a class of subsets of X (called *blocks*), each containing at least two points, (iv) no block meets a group in more than one point, and (v) each pairset $\{x,y\}$ of points not contained in a group is contained in precisely one block.

At least in the case where all groups $G \in \mathcal{S}$ have size $|G| \geq 2$, a GDD can be thought of as a PBD $(X,\mathcal{S} \cup \mathcal{A})$ in which a class of blocks which partition X has been distinguished. But it seems important to make the distinction between PBD's and GDD's, as GDD's are clearly the right concept for the Fundamental Construction (F.C.) 3.1 below. We preface the F.C. with several remarks concerning the relation between PBD's and GDD's.

REMARK 3.1. If $\mathcal{S}$ consists of all singleton subsets of X, then $(X,\mathcal{A})$ is a PBD if and only if $(X,\mathcal{S},\mathcal{A})$ is a GDD.

REMARK 3.2. (ADJOINING AND DELETING POINTS). If $(X,\mathcal{S},\mathcal{A})$ is a GDD, we may *adjoin a point* $\theta \notin X$ to obtain a PBD $(X',\mathcal{A}')$ where

$$X' = X \cup \{\theta\} ,$$

$$\mathcal{A}' = \mathcal{A} \cup \{G \cup \{\theta\} : G \in \mathcal{S}\} .$$

Conversely, given a PBD $(X',\mathcal{A}')$ we may *delete a point* $\theta \in X'$ to obtain a GDD $(X,\mathcal{S},\mathcal{A})$, where

$$X = X' - \{\theta\} ,$$

$$\mathcal{S} = \{A - \{\theta\} : A \in \mathcal{A}', \theta \in A\} ,$$

$$\mathcal{A} = \{A : A \in \mathcal{A}', \theta \notin A\}.$$

By a *subdesign* of a PBD $(X,\mathcal{B})$, we mean a PBD $(Y,\mathcal{C})$ such that $Y \subseteq X$ and $\mathcal{C} \subseteq \mathcal{B}$. Evidently, the blocks $\mathcal{B}-\mathcal{C}$ cover exactly the pairs x,y of distinct points of X with not both $x,y \in Y$. We admit $|Y| \leq 1$, in which case $\mathcal{C}$ is empty.

REMARK 3.3. (ADJOINING SUBDESIGNS). Let $(X,\mathcal{S},\mathcal{A})$ be a GDD and F a set, $F \cap X = \emptyset$. Let $(F,\mathcal{D})$ be a PBD, and for each group $G \in \mathcal{S}$, let $(G \cup F, \mathcal{B}_G)$ be a PBD containing a PBD $(F,\mathcal{C}_G)$ as a subdesign. Then with

$$X' = X \cup F ,$$

$$\mathcal{A}' = \mathcal{A} \cup \mathcal{D} \cup (\bigcup_{G\in\mathcal{S}} \mathcal{B}_G - \mathcal{C}_G) ,$$

$(X',\mathcal{A}')$ is a PBD.

REMARK 3.4. (BREAKING UP BLOCKS). Let $(X,\mathcal{A})$ be a PBD and for each block $A \in \mathcal{A}$, let $(A,\mathcal{B}_A)$ be a PBD. Then with

$$\mathcal{B} = \bigcup_{A\in\mathcal{A}} \mathcal{B}_A ,$$

$(X,\mathcal{B})$ is a PBD.

THE FUNDAMENTAL CONSTRUCTION (F.C.) 3.1. Let $(X,\mathcal{S},\mathcal{A})$ be a "master" GDD and let a positive integral weight s_x be assigned to each point $x \in X$. Let $(S_x : x \in X)$ be pairwise disjoint sets with $|S_x| = s_x$. With the notation $S_Y = \bigcup_{x\in Y} S_x$ for $Y \subseteq X$, put

$$X^* = S_X$$

$$\mathcal{S}^* = \{S_G : G \in \mathcal{S}\} .$$

For $A \in \mathcal{A}$, we have a natural partition $\pi_A = (S_A,\{S_x : x \in A\})$; we suppose that for each block $A \in \mathcal{A}$, a GDD

$$(S_A , \{S_x : x \in A\}, \mathcal{B}_A)$$

is given, and put

$$\mathcal{A}^* = \bigcup_{A\in\mathcal{A}} \mathcal{B}_A .$$

Then $(X^*,\mathcal{S}^*,\mathcal{A}^*)$ is a GDD.

We point out that in view of remark 3.1, remark 3.4 is a special case of the F.C. 3.1 (where the master GDD has groups of size 1 and we weight each of its points with 1).

We give below some of the neater corollaries of our remarks and the F.C. In the various applications here and in the following section, all points of the master GDD in the F.C. will be weighted equally. But instances of non-uniform weighting were essential for the proof of theorem 1.1.

A GDD $(X,\mathcal{S},\mathcal{A})$ will be said to have block sizes from K and *type* $(1^{l_1}2^{l_2}3^{l_3}\ldots)$ when $|A| \in K$ for each $A \in \mathcal{A}$ and $(1^{l_1}2^{l_2}3^{l_3}\ldots)$ is the partition of the integer $|X|$ arising from the partition $(X,\mathcal{S})$ of the set X, i.e. there are l_i groups of size i, i = 1,2,3,... . We say a class $\mathcal{B}$ of sets is k-*uniform* when $|B| = k$ for each $B \in \mathcal{B}$.

Transversal designs TD(k,n) are GDD's with type (n^k), i.e. k groups of size n, and block size k. The existence of a TD(k,n) is equivalent to the existence of a set of k-2 mutually orthogonal Latin squares [8].

THEOREM 3.1. (MACNEISH). *If there exists a* TD(k,n) *and a* TD(k,m), *then there exists a* TD(k,mn).

PROOF. Take the TD(k,n) as the master GDD in the F.C., and weight each of its points with m. The type of each partition π_A is then (m^k), and as there exists a TD(k,m), we may choose $\mathcal{B}_A$ to be k-uniform. The F.C. then produces a TD(k,mn). □

From the existence of TD(q+1,q) for prime powers q (obtainable by deleting a point from a $B[q+1, q^2+q+1]$ by remark 3.2) follows

THEOREM 3.2. *If* $n = p_1^{\alpha_1}p_2^{\alpha_2}\cdots p_r^{\alpha_r}$ *is the factorization of* $n > 1$ *into powers of distinct primes, then there exists a* TD(k,n) *whenever* $k \le 1 + \min_{1\le i\le r} p_i^{\alpha_i}$.

The following theorem was first proved in the case k = 3 by E.H. MOORE [15] and in 1893. The generalization was pointed out to the author by D.K. RAY-CHAUDHURI.

THEOREM 3.3. *If*

(i) *there exists a* B[k,u] *containing a* B[k,w] *as a subdesign (*w = 0,1, *or* k *is permitted),* *and*

(ii) *there exists a* TD(k,u-w),

then the existence of a B[k,v] *implies the existence of a* B[k,v(u-w)+w].

PROOF. As the master GDD in the F.C., take a B[k,v] with all singletons as groups, and weight each point with u-w. The type of each partition π_A is

$((u-w)^k)$ and $\mathcal{B}_A$ may be taken to be k-uniform by (ii). The F.C. produces a GDD of type $((u-w)^v)$, to which we may adjoin a subdesign on w points by (i) and remark 3.3 to obtain a B[k,v(u-w)+w]. □

THEOREM 3.4. *If*

(i) *there exists a* B[k,u] *containing a* B[k,w] *as a subdesign* $w \geq 1$, *and*

(ii) *there exists a* TD(k,(u-w)/(k-1)),

then the existence of a B[k,v] *implies the existence of a* B[k,(v-1)(u-w)/(k-1)+w].

PROOF. Delete a point (remark 3.2) from a B[k,v] to obtain a GDD with group type $((k-1)^r)$, $r = (v-1)/(k-1)$, and block size k. Weight each point with (u-w)/(k-1) (which is an integer since $u \equiv w \equiv 1 \pmod{k-1}$). By (ii), $\mathcal{B}_A$ may be chosen k-uniform. The F.C. produces a GDD of type $((u-w)^r)$ to which we adjoin a subdesign on w points by (i) to obtain a B[k,r(u-w)+w]. □

The next theorem is an elegant and powerful construction due to HANANI [9]. For each $k \geq 2$, we let R_k denote the set of positive integers r for which there exists a B[k,r(k-1)+1].

THEOREM 3.5. *If there exists a* PBD[R_k,v], *then* $v \in R_k$.

PROOF. By remark 3.2, R_k is exactly the set of positive integers r for which there exists a GDD of type $((k-1)^r)$ with block size k.

As the master GDD in the F.C., take a PBD[R_k,v] considered as a GDD of type (k^v) with block sizes from R_k. Weight each point with k-1. The type of a partition π_A is $((k-1)^r)$ where $|A| = r \in R_k$, and so $\mathcal{B}_A$ may be chosen k-uniform. The F.C. produces a GDD of type $((k-1)^v)$, block size k, which by the observation of the previous paragraph means $v \in R_k$. □

A PBD (X,$\mathcal{A}$) is said to be *resolvable* iff there exists a partition

$$\mathcal{A} = \mathcal{A}_1 \cup \mathcal{A}_2 \cup \ldots \cup \mathcal{A}_r$$

such that each set $\mathcal{A}_i$ of blocks is a partition of X (a *parallel class* of blocks). If we let R_k^* denote the set of positive integers r for which there exists a B^*[k,r(k-1)+1] (i.e., a resolvable B[k,r(k-1)+1]), then we have

THEOREM 3.6. *If there exists a* PBD[R_k^*,v], *then* $v \in R_k^*$.

This theorem, due to RAY-CHAUDHURI and the author, may be found in [17]. We remark only that it can be proved with the same construction as in theorem 3.5, and that it was an important step in the proofs that $B^*[3,6t+3]$ and $B^*[4,12t+4]$ exist for all t (see [17,13]) and that $B^*[k,k(k-1)t+k]$ exist for all t sufficiently large with respect to k (see [18]).

Resolvable designs $B^*[k,v]$ (with fixed parallelism, i.e. partition of the blocks into parallel classes) are also known as Sperner spaces.

Instances and/or extensions of the constructions discussed in this section can be found in [3,9,11,12,17,22,23].

We conclude this section with two remarks to be used in section 5. We do not attempt to state these in the most general form.

REMARK 3.5. (COMPLETION). If $(X,\mathcal{A})$ is a resolvable B[k,v], say $\mathcal{A} = \mathcal{A}_1 \cup \ldots \cup \mathcal{A}_r$ is a partition of $\mathcal{A}$ into parallel classes ($r = (v-1)/(k-1)$), then we may *complete* $(X,\mathcal{A})$ to a PBD[{k+1,r},v+r] $(X',\mathcal{A}')$ by adding r "points at infinity" $\theta_1,\ldots,\theta_r$ and a "block at infinity" $B = \{\theta_1,\ldots,\theta_r\}$. That is, we put

$$X' = X \cup \{\theta_1,\ldots,\theta_r\} ,$$

$$\mathcal{A}' = \{B\} \cup \{A \cup \{\theta_i\} : A \in \mathcal{A}_i;\ i=1,2,\ldots,r\} .$$

For $l \le r$, we may partially complete $(X,\mathcal{A})$ by adding a block of size l "at infinity" (adjoining "points at infinity" only to l of the parallel classes $\mathcal{A}_i$) to obtain a PBD[{k,k+1,l},v+l].

REMARK 3.6. (TRUNCATION). Given a transversal design TD[k+1,t], we obtain a GDD of type (s^1t^k) with block sizes from {k,k+1} by deleting t-s points from one of the groups (and from the blocks which contain them) of the transversal design.

4. A CLOSURE OPERATION

Given a set K (finite or infinite) of positive integers, we denote by $\mathbb{B}(K)$ the set of positive integers v for which there exists a PBD[K,v]. The mapping $K \to \mathbb{B}(K)$ is a *closure operation* on the subsets of the positive integers; that is, it enjoys the properties

(i) $K \subseteq \mathbb{B}(K)$,

(ii) $K_1 \subseteq K_2 \Rightarrow \mathbb{B}(K_1) \subseteq \mathbb{B}(K_2)$,

(iii) $\mathbb{B}(\mathbb{B}(K)) = \mathbb{B}(K)$.

These are easily verified. (Property (iii) is a consequence of remark 3.4.)

We call a set K of positive integers *PBD-closed* (or simply *closed*) when $\mathbb{B}(K) = K$. Theorem 1.1 asserts that every closed set K contains all sufficiently large integers v with $v \equiv 1 \pmod{\alpha(K)}$ and $v(v-1) \equiv 0 \pmod{\beta(K)}$. Thus there are only countably many closed sets.

Using the fact that the greatest common divisor of any set is equal to the g.c.d. of a finite subset, a consequence of theorem 1.1 is (see [23])

THEOREM 4.1. *If* K *is a closed set, then there exists a finite subset* $J \subseteq K$ *such that* $K = \mathbb{B}(J)$.

Examples of closed sets seem to arise naturally in certain existence problems. We give several such examples below. The proof that a set defined combinatorially is closed invariably involves a construction (as in remark 3.4, which shows that $\mathbb{B}(K)$ is closed for any set K). Of course, the construction contains far more information than simply the assertion that a set is closed.

Theorems 3.5 and 3.6 can be rephrased as

EXAMPLE 4.1. For any $k \geq 2$, the sets R_k and R_k^* are closed.

To apply theorem 1.1 to a closed set K, it is necessary to know something about the parameters $\alpha(K)$ and $\beta(K)$. This involves exhibiting some elements of K. For if we know $k \in K$, then already we can say $\beta(K)$ divides $k(k-1)$ and hence all large $v \equiv 1 \pmod{k(k-1)}$ belong to K. More generally, if we know $\beta(K)$ divides b and $u \in K$, K a closed set, then K contains all large $v \equiv u \pmod{b}$.

We point out some simple arithmetic examples of closed sets.

EXAMPLE 4.2. Let a be a positive integer. Then $H^a = \{v : v \equiv 1 \pmod{a}\}$ is closed.

For a clearly divides $\alpha(H_a)$; hence $v \in \mathbb{B}(H^a)$ implies $v \in H^a$.

EXAMPLE 4.3. Let b be a positive integer. Then $H_b = \{v : v(v-1) \equiv 0 \pmod{b}\}$ is closed.

For b divides $\beta(H_b)$; hence $v \in \mathbb{B}(H_b)$ implies $v \in H_b$.

EXAMPLE 4.4. Let m be a positive integer. Then $\{1\} \cup \{v : v \geq m\}$ is closed.

This is easily seen. Since the intersection of closed sets is closed, we have

EXAMPLE 4.5. Let a,b,m be positive integers. Then

$$H_b^a = \{v : v \equiv 1 \pmod a, v(v-1) \equiv 0 \pmod b\}$$

and

$$H_b^a \cap \{v : v = 1 \quad \text{or} \quad v \geq m\} \quad \text{are closed.}$$

We remark that $\alpha(H_b^a) = a$ and $\beta(H_b^a) = b$ if and only if a divides b, b is even, and b/a is relatively prime to a (allowing a = b = 0) [23].

The next several examples of closed sets arise from the following observation that PBD's can be used to combine idempotent quasigroups to form another, larger, idempotent quasigroup.

A *quasigroup* on a finite set X is a binary operation $Q: X \times X \to X$ which satisfies both cancellation laws, i.e. the values of any two of $x,y,z \in X$ uniquely determine the value of the third so that the equation $xQy = z$ is valid. The quasigroup Q is *idempotent* when $xQx = x$ for all $x \in X$.

Let $(X,\mathcal{A})$ be a PBD and for each block $A \in \mathcal{A}$, let Q_A be an idempotent quasigroup on A. Then define Q on X by $xQx = x$, and for distinct $x,y \in X$, $xQy = xQ_Ay$ where A is the unique block of $\mathcal{A}$ containing x and y. It is easily checked that Q is an idempotent quasigroup on X.

EXAMPLE 4.6. Given k, let L_k' denote the set of positive integers n for which there exists a set of k mutually orthogonal Latin squares of order n which admit a common transversal (we may suppose, e.g., that all symbols occur on the diagonal of each square). Then L_k' is a closed set.

This is an observation of BOSE, SHRIKHANDE & PARKER [4] and was instrumental in their disproof of EULER's conjecture (see also [8,14]). We find it convenient here to explain their construction in terms of quasigroups.

Two quasigroups Q_1, Q_2 on the same set X are *orthogonal* iff for any $a,b \in X$, the system of equations

$$xQ_1y = a, \qquad xQ_2y = b$$

has a unique simultaneous solution (x,y). The assertion $n \in L_k'$ is equivalent to the existence of k mutually orthogonal idempotent quasigroups of order n.

Now given $n \in \mathbb{B}(L_k')$, there exists a PBD $(X,\mathcal{A})$ with $|X| = n$ and $|A| \in L_k'$ for all $A \in \mathcal{A}$. For each block A, we can find k mutually orthogonal idempotent quasigroups $Q_A^{(1)},\ldots,Q_A^{(k)}$ on A. Then the above construction produces idempotent quasigroups $Q^{(1)},\ldots,Q^{(k)}$ on X, which are readily seen to be mutually orthogonal. Hence $n \in L'_k$ and we have shown that L'_k is closed.

There is no point in applying theorem 1.1 to the existence problem for orthogonal Latin squares, since these results are used heavily in the proof of theorem 1.1. However, one can easily see that the number of non-isomorphic sets of k mutually orthogonal Latin squares of order n goes to infinity with n (use theorem 4.1 in a manner analogous to the proof of theorem 2 in [24]).

A quasigroup Q is *self-orthogonal* when it is orthogonal to its transpose Q' defined by xQ'y = yQx.

EXAMPLE 4.7. The set S of positive integers n for which there exists a self-orthogonal quasigroup of order n is closed.

The same construction works, as noticed by J.F. LAWLESS [14]. Using $S = \mathbb{B}(S)$ and some special constructions, BRAYTON, COPPERSMITH & HOFFMAN [5] have recently shown that S contains all positive integers except 2,3 and 6. Again, we claim that the number of non-isomorphic such quasigroups tends to infinity with n.

A *Room pair* of quasigroups is a pair Q_1,Q_2 of commutative idempotent quasigroups on the same set X such that for any $a,b \in X$, there is at most one unordered pair {x,y} with $xQ_1y = a$ and $xQ_2y = b$.

EXAMPLE 4.8. The set R of positive integers n such that there exists a Room pair of quasigroups of order n is closed.

This was first noticed by J.F. LAWLESS [14]. The construction is also discussed in [19].

We conclude our remarks on quasigroups by observing that GDD's can be used to construct (not necessarily idempotent) quasigroups: if each block A of a GDD $(X,\mathcal{S},\mathcal{A})$ is equipped with an idempotent quasigroup Q_A and each group G is equipped with any quasigroup Q_G, then there is a natural quasigroup Q on X, where $xQx = xQ_Gx$ for $x \in G$. When $\mathcal{S}$ consists of all singleton subsets of X, this construction degenerates into the basic construction with PBD's (cf. remark 3.1).

We introduce designs $PBD_\lambda[K,v]$, where each pair of distinct points is contained in exactly λ blocks, just for the purpose of stating

EXAMPLE 4.9. For any set K of positive integers and a positive integer λ, the set $\mathbb{B}_\lambda(K) = \{v : \text{there exists a } PBD_\lambda[K,v]\}$ is a closed set.

Other examples of closed sets are

EXAMPLE 4.10. The set of positive integers r for which there exists a reverse Steiner triple system of order 2r+1 (J. DOYEN [6]).

EXAMPLE 4.11. The set of v for which there exists a pair of orthogonal Steiner triple systems of order v (J.F. LAWLESS [14]).

EXAMPLE 4.12. The set of integers n for which there exists a GDD of type (m^n) and block sizes from K (see [22,20]).

One further example is given in section 6.

5. GENERATING CLOSED SETS

We observe that each closed set K has a unique minimal generating set. Let us call an element $x \in K$ *essential* in K iff $x \notin \mathbb{B}(K - \{x\})$, or equivalently, $x \notin \mathbb{B}(\{y \in K : y < x\})$. Letting E_K denote the set of all essential elements of K, we have

PROPOSITION 5.1. *Let* J *be a subset of a closed set* K. *Then* $\mathbb{B}(J) = K$ *if and only if* $E_K \subseteq J$.

PROOF. Clearly, $\mathbb{B}(J) = K$ implies $E_K \subseteq J$. We claim now that $\mathbb{B}(E_K) = K$. If not, let x be the least element of $K-\mathbb{B}(E_K)$. Then x is not essential, so $x \in \mathbb{B}(\{y \in K : y < x\}) \subseteq \mathbb{B}(\mathbb{B}(E_K)) = \mathbb{B}(E_K)$, a contradiction. □

Note that theorem 4.1 implies E_K is finite.

We may ask for the set of essential elements, or at least a reasonably small generating set, for the arithmetic examples H_b^a of section 4. HANANI [9,10] establishes that

$$H_3^1 = \{0,1 \pmod 3\} = \mathbb{B}\{3,4,6\} ,$$

$$H_4^1 = \{0,1 \pmod 4\} = \mathbb{B}\{4,5,8,9,12\} ,$$

$$H_5^1 = \{0,1 \pmod 5\} = \mathbb{B}\{5,6,10,11,15,16,20,35,36,40,70,71,75,76\} ,$$

as preliminary results before proving

$$H_6^2 = \{1,3 \pmod 6\} = \mathbb{B}\{3\} ,$$
$$H_{12}^3 = \{1,4 \pmod{12}\} = \mathbb{B}\{4\} ,$$
$$H_{20}^4 = \{1,5 \pmod{20}\} = \mathbb{B}\{5\} .$$

The reader may check that the essential elements of H_3^1 and H_4^1 are exactly those listed. We can improve the result for H_5^1 to

PROPOSITION 5.2. $H_5^1 = \mathbb{B}\{5,6,10,11,16,20,35,40\}$.

PROOF. We assume HANANI's result above and proceed to show that 36,70,71, 75,76 are not essential in H_5^1.

R.C. BOSE [2] has shown the existence of resolvable $B[q+1, q^3+1]$ for all prime powers q. In particular, we have the existence of a resolvable B[5,65], which we may partially complete (remark 3.5) by adding a block at infinity of size 5,6,10,11, respectively, to find $70,71,75,76 \in \mathbb{B}\{5,6,10,11\}$.

To see 36 is non-essential in H_5^1, we remove two intersecting lines (and their points) from a B[7,49] (affine plane of order 7) to obtain a PBD on 36 points with 18 blocks of size 6 and 36 of size 5 (i.e., $36 \in \mathbb{B}\{5,6\}$). □

The author does not know whether 35 and 40 are essential in H_5^1.

HANANI's paper [9] also exhibits finite generating sets for the closed sets $\{1\} \cup \{v : v \geq k\}$ for k = 3,4,5. And the result $\{1\} \cup \{v : v \geq k\} = \mathbb{B}\{k,k+1,k+2,\ldots,kq_k-1\}$ where q_k is the smallest prime power such that $q_k \geq 2k-1$, can be found in [12].

THEOREM 5.1.

(i) $H_2^2 = \{1 \pmod 2\} = \mathbb{B}\{3,5\}$,

(ii) $H_3^3 = \{1 \pmod 3\} = \mathbb{B}\{4,7,10,19\}$,

(iii) $H_4^4 = \{1 \pmod 4\} = \mathbb{B}\{5,9,13,17,29,33,49,57,89,93,129,137\}$.

LEMMA 5.1. *If* k *and* k+1 *are prime powers, the existence of a* GDD *on* v *points with block sizes from* {k+1,k+2} *and at least two groups implies that* vk+1 *is not essential in* $H_k^k = \{1 \pmod k\}$.

PROOF. We take such a GDD as the master GDD in the Fundamental Construction 3.1. Each point is to be weighted with k. The type of any partition π_A is (k^{k+1}) or (k^{k+2}). The classes $\mathcal{B}_A$ may be taken to be (k+1)-uniform as GDD's with these types and block size k+1 may be obtained by deleting points (remark 3.2) from $B[k+1, k^2+k+1]$ and $B[k+1,(k+1)^2]$, respectively. The F.C. produces a GDD on vk points with groups of size divisible by k, and blocks of size k+1. Adjoining a point by remark 3.2, we obtain a PBD on vk+1 points with block sizes from $H_k^k = \{1 \pmod k\}$ (and less than vk+1). This proves the lemma. □

<u>PROOF OF THEOREM 5.1</u>. If there exists a TD(k+2,t), remark 3.6 shows that GDD's with v points satisfying the hypothesis of lemma 5.1 exist for $(k+1)t \le v \le (k+2)t$. With this observation and theorem 3.2, it can be seen (we omit the details) that with E_k denoting the essential elements of H_k^k,

$$E_2 \subseteq \{3,5,11,13,15,17\} ,$$

$$E_3 \subseteq \{4,7,10\} \cup \{19,22,\ldots,43,46\} \cup \{79,82\} ,$$

$$E_4 \subseteq \{5,9,13,17\} \cup \{29,33,\ldots,93,97\} \cup \{125,129,133,137\} .$$

We proceed to show that some of the indicated integers are not essential in the respective sets H_k^k.

<u>Case k = 2.</u> 13 and 15 belong to $\mathbb{B}\{3\}$ and hence are certainly non-essential in H_2^2. Completing a resolvable design $B^*[2,6]$ by remark 3.5 shows $11 \in \mathbb{B}\{3,5\}$. Deleting a point (remark 3.2) from a B[3,9] gives a GDD satisfying the hypothesis of lemma 5.1; hence $17 = 2\cdot 8+1$ is non-essential in H_2^2. Thus $E_2 \subseteq \{3,5\}$.

<u>Case k = 3.</u> $25,28,37,40 \in \mathbb{B}\{4\}$. Completing $B^*[3,15]$ and $B^*[3,21]$ shows $22 \in \mathbb{B}\{4,7\}$ and $31 \in \mathbb{B}\{4,10\}$. We show 43,46,79,82 are non-essential in H_3^3 by exhibiting GDD's on 14,15,26,27 points, respectively, with blocks of size 4 and using lemma 5.1. For 15 and 27, delete points (remark 3.2) from B[4,16] and B[4,28]. For 14, take the 14 points to be Z_{14}, groups {i,i+7} (i=0,1,...,6), and blocks {2+i,4+i,8+i,7+i} (i=0,1,...,13). For 26, take points Z_{26}, groups {i,i+13} (i=0,1,...,12), and blocks {2+i,6+i,18+i, 13+i}, {4+i,12+i,10+i,13+i} (i=0,1,...,25). Finally, $34 \in \mathbb{B}\{4,7,10\}$ follows from lemma 5.2 below (q = 3). We have proved $E_3 \subseteq \{4,7,10,19\}$ (and it is easily checked that equality holds).

<u>Case k = 4.</u> $41,45,61,65,81,125 \in \mathbb{B}\{5\}$ (cf.[9]) and $73 \in \mathbb{B}\{9\}$ (projective plane of order 8), hence these values are non-essential in H_4^4. Completing (by remark 3.5) resolvable designs $B^*[4,v]$ for v = 28,40,52,100 (which exist [13]), we see that 37,53,69,133 are

non-essential in H_4^4. Deleting a point from a $B[5,25]$ gives a GDD on 24 points and lemma 5.1 shows $97 = 4\cdot 24+1$ is non-essential. Finally, lemma 5.2 below ($q = 4$) shows $77 \in \mathbb{B}\{5,13,17\}$. We have proved that all $x \equiv 1 \pmod 4$ are non-essential in H_4^4 with the possible exceptions of $x = 5,9,13,17,29,33,49,57,89,93,129,137$. □

LEMMA 5.2. *If* q *is a prime power, then* $q^3+q^2-q+1 \in \mathbb{B}\{q+1,q^2-q+1,q^2+1\}$.

PROOF. Let Π be a projective plane of order q^2 with a Baer subplane Π_0 (of order q). Let θ be a point of Π_0, $L_0,L_1,\ldots,L_q$ the lines of Π which contain θ and belong to the subplane Π_0, and L^* another line of Π containing θ. Let L_i' denote the set of the q^2-q points of L_i which do not belong to Π_0. The set $X = L^* \cup (\bigcup_{i=0}^{q} L_i')$ of $(q^2+1) + (q+1)(q^2-q)$ points of Π, together with the non-trivial intersections of lines of Π with X, provides a PBD with one block (namely L^*) of size q^2+1, $q+1$ blocks (namely, $L_i' \cup \{\theta\}$, $i=0,1,\ldots q$) of size q^2-q+1, and the remaining blocks have size $q+1$. This last assertion follows from the fact that each line of Π contains exactly one or $q+1$ points of Π_0. □

We indicate how theorem 5.1 (ii) can be used to derive two known results.

Kirkman Designs $B^*[3,v]$: Simple direct constructions [17] show that $B^*[3,v]$ exist for $v = 9,15,21,39$, i.e. $\{4,7,10,19\} \subseteq R_3^*$. By theorems 5.1 (ii) and 3.6, $H_3^3 = \mathbb{B}\{4,7,10,19\} \subseteq \mathbb{B}(R_3^*) = R_3^*$; which means that resolvable designs $B^*[3,6t+3]$ exist for all t.

Designs $B_2[4,3t+1]$ ($\lambda = 2$): If we establish that designs $B_2[4,v]$ exist for $v = 4,7,10,19$, it follows from theorem 5.1 (ii) and example 4.9 that $B_2[4,3t+1]$ exist for all t.

6. EDGE-DECOMPOSITIONS OF COMPLETE GRAPHS

Let Γ be a graph. By an *edge-decomposition* (or simply a *decomposition*) of a graph Γ^* into Γ-graphs we mean a set $\Gamma_1,\Gamma_2,\ldots,\Gamma_t$ of subgraphs of Γ^*, each isomorphic to Γ, such that each edge of Γ^* occurs in exactly one of the subgraphs Γ_i.

A design $B[k,v]$ may be considered as a decomposition of the complete graph on v vertices into k-cliques (complete graphs on k vertices). (More generally, a PBD can be viewed as a decomposition of a complete graph into cliques; and a GDD should perhaps be viewed as a decomposition of a complete multipartite graph into cliques.)

Given a simple graph Γ, we denote by K_Γ the set of positive integers v for which the complete graph on v vertices admits a decomposition into Γ-graphs. The purpose of this last section is to provide a proof for

THEOREM 6.1. *Let* Γ *be a simple graph with* m *edges. Then* K_Γ *is a closed set and* $\beta(K_\Gamma) = 2m$.

That K_Γ is closed is easily verified; a consequence of theorem 6.1 and theorem 1.1 is that K_Γ contains all sufficiently large integers $v \equiv 1 \pmod{2m}$.

We observe that if $v \in K_\Gamma$, then surely the number m of edges of Γ divides $\binom{v}{2}$, i.e. $v(v-1) \equiv 0 \pmod{2m}$. Thus 2m is a divisor of $\beta(K_\Gamma)$. We show below that K_Γ contains the set Q of sufficiently large prime powers $q \equiv 1 \pmod{2m}$ (lemma 6.1), and observe that $\beta(Q) = 2m$ (lemma 6.2). $Q \subseteq K_\Gamma$ implies $\beta(K_\Gamma)$ divides $\beta(Q) = 2m$, and the proof of theorem 6.1 will then be complete. □

LEMMA 6.1. *If* Γ *is a simple graph with* m *edges and* k *vertices, then the complete graph on* q *vertices can be decomposed into* Γ*-graphs whenever* q *is a prime power of the form* $q = 2mt+1$ *and* $q > m^{k^2}$.

PROOF. Let $C_0, C_1, \ldots, C_{m-1}$ be the cyclotomic classes of index m in GF(q) and let S_i be the set of pairs $\{x,y\}$ of distinct field elements such that $x-y \in C_i$ (since $-1 \in C_0$ in our case, x-y and y-x belong to the same class).

Theorem 3 of [21] asserts that if $q > m^{k^2}$, then for any choice of $l_{ij} \in \{0,1,\ldots,m-1\}$ there exist field elements $a_1, a_2, \ldots, a_k$ such that $\{a_i, a_j\} \in S_{l_{ij}}$ for all i,j $(1 \le i < j \le k)$. (This can also be proved with lemmas 2.1 and 2.2). If we think of the edges of S_i as being colored with color i, then the claim is that all possible m-colorings are found among the edge colorings induced on k element subsets. So surely we may find a subgraph Γ_0 of the complete graph with vertex set GF(q) which is isomorphic to Γ and such that its m edges receive distinct colors.

Let S be a system of representatives for the cosets of the factor group $C_0/\{1,-1\}$. Then the set (not necessarily group) of permutations $\{x \to ax+b : a \in S, b \in GF(q)\}$ of GF(q) is sharply transitive on the edges of color i for each i. Applying these permutations to Γ_0 produces a set of isomorphic subgraphs which partition the edges of the complete graph on GF(q). □

LEMMA 6.2. *Let* m *and* c *be positive integers and let* Q *be the set of prime powers* q *such that* $q \equiv 1 \pmod{2m}$ *and* $q > c$. *Then* $\beta(Q) = 2m$.

PROOF. Clearly $\beta(Q)$ is divisible by $2m$, say $\beta(Q) = 2mt$. We claim $t = 1$. If not, let p be a prime divisor of t. Now for at least one choice of sign, $2mp$ and $2m(p\pm1) + 1$ are relatively prime, and by DIRICHLET's theorem on primes in arithmetic progressions, there exists a prime $q = 2mpl + 2m(p\pm1)+1 >$ $> \max(c,\beta(Q))$. Then $q \in Q$, so $\beta(Q)$ divides $q(q-1)$. Since q is prime and larger than $\beta(Q)$, $\beta(Q) = 2mt$ divides $q-1$; hence $2mp$ divides $q-1 = 2m(pl+p\pm1)$. This contradiction shows that t has no prime divisors and completes the proof of the lemma. □

We close by remarking that theorem 6.1 can be used to prove a recent theorem of B. GANTER [7]. A partial PBD[K,v] is a system $(X,\mathcal{A})$ such that $|X| = v$, $|A| \in K$ for all $A \in \mathcal{A}$, and each pair of distinct points $x,y \in X$ is contained in at most one block $A \in \mathcal{A}$. GANTER's theorem asserts that for any finite partial PBD[K,v], there exists a finite PBD[K,v^*] $(X^*,\mathcal{A}^*)$ such that $X \subseteq X^*$ and $\mathcal{A} \subseteq \mathcal{A}^*$.

Now given a partial PBD $(X,\mathcal{A})$, we may consider the graph Γ with vertex set X and edge set consisting of those pairs $\{x,y\}$ of points which do occur in some block $A \in \mathcal{A}$. If the complete graph with vertex set X^* can be decomposed into Γ-graphs, then it is clear that we may find a PBD $(X^*,\mathcal{A}^*)$ where $\mathcal{A}^*$ is a union of isomorphic copies of $\mathcal{A}$.

REFERENCES

[1] BOSE, R.C., *On the construction of balanced incomplete block designs*, Annals of Eugenics, 9 (1939) 353-399.

[2] BOSE, R.C., *On the application of finite projective geometry for deriving a certain series of balanced Kirkman arrangements*, Calcutta Math. Soc. Golden Jubilee Vol., 1959, pp. 341-354.

[3] BOSE, R.C. & S.S. SHRIKHANDE, *On the composition of balanced incomplete block designs*, Canad. J. Math., 12 (1960) 177-188.

[4] BOSE, R.C., S.S. SHRIKHANDE & E.T. PARKER, *Further results on the construction of mutually orthogonal Latin squares and the falsity of a conjecture of Euler*, Canad. J. Math., 12 (1960) 189-203.

[5] BRAYTON, R.K., D. COPPERSMITH & A.J. HOFFMAN, *Self-orthogonal Latin squares of all orders $n \neq 2,3,6$*, Bull. Amer. Math. Soc., 80 (1974) 116-118.

[6] DOYEN, J., *A note on reverse Steiner triple systems*, Discrete Math., 1 (1971-72) 315-319.

[7] GANTER, B., *Partial pairwise balanced designs*, Technische Hochschule Darmstadt Preprint No. 99, November, 1973.

[8] HALL JR., M., *Combinatorial theory*, Blaisdell, Waltham, Mass., 1967.

[9] HANANI, H., *The existence and construction of balanced incomplete block designs*, Ann. Math. Statist., 32 (1961) 361-386.

[10] HANANI, H., *A balanced incomplete block design*, Ann. Math. Statist., 36 (1965) 711.

[11] HANANI, H., *On balanced incomplete block designs with blocks having five elements*, J. Combinatorial Theory A, 12 (1972) 184-201.

[12] HANANI, H., *On balanced incomplete block designs and related designs*, to appear.

[13] HANANI, H., D.K. RAY-CHAUDHURI & R.M. WILSON, *On resolvable designs*, Discrete Math., 3 (1972) 343-357.

[14] LAWLESS, J.F., *Pairwise balanced designs and the construction of certain combinatorial systems*, in: Proceedings of the Second Louisiana Conference on Graph theory, Combinatorics and Computing, 1971.

[15] MOORE, E.H., *Concerning triple systems*, Math. Ann., 43 (1893) 271-285.

[16] PELTESOHN, R., *Eine Lösung der beiden Heffterschen Differenzenprobleme*, Compositio Math., 6 (1939) 251-257.

[17] RAY-CHAUDHURI, D.K. & R.M. WILSON, *Solution of Kirkman's school girl problem*, in: Proceedings of Symposia in Pure Mathematics, Vol. 19, *Combinatorics*, T.S. MOTZKIN (ed.), Amer. Math. Soc., Providence, R.I., 1971, pp. 187-204.

[18] RAY-CHAUDHURI, D.K. & R.M. WILSON, *The existence of resolvable designs*, in: *A Survey of Combinatorial Theory*, J.N. SRIVASTAVA a.o., (ed.), North-Holland/American Elsevier, Amsterdam/New York, 1973, pp. 361-376.

[19] WALLIS, W.D., A.P. STREET & J.S. WALLIS, *Combinatorics: Room squares, sum-free sets, Hadamard matrices*, Lecture Notes in Mathematics 292, Springer-Verlag, Berlin etc., 1972.

[20] WILSON, R.M., *The construction of group divisible designs and partial planes having the maximum number of lines of a given size*, in: Proceedings of the Second Chapel Hill Conference on Combinatorial Mathematics and its Applications, University of North Carolina at Chapel Hill, 1970, pp. 488-497.

[21] WILSON, R.M., *Cyclotomy and difference families in elementary abelian groups*, J. Number Theory, 4 (1972) 17-47.

[22] WILSON, R.M., *An existence theory for pairwise balanced designs, I: Composition theorems and morphisms*, J. Combinatorial Theory A, 13 (1972) 220-245.

[23] WILSON, R.M., *An existence theory for pairwise balanced designs, II: The structure of PBD-closed sets and the existence conjectures*, J. Combinatorial Theory A, 13 (1972) 246-273.

[24] WILSON, R.M., *An existence theory for pairwise balanced designs, III: Proof of the existence conjectures*, J. Combinatorial Theory, to appear.

ON TRANSVERSAL DESIGNS

H. HANANI

University of the Negev, Beer Sheva, Israel

1. BASIC LEMMAS

A *design* is a pair $(X,\mathcal{B})$ where X is a finite set of *points* and $\mathcal{B}$ is a family of -not necessarily distinct- subsets B_i (called *blocks*) of X.

A *parallel class* of blocks of a design $(X,\mathcal{B})$ is a subfamily $\mathcal{B}_1 \subset \mathcal{B}$ of disjoint blocks which cover X.

In a design $(X,\mathcal{B})$ let the family $\mathcal{B}$ of blocks be composed of two subfamilies $\mathcal{B} = \mathcal{G} \cup \mathcal{P}$ where $\mathcal{G}$ is a parallel class of blocks. The elements (blocks) of $\mathcal{G}$ will be called *groups* and the elements (blocks) of $\mathcal{P}$ -*proper blocks* or for short- *blocks*. A design $(X,\mathcal{G} \cup \mathcal{P})$ is a *transversal design* $T[s,\lambda,r]$ iff

(i) $|G_i| = r$ for every $G_i \in \mathcal{G}$,

(ii) $|\mathcal{G}| = s$,

(iii) $|G_i \cap B_j| = 1$ for every $G_i \in \mathcal{G}$ and every $B_j \in \mathcal{P}$,

(iv) every pairset $\{x,y : x \in G_i,\ y \in G_j,\ G_i \neq G_j\}$ is contained in exactly λ blocks of $\mathcal{P}$.

It follows immediately that in $T[s,\lambda,r]$, $|X| = sr$, $|B_j| = s$ for every $B_j \in \mathcal{P}$, and $|\mathcal{P}| = r^2\lambda$.

Let us denote by $T(s,\lambda)$ the set of integers r for which designs $T[s,\lambda,r]$ exist. The following lemmas are evident.

LEMMA 1. $T(s,\lambda) \subset T(s',\lambda)$ *for every* $s' \le s$.

LEMMA 2. $T(s,\lambda) \subset T(s,n\lambda)$ *for every positive integer* n.

Lemma 2 may be generalized as follows.

LEMMA 3. *If* $r \in T(s,\lambda)$ *and* $r \in T(s,\lambda')$, *then* $r \in T(s,n\lambda+n'\lambda')$ *for all non-negative integers* n *and* n'.

M. Hall, Jr. and J. H. van Lint (eds.), Combinatorics, 43-53.

The following lemma has been proved by MACNEISH [5] for $\lambda = \lambda' = 1$. However, this, more general wording, does not involve any change in the proof.

LEMMA 4. *If* $r \in T(s,\lambda)$ *and* $r' \in T(s,\lambda')$, *then* $rr' \in T(s,\lambda\lambda')$.

We shall now prove

LEMMA 5. *In a transversal design* $T[s,\lambda,r]$, $s \le (r^2\lambda-1)/(r-1)$ *holds.*

PROOF. Let $X = I_r \times I_s$. We may assume that one of the blocks is $\{(0;\sigma) : \sigma \in I_s\}$. There are exactly $r\lambda-1$ additional blocks containing the point $(0;0)$. Each of the points $(0;\sigma')$, $\sigma'=1,2,\ldots,s-1$ occurs in these blocks exactly $\lambda-1$ times and accordingly the total number of points having as first index 0 in those $r\lambda-1$ blocks is $r\lambda-1+(s-1)(\lambda-1)$. The total number of pairs $\{(0;\sigma),(0;\sigma') : \sigma,\sigma' \in I_s, \sigma \ne \sigma'\}$ occurring in the blocks is minimized if each of the blocks has the same number of points having first index 0, namely $1+(s-1)(\lambda-1)/(r\lambda-1)$ and then the total number of the said pairs is $\frac{1}{2}(r\lambda-1)[(s-1)(\lambda-1)/(r\lambda-1)+1][(s-1)(\lambda-1)/(r\lambda-1)]$.

There are exactly $(r-1)r\lambda$ blocks not containing the point $(0;0)$. The number of points with first index 0 in those blocks is $(r-1)\lambda(s-1)$ and the total number of pairs of such points is minimized if in each block there is an equal number, i.e. $(s-1)/r$, of points with first index 0.

Summing up we have for the total number P of pairs of points with first index 0

$$P = \tfrac{1}{2}s(s-1)\lambda \ge$$

$$\ge \tfrac{1}{2}s(s-1) + \tfrac{1}{2}(r\lambda-1)\left[\frac{(s-1)(\lambda-1)}{r\lambda-1} + 1\right]\frac{(s-1)(\lambda-1)}{r\lambda-1} +$$

$$+ \tfrac{1}{2}(r-1)r\lambda\frac{s-1}{r}\left[\frac{s-1}{r} - 1\right],$$

which gives $s \le (r^2\lambda-1)/(r-1)$. □

Transversal designs satisfying $s = (r^2\lambda-1)/(r-1)$ will be called *complete transversal designs*. Transversal designs with $s < (r^2\lambda-1)/(r-1)$ will be called *incomplete transversal designs*.

If the blocks of a transversal design $T[s,\lambda,r]$ can be partitioned into $r\lambda$ parallel classes of blocks then the design will be called a *resolvable transversal design* $RT[s,\lambda,r]$. As usual the set of integers r for which designs $RT[s,\lambda,r]$ exist will be denoted by $RT(s,\lambda)$.

By adding an additional group to a resolvable transversal design $RT[s,\lambda,r]$ and adjoining each element of this group to λ distinct parallel classes of blocks we obtain

LEMMA 6. $RT(s,\lambda) \subset T(s+1,\lambda)$.

For $\lambda = 1$ the stronger result is known

LEMMA 7. $RT(s,1) = T(s+1,1)$.

Transversal designs are also known as *Orthogonal arrays*. A full description of such arrays including bibliography may be found in the book of DAMARAJU RAGHAVARAO [8, p.9-31].

2. COMPLETE TRANSVERSAL DESIGNS

2.1. Hadamard matrices

The complete transversal designs with r = 2, namely the designs $T[4\lambda-1,\lambda,2]$, are equivalent to the Hadamard matrices. To see this, write the groups of the design in any fixed order and in each group denote one point by +1 and the other by -1. Further write the blocks of the design as rows of a matrix and add a column of +1's. The obtained matrix is a $4\lambda \times 4\lambda$ Hadamard matrix.

The designs $T[4\lambda-1,\lambda,2]$ with $\lambda=2,3,\ldots,8$ are given herewith.

$2 \in T(7,2)$ $\quad X = I_2 \times Z_7$

$\{(0;0),(0;1),(0;2),(0;3),(0;4),(0;5),(0;6)\}$

$\{(0;1),(0;2),(0;4),(1;0),(1;3),(1;5),(1;6)\}$ mod $(-;7)$

$2 \in T(11,3)$ $\quad X = I_2 \times Z_{11}$

$\{(0;\zeta) : \zeta \in Z_{11}\}$

$\{(1;0),(0;2^{2\alpha}),(1;2^{2\alpha+1}) : \alpha=0,1,2,3,4\}$ mod $(-;11)$

$2 \in T(15,4)$ $\quad X = I_2 \times Z_{15}$

$\{(0;\zeta) : \zeta \in Z_{15}\}$

$\{(0;1),(0;2),(0;3),(0;5),(0;6),(0;9),(0;11),(1;0),(1;4),(1;7),(1;8),(1;10),$
$(1;12),(1;13),(1;14)\}$ mod $(-;15)$

$2 \in T(19,5)$ $\quad X = I_2 \times Z_{19}$

$\{(0;\zeta) : \zeta \in Z_{19}\}$

$\{(1;0),(0;2^{2\alpha}),(1;2^{2\alpha+1}) : \alpha=0,1,\ldots,8\}$ mod $(-;19)$

$2 \in T(23,6)$ $\quad X = I_2 \times Z_{23}$

$\{(0;\zeta) : \zeta \in Z_{23}\}$

$\{(1;0),(0;5^{2\alpha}),(1;5^{2\alpha+1}) : \alpha=0,1,\ldots,10\} \bmod (-;23)$

$2 \in T(27,7)$ $\quad X = I_2 \times GF(27)$ $\quad x^3 = x + 2$

$\{(0;\zeta) : \zeta \in GF(27)\}$

$\{(1;0),(0;x^{2\alpha}),(1;x^{2\alpha+1}) : \alpha=0,1,\ldots,12\} \bmod (-;27)$

$2 \in T(31,8)$ $\quad X = I_2 \times Z_{31}$

$\{(0;\zeta) : \zeta \in Z_{31}\}$

$\{(1;0),(0;3^{2\alpha}),(1;3^{2\alpha+1}) : \alpha=0,1,\ldots,14\} \bmod (-;31)$.

2.2. Projective planes

The complete transversal designs with $\lambda = 1$, namely the designs $T[q+1,1,q]$ are equivalent to the finite projective planes $PG(2,q)$. As lines in the plane may serve the blocks as well as the groups with an additional point (∞) adjoint.

It is known that finite projective planes exist whenever q is a power of a prime and accordingly we have

LEMMA 8. *If* q *is a power of a prime, then* $q \in T(q+1,1)$.

2.3. Projective geometries

It is known that finite projective geometries $PG(d,q)$ of any dimension d exist if q is a power of a prime. Such geometry enables us to construct a complete transversal design $T[(q^d-1)/(q-1),q^{d-2},q]$. To this end fix any point A of $PG(d,q)$ and define as groups all the $(q^d-1)/(q-1)$ lines through A, with A deleted. Every $(d-1)$-dimensional hyperplane not incident with A intersects every group in exactly one point and we may define those $(d-1)$-dimensional hyperplanes as blocks of the design. Through every two points of distinct groups goes exactly one line of the geometry. This line is contained in $(q^{d-1}-1)/(q-1)$ $(d-1)$-dimensional hyperplanes; out of these hyperplanes $(q^{d-2}-1)/(q-1)$ are incident with A, the other q^{d-2} are blocks of the design. Accordingly every pair of points in distinct groups is contained in exactly q^{d-2} blocks. Consequently we obtain

THEOREM 1. *If* q *is a power of a prime, then* $q \in T((q^d-1)/(q-1), q^{d-2})$ *for every integer* $d > 1$.

Two designs of the described form are given herewith.

$3 \in T(13,3)$ $\quad X = (Z_2 \cup \{\infty\}) \times Z_{13}$

$\{(\infty;\zeta) : \zeta \in Z_{13}\}$

$\{(\infty;0),(\infty;2^{4\alpha+1}),(0;2^{4\alpha}),(1;2^{4\alpha+2}),(1;2^{4\alpha+3}) : \alpha=0,1,2\} \bmod (2;13)$

$5 \in T(31,5)$ $\quad X = (Z_4 \cup \{\infty\}) \times Z_{31}$

$\{(\infty;\zeta) : \zeta \in Z_{31}\}$

$\{(\infty;3^{10\alpha}),(\infty;3^{10\alpha+3}),(0;0),(0;3^{10\alpha+1}),(0;3^{10\alpha+4}),(1;3^{10\alpha+5}),(1;3^{10\alpha+7}),$
$(1;3^{10\alpha+9}),(2;3^{10\alpha+8}),(3;3^{10\alpha+2}),(3;3^{10\alpha+6}) : \alpha=0,1,2\} \bmod (4;31).$

3. INCOMPLETE TRANSVERSAL DESIGNS

3.1. Affine geometries

In the case $\lambda = 1$ an incomplete resolvable transversal design $RT[q,1,q]$ representing an affine plane $AG(2,q)$ is obtained from a complete transversal design $T[q+1,1,q]$ representing a projective plane $PG(2,q)$ by omitting one group. Such a design exists whenever q is a power of a prime.

In the general case, finite affine geometries $AG(d,q)$ of any dimension d exist if q is a power of a prime. Such geometry enables us to construct a resolvable transversal design $RT[q^{d-1},q^{d-2},q]$ as follows:
Take any class of parallel lines of the geometry as groups and every (d-1)-dimensional hyperplane which does not contain a group, as a block. In this construction we obtain q^{d-1} classes of q parallel blocks each. Consequently

THEOREM 2. *If* q *is a power of a prime then for every integer* d, $q \in RT(q^{d-1},q^{d-2})$.

A design of the described form is given herewith

$3 \in T(9,3)$ $\quad X = Z_3 \times GF(9)$ $\quad x^2 = 2x + 1$

$\{(0;0),(1;x^{2\alpha}),(2;x^{2\alpha+1}) : \alpha=0,1,2,3\} \bmod (3;9)$

3.2. Latin squares

It has been observed long ago, that the existence of a transversal design T[s,1,r] is equivalent to the existence of s-2 mutually orthogonal Latin squares of order r. As the Latin squares have become more popular, most of the results are stated in a form convenient in that field. We shall word these results in a way accepted for transversal designs.

From lemmas 4 (with $\lambda = \lambda' = 1$) and 8 follows the result of MACNEISH [6].

LEMMA 9. *If* $r = \Pi p_i^{\alpha_i}$ *is the factorization of* r *into powers of distinct primes, then* $r \in T(s,1)$, *where* $s-1 = \min p_i^{\alpha_i}$.

Some other most important results in this field are listed below.

(1) $r \in T(s,1)$ whenever $r > (3s)^{91}$ (CHOWLA, ERDÖS & STRAUSS [3]).
This result has been improved by ROGERS [9] and lately by WILSON who proved

(2) for sufficient large s, $r \in T(s,1)$ whenever $r > s^{17}$ (WILSON [10]).
Further it has been proved

(3) for every $r > 6$, $r \in T(4,1)$ holds (BOSE, PARKER and SHRIKHANDE [7,2,1]),

(4) for every $r > 51$, $r \in T(5,1)$ holds (HANANI [5]),

(5) for every $r > 62$, $r \in T(7,1)$ holds (HANANI [5]),

(6) for every $r > 90$, $r \in T(8,1)$ holds (WILSON [10]).

For further reference we mention a result by DULMAGE, JOHNSON & MENDELSOHN [4] who proved the existence of 5 mutually orthogonal Latin squares of order 12, or, in our notation

LEMMA 10. $12 \in T(7,1)$.

3.3. General transversal designs

Let a transversal design $T[\sigma,\lambda,\rho]$ exist. Delete some elements from one of the groups so that in this group $\rho' \leq \rho$ elements are left. The blocks will be of size σ and $\sigma-1$. It follows immediately

LEMMA 11. *If* $\rho \in T(\sigma,\lambda)$ *and* $\rho' \leq \rho$, *and if for a given* s, $\{\sigma,\sigma-1\} \subset RT(s,1)$ *and* $\{\rho,\rho'\} \subset T(s,\lambda)$, *then* $r = \rho(s-1) + \rho' \in T(s,\lambda)$.

We shall now prove

THEOREM 3. *For every* $r \geq 1$ *and every* $\lambda > 1$, $r \in T(7,\lambda)$ *holds.*

PROOF. For $r = 1$ the lemma is trivial. Further, it follows from lemmas 4 and 11 that in order to prove our lemma for every $r > 1$ it is sufficient to prove it for the factors of 60. This is done presently.

For $r = 2$ we proved in section 2.1 that $2 \in T(7,2)$ and $2 \in T(11,3)$. By lemmas 1 and 3 it follows

(3.1) $\quad 2 \in T(7,\lambda)$ for every $\lambda > 1$.

For $r = 3$ it follows from theorem 1 (with $d = 3$) that $3 \in T(13,3)$. Further we have

$3 \in RT(6,2) \qquad X = Z_3 \times (Z_3 \times I_2)$

$\{(0;0,0),(0;1,0),(1;2,0),(2;0,1),(2;1,1),(1;2,1)\}$ mod $(3;3,-)$

$\{(0;2\alpha,0),(\alpha;2\alpha+1,0),(2\alpha;2\alpha+2,0),(0;\alpha,1),(\alpha;\alpha+1,1),(2\alpha;\alpha+2,1)\}$ mod $(3;-,-)$, $\alpha=0,1,2$.

Consequently by lemmas 6, 1 and 3 we have

(3.2) $\quad 3 \in T(7,\lambda)$ for every $\lambda > 1$.

For $r = 4$ we have

$4 \in RT(8,2) \qquad X = GF(4) \times (Z_7 \cup \{\infty\}) \qquad x^2 = x + 1$

$\{(0;\zeta) : \zeta \in (Z_7 \cup \{\infty\})\}$ mod $(4;-)$

$\{(0;\infty),(0;0),(x^0;3^\alpha),(x^1;3^{\alpha+2}),(x^2;3^{\alpha+4}) : \alpha=0,1\}$ mod $(4;7)$

$4 \in RT(8,3) \qquad X = GF(4) \times I_8 \qquad x^2 = x + 1$

$\{(0;0),(0;1),(x^\alpha;2),(x^{\alpha+1};3),(0;4),(x^{\alpha+1};5),(x^\alpha;6),(x^\alpha;7)\}$ mod $(4;-)$, $\alpha=0,1,2$

$\{(0;0),(x^\alpha;1),(0;2),(x^{\alpha+1};3),(x^{\alpha+1};4),(0;5),(x^\alpha;6),(x^{\alpha+1};7)\}$ mod $(4;-)$, $\alpha=0,1,2$

$\{(0;0),(x^\alpha;1),(x^{\alpha+1};2),(0;3),(x^{\alpha+1};4),(x^\alpha;5),(0;6),(x^\alpha;7)\}$ mod $(4;-)$, $\alpha=0,1,2$

$\{(0;0),(x^{\alpha+1};1),(x^{\alpha+1};2),(x^{\alpha+1};3),(x^\alpha;4),(x^\alpha;5),(x^\alpha;6),(0;7)\}$ mod $(4;-)$, $\alpha=0,1,2$.

Considering lemmas 1, 3 and 6 we have

(3.3) $\quad 4 \in RT(8,\lambda)$ and $4 \in T(9,\lambda)$ for every $\lambda > 1$.

For $r = 5$

$5 \in RT(10,2)$ $X = Z_5 \times (Z_5 \times I_2)$

$\{(0;0,\alpha),(4;2^{2\alpha},0),(1;2^{2\alpha+1},0),(2;2^{2\alpha},1),(3;2^{2\alpha+1},1) : \alpha=0,1\} \bmod (5;5,-)$

$\{(0;0,\alpha),(2;2^{2\alpha},0),(3;2^{2\alpha+1},0),(3;2^{2\alpha},1),(2;2^{2\alpha+1},1) : \alpha=0,1\} \bmod (5;5,-)$.

$5 \in RT(7,3)$ $X = Z_5 \times Z_7$

$\{(0;\zeta) : \zeta \in Z_7\} \bmod (5;-)$

$\{(0;0),(2^{\beta};3^{3\alpha}),(2^{\beta+1};3^{3\alpha+2}),(2^{\beta+3};3^{3\alpha+1}) : \alpha=0,1\} \bmod (5;7)$, $\beta=0,1$.

Consequently by lemmas 1, 3 and 6

(3.4) $5 \in RT(7,\lambda)$ and $5 \in T(8,\lambda)$ for every $\lambda > 1$.

For $r = 6$

$6 \in T(7,2)$ $X = (Z_5 \cup \{\infty\}) \times Z_7$

$\{(\infty;\zeta) : \zeta \in Z_7\}$ 2 times

$\{(\infty;0),(0;3^{3\alpha}),(2^{\beta};3^{3\alpha+2}),(2^{\beta+2};3^{3\alpha+1}) : \alpha=0,1\} \bmod (5;7)$, $\beta=0,1$.

$6 \in T(7,3)$ $X = (Z_5 \cup \{\infty\}) \times Z_7$

$\{(\infty;\zeta) : \zeta \in Z_7\}$ 3 times

$\{(\infty;0),(0;2),(0;3),(0;5),(2^{2\alpha};4),(2^{2\alpha+1};1),(2^{2\alpha+3};6)\} \bmod (5;7)$, $\alpha=0,1$

$\{(\infty;0),(0;3^{3\alpha}),(1;3^{3\alpha+2}),(4;3^{3\alpha+1}) : \alpha=0,1\} \bmod (5;7)$.

Consequently, by lemma 3, we have

(3.5) $6 \in T(7,\lambda)$ for every $\lambda > 1$.

For $r = 10$

$10 \in T(7,2)$ $X = (Z_9 \cup \{\infty\}) \times Z_7$

$\{(\infty;\zeta) : \zeta \in Z_7\}$ 2 times

$\{(0;\zeta) : \zeta \in Z_7\} \bmod (9;-)$

$\{(\infty;0),(0;3^{3\alpha}),(1;3^{3\alpha+2}),(2;3^{3\alpha+4}),(4;3^{3\alpha+5}),(6;3^{3\alpha+1}),(7;3^{3\alpha+3})\} \bmod (9;7)$, $\alpha=0,1$

$\{(0;0),(1;3^{3\alpha+2}),(2;3^{3\alpha}),(7;3^{3\alpha+1}) : \alpha=0,1\} \bmod (9;7)$.

$10 \in T(8,3)$ $\quad X = (Z_3 \times Z_3 \cup \{\infty\}) \times (Z_2 \times Z_2 \times Z_2)$

$\{(\infty;\zeta) : \zeta \in Z_2 \times Z_2 \times Z_2\}$ 3 times

$\{(0,0;\zeta) : \zeta \in Z_2 \times Z_2 \times Z_2\}$ mod $(3,3;-)$

$\{(\infty;0,0,0),(0,0;0,0,1),(0,0;0,1,0),(0,1;0,1,1),(0,2;1,1,0),(1,2;1,0,1),$
$(1,2;1,0,0),(2,0;1,1,1)\}$ mod $(3,3;2,2,2)$

$\{(\infty;0,0,0),(0,0;0,1,0),(0,0;1,0,0),(0,1;1,1,0),(0,2;1,0,1),(1,1;0,1,1),$
$(1,1;0,0,1),(2,1;1,1,1)\}$ mod $(3,3;2,2,2)$

$\{(\infty;0,0,0),(0,0;1,0,0),(0,0;0,0,1),(0,1;1,0,1),(0,2;0,1,1),(1,0;1,1,0),$
$(1,0;0,1,0),(2,2;1,1,1)\}$ mod $(3,3;2,2,2)$

$\{(0,0;0,1,1),(0,1;1,0,1),(0,2;1,1,0),(1,0;0,1,0),(1,1;1,0,0),(1,2;0,0,1),$
$(2,0;0,0,0),(2,0;1,1,1)\}$ mod $(3,3;2,2,2)$.

By lemmas 1 and 3 it follows

(3.6) $\quad 10 \in T(7,\lambda)$ for every $\lambda > 1$.

For $r = 15$

$15 \in RT(7,2)$ $\quad X = (Z_3 \times Z_5) \times Z_7$

$\{(0,0;\zeta) : \zeta \in Z_7\}$ mod $(3,5;-)$, 2 times

$\{(0,2^{\gamma+3};0),(2^{\alpha+\beta},0;3^{3\alpha}),(2^{\alpha+\beta},2^{\gamma};3^{3\alpha+2}),(2^{\alpha+\beta},2^{\gamma+2};3^{3\alpha+4}) : \alpha=0,1\}$
mod $(3,5;7)$, $\beta=0,1$, $\gamma=0,1$.

$15 \in RT(7,3)$ $\quad X = Z_{15} \times Z_7$

$\{(0;\zeta) : \zeta \in Z_7\}$ mod $(15;-)$, 3 times

$\{(0;0),(1;3^{\alpha}),(2;3^{\alpha+1}),(4;3^{\alpha+2}),(5;3^{\alpha+3}),(8;3^{\alpha+4}),(10;3^{\alpha+5})\}$ mod $(15;7)$,
$\alpha=0,1,\ldots,5$.

By lemmas 3 and 6 we have

(3.7) $\quad 15 \in RT(7,\lambda)$ and $15 \in T(8,\lambda)$ for every $\lambda > 1$.

For r = 20

$20 \in T(7,2) \qquad X = (Z_{19} \cup \{\infty\}) \times Z_7$

$\{(\infty;\zeta) : \zeta \in Z_7\}$ 2 times

$\{(\infty;0),(2^{6\alpha+\mu};3^{2\alpha}),(2^{6\alpha+\mu};3^{2\alpha+3}) : \alpha=0,1,2\} \bmod (19;7),\ \mu=2,12$

$\{(0;0),(2^{6\alpha+9\beta};3^{2\alpha+4\gamma}),(2^{6\alpha+9\beta+1};3^{2\alpha+4\gamma+3}) : \alpha=0,1,2\} \bmod (19;7),\ \beta=0,1,$
$\gamma=0,1.$

$20 \in T(7,3) \qquad X = (Z_{19} \cup \{\infty\}) \times Z_7$

$\{(\infty;\zeta) : \zeta \in Z_7\}$ 3 times

$\{(\infty;0),(2^{6\alpha+\mu};3^{2\alpha}),(2^{6\alpha+\mu};3^{2\alpha+3}) : \alpha=0,1,2\} \bmod (19;7),\ \mu=6,14,16$

$\{(0;0),(2^{6\alpha+2\beta+9\gamma};3^{2\alpha}),(2^{6\alpha+2\beta+9\gamma+1};3^{2\alpha+3}) : \alpha=0,1,2\} \bmod (19;7),\ \beta=0,1,2,$
$\gamma=0,1.$

By lemma 3 we have

(3.8) $\qquad 20 \in T(7,\lambda)$ for every $\lambda > 1$.

For r = 12 see lemma 10 and for r = 30 and r = 60 apply lemma 11: in the case of r = 30 with parameters $\sigma = 8$, $\rho = 4$ and $\rho' = 2$, and in the case r = 60 with parameters $\sigma = 8$, $\rho = 8$ and $\rho' = 4$. □

REFERENCES

[1] BOSE, R.C., E.T. PARKER & S.S. SHRIKHANDE, *Further results on the construction of mutually orthogonal Latin squares and the falsity of Euler's conjecture*, Canad. J. Math., 12 (1960) 189-203.

[2] BOSE, R.C. & S.S. SHRIKHANDE, *On the construction of sets of mutually orthogonal Latin squares and the falsity of a conjecture of Euler*, Trans. Amer. Math. Soc., 95 (1960) 191-209.

[3] CHOWLA, S., P. ERDÖS & E.G. STRAUSS, *On the maximal number of pairwise orthogonal Latin squares of a given order*, Canad. J. Math., 12 (1960) 204-208.

[4] DULMAGE, A.L., D.M. JOHNSON & N.S. MENDELSOHN, *Orthomorphisms of groups and orthogonal Latin squares, I*, Canad. J. Math., 13 (1961) 356-372.

[5] HANANI, H., *On the number of orthogonal Latin squares*, J. Combinatorial Theory, 8 (1970) 247-271.

[6] MACNEISH, H.F., *Euler squares*, Ann. of Math., 23 (1922) 221-227.

[7] PARKER, E., *Construction of some sets of mutually orthogonal Latin squares*, Proc. Amer. Math. Soc., 10 (1959) 946-949.

[8] RAGHAVARAO, D., *Constructions and combinatorial problems in design of experiments*, J. Wiley & Sons, New York, 1971.

[9] ROGERS, K., *A note on orthogonal Latin squares*, Pacific J. Math., 14 (1964) 1395-1397.

[10] WILSON, R.M., *Concerning the number of mutually orthogonal Latin squares*, in print.

COMBINATORICS OF FINITE GEOMETRIES

A. BARLOTTI

Università di Bologna, Bologna, Italy

In this lecture we intend to present a brief survey of very recent results. We shall be interested in the development of some topics considered in section 3.2 of DEMBOWSKI's book [21] (Combinatorics of finite planes) and in problems connected with the existence of finite geometrical structures.

1. THE STUDY OF SYSTEMS AXIOMATIZING FINITE PLANES AND DESIGNS*)

1.1. If $P = (p, L, I)$ is a projective plane of order n, then the following properties hold:

(1) $|p| = n^2 + n + 1$;

(2) $|L| = n^2 + n + 1$;

(3) $[p] = n + 1$, for every $p \in p$;

(4) $[L] = n + 1$, for every $L \in L$;

(5) $[p,q] = 1$, for $p,q \in p$ and $p \neq q$;

(6) $[L,M] = 1$, for $L,M \in L$ and $L \neq M$.

Conversely, if in a structure P consisting of a set p of points, a set L of lines with an incidence defined in $p \times L$, properties (1) to (6) hold, with $n \geq 2$, then P is a (non-degenerate) projective plane. Moreover, the above properties (1) to (6) are a redundant set of conditions to ensure that P is a projective plane. A subset of (1) - (6) is called a *complete* system when its properties are sufficient to imply that P is a projective plane. A complete system is called *minimal* if none of its proper subsets is also complete.

*) See also DEMBOWSKI [21], pp. 138-139. Here, and in what follows we shall use the symbols used in this book.

M. Hall, Jr. and J. H. van Lint (eds.), Combinatorics, 55-63.

In M. HALL [30] it is proved that (2), (4), (5) and its "dual"*) (1), (3), (6) are complete systems. All complete minimal systems of (1) to (6) were found in [1] and the following list was given:

(1) (3) (6) , (2) (4) (5),
(1) (4) (5) , (2) (3) (6),
(1) (2) (3) (5), (1) (2) (4) (6).

If $\mathcal{p} \cup \mathcal{L}$ is not empty, the systems

(3) (4) (5) (3) (4) (6)

are complete and minimal.

Further, if $|\mathcal{p}| \geq 2$ the system (3) (5) (6) is also complete and minimal, and the dual of this statement also holds.

G. CORSI [19] considered a refinement of this problem using, instead of properties (1),...,(6) the twelve properties obtained by replacing property (i) by the properties (i'), in which " = " is replaced by " ≤ ", and (i"), in which " = " is replaced by " ≥ ".**) The full list of complete minimal subsystems of (1'),...,(6") is given in DEMBOWSKI [21] pp. 138-139. Notice, however, that in the statement of the result given there we must replace the words "nondegenerate incidence structure" by "incidence structure containing at least two points or two lines".

Clearly, if we add some further hypotheses to (1') - (6") the list of complete minimal subsystems of these will be modified. We list here some results in this direction.

a) A. BASILE [4] added the hypothesis that the structure P is *nondegenerate*.
b) C. BERNASCONI [5] studied the problem for a *connected* and nondegenerate structure P.

The fact that P is connected adds to the list of complete minimal systems given in [4] the following:

(1") (3') (4') (5"), (2") (3') (4') (6"),
(1") (3') (5), (2") (4') (6),

and requires the cancellation of four systems which are no longer minimal, viz.:

---

*) Clearly, properties (2),(4),(6) are respectively the duals of (1),(3),(5).

**) Logical questions connected with the replacement of condition (i) by (i') or (i") are considered in R. MAGARI [34].

(1") (3') (4) (5"), (2") (3) (4') (6"),

(1") (3') (4') (5") (6"), (2") (3') (4') (5") (6").

c) M. CRISMALE [20] found all the complete minimal subsets of the system obtained from (1') - (6") by replacing (3'), (3"), (4'), (4") respectively by the following properties ($\hat{3}$'), ($\hat{3}$"), ($\hat{4}$'), ($\hat{4}$"):

($\hat{3}$') $b_1 \le n+1$, ($\hat{3}$") $b_1 \ge n+1$,

($\hat{4}$') $v_1 \le n+1$, ($\hat{4}$") $v_1 \ge n+1$,

where b_1 is the average number of lines on a point and v_1 is the average number of points on a line. The list of the complete minimal subsets is the following:

(1") (2') ($\hat{3}$") (6'), (1') (2") ($\hat{4}$") (5'),

(1") (2') ($\hat{3}$") (5'), (1') (2") ($\hat{4}$") (6'),

(1") ($\hat{3}$) ($\hat{4}$") (5") (6'), (2") ($\hat{3}$") ($\hat{4}$) (5') (6"),

(1") ($\hat{3}$) ($\hat{4}$") (5), (2") ($\hat{3}$") ($\hat{4}$) (6),

(1) ($\hat{3}$") ($\hat{4}$") (6'), (2) ($\hat{3}$") ($\hat{4}$") (5'),

(1) ($\hat{3}$") ($\hat{4}$") (5'), (2) ($\hat{3}$") ($\hat{4}$") (6'),

(1") ($\hat{3}$") ($\hat{4}$) (6), (2") ($\hat{3}$) ($\hat{4}$") (5),

(1") ($\hat{3}$") ($\hat{4}$) (5') (6"), (2") ($\hat{3}$) ($\hat{4}$") (5") (6').

1.2. Similar questions can be studied for affine planes considering together with properties (3) and (5) above the following:

(7) $|\mathcal{P}| = n^2$,

(8) $|\mathcal{L}| = n(n+1)$,

(9) $[L] = n$ for every $L \in \mathcal{L}$.

Partial results, needed as lemmas to obtain other results, were obtained by OSTROM [37] and DEMBOWSKI & OSTROM [22]. The systematic search for the complete minimal subsystems was done by U. OLIVERI [36] for the set (3) (5) (6') (7) (8) (9) and by P. BRUTTI [15] for (3') (3") (5') (5") (7') (7") (8') (8") (9') (9") with the additional condition that the structure is nondegenerate. C. BERNASCONI [5] studied how the hypothesis that the structure is connected modifies the list given in [15].

1.3. Similar questions may be studied for other finite structures; this was done for inversive planes by R. BUMCROT & D. KNEE [17] and for projective designs by C. BERNASCONI [6]. We shall give here the results obtained in [6].

Let $\mathcal{D}$ be a structure consisting of a set $\mathcal{P}$ of points, a set $\mathcal{B}$ of blocks and an incidence defined in $\mathcal{P} \times \mathcal{B}$ for which the following properties hold:

(D1) $|\mathcal{P}| = v = 1 + \frac{k(k-1)}{\lambda}$,

(D2) $|\mathcal{B}| = b = 1 + \frac{k(k-1)}{\lambda}$,

(D3) $[p] = r = k$ for every $p \in \mathcal{P}$,

(D4) $[B] = k$ for every $B \in \mathcal{B}$,

(D5) $[pp'] = \lambda$ for $p,p' \in \mathcal{P}$ and $p \neq p'$,

(D6) $[BB'] = \mu = \lambda$ for $B,B' \in \mathcal{B}$ and $B \neq B'$,

with the hypothesis $\lambda > 0$ and $k - \lambda \geq 2$. The above list is the extension of (1) to (6) of 1.1 to the case of projective designs, and in [6] it is proved that the list of complete minimal systems of (D1) to (D6) is the same (and with the same additional conditions) as the list given in 1.1 for the case of projective planes. This result shows that projective designs can be characterized by cardinality conditions, without any symmetry assumptions.

In connection with this fact we wish to observe that D.A. DRAKE [29] has proved that in a finite affine Hjelmslev plane, cardinality assumptions cannot replace the parallel axiom.

Other papers in connection with the above topics are [7], [40] and [41]. The following theorem (N.G. DE BRUIJN & P. ERDÖS [14]) represents a seminal result in the study of the problems considered in the first section:

Let there be given n points $a_1,\ldots,a_n$ and denote by $A_1,\ldots,A_m$ blocks of points such that we have:

(i) $|A_i| \geq 2$;

(ii) each pair (a_i,a_j) is contained in one and only one block.

Then we have $m \geq n$, with equality occurring only if either the system is of the type:

$A_1 = (a_1,a_2,\ldots,a_{n-1}),\ A_2 = (a_1,a_n),\ A_3 = (a_2,a_n),\ldots,\ A_n = (a_{n-1},a_n)$

or if the system is a projective plane, with the lines given by $A_1,\ldots,A_n$.

2. REFERENCES TO RECENT RESULTS IN OTHER TOPICS

Representation of nets and planes by sets of mutually orthogonal Latin squares is one of the central topics in combinatorics of finite geometric structures. For recent results and problems in this field we refer to A. BARLOTTI [3], G. DÉNES & A.D. KEEDWELL [23] and P. HOHLER [33].

The purpose of Galois geometries is to study geometry of finite spaces. Non-linear properties are of particular interest. Characterizations of algebraic varieties in finite spaces are illustrated in G. TALLINI [43]. Problems related to (k;n)-arcs and (k;n)-caps of given "kind" are considered in M. TALLINI SCAFATI [44].

In connection with the "packing problem" for k-caps, we quote the following results:

$m(4,3) = 20$ due to G. PELLEGRINO [38],
$m(5,3) = 56$ due to R. HILL [31].

We wish also to point out how a representation of the plane in higher dimensional space may sometimes simplify the study of the arcs in the plane. (See R.C. BOSE [8].)

Combinatorial arguments may be of big help in problems connected with the existence or non-existence and characterization of finite geometric structures. See R.H. BRUCK [10], A. BRUEN [11], A. BRUEN & J.C. FISHER [12], J. COFMAN [18] and R.H.F. DENNISTON [24 to 28].

REFERENCES

[1] BARLOTTI, A., *Un'osservazione sulle proprietà che caratterizzano un piano grafico finito*, Boll. Un. Mat. Ital., 17 (1962) 394-398.

[2] BARLOTTI, A., *Some classical and modern topics in finite geometrical structures*, in: *A survey of combinatorial theory*, J.N. SRIVASTAVA a.o. (eds.), North-Holland Publ. Cy., Amsterdam, 1973.

[3] BARLOTTI, A., *Alcune questioni combinatorie nello studio delle strutture geometriche*, in: Atti Convegno Teorie Combinatorie, Acc. Lincei, Rome 1973, to appear.

[4] BASILE, A., *Sugli insiemi di proprietà che definiscono un piano grafico finito*, Le Matematiche, 25 (1970) 84-95.

[5] BERNASCONI, C., *Strutture di incidenza connesse e definizione assiomatica di piani grafici e affini*, Ann. Univ. Ferrara, to appear.

[6] BERNASCONI, C., *Sistemi di assiomi che caratterizzano i disegni proiettivi*, to appear.

[7] BISCARINI, P., *Sets of axioms for finite inversive planes*, to appear.

[8] BOSE, R.C., *On a representation of Hughes planes*, in: Proc. Internat. Conf. on Projective Planes, M.J. KALLAHER & T.G. OSTROM (eds.), Washington State Univ. Press, 1973, pp.27-57.

[9] BRAMWELL, D.L. & B.J. WILSON, *The (11,3)-arcs of the Galois plane of order 5*, Proc. Cambridge Philos. Soc., 74 (1973) 247-250.

[10] BRUCK, R.H., *Construction problems in finite projective spaces*, in: *Finite geometric structures and their applications*, C.I.M.E. II ciclo 1972, Ed. Cremonese, Rome, 1973, pp.105-188.

[11] BRUEN, A., *Blocking sets in finite projective planes*, SIAM J. Appl. Math., 21 (1971) 380-392.

[12] BRUEN, A. & J.C. FISHER, *Arcs and ovals in derivable planes*, Math. Z., 125 (1972) 122-128.

[13] BRUEN, A. & J.C. FISHER, *Blocking-sets, k-arcs and nets of order ten*, Advances in Math., 10 (1973) 317-320.

[14] BRUIJN, N.G. DE & P. ERDÖS, *On a combinatorial problem*, Kon. Nederl. Akad. Wetensch. Proc. A, 51 (1948) 1277-1279 (= Indag. Math., 10 (1948) 421-423).

[15] BRUTTI, P., *Sistemi di assiomi che definiscono un piano affine di ordine n*, Ann. Univ. Ferrara Sez VII, 14 (1969) 109-118.

[16] BUEKENHOUT, F. & R. METZ, *On circular spaces having no disjoint circles*, to appear.

[17] BUMCROT, R. & D. KNEE, private communication.

[18] COFMAN, J., *On combinatorics of finite projective spaces*, in: Proc. Internat. Conf. on Projective Planes, M.J. KALLAHER & T.G. OSTROM (eds.), Washington State Univ. Press, 1973, pp.59-70.

[19] CORSI, G., *Sui sistemi minimi di assiomi atti a definite un piano grafico finito*, Rendic. Sem. Mat. Padova, 34 (1964) 160-175.

[20] CRISMALE, M., *Sui sistemi minimi di assiomi atti a definite un piano proiettivo finito*, to appear.

[21] DEMBOWSKI, P., *Finite geometries*, Ergebnisse der Mathematik 44, Springer-Verlag, Berlin etc., 1968.

[22] DEMBOWSKI, P. & T.G. OSTROM, *Planes of order n with collineation groups of order n^2*, Math. Z., 103 (1968) 239-258.

[23] DÉNES, J. & A.D. KEEDWELL, *Latin squares and their applications*, Acad. Press, New York and London, 1974.

[24] DENNISTON, R.H.F., *Some packings of projective spaces*, Rend. Acc. Naz. Lincei (8), 52 (1972) 36-40.

[25] DENNISTON, R.H.F., *Cyclic packings of the projective space of order 8*, Rend. Acc. Naz. Lincei (8), 54 (1973) 373-377.

[26] DENNISTON, R.H.F., *Packings of PG $(3,q)$*, in: *Finite geometric structures and their applications*, C.I.M.E. II ciclo 1972, Ed. Cremonese, Rome, 1973, pp.193-199.

[27] DENNISTON, R.H.F., *Spreads which are not subregular*, Glasnik Mat. Ser. III, 8 (1973) 3-5.

[28] DENNISTON, R.H.F., *Some spreads which contain reguli without being subregular*, to appear.

[29] DRAKE, D.A., *Near affine Hjelmslev planes*, J. Comb. Theory, 16 (1974) 34-50.

[30] HALL, M. Jr., *The theory of groups*, Mac Millan, New York, 1959.

[31] HILL, R., *On the largest size of cap in $S_{5,3}$*, Rend. Acc. Naz. Lincei, to appear.

[32] HILL, R., *Caps and groups*, to appear.

[33] HOHLER, P., *Eigenschaften von vollständigen Systemen orthogonaler Lateinischer Quadrate, die bestimmte affine Ebenen repräsentieren*, J. of Geometry 2 (1972) 161-174.

[34] MAGARI, R., *Sui sistemi di assiomi "minimali" per una data teoria*, Boll. Un. Mat. Ital., 19 (1964) 423-435.

[35] MENICHETTI, G., *q-archi completi nei piani di Hall di ordine* $q = 2^k$, to appear.

[36] OLIVERI, U., *Alcune proprietà che caratterizzano un piano affino finito*, Le Matematiche, 22 (1967) 397-402.

[37] OSTROM, T.G., *Semi translation planes*, Trans. Amer. Math. Soc., 111 (1964) 1-18.

[38] PELLEGRINO, G., *Sul massimo ordine delle calotte in* $S_{4,q}$, Le Matematiche, 25 (1970) 149-157.

[39] PELLEGRINO, G., *Procedimenti geometrici per la construzione di alcune classi di calotte complete in* $S_{r,3}$, Boll. Un. Mat. Ital. (4), 5 (1972) 109-115.

[40] REIMAN, I., *Su una proprietà dei piani grafici finiti*, Rend. Acc. Naz. Lincei, 35 (1963) 279-281.

[41] REIMAN, I., *Su una proprietà dei due disegni*, Rend. Mat. e Appl., 1 (1968) 75-81.

[42] SEGRE, B., *Proprietà elementari relative ai segmenti ed alle coniche sopra un campo qualsiasi ed una congettura di Seppa Ilkka per il caso dei campi di Galois*, Ann. Mat. Pura Appl. (4), 96 (1973) 289-337.

[43] TALLINI, G., *Graphic characterization of algebraic varieties in a Galois space*, in: Atti Convegno Teorie Combinatorie, Acc. Lincei, Rome 1973, to appear.

[44] TALLINI SCAFATI, M., *The k-sets of type (m,n) in a Galois space* $S_{r,q}$ $(r \geq 2)$, in: Atti Convegno Teorie Combinatorie, Acc. Lincei, Rome 1973, to appear.

[45] THAS, J.A., *Connection between the n-dimensional affine space* $A_{n,q}$ *and the curve C, with equation* $y = x^q$, *of the affine plane* A_{2,q^n}, Rend. Trieste, 2 (1970) 146-151.

[46] THAS, J.A., *A combinatorial problem*, Geometriae Dedicata, 1 (1973) 236-240.

[47] THAS, J.A., *4-gonal configurations*, in: *Finite geometric structures and their applications*, C.I.M.E. II ciclo 1972, Ed. Cremonese, Rome, 1973, pp.249-263.

[48] THAS, J.A., *On 4-gonal configurations*, Geometriae Dedicata, 2 (1973) 317-326.

[49] THAS, J.A., *Flocks of finite egglike inversive planes*, in: *Finite geometric structures and their applications*, C.I.M.E. II ciclo 1972, Ed. Cremonese, Rome, 1973, pp.189-191.

[50] THAS, J.A., *Some results concerning* $\{(q+1)(n-1);n\}$*-arcs and* $\{(q+1)(n-1)+1;n\}$*-arcs in finite projective planes of order* q, to appear.

[51] THAS, J.A., *On 4-gonal configurations with parameters* $r = q^2+1$ *and* $k = q+1$, to appear.

ON FINITE NON-COMMUTATIVE AFFINE SPACES

J. ANDRÉ

Universität des Saarlandes, D 66 Saarbrücken, GFR

INTRODUCTION

We consider incidence structures (cf. DEMBOWSKI [10,p.1]) consisting of a non-void set of elements called *points*, certain subsets of points called *lines*, an operation sometimes called *join*, mapping every ordered pair of different points surjectively onto the set of all lines and finally a binary relation on the lines called *parallelism*, all these things with additional properties gained in a natural way. The most known structures of this type are the well-known affine spaces (e.g. in the sense of TAMASCHKE [20,21]) but also many other examples have been developed in recent times (see e.g. DEMBOWSKI [10], PICKERT [18], SPERNER [19], ARNOLD [6], WILLE [26], also ANDRÉ [1], BACHMANN [8]).

In very general spaces with a *commutative* join many results are known in the meantime, especially by the extensive work of WILLE [26]. For spaces with a non-commutative join, however, almost nothing is known up to now. It is the purpose of this paper to develop some results just for such "non-commutative spaces". But we do not want to consider too general spaces of such a type. For this reason we introduce some further axioms in such a way that the space under consideration becomes an affine space (in the usual sense) if additionally the join is commutative, thus obtaining "non-commutative affine spaces". If we do so in a suitable way we get the so-called *quasi-affine spaces* and some of their particular types, especially the *near-affine spaces (fastaffine Räume)*.

As we shall see in this note these geometries can be applied to the theory of groups, esp. permutation groups (see also ANDRÉ [3]), to algebra, esp. nearfields (see also ANDRÉ [4], BACHMANN [8]), to the theory of combinatorial designs (see also ANDRÉ [2]) and to the foundations of geometry (see also ANDRÉ [5] and BACHMANN [8]).

There may possibly be other applications. For instance, by considering the set of all countable sequences with elements of a Kalscheuer-nearfield (cf. KALSCHEUER [16]) and with a convergent series of the squares

M. Hall, Jr. and J. H. van Lint (eds.), Combinatorics, 65-113.

of their absolute values we get a generalization of Hilbert space which could give new aspects on functional analysis, perhaps also new deep insights in quantum mechanics and generalizations.*)

In the first chapter the basic definitions, esp. of the quasi-affine and nearaffine spaces, will be stated (cf. sections 1 and 3). A simple but, as I suppose, far reaching application to transitive permutation groups (not necessarily of finite degree) is contributed in section 2. In the last section of this chapter axioms of configurations, some of them analogous to that of DESARGUES in affine spaces, are described; their significance for geometric structures will be given in the last chapter. It may also be possible to generalize e.g. the Reidemeister-, Thomsen- and Klingenberg-configurations (see e.g. KLINGENBERG [17], HUGHES & PIPER [12], PICKERT [18]) to nearaffine spaces and then to characterize such spaces in some other ways (e.g. algebraically). All results stated in this first chapter are valid for both finite and infinite spaces.

For the second chapter, however, the condition of finiteness for all spaces under consideration is essential. We introduce there the concepts of order, dimension, subspaces and others for finite nearaffine spaces in a natural way. We gain many theorems concerning their detailed structure some of them being analogous to well-known properties in affine spaces. One of the main results is that two different points are incident with either exactly one or exactly n lines where n is the order (i.e. the number of points on a line) of the space under consideration (cf. chapter II, theorem 6.1). This leads to another type of join introduced in definition 6.1. It is commutative but deeper properties of it are not yet known.

In the last chapter we consider finite nearaffine spaces with further properties which, roughly spoken, assume the existence of "sufficiently many" points, lines etc. with special properties. Under such conditions one can prove the little Desargues condition (cf. chapter III, section 1) and as a consequence the existence of sufficiently many translations. Nothing, however, is known to the author which hypotheses imply the general Desargues theorem. But if this is true then the nearaffine space can be described as a space over a nearfield, even in the infinite case; this will be sketched in the last section.

---

*) For a "geometrical" theory of quantum mechanics see e.g. VARADARAJAN [22,23], with further references. For perhaps possible generalizations in this direction see e.g. JORDAN, VON NEUMANN & WIGNER [15], JORDAN [13] or (with respect to non-commutative lattices) JORDAN [14]. (See also problem (7) of the appendix.)

Finally we shall list some unsolved problems related to quasi-affine and nearaffine spaces whose solutions might possibly lead to a deeper understanding of those structures and their relations to other fields of mathematics.

I. BASIC CONCEPTS

1. QUASI-AFFINE SPACES

We consider structures of the type

(1.1) $$S = (X, \mathsf{L}, \sqcup, \parallel)$$

with the following *basic hypotheses*:

(1) X is an arbitrary non-void set whose elements are called *points*. They are denoted by small latin letters.

(2) L is a subset of the powerset $\mathsf{P}(X)$ of X; their elements are called *lines* (*Linien*). They are denoted by capital latin letters.

(3) $\sqcup$ is a surjective mapping of the set of all ordered pairs (x,y) of different points onto L:

(1.2) $$\begin{cases} \sqcup: X^2\setminus\Delta_X \twoheadrightarrow \mathsf{L}, \\ (x,y) \longmapsto x\sqcup y, \qquad (x\neq y)\,. \end{cases}$$

The line $x\sqcup y$ is called the *connection-line* or *join from* x *to* y (*in this order*).*)

(4) $\parallel$ is a binary equivalence relation on L called *parallelism*.

In this way S becomes an incidence structure (in the sense of DEMBOWSKI [10] **)) where $\in$ is the incidence relation between X and L.

We shall give some further conditions on S in such a way that S becomes an affine space (in the usual sense) if we additionally assume the commutativity of $\sqcup$. We subdivide these axioms into four classes: (L) *conditions on lines*, (P) *conditions on parallelism*, (R) *a condition of rich-*

*) It is sometimes useful to define also $x\sqcup x =: \{x\}$.

**) There the lines are called *blocks*.

ness, and (T) *a condition on similar triangles* (see also ANDRÉ [2]).

DEFINITION 1.1. A structure $S = (X, \mathsf{L}, \sqcup, \parallel)$ with the basic hypotheses (1) to (4) is called a *quasi-affine space* if the following conditions hold:

(L1) $x, y \in x \sqcup y$ *for all* $x, y \in X$ *with* $x \neq y$.

(L2) $z \in (x \sqcup y) \setminus \{x\} \Leftrightarrow x \sqcup y = x \sqcup z$.

(L3) $x \sqcup y = y \sqcup x = x \sqcup z \Rightarrow x \sqcup z = z \sqcup x$.

(P1) *For* $L \in \mathsf{L}$ *and* $x \in X$ *there exists exactly one line* $L' \parallel L$ *such that* $L' = x \sqcup y$ *for a suitable* $y \in X$ (Euclid's axiom of parallelism). This line is denoted by

$$(1.3) \qquad (x \parallel L).$$

(P2) *If* $G = x \sqcup y = y \sqcup x$ *and* $L \parallel G$ *then* $x', y' \in L$ *with* $x' \neq y'$ *imply* $x' \sqcup y' = y' \sqcup x'$.

(R) *There exist at least two lines in* S.

(T) *If* x, y, z *are pairwise different and* x', y' *are different points with* $x \sqcup y \parallel x' \sqcup y'$ *then*

$$(1.4) \qquad (x' \parallel x \sqcup z) \cap (y' \parallel y \sqcup z) \neq \emptyset. \text{ *)}$$

Consequences and remarks

(a) Condition (L1) implies that every line contains at least two points.

(b) A point x on a line L such that $L = x \sqcup y$ for a suitable $y \in X$ is called a *base point* (*Aufpunkt*) of L.

A simple consequence of (L3) is

(c) The following conditions are equivalent:

 (α) L has at least two base points.

 (β) Every point of L is a base point of L.

 (γ) If x, y are different points on L then $L = x \sqcup y = y \sqcup x$.

(d) A line with a condition given in (c) is called a *straight line* (*Gerade*). The set of all straight lines of the space S is denoted by G. Lines not being straight are called *proper lines*.

(e) Condition (P2) means that every line parallel to a straight line is also straight.

---

*) See TAMASCHKE [21,p.319] for an analogous condition on affine spaces.

(f) If we specialize (T) by $x = x'$ we get the following affine version of a *Veblen-condition*:

(V) *Let* x,y,z *be pairwise different points,* $y' \in x \sqcup y$, *then*

(1.4') $$(x \sqcup z) \cap (y' \parallel y \sqcup z) \neq \emptyset.$$

Note that this intersection [as well as (1.4)] need not consist of only one point.

The following theorem is easy to prove (see also ANDRÉ [2,theorem 2.1]).

THEOREM 1.1. *A quasi-affine space* $S = (X, L, \sqcup, \parallel)$ *with a commutative join* $\sqcup$ *is an affine space (of dimension* ≥ 2).

DEFINITION 1.2. Let $A \subseteq X$. Two points $x,y \in A$ are called *joinable (verbindbar) with respect to* A, denoted by

$$x \underset{A}{\sim} y,$$

if either $x = y$ or there exist finitely many points $x = x_0, x_1, \ldots, x_n = y$ such that all $x_i \sqcup x_{i+1}$ $(i \in \{0,1,\ldots,n-1\})$ are straight lines totally contained in A (i.e. x and y can be connected by a finite chain of straight lines all completely belonging to A).

It is straightforward that $\underset{A}{\sim}$ is an equivalence relation on X. For $A = X$ we simply say that x and y are *joinable* and simplify the sign $\underset{A}{\sim}$ to $\sim$. The equivalence classes with respect to $\sim$ all reduce to single points iff S contains no straight lines.

2. EXAMPLES RELATED TO PERMUTATION GROUPS

Let X be a non-void set (not necessarily finite) and Γ a permutation group acting on X. For $x \in X$ let Γ_x be, as usual, the *stabilizer*, i.e. the subgroup of all those $\gamma \in \Gamma$ leaving x fixed. Moreover, assume always that $x \neq y$ implies $\Gamma_x \neq \Gamma_y$. We define a *join* $\sqcup$ on X by

(2.1) $$x \sqcup y := \{x\} \cup \Gamma_x(y),$$

thus obtaining a set L of lines on X. Such a line with a base point x therefore consists of x and an orbit of Γ_x different from x. A parallelism on L is defined by

(2.2) $\quad L \parallel L' :\Longleftrightarrow L' = \gamma(L)$ for a suitable $\gamma \in \Gamma$.

In this way we get a structure $I_\Gamma = (X, L, \sqcup, \parallel)$ called the *group space* of Γ possessing all basic hypotheses [cf. section 1, (1) to (4)]. Moreover, the following property is easy to verify.

PROPOSITION 2.1. *If Γ is a transitive but not doubly transitive permutation group on X then the space I_Γ defined by* (2.1) *and* (2.2) *is a quasi-affine space. Moreover, for all $\gamma \in \Gamma$ we have*

(2.3) $\quad \gamma(x \sqcup y) = \gamma(x) \sqcup \gamma(y)$

for all $x, y \in X$, thus γ being an automorphism of I_Γ.

We give some characterizations for additional conditions on Γ or I_Γ.

PROPOSITION 2.2. *The condition*

(2.4) $\quad x \sqcup y \parallel y \sqcup x, \qquad (x, y \in X,\ x \neq y)$

holds iff Γ contains a γ exchanging x and y. *)

PROOF. $\gamma(x) = y$, $\gamma(y) = x$ and (2.3) imply $x \sqcup y \parallel \gamma(x \sqcup y) = \gamma(x) \sqcup \gamma(y) = y \sqcup x$, hence (2.4).

Conversely assume now (2.4). By (2.2) there exists a $\delta \in \Gamma$ with $y \sqcup x = \delta(x \sqcup y) = \delta(x) \sqcup \delta(y)$. Using (2.1) we have $\delta(x) = y$ and $\delta(y) \in \Gamma_y(x)$. This yields the existence of an $\eta \in \Gamma_y$ mapping $\delta(y)$ on x. The permutation $\gamma := \eta\delta \in \Gamma$ exchanges x and y.

PROPOSITION 2.3. *The equation*

(2.5) $\quad x \sqcup y = y \sqcup x, \qquad (x, y \in X,\ x \neq y),$

holds iff for any $\gamma \in \Gamma_x \backslash \Gamma_y$ there exists a $\gamma' \in \Gamma_y \backslash \Gamma_x$ such that

---

*) Another interesting characterization of this exchanging condition is given in WIELANDT [25, theorem 16.4, p.45]. Moreover, it is easy to prove by similar methods that the orbit $\neq \{x\}$ of Γ_x given by $(x \parallel y \sqcup x)$ is the reflection (cf. [25, §16, p.44]) of the orbit determined by $x \sqcup y$, independently of the special choice of y. Other relations between Γ and I_Γ will be published elsewhere.

$$\gamma(y) = \gamma'(x)$$

and a corresponding relation holds if the roles of x *and* y *are interchanged.*

PROOF. See ANDRÉ [3,Satz 1.1]. □

THEOREM 2.1. *Let* Γ *be a transitive but not doubly transitive permutation group on* X *such that for all different points* x, y *and any* $\gamma \in \Gamma_x \setminus \Gamma_y$ *there exists a* $\gamma' \in \Gamma_y \setminus \Gamma_x$ *with* $\gamma(y) = \gamma'(x)$. *Then* Γ *is isomorphic (as permutation group) to the group of all dilatations (i.e. translations or homotheties) of a desarguesian affine space (of dimension* ≥ 2*).*

This is a consequence of theorem 1.1 and proposition 2.3. For further details cf. ANDRÉ [3].

PROPOSITION 2.4. *The equivalence classes on* X *with respect to the joinable relation* $\sim$ (cf. definition 1.2) *form a complete block system of* Γ (see e.g. WIELANDT [25,p.12]), *i.e.*

(1) *for any equivalence class* X_i *and any* $\gamma \in \Gamma$ *we have either* $\gamma(X_i) = X_i$ *or* $X_i \cap \gamma(X_i) = \emptyset$, *and*

(2) *for any two classes* X_i *and* X_j *there exists a* $\gamma \in \Gamma$ *with* $X_j = \gamma(X_i)$.

This follows because the image $\gamma(x \sqcup y)$ $[\gamma \in \Gamma]$ of a straight line is again straight.

THEOREM 2.2. *A transitive group* Γ *whose group space* I_Γ *contains at least one straight line is either doubly transitive or imprimitive.*

PROOF. Assume Γ to be transitive but not doubly transitive. Let G be a straight line of I_Γ then $\emptyset \neq G \subset X$. Let $x \in X\setminus G$ and $y \in X\setminus\{x\}$ such that $x \sqcup y \parallel G$. Then $\gamma(x) \in G$ implies $\gamma(y) \in G$ for all $\gamma \in \Gamma$. Due to Rudio's theorem (see e.g. WIELANDT [25, theorem 8.1]) Γ is imprimitive. □

For considerations occurring in the next section and in chapter III the following concept may be useful. Again let Γ be a group operating transitively but not doubly transitively on X.

DEFINITION 2.1. A $\tau \in \Gamma$ is called a *translation* if either $\tau = 1$ (the identity) or τ acts fixpointfree on X and

(2.6) $\quad x \sqcup \tau(x) \parallel y \sqcup \tau(y)$

holds for all $x, y \in X$.

The set T of all translations need not be a subgroup of Γ but trivially $\tau \in T$ and $\gamma \in \Gamma$ imply $\gamma\tau\gamma^{-1} \in T$, and if additionally (2.4) holds for all $x, y \in X$ then $\tau \in T$ implies $\tau^{-1} \in T$ due to

$$\begin{aligned} & x \sqcup \tau^{-1}(x) = \\ & = \tau^{-1}(\tau(x) \sqcup x) \parallel \tau(x) \sqcup x \parallel x \sqcup \tau(x) \parallel y \sqcup \tau(y) \parallel \tau(y) \sqcup y \parallel \tau^{-1}(\tau(y) \sqcup y) = \\ & = y \sqcup \tau^{-1}(y). \end{aligned}$$

<u>PROPOSITION 2.5.</u> *A fixpointfree permutation* $\tau \in \Gamma$ *belongs to* T *iff for all* $x, y \in X$ *there exists a* $\gamma \in \Gamma$ *such that*

(2.7) $\quad y = \gamma(x) \quad$ *and* $\quad y = \tau^{-1}\gamma\tau(x)$

hold.

<u>PROOF</u>. Assume (2.7); then

$$y \sqcup \tau(y) = \gamma(x) \sqcup \tau\tau^{-1}\gamma\tau(x) = \gamma(x \sqcup \tau(x)) \parallel x \sqcup \tau(x).$$

Conversely suppose (2.6) for τ. According to the transitivity of Γ there exists a $\delta \in \Gamma$ with $\delta(x) = y$, Now (2.6) and (P1) yield

$$y \sqcup \delta\tau(x) = \delta(x \sqcup \tau(x)) = y \sqcup \tau(y),$$

hence $\tau(y) \in \Gamma_y \delta\tau(x)$ due to (2.1). Thus we have

$$y \in \tau^{-1}\Gamma_y \delta\tau(x),$$

so that we can find an $\eta \in \Gamma_y$ with $y = \tau^{-1}\eta\delta\tau(x)$. The permutation $\gamma := \eta\delta \in \Gamma$ has all properties of (2.7). □

<u>DEFINITION 2.2.</u> A translation τ is called *straight* (*strikt*) if either $\tau = 1$ or $x \sqcup \tau(x)$ is straight.

REMARK. Because of (P2) this condition does not depend on the special choice of x.

The straight translations will play an important role in some special type of quasi-affine spaces considered in the next section. It should be important to characterize those groups Γ whose translations form a subgroup, especially a (sharply) transitive subgroup. It is an interesting but up to now unsolved problem to characterize the spaces I_Γ in a purely geometric way. This problem is only solved for some very special types of permutation groups Γ. (See also problem (2) of the appendix in this paper.)

3. NEARAFFINE SPACES

In this section we specialize the concept of a quasi-affine space in such a way that we essentially assume the existence of "sufficiently many" straight lines.

DEFINITION 3.1. A quasi-affine space $F = (X, L, \sqcup, \|)$ is called a *nearaffine space (fastaffiner Raum)* if the following additional conditions hold in F (they are subdivided into two classes: one *condition concerning parallelism* (P) and two *conditions* (G) *concerning straight lines*):

(P3) *For all* $x,y \in X$ *with* $x \neq y$ *we have* *)

$$x \sqcup y \ \| \ y \sqcup x \, .$$

(G1) *Every line* L *meets every straight line* $G \neq L$ *in at most one point, i.e.* $|G \cap L| \leq 1$.

(G2) Chain condition. *All points of* X *are joinable in the sense of definition 1.2.*

As a consequence of (T) [cf. definition 1.1] and (P3) the following condition holds.

THEOREM 3.1. Condition for closed parallelograms (Pa). *Let* x,y,z *be pairwise different points with* $x \sqcup y \neq x \sqcup z$; *then*

$$(3.1) \qquad (z \ \| \ x \sqcup y) \cap (y \ \| \ x \sqcup z) \neq \emptyset$$

---

*) This is exactly the condition (2.4).

(but the intersection need not consist of only one point).

PROOF. Consider the diagonal $y \sqcup z$. Using (P3) we have $x \sqcup z \parallel z \sqcup x$, $x \sqcup y \parallel y \sqcup x$ and $y \sqcup z \parallel z \sqcup y$. If we apply (T) on the triangle x,y,z and on z,y we obtain (3.1). □

REMARK. For nearaffine spaces (V) and (Pa) are consequences of (T). It is an unsolved problem, even in the finite case, whether conversely (T) follows from (V) and (Pa). For the analysis of finite nearaffine spaces as done in the following chapters, however, we *only* need the conditions (V) and (Pa).

A fundamental property of nearaffine spaces is given by the following theorem.

THEOREM 3.2. *All lines of a nearaffine space have the same number of points.*

PROOF. The proof will be given in several steps: First using (V) and (G1) we show that two intersecting straight lines have the same number of points. By the use of (G2) we secondly prove that any two straight lines have equally many points. The third step is to verify that two lines with a common base point have the same number of points. For that we heavily use (G2) in the form of an induction on the number of straight lines connecting two points of both lines, different from their common base point. The complete theorem is then a simple consequence of this result. For further details cf. ANDRÉ [2,I,theorem 3.1]. □

Note that for the proof of this theorem we use the axioms (G) and (V) but not (Pa) and instead of (P3) only the weaker condition $x \sqcup y \parallel x' \sqcup y' \Rightarrow y \sqcup x \parallel y' \sqcup x'$ (by using prop. 3.3).

DEFINITION 3.2. The (common) number of points on a line (which may of course be infinite) is called the *order* of the nearaffine space. It is usually denoted by n.

REMARKS.

(1) By (L1) we always have $n \geq 2$. For $n = 2$ we get $x \sqcup y = \{x,y\} = y \sqcup x$, i.e. an affine space. Hence $n \geq 3$ holds for any proper nearaffine space.

(2) It is well-known that the numbers of points of the orbits of Γ_x different from x differ in general (see e.g. WIELANDT [25,chapter III]). Therefore theorem 3.2 does not hold in a general quasi-affine space.

EXAMPLES OF NEARAFFINE SPACES. We consider the following permutation group Γ acting on the set X. Let I be an index set consisting of at least two elements. Moreover, let F be a set in which a multiplication and for every $i \in I$ an addition $+_i$ is defined in such a way that $(F,+_i,\cdot)$ becomes a (not necessarily planar) nearfield (with the distributive law $\alpha(\beta +_i \gamma) = \alpha\beta +_i \alpha\gamma$ for all $\alpha,\beta,\gamma \in F$) *) and all these nearfields have the same neutral element 0 with respect to their additions $+_i$. Let X be defined by

$$\text{(3.2)} \qquad X := \{(\xi_i)_{i\in I} \mid \xi_i \in F,\ \xi_i \neq 0 \text{ for only finitely many } i \in I\}.$$

Define addition and multiplication by

$$\text{(3.3)} \qquad (\xi_i)_{i\in I} + (\eta_i)_{i\in I} := (\xi_i +_i \eta_i)_{i\in I}$$

and

$$\text{(3.4)} \qquad \alpha(\xi_i)_{i\in I} := (\alpha\xi_i)_{i\in I}$$

resp., $(\alpha,\xi_i,\eta_i \in F)$. Then $(X,+)$ is an abelian group and every $\alpha \in F\setminus\{0\}$ becomes an automorphism $x \mapsto \alpha x$ of $(X,+)$. Moreover, $\alpha x = \beta x$ implies $\alpha = \beta$ or $x = 0$.

The group Γ of all permutations

$$\text{(3.5)} \qquad x \mapsto \alpha x+v, \qquad (x,v \in X,\ \alpha \in F\setminus\{0\})$$

is transitive but (due to $|I| \geq 2$) not doubly transitive on X. Their translations (cf. definition 2.1) are exactly the mappings (3.5) with $\alpha = 1$; they form a sharply transitive normal subgroup of Γ. A translation $x \mapsto x+v$ is straight (cf. definition 2.2) iff v belongs to the so-called *quasi-nucleus* (*Quasikern*) Q of X defined by

$$\text{(3.6)} \qquad Q := \{v \in X \mid \forall\alpha,\beta \in F \ \ \exists\gamma \in F \text{ with } \alpha v+\beta v = \gamma v\}.$$

The lines $x \sqcup y$ of I_Γ can also be described by

*) For the definition and properties of nearfields see e.g. DEMBOWSKI [10,p.33] (with the left distributive law instead of the right one quoted here).

(3.7) $\quad x \sqcup y := F(y-x) + x := \{\alpha(y-x) + x \mid \alpha \in F\}.$

Such a line is straight iff $y-x \in Q$. The parallelism of two lines is given by

(3.8) $\quad Fz + v \parallel Fz' + v' :\Longleftrightarrow Fz = Fz', \qquad (z,z' \in X\setminus\{0\},\ v,v' \in X).$

The cardinality $|I|$ of the index set I is sometimes called the *dimension* of I_Γ; it is denoted by

(3.9) $\quad \mathrm{Dim}\ I_\Gamma := |I|$

and does only depend on the geometrical structure of I_Γ.

DEFINITION 3.3. The spaces I_Γ defined by the group (3.5) are sometimes called *nearfield spaces*.

Due to proposition 2.1 all nearfield spaces are quasi-affine. But we can prove more.

PROPOSITION 3.1. *A nearfield space is nearaffine.*

The proofs of (P3) and (G2) are straightforward. For the proof of (G1) we essentially have to use that

$$\alpha v + \beta x = \alpha' v + \beta' x, \qquad (v \in Q\setminus\{0\},\ x \in X\setminus Fv)$$

implies $\alpha = \alpha'$ and $\beta = \beta'$. For the proof of this proposition cf. ANDRÉ [4, Satz 4.17].*) Also the following proposition is easy to verify.

PROPOSITION 3.2. *Let* I_Γ *be a nearfield space. Then the additions* $+_i$ *given in* (3.3) *do not depend on* i *(so that we can put* $+_i =: +$*) iff the following additional condition holds in* I_Γ.

(G3) Triangle condition. *For any two different straight lines* G *and* G' *with the common point* x *there exist* $y \in G\setminus\{x\}$ *and* $y' \in G'\setminus\{x\}$ *such that* $y \sqcup y'$ *is straight.*

*) In that paper one can find an algebraic theory of the structures defined by (3.2) to (3.4), there called *Fastvektorräume* (*nearvectorspaces*). The geometric consequences will be given in ANDRÉ [5].

We conclude this section with a proposition valid for general near-affine planes which will be useful in the next chapter.

PROPOSITION 3.3. *Let* x *and* y *be different points on a straight line* G *and let* L *be a line different from* G *with* x *as a base point. Then* $L \cap (y \parallel L) = \emptyset$.

PROOF. Assume $z \in L \cap (y \parallel L)$. Then $L = x \sqcup z$ and $(y \parallel L) = y \sqcup z$ by (L2). Using (P3) we get

$$z \sqcup x \parallel L \parallel (y \parallel L) \parallel z \sqcup y ,$$

by (P1) thus $z \sqcup x = z \sqcup y$, hence $x, y \in G \cap (z \sqcup x)$ contradicting (G1) because G is straight. $\square$

REMARK. It is easy to see that for the proof of proposition 3.3 we only need the condition $x \sqcup y \parallel x' \sqcup y' \Rightarrow y \sqcup x \parallel y' \sqcup x'$ instead of the stronger one (P3).

4. CONFIGURATIONS

In order to get deeper properties of nearaffine spaces it is often necessary to add further hypotheses. They may be of group-theoretical nature, e.g. by assuming the existence of "sufficiently many" collineations (automorphisms) with certain properties. On the other hand it is sometimes useful to make hypotheses of geometrical nature, i.e. by requiring geometrical configurations analogous to the Desargues or Pappos configurations in affine spaces. We first formulate three types of Desargues conditions.

(D1) Little Desargues configuration. *If* $x, x', y, y', z, z' \in X$ *are pairwise different,* $x \sqcup x' \parallel y \sqcup y' \parallel z \sqcup z'$ *are straight and pairwise different, then* $x \sqcup y \parallel x' \sqcup y'$ *and* $x \sqcup z \parallel x' \sqcup z'$ *imply* $y \sqcup z \parallel y' \sqcup z'$.

(D2) Desargues configuration (first kind). *If* $u, x, x', y, y', z, z' \in X$ *are pairwise different,* $u \sqcup x$ *is straight and* $u \sqcup y$, $u \sqcup z$ *are lines different from each other and from* $u \sqcup x$, *then* $x' \in u \sqcup x$, $y' \in u \sqcup y$, $z' \in u \sqcup z$, $x \sqcup y \parallel x' \sqcup y'$ *and* $x \sqcup z \parallel x' \sqcup z'$ *imply* $y \sqcup z \parallel y' \sqcup z'$.

(D3) Desargues configuration (second kind). *If* $u, x, x', y, y', z, z' \in X$ *are pairwise different,* $u \sqcup x$, $u \sqcup y$ *and* $u \sqcup z$ *are pairwise different lines, and if* $x \sqcup y \parallel x' \sqcup y'$ *and* $x \sqcup z \parallel x' \sqcup z'$ *are straight then* $y \sqcup z \parallel y' \sqcup z'$.

The following configuration trivially holds in any affine space.

(Di) Diagonal condition for quadrilaterals. *If* $x,y,z,w \in X$ *are pairwise different such that* $x \sqcup y$, $y \sqcup z$, $z \sqcup w$, $w \sqcup x$ *and the one diagonal* $x \sqcup z$ *are straight then also the other diagonal* $y \sqcup w$ *is straight.*

Nearaffine spaces with (D1), (D2) and (Di) are called *desarguesian.**) The following theorem is fundamental.

THEOREM 4.1. *Any nearfield space* (cf. definition 3.3) *is desarguesian.*

For the proof see ANDRÉ [5]. Conversely all desarguesian nearaffine spaces are essentially of that type; for more details see section 4 of chapter III and ANDRÉ [5].

There exist non-desarguesian nearaffine spaces of arbitrarily high dimension. They can be constructed from affine spaces over the reals by "refracting" lines in a suitable way, cf. ANDRÉ [2,chapter I]. In chapter III of this paper we shall see that in all finite nearaffine spaces, being no planes, the little Desargues condition (D1) holds. In the infinite case this problem remains open.

It is also possible to state a condition analogous to Pappos' configuration (ANDRÉ [2,5]). A nearfield space is pappian iff the multiplication of the nearfields $(F,+_i,\cdot)$ is commutative, hence all those nearfields become (commutative) fields. If additionally (G3) holds then by proposition 3.2 the set X forms a vectorspace over a field and $\mathcal{F}$ becomes an affine pappian space.

II. GENERAL THEORY OF FINITE NEARAFFINE SPACES

In this and the following chapter $\mathcal{F} := (X, \mathcal{L}, \sqcup, \|)$ is always a fixed *finite* nearaffine space (see chapter I, definition 3.1), i.e. X is a finite set. Hence the order n of $\mathcal{F}$ (cf. chapter I, definition 3.2) is a natural number. The case $n = 2$ is trivial (cf. the first remark after definition 3.2 in chapter I), hence *we always assume* $n \geq 3$. Instead of (T) we only suppose (V) and (Pa) (cf. chapter I, sections 1 and 3, esp. the remark after theorem 3.1).

---

*) (D3) is then a conclusion of these conditions as will be shown in ANDRÉ [5]. But it is unknown whether (D2) is a consequence of (D1) and (D3), or whether (D1) follows from (D2) and (D3).

1. SUBSPACES

DEFINITION 1.1. A subset U of points of a nearaffine space $\mathcal{F}$ is called an *affine subspace* or simply *subspace* of $\mathcal{F}$ if the following two conditions hold.

(S1) $x,y \in U$, $x \neq y$ *imply* $x \sqcup y \subseteq U$.

(S2) *Any two points of* U *are joinable with respect to* U (cf. chapter I, definition 1.2).

Trivially $\emptyset$, the single points, the straight lines and the whole set X are all subspaces. The property of U being a subspace will be denoted by

$$U \leq X. \tag{1.1}$$

PROPOSITION 1.1. *Let* U *be a subspace and* G *a straight line. Then either* $G \leq U$ *or* $|U \cap G| \leq 1$.

PROPOSITION 1.2. *Let* U *be a subspace,* $x \in U$ *and* L *a line totally contained in* U*, then also* $(x \parallel L) \subseteq U$. (For the definition of $(x \parallel L)$ see chapter I, (1.3).)

PROOF. Let y be a base point (cf. chapter I, section 1, (b)) of L. If $x = y$ or $L = x \sqcup y$ then $(x \parallel L) = L \subseteq U$. Assume therefore $x \neq y$ and $L \neq x \sqcup y$. We first suppose in addition that $x \sqcup y$ is straight. Due to $n \geq 3$ there exists a $z \in (x \sqcup y) \setminus \{x,y\}$, obviously $z \notin L$. Let x' be any point of $(x \parallel L) \setminus \{x\}$, then $x' \notin x \sqcup y$. By the Veblen condition (V) there exists a $y' \in (z \sqcup x') \cap L \subseteq U$. This gives $x' \in U$, hence $(x \parallel L) \subseteq U$ if $x \sqcup y$ is straight. To complete the proof we apply (S2) for an induction on the number of straight lines in U connecting x with y. □

COROLLARY. *If* U *is a subspace of* $\mathcal{F}$ *then all axioms of a nearaffine space, possibly except* (R) (cf. chapter I, sections 1 and 3) *hold in the space*

$$\mathcal{F}_U := (U, \mathcal{L}_U, \sqcup_{|U}, \parallel_{|U}), \tag{1.2}$$

where $\mathcal{L}_U := \{L \in \mathcal{L} \mid L \subseteq U\}$ *and* $\sqcup_{|U}$, $\parallel_{|U}$ *are the restrictions of* $\sqcup$, $\parallel$ *resp., to* U *(in the usual sense). If* (G3) *holds in* $\mathcal{F}$ *then also in* $\mathcal{F}_U$.

PROPOSITION 1.3. *Let* U *be a subspace and* G *a straight line with* $G \cap U \neq \emptyset$. *Then*

$$(1.3) \qquad [U,G] := \bigcup_{y \in U} (y \parallel G)$$

is also a subspace.

PROOF.

(1) If $G \leq U$ then $[U,G] = U$ is a subspace. Assume now $G \nleq U$, i.e. $|G \cap U| = 1$ by proposition 1.1.

(2) Let $z \in V := [U,G]$. Then by (1.3) and proposition 1.1 we have $|(z \parallel G) \cap U| = 1$. Define $\bar{z}_G$ or simply $\bar{z}$, the *projection* of z on U in the direction G, by

$$(1.4) \qquad \{\bar{z}_G\} := \{\bar{z}\} := (z \parallel G) \cap U.$$

(3) Let z, z' be two different points on V. If $\bar{z} = \bar{z}'$ then $z \sqcup z' = (z \parallel G) \subseteq V$. Assume now $\bar{z} \neq \bar{z}'$. Then $\bar{z} \sqcup \bar{z}' \subseteq U$ according to (S1) for U. By (Pa) (cf. chapter I, theorem 3.1) and the definition of V we obtain $(z \parallel \bar{z} \sqcup \bar{z}') \subseteq V$. Applying now (V) to this line and $z \sqcup z'$ we finally conclude $z \sqcup z' \subseteq V$. This gives (S1) for V.

(4) By (S2) for U there is a chain of straight lines in U connecting $\bar{z}$ with $\bar{z}'$. Because $z \sqcup \bar{z}$ and $z' \sqcup \bar{z}'$ are straight (or $z = \bar{z}$, or $z' = \bar{z}'$) there also exists a chain of straight lines totally contained in V connecting z with z'. This proves (S2) for V. □

COROLLARY. *If* $|U \cap G| = 1$ *then*

$$(1.5) \qquad |[U,G]| = |U|\,|G| = |U|n\,,$$

if n *is the order of the space under consideration.*

THEOREM 1.1. *The number of points of a non-void subspace of a finite near-affine space of order* n *is a power of* n.

PROOF. Let V be a non-void subspace. The theorem is true if $|V| = 1$, i.e. V is a point. Use induction on $|V|$ and assume $|V| > 1$. Let U be a *proper* subspace of V with a maximal number of points. Let $x \in U$ and $y \in V \setminus U$. Due to (S2) for V a chain of straight lines completely contained in V connects x with y. Hence there is a straight line G through x contained in V but not in U. Thus $U < [U,G] \leq V$, hence $[U,G] = V$ because of the maximality of U. By (1.5) the induction hypothesis yields the theorem. □

DEFINITION 1.2. Let U be a non-void subspace. Then the exponent d in

$$|U| = n^d \tag{1.6}$$

(cf. theorem 1.1) is called the *dimension* *) of U; it is denoted by

$$d := \text{Dim } U. \tag{1.7}$$

By convention

$$\text{Dim } \emptyset := -1. \tag{1.7'}$$

The dimension of the whole space X of $\mathcal{F}$ is sometimes called the *dimension* Dim $\mathcal{F}$ *of* $\mathcal{F}$.

DEFINITION 1.3. A *hyperplane* of a nearaffine space $\mathcal{F}$ is a proper maximal subspace.

Hence, if H is a hyperplane then

$$\text{Dim } H = \text{Dim } \mathcal{F} - 1. \tag{1.8}$$

Finally we remark

$$U \leq V \leq X, \quad \text{Dim } U = \text{Dim } V \Rightarrow U = V. \tag{1.9}$$

2. SOME NUMBER PROPERTIES

Let $\mathcal{F}$ be a nearaffine space of order n and dimension d. Then $\mathcal{F}$ contains n^d points. We shall state further number-theorems for such spaces.

---

*) As we shall see in chapter III this concept of dimension coincides with that given in chapter I, (3.9).

PROPOSITION 2.1. *The number of lines parallel to a given straight line is* n^{d-1}.

PROOF. Let G be a fixed straight line and g the number of lines parallel to G. By (P1) and (P2) (cf. chapter I, section 1) every point is contained in exactly one line parallel to G. Hence we have $|G|g = ng = |X| = n^d$. □

PROPOSITION 2.2. *The number of lines possessing a given point as base point is*

$$\frac{n^d - 1}{n - 1} = 1 + n + \ldots + n^{d-1}.$$

PROOF. Let x be the given point. By (L2) every point $y \neq x$ contains exactly one line with x as a base point. This proves the proposition by counting arguments. □

The axioms (P1) and (P2) yield:

PROPOSITION 2.3. *The number* γ *of straight lines through a fixed point does not depend on the special choice of this point.*

REMARK. We call this number γ the *class* of the nearaffine space under consideration.

PROPOSITION 2.4. *The number of all lines incident with a given fixed point is*

$$r := \gamma + n\left(\frac{n^d - 1}{n - 1} - \gamma\right) = n\,\frac{n^d - 1}{n - 1} - (n-1)\gamma. \tag{2.1}$$

PROOF. Let $x \in X$ be fixed. The number of straight lines through x is γ by proposition 2.3. Hence, due to proposition 2.2, the number of proper lines with x as base point is $(n^d-1)/(n-1) - \gamma$. The number of points $y \neq x$ such that $x \sqcup y$ is a proper line is

$$(n-1)((n^d-1)/(n-1) - \gamma).$$

Now (2.1) follows by the fact that proper lines with different base points are also different (by consequence (c) of chapter I, section 1). □

THEOREM 2.1. *A finite nearaffine space* $\mathcal{F}$ *considered as incidence structure* (with respect to points and lines and with the incidence relation ϵ, cf. DEMBOWSKI [10,p.1]) *is a tactical configuration, i.e. every line is incident with the same number* k *of points and dually every point is incident with the same number* r *of lines* (see e.g. DEMBOWSKI [10,p.4-5]).

If $\mathcal{F}$ *has order* n, *dimension* d *and class* γ, *if, moreover,* v *and* b *are the number of points and lines resp., then*

$$(2.2)\qquad \begin{cases} v = n^d, & b = n^{d-1}\gamma + n^d\left(\frac{n^d - 1}{n - 1} - \gamma\right), \\ k = n, & r = \gamma + n\left(\frac{n^d - 1}{n - 1} - \gamma\right). \end{cases}$$

PROOF. By theorem 3.2 (chapter I) we have k = n. By definition 1.2 (cf. also theorem 1.1) the number of points is $v = n^d$. The formula for r is (2.1), that for b follows from

$$(2.3)\qquad bk = vr$$

being valid for all tactical configurations (cf. DEMBOWSKI [10,p.5(9')]). □

PROPOSITION 2.5. *In a finite nearaffine space of order* n, *dimension* d *and class* γ *the number of all straight lines is* $n^{d-1}\gamma$.

PROOF. By the propositions 2.2 and 2.3 the number of all proper lines with a given point as base point is $(n^d-1)/(n-1) - \gamma$, hence by consequence (c) (cf. chapter I, section 1) the number of all proper lines is $n^d((n^d-1)/(n-1) - \gamma)$. Our proposition follows from theorem 2.1. □

3. A DEPENDENCE RELATION ON THE STRAIGHT LINES OF A BUNDLE

Let $\mathcal{F}$ be a finite nearaffine space, and $\mathcal{L}$ and $\mathcal{G}$ the set of all lines and straight lines resp. Let x be a fixed point in $\mathcal{F}$. Then $\mathcal{L}_x$ is by definition the set of all lines with x as a base point, the *bundle* generated by x. Morover, let $\mathcal{G}_x := \mathcal{G} \cap \mathcal{L}_x$.

Let $G_1,\dots,G_k \in \mathcal{G}_x$; then define

$$(3.1)\qquad \begin{cases} [G_1] := G_1 \\ [G_1,\dots,G_k] := [[G_1,\dots,G_{k-1}],G_k]. \end{cases}$$

Obviously $[G_1,\dots,G_k]$ is a subspace. We shall prove that it does not depend on the order of the G_i's.

LEMMA 3.1. $[G_1,\dots,G_{k-2},G_{k-1},G_k] = [G_1,\dots,G_{k-2},G_k,G_{k-1}]$.

PROOF. Define $[G_1,\dots,G_{k-2}] =: U$. First by (1.3) we get

$$[U,G_{k-1}] \le [[U,G_k],G_{k-1}] = [U,G_k,G_{k-1}].$$

Moreover,

$$G_k \le [U,G_k] \le [U,G_k,G_{k-1}].$$

Due to proposition 1.2 we get

$$(v \parallel G_k) \le [U,G_k,G_{k-1}]$$

for all $v \in [U,G_{k-1}]$. Hence

$$[U,G_{k-1},G_k] = \bigcup_{v\in[U,G_{k-1}]} (v \parallel G_k) \le [U,G_k,G_{k-1}].$$

By symmetry we also have the converse inequality, thus the lemma. □

LEMMA 3.2. $[G_1,\dots,G_{i-1},G_i,\dots,G_k] = [G_1,\dots,G_i,G_{i-1},\dots,G_k]$.

PROOF. This follows from lemma 3.1 because of

$$[G_1,\dots,G_{i-1},G_i,\dots,G_k] = [[G_1,\dots,G_i],\dots,G_k]. \quad \square$$

Every permutation of $1,\dots,k$ can be generated by transpositions of neighboured numbers. This gives

PROPOSITION 3.1. *The subspace* $[G_1,\dots,G_k]$ *does not depend on the order of the* G_i*'s.*

DEFINITION 3.1. Let $G,G_1,\dots,G_k \in \mathbf{G}_x$. We call G *dependent on* $G_1,\dots,G_k$, denoted by

(3.2) $\quad G \text{ dep } \{G_1,\dots,G_k\}$,

iff $G \le [G_1,\dots,G_k]$.

THEOREM 3.1. *The set* G_x *together with the dependence relation given in definition 3.1 forms a dependence space* (see e.g. VAN DER WAERDEN [24,§20]), *i.e. the relation* "dep" *has the following three properties*:

(Dep 1) G_i dep $\{G_1,\ldots,G_k\}$,

(Dep 2) G dep $\{G_1,\ldots,G_k\}$, G_i dep $\{H_1,\ldots,H_r\}$ *imply* G dep $\{H_1,\ldots,H_r\}$,

(Dep 3) Exchange condition. G dep $\{G_1,\ldots,G_k\}$, *not* G dep $\{G_1,\ldots,G_{k-1}\}$ *imply* G_k dep $\{G_1,\ldots,G_{k-1},G\}$.

PROOF. The first two conditions are straightforward. For the proof of (Dep 3) we first remark $[G_1,\ldots,G_{k-1}] < [G_1,\ldots,G_k]$, hence $\mathrm{Dim}[G_1,\ldots,G_k] = \mathrm{Dim}[G_1,\ldots,G_{k-1}] + 1$. For the same reason $\mathrm{Dim}[G_1,\ldots,G_{k-1},G] = \mathrm{Dim}[G_1,\ldots,G_{k-1}] + 1$. Due to $[G_1,\ldots,G_{k-1},G] \leq [G_1,\ldots,G_{k-1},G_k]$ and (1.9) we have equality for these spaces, hence $G_k \leq [G_1,\ldots,G_{k-1},G]$. □

DEFINITION 3.2. As usual we say $\{G_1,\ldots,G_k\} \subseteq G_x$ is *independent* iff G_i does not depend on $\{G_1,\ldots,G_{i-1},G_{i+1},\ldots,G_k\}$ for all $i \in \{1,\ldots,k\}$. Otherwise the set is called *dependent*.

THEOREM 3.2. *Let* F *be a finite nearaffine space and* $G_1,\ldots,G_k$ *straight lines of* F *going through a fixed point. Then*

$$\mathrm{Dim}[G_1,\ldots,G_k] \leq k \tag{3.3}$$

and equality holds iff $\{G_1,\ldots,G_k\}$ *is independent.*

PROOF. Formula (3.3) follows easily from (3.1), (1.5) and definition 1.2 by induction on k. Equality in (3.3) is equivalent with

$$G_{i+1} \not\leq [G_1,\ldots,G_i] \tag{3.4}$$

for all $i \in \{1,\ldots,k-1\}$, hence by proposition 3.1, equivalent with

$$G_{s_{i+1}} \not\leq [G_{s_1},\ldots,G_{s_i}] \tag{3.4'}$$

for any permutation $\{s_1,\ldots,s_k\}$ of $\{1,\ldots,k\}$. Now it is easy to see that the specialization i = k is really equivalent to (3.4'). This completes the proof. □

COROLLARY. *In a finite nearaffine space of dimension* d *the set* G_x *of all straight lines through* x *contains* d *independent straight lines, but more than* d *straight lines of* G_x *are always dependent.*

THEOREM 3.3. *Any* k*-dimensional* (k ≥ 1) *subspace* U *of a finite nearaffine space containing the point* x *can be represented by*

$$U = [G_1,\dots,G_k],$$

where $G_1,\dots,G_k$ *are independent straight lines of* G_x.

PROOF. We prove this theorem by induction on k = Dim U. For k = 1 we have U = [U]. For k > 1 let V be a maximal proper subspace of U. It has the dimension k-1. Let $G_k \le U$, $G_k \not\le V$ a straight line, then $U = [V,G_k]$ and the theorem is proved by using the induction hypothesis. □

PROPOSITION 3.2. *Let* $U = [G_1,\dots,G_k]$, $G_i \in \mathsf{G}_x$ *and* $y \in U$. *Define* $G_i' := (y \parallel G_i) \in \mathsf{G}_y$. *Then also* $U = [G_1',\dots,G_k']$.

PROOF. Due to $y \in U$ and $G_i \le U$ we have $G_i' \le U$ by proposition 1.2. This implies $[G_1',\dots,G_k'] \le U$. But otherwise $G_i = (x \parallel G_i')$ and the same argument yields the converse inequality. □

PROPOSITION 3.3. *Let* U, G_i, y *and* G_i' *be defined as in proposition 3.2 and let*

$$U_y := [G_1',\dots,G_k'].$$

Then either $U_y = U$ *or* $U \cap U_y = \emptyset$.

PROOF. Assume $z \in U \cap U_y$. Then proposition 3.2 yields $U = U_z = U_y$. □

4. PENCILS OF PARALLEL HYPERPLANES

Let $F = (X,L,\sqcup,\parallel)$ be a finite nearaffine space of order n and dimension d. Let H be an arbitrary hyperplane (cf. definition 1.3) of F; it has the dimension d-1. By (G2) (cf. chapter I, section 3) there is a straight line G such that $H \cap G = \{x\}$ is a point. Using theorem 3.3 and proposition

3.2 it is easy to see that the hyperplanes H_y (for the definition cf. proposition 3.3) are pairwise disjoint for different $y \in G$. By counting arguments we thus obtain

$$(4.1) \qquad X = \dot{\bigcup_{y \in G}} H_y$$

where the dot on the union sign means that the sets whose union will be formed are pairwise disjoint. The set

$$(4.2) \qquad \mathcal{H} := \{H_y \mid y \in G\}$$

is called the *pencil of* (parallel) *hyperplanes generated by* H *) or more briefly a *pencil*. We shall give some properties of such a pencil. For most of them we essentially use the finiteness of the space $\mathcal{F}$. The first proposition, however, is also valid in the infinite case provided of course that hyperplanes do exist.

LEMMA 4.1. *Let* H *be a hyperplane,* $x \notin H$ *and* L *a line with* x *as a base point and* $y \in L \setminus \{x\}$. *Moreover, let* $G \not\subset H$ *be straight and* $\bar{x}, \bar{y}$ *the projections of* x, y *resp. in the direction* G (cf. (1.4)). *Assume* $\bar{x} = \bar{y}$ *or* $L \nparallel \bar{x} \sqcup \bar{y}$. *Then* $L \cap H \neq \emptyset$.

PROOF. This is trivial if $L = G$. Assume therefore $L \neq G = x \sqcup \bar{x}$. Then $\bar{x} \neq \bar{y}$. Using (Pa) (cf. chapter I, theorem 3.1), we obtain

$$(y \parallel \bar{y} \sqcup \bar{x}) \cap (x \sqcup \bar{x}) \neq \emptyset;$$

hence by (G1)

$$(y \parallel \bar{y} \sqcup \bar{x}) \cap (x \sqcup \bar{x}) = \{x'\}.$$

Now $L \neq G$ implies $x' \neq y$ and $x \sqcup y \nparallel \bar{x} \sqcup \bar{y}$, hence $y \sqcup x \nparallel \bar{y} \sqcup \bar{x}$ by (P3); this implies $x' \neq x$. Thus we can apply the Veblen condition (V) and we get

$$(\bar{x} \sqcup \bar{y}) \cap (x \sqcup y) \neq \emptyset.$$

---

*) It is easy to verify that the pencil is independent of the special position of G provided only that $G \not\subseteq H$.

This proves the lemma because of $\bar{x}\sqcup\bar{y} \subseteq H$. □

PROPOSITION 4.1. *Let* H *and* H' *be hyperplanes of the same pencil and* L *be a line contained in* H. *Then* $y \in H'$ *implies* $(y \parallel L) \subseteq H'$.

PROOF. This is clear for $H = H'$ by proposition 1.2. Assume $H \neq H'$. Let x be a base point of L and let $x' \in H'$ be such that $x\sqcup x'$ is straight. Assume $z \in L\backslash\{x\}$ and let $\bar{z}$ be the projection of z onto H' in the direction G (cf. (1.4)). Obviously $x\sqcup\bar{z} \subseteq H'$. If $x'\sqcup\bar{z} \nparallel x\sqcup z = L$ then $(x'\sqcup\bar{z}) \cap H \neq \emptyset$ according to lemma 4.1, hence $H \cap H' \neq \emptyset$. But this contradicts $H \neq H'$ and (4.1). Thus $H' \supseteq (x' \parallel L) = x'\sqcup z \parallel L$ and also $(y \parallel L) \subseteq H'$. □

PROPOSITION 4.2. *Let* L *be a line and* $\mathcal{H}$ *a pencil of hyperplanes. Then either* $L \subseteq H$ *for exactly one* $H \in \mathcal{H}$ *or* $|L \cap H| = 1$ *for all* $H \in \mathcal{H}$.

PROOF. Assume $L \nsubseteq H$ for all $H \in \mathcal{H}$. Let x be a base point of L and H_0 the (by (4.1)) uniquely determined hyperplane of $\mathcal{H}$ containing x. By our hypothesis there exists an $H_1 \in \mathcal{H}\backslash\{H_0\}$ and a point $y \in L \cap H_1$, consequently $L = x\sqcup y$. Let G be a straight line containing x but not contained in H_0. Using (4.1) we see that $\{x'\} := G \cap H_1$ defines a point. Let H_2 be an other hyperplane of $\mathcal{H}$ and x'' defined by $\{x''\} := G \cap H_2$. If $y = x'$ then $L = G$ has exactly one point in common with H_2. If $y \neq x'$ then $x'\sqcup y \subseteq H_1$. According to proposition 4.1 we get $(x'' \parallel x'\sqcup y) \subseteq H_2$. Now (V) implies $(x\sqcup y) \cap (x'' \parallel x'\sqcup y) \neq \emptyset$. Hence $L \cap H_2 \neq \emptyset$ for all $H_2 \in \mathcal{H}$. But both $\mathcal{H}$ and L contain n elements. By counting arguments thus $|L \cap H_2| = 1$ for all $H_2 \in \mathcal{H}$. This proves the proposition. □

COROLLARY. *Any hyperplane* H *of a finite nearaffine space* $\mathcal{F} = (X, \mathcal{L}, \sqcup, \parallel)$ *is a flat* (cf. ANDRÉ [1]), *i.e. any line incident with* $x, y \in H$, $x \neq y$ *(not necessarily as base points) lies completely in* H.

We now generalize this property.

THEOREM 4.1. *Every subspace of a finite nearaffine space is a flat.*

PROOF. Let U be a subspace. We use induction on $s_U := |X| - |U|$. For $s_U = 0$ we have $X = U$ and the theorem is true. Assume $s_U > 0$ and select V in such a way that U is a maximal subspace of V, i.e. a hyperplane of the space $\mathcal{F}_V$ (cf. (1.2)); this can be done e.g. by using proposition 1.3. Obviously $s_V < s_U$ so that we can apply the induction hypothesis to V. Let x,y be different points of U and L a line incident with both x and y. The induction hypothesis and $x,y \in V$ imply $L \subseteq V$, hence L is a line belonging to the space F_V. The corollary to proposition 4.2 now proves the theorem. □

We apply this theorem and obtain a generalization of proposition 3.3 of chapter I.

PROPOSITION 4.3. *Let* U *be a subspace,* $x \in U$ *and* L *a line having* x *as a base point such that* $L \not\subseteq U$. *Then* $y \in U\setminus\{x\}$ *implies* $L \cap (y \parallel L) = \emptyset$.

PROOF. Assume $z \in L \cap (y \parallel L)$. Then $z \notin U$ and $L = x\sqcup z$ and $(y \parallel L) = y\sqcup z$. Due to (P3) we get

$$z\sqcup x \parallel L \parallel (y \parallel L) \parallel z\sqcup y\,,$$

hence $z\sqcup x = z\sqcup y$ by (P1). Because of $x,y \in U$, $x \neq y$, theorem 4.1 now yields $z\sqcup x \subseteq U$, hence $z \in U$ contradicting our hypothesis. □

Now we give another description of the hyperplanes of a pencil.

PROPOSITION 4.4. *Let* H *be a hyperplane and* y *a point. Then that hyperplane of the pencil defined by* H *and passing through* y *can be described by*

$$\bigcup_{H \supseteq L \in \mathcal{L}} (y \parallel L)\,.$$

PROOF. Let H' be the (uniquely determined) hyperplane going through y and belonging to the pencil generated by H. Due to proposition 4.1 we have

$$\bigcup_{H \supseteq L \in \mathcal{L}} (y \parallel L) \subseteq H'.$$

But on the other hand both point sets have the same number of points, namely n^{d-1}. Hence they coincide. □

We conclude this section with a characterization of two-dimensional subspaces.

DEFINITION 4.1. A nearaffine space is called a *nearaffine plane* if every line L intersects every straight line $G \nparallel L$ in exactly one point (cf. also ANDRÉ [2]).

PROPOSITION 4.5. *A nearaffine space is a nearaffine plane iff its dimension is two.*

PROOF.

(1) Dim X = 2. Then any straight line G is a hyperplane. Using proposition 4.2 we see that any line $L \not\parallel G$ has exactly one intersection point with G.

(2) $L \nparallel G$ *implies* $|L \cap G| = 1$. Assume Dim X ≥ 3. The corollary of theorem 3.2 then implies the existence of three independent straight lines G_1, G_2, G_3 going through a point x. According to the dependence conditions stated in section 3 we assume $G = G_1$ without restriction of generality. But then e.g. $(y \parallel G_2)$ has no intersection with G whenever $y \in G_3 \setminus \{x\}$. □

5. WEAK SUBSPACES

Let $F = (X, L, \sqcup, \parallel)$ be a finite nearaffine space of order n.

DEFINITION 5.1. A subset $S \subseteq X$ such that $x,y \in S$, $x \neq y$, imply $x \sqcup y \subseteq S$ is called a *weak subspace*.

Trivially any subspace is a weak subspace. It is our aim to prove the converse relation, i.e. that both concepts coincide.

By proposition 1.2 any subspace is also a weak subspace. It is our aim to prove the converse relation, i.e. that both concepts coincide.

PROPOSITION 5.1. *The intersection of weak subspaces is also a weak subspace.*

For the sake of simplicity we shall give the two following definitions.

DEFINITION 5.2. Let U be a subspace and L a line. We say L is *parallel* to U if for any $x \in U$ we have $(x \parallel L) \subseteq U$. The relation thus defined is denoted by

$$(5.1) \qquad L \parallel U.$$

Moreover, a weak subspace satisfying (S2) (cf. definition 1.1), i.e. a subspace, is sometimes also called a *strong subspace*.

THEOREM 5.1. *Let* $F = (X, L, \sqcup, \|)$ *be a finite nearaffine space of order* $n \geq 3$ *and let* U *be a set of points. Then the following conditions are equivalent:*

(A) $x, y \in U$, $x \neq y$ *imply* $x \sqcup y \subseteq U$ *(i.e.* U *is a weak subspace).*

(B) U *is a subspace* (cf. def. 1.1).

(C) U *is a flat, i.e. any line incident with* $x, y \in U$, $x \neq y$ *completely lies in* U.

PROOF. (C) ⇒ (A) is trivial and (B) ⇒ (C) is theorem 4.1. It remains to prove (A) ⇒ (B). We subdivide this proof into several steps*).

(1) Minimal counterexample. Assume there is a finite nearaffine space possessing a proper weak subspace (i.e. a set of points being a weak subspace but not a strong one). Then there exists a nearaffine space say $F = (X, L, \sqcup, \|)$ of minimal dimension d with this property. In a plane the only weak subspaces are the empty set, the points and the straight lines, all of them being strong. Hence $d \geq 3$.

(2) Minimal weak subspaces. Let S be a fixed proper weak subspace of F with minimal $|S|$. Moreover, let U be a fixed proper subspace of S with maximal $|U|$. By the minimality of S and $U \subset S$ we see U is strong.

(3) *If* G *is straight,* $G \subset S$, *then* $G \| U$; *especially* U *is not a point.* Assume $G \nparallel U$ and let be $x \in U$ and $G' := (x \| G)$. Then $G' \cap U = \{x\}$. We have to prove $G' \subseteq S$. If $x \in G$ we have $G' = G \subseteq S$; assume therefore $x \notin G$. Since $n \geq 3$ we can chose three pairwise different points $r, s, t \in G$. Now (Pa) implies the existence of a $p \in (t \| s \sqcup x) \cap (x \| G)$. By proposition 3.3 of chapter I we have $p \neq x$. Now (V) implies that $q \in (r \sqcup x) \cap (t \| s \sqcup x)$ exists. Moreover, $q \in S$ by $r \sqcup x \subseteq S$. If $q = t$ then $x \in G$, hence $q \neq t$ and $(t \| s \sqcup x) = t \sqcup q \subseteq S$, whence $p \in S$ and thus $G' = x \sqcup p \subseteq S$. Because U is strong also $[U, G']$ is strong by proposition 1.3. On the other hand $U \subset [U, G'] \subseteq S$, hence $[U, G'] = S$ due to the maximality of $|U|$ assumed in (2). But then S is strong, contradicting our hypothesis on S.

(4) *Let* V *be strong,* $V \supseteq U$. *Then either* $V \cap S = U$ *or* $V = X$. If $V \cap S \neq U$ then $U \subset V \cap S \subseteq S$, hence $V \cap S$ is not strong by (2). This implies $S \subseteq V$ and by (1) it must be $V = X$.

---

*) I could only prove this theorem with the further hypothesis $L \subseteq U$, $x \in U \Rightarrow (x \| L) \subseteq U$ (cf. the 1st ed.). I owe to O. BACHMANN this proof working with weaker hypotheses.

(5) *Let* V *be strong,* $V \supseteq U$ *and* $V \cap S \subseteq U$ *such that* V *has maximal dimension. Then* V *is a hyperplane.* For $V \cap S \subseteq U$ first implies $V < X$. Let $x \in U$. Then we can find a straight line G with $G \cap V = \{x\}$, especially $G \nparallel U$. Hence $G \nsubseteq S$ by (3). Due to $U \leq V < [V,G]$ and the maximality of Dim V we have $[V,G] \cap S \nsubseteq U$, hence $[V,G] = X$ by (4). Thus V is a hyperplane.

(6) *Let* G *be straight and* $x \in U \cap G$, $G \cap V = \{x\}$. *Then there exists a hyperplane* W *such that* $G \subseteq W$ *and* $U \nsubseteq W$. Step (1) implies $d = \text{Dim } X \geq 3$, hence Dim $V \geq 2$ by (5). Let G' be straight such that $x \in G' \subseteq U$. Using the results of section 3, esp. theorems 3.1 and 3.3, there exist independent straight lines $G' = G_1, \ldots, G_{d-1}$ such that $V = [G_1, \ldots, G_{d-1}]$. Then $W := [G_2, \ldots, G_{d-1}, G]$ is a hyperplane with the desired properties.

(7) *Let* x, U, G, V, W *and* S *be as before. If* L *is a line such that* $L \subseteq S$ *but* $L \nsubseteq V$ *and such that a base point* b *of* L *is in* U, *then* $L \parallel W$. Assume $L \nparallel W$ and $y \in U \setminus W$, hence also $S \nsubseteq W$. The propositions 4.1 and 4.2 imply

$$(y \parallel L) \cap W = \{z\}$$

for one point z. We have to show $z \in S$. For this select a point $y' \in U$ such that $y' \sqcup b$ is straight; this is possible because U is not a point by (3). Chose $h \in (y' \sqcup b) \setminus \{y', b\}$ and $b' \in L \setminus \{b\}$. Then (V) implies the existence of a point $d \in (y' \parallel L) \cap (h \sqcup b')$ with $d \neq y'$. From $h, b' \in S$ it follows $d \in S$, hence $(y' \parallel L) = y' \sqcup d \subseteq S$. Applying (G2) on U we conclude $(y \parallel L) \subseteq S$ for all $y \in U$, hence $z \in S$ and then $z \in S \setminus V$ due to $L \nsubseteq V$. Because of $S \cap W \subset S$ we know that $S \cap W$ is strong, hence there exists a chain of straight lines totally contained in $S \cap W$ and connecting x with z. But this contradicts (3). The hypothesis $L \nparallel W$ is thus wrong.

(8) <u>Conclusion of the proof</u>. Let again be $y \in U \setminus W$. As before let be $x \in U \cap W$ and let L be a line in S having x as a base point such that $L \nsubseteq U$. By (7) with $b = x$ we have $L \parallel W$, hence $L \subseteq W$ by (P1). Now let be $z \in L \setminus \{x\} \subseteq W$ and $L' := y \sqcup z \subseteq S$, hence $L' \parallel W$ by (7). Using (P3) we obtain

$$z \sqcup y = (z \parallel L') \subseteq W,$$

again by (7). But this gives $y \in W$, contradicting the hypothesis $y \notin W$. □

COROLLARY. *The set* $\mathcal{U}$ *of all subspaces forms a lattice with respect to the set-theoretical intersection* $\cap$ *and the generating union given by*

$$[U,V] = \bigcap_{U,V \subseteq W \in \mathcal{U}} W, \qquad (U,V \in \mathcal{U}).$$

The same is also true for the set $\mathcal{U}_x$ *of all subspaces containing the point* x.

PROPOSITION 5.2. *Let* U *be a non-void subspace. Then*

$$(5.2) \qquad U = \bigcap\{H \mid U \subseteq H, H \text{ hyperplane}\}.$$

PROOF. Let D be the intersection of all hyperplanes $H \supseteq U$. Then trivially $D \supseteq U$. Assume now $x \in U$ and $y \in D\backslash U$. Then $V := [U,y] \supset U = [G_1,\ldots,G_s]$ where the straight lines $G_i \in \mathcal{G}_x$ are assumed to be independent. There exists a straight line $G_{s+1} \leq V$ not lying in U. Enlarge the set $\{G_1,\ldots,G_{s+1}\}$ to a set $\{G_1,\ldots,G_d\}$ of d independent straight lines of $\mathcal{G}_x$ ($d = \text{Dim}\, F$). Then $H := [G_1,\ldots,G_s,G_{s+2},\ldots,G_d]$ is a hyperplane with $U \leq H$ but $V \nleq H$, hence $D \nsubseteq H$, contradicting the construction of D. □

6. NEARAFFINE SPACES AS COMBINATORIAL DESIGNS

Let $F = (X, L, \sqcup, \|)$ be a finite nearaffine space of order n. We shall prove in this section that given any two different points x and x' such that $x \sqcup x'$ is not straight there exist exactly n different lines incident with both x and x' (cf. also ANDRÉ [2], BACHMANN [8]).

PROPOSITION 6.1. *Let* x *and* x' *be different points. Then there exists a pencil* $\mathcal{H}$ *of parallel hyperplanes* (cf. section 4) *such that* $x \in H \in \mathcal{H}$, $x' \in H' \in \mathcal{H}$ *and* $H \neq H'$.

PROOF. According to proposition 5.2 applied to $U = \{x\}$ there exists a hyperplane H containing x but not x'. The pencil generated by H (cf. section 4) has the desired property. □

PROPOSITION 6.2. *Let* U *be a subspace and* x,x' *two different points in it. Let* $L,L' \nsubseteq U$ *be lines having* x,x' *resp. as base points. Then* $|L \cap L'| \leq 1$.

PROOF. First $L, L' \not\subseteq U$ imply $L \cap U = \{x\}$ and $L' \cap U = \{x'\}$. Assume $L \cap L'$ contains two different points y and y'. Both these points do not belong to U. Using (P3) we obtain

$$y \sqcup x' \parallel x' \sqcup y = x' \sqcup y' \parallel y' \sqcup x' .$$

By counting arguments we conclude that the set S of all those points $z \in (x \sqcup x') \setminus \{x\}$ for which there exists a line L^* with a base point on L and parallel to $y \sqcup x'$ such that $L^* \cap (x \sqcup x') = \{z\}$ is a proper subset of $x \sqcup x'$. For $z^* \in (x \sqcup x') \setminus S$ we then have

$$(z^* \parallel x' \sqcup y) \cap L = \emptyset$$

contradicting the Veblen condition (V). □

PROPOSITION 6.3. *Let* x *and* x' *be two different points of a finite nearaffine space of order* n. *Then there exist at most* n *lines incident with both* x *and* x'.

PROOF. Let $\mathfrak{H}$ be a pencil as in proposition 6.1 and H, H' those (uniquely determined) hyperplanes of $\mathfrak{H}$ going through x, x' resp. For any $H^* \in \mathfrak{H} \setminus \{H, H'\}$ there exists at most one point y such that $y \sqcup x = y \sqcup x'$; this is a consequence of the preceding proposition. But $\mathfrak{H}$ possesses exactly n hyperplanes (cf. (4.1),(4.2)). This proves the proposition. □

THEOREM 6.1. *Two different points on a finite nearaffine space of order* n *not being on a straight line are incident with exactly* n *lines.*

PROOF. Let x be a fixed point of the nearaffine space $\mathfrak{F}$ considered as incidence structure (with respect to ϵ). Denote by $\mathfrak{F}_x$ the internal structure (cf. DEMBOWSKI [10,p.3]) obtained from $\mathfrak{F}$ by removing x and all lines not incident with x. Let v', k' and b' be the number of all points, all points on a line and all lines resp. of $\mathfrak{F}_x$. Then theorem 2.1 gives

$$(6.1) \qquad \begin{cases} v' = v-1 = n^d-1, \\ k' = k-1 = n-1, \\ b' = r = \gamma + n\left(\dfrac{n^d-1}{n-1} - \gamma\right) = n\dfrac{n^d-1}{n-1} - (n-1)\gamma; \end{cases}$$

here γ is the number of all straight lines of $\mathfrak{F}$ incident with a given point (cf. proposition 2.3 and the remark thereafter). Let v_i' ($i \in \mathbb{N}$) be the number of all points of F_x incident with exactly i lines of F_x. Clearly

(6.2) $$v_1' = \gamma(n-1)$$

is the number of all those points y of F_x for which $x \sqcup y$ is straight. By proposition 6.3 we have $v_i' = 0$ for $i > n$. Trivially

(6.3) $$n^d - 1 = v' = \sum_{i=1}^{n} v_i'.$$

Using the principle of *double counting* (see e.g. DEMBOWSKI [10,p.4,(10)]) we get

(6.4) $$b'k' = \sum_{i=1}^{n} i v_i'.$$

Assume now $v_i' > 0$ for some $i \in \{2,\dots,n-1\}$. Using (6.1) to (6.4) this hypothesis would give the contradiction

$$b'k' < v_1' + n \sum_{i=2}^{n} v_i' = n \sum_{i=1}^{n} v_i' - (n-1)v_1' =$$

$$= n(n^d-1) - \gamma(n-1)^2 = b'k'.$$

This proves the theorem. □

DEFINITION 6.1. Let x,x' be two different points of a finite nearaffine space $F = (X,L,\sqcup,\|)$. Define $x\nabla x'$ as the set of all points being base points of lines through x and x', i.e.

(6.5) $$x\nabla x' := \{x,x'\} \cup \{y \in X \mid y \sqcup x = y \sqcup x'\}.$$

Point sets of this type are called *gravity curves* *) *(Schwerpunktkurven)* or *blocks* (cf. ANDRÉ [2,II,§3]). The points x,x' are called the *nodes* *(Knoten)* of $x\nabla x'$.

---

*) This name has been chosen because in the case of a nearfield space (cf. chapter I,definition 3.3) the set defined by (6.5) takes the form

(6.5') $$x\nabla x' = \{y \mid \alpha y + \alpha' y = \alpha x + \alpha' x' \text{ for some } (\alpha,\alpha') \in F^2 \setminus \{(0,0)\}\}.$$

Here $y \in x\nabla x'$ looks like the centre of gravity (Schwerpunkt) of x and x' provided that x,x' are covered by suitable "masses" α,α' resp. (not both 0). Of course this interpretation here is only a formal one and has no physical reality

We note the following obvious results on gravity curves (see also ANDRÉ [2]).

THEOREM 6.2. *Let* $F = (X, L, \sqcup, \|)$ *be a finite nearaffine space of order* n. *Then the join* ∇ *defined by* (6.5) *has the following properties:*

(1) $x, x' \in x\nabla x'$,

(2) $x\nabla x' = x'\nabla x$,

(3) $x\nabla x'' = x\nabla x'$, $x'' \neq x' \Rightarrow x\nabla x' = x'\nabla x''$,

(4) $|x\nabla x'| = n$,

(5) $x\nabla x' = x\sqcup x'$ *iff* $x\sqcup x'$ *is straight,*

(6) *if* H *is a hyperplane not containing* $x\sqcup x'$ *then* $|H \cap (x\nabla x')| = 1$,

(7) *if* U *is a subspace,* $x, x' \in U$ *and* $x \neq x'$, *then* $x\nabla x' \subseteq U$.

III. FINITE NEARAFFINE SPACES WITH SPECIAL PROPERTIES

In this chapter we expand the theory developed in the last chapter to special types of finite nearaffine spaces.

1. THE LITTLE DESARGUES CONFIGURATION AND TRANSLATIONS

It is our aim to prove the validity of the little Desargues configuration (D1) (cf. chapter I, section 4) for all finite nearaffine spaces of dimension ≥ 3. To do so it is useful to generalize (D1) in the following way.

(D1') *If* $x, x', y, y', z, z' \in X$ *are pairwise different and* $x\sqcup x' \| y\sqcup y' \| z\sqcup z'$ *are pairwise different lines then* $x\sqcup y \| x'\sqcup y'$ *and* $x\sqcup z \| x'\sqcup z'$ *imply* $y\sqcup z \| y'\sqcup z'$.

REMARK. (D1') yields (D1) by the additional hypothesis $x\sqcup x'$ is straight.

PROPOSITION 1.1. (D1') *holds provided that there exists a hyperplane* H *containing* x,y,z *but not* $x\sqcup x'$ *(and hence also not* $y\sqcup y'$ *and* $z\sqcup z'$*).*

PROOF. The hyperplane H' through x' belonging to the pencil (cf. chapter II, section 4) generated by H is different from H due to the hypothesis. Using

$x \sqcup y \parallel x' \sqcup y'$, $x \sqcup z \parallel x' \sqcup z'$ and proposition 4.4 from chapter II, we get $y',z' \in H'$. Applying now propositions 4.1 and 4.2 (chapter II) we get $x \sqcup z \parallel x' \sqcup z'$, hence (D1'). □

For the proof of the general validity of (D1) we need the following lemma.

LEMMA 1.1. *Let* G *be a straight line and* $x,y \notin G$. *Then there exists a hyperplane* $H \geq G$ *not containing* x *and* y.

PROOF. Define $U := [G,x,y]$. Applying chapter II, proposition 5.2 to G on the one hand and to U on the other hand we immediately obtain a hyperplane with the desired properties. □

THEOREM 1.1. *In a finite nearaffine space* F *with dimension* $d \geq 3$ *the little Desargues configuration* (D1) *generally holds.*

PROOF. This is true if x,y,z and x',y',z' lie on different hyperplanes by proposition 1.1. Assume now that all points occurring in (D1) belong to the same hyperplane H which due to $d \geq 3$ is at least two-dimensional. Applying lemma 1.1 we can find a hyperplane S of H containing x and x' but not y and z. Let G be a straight line through x with $G \nleq H$. Then $[S,G] =: H_1$ is another hyperplane of our space F. Fix $x'' \in G \setminus \{x\}$ and let H' be that hyperplane of the pencil generated by H which contains x". Using results of section 4 from chapter II we obtain y" and z" by

$$\{y''\} := (y \parallel G) \cap H', \qquad \{z''\} := (z \parallel G) \cap H'$$

resp. and $x \sqcup y \parallel x'' \sqcup y''$, $x \sqcup z \parallel x'' \sqcup z''$ and $y \sqcup z \parallel y'' \sqcup z''$. Now we use proposition 1.1 for the triangles x,x',x" and y,y',y" and conclude $x'' \sqcup x' \parallel y'' \sqcup y'$, similarly $x'' \sqcup x' \parallel z'' \sqcup z'$. Again applying proposition 1.1, now to the triangles x",y",z" and x',y',z', we get the desired result $y \sqcup z \parallel y' \sqcup z'$. □

We apply this theorem in view of group-theoretical properties.

THEOREM 1.2. *The translations* (cf. chapter I, definition 2.1) *of a finite at least three-dimensional nearaffine space form an abelian group generated by the straight translations* (cf. chapter I, definition 2.2) *and acting sharply transitive on the points of the space.*

SKETCH OF PROOF.*) If x,x' are points such that $x \sqcup x'$ is straight then due to (D1) there exists a straight translation mapping x onto x'. The proof goes just as in the case of the affine space (see e.g. ARTIN [1,chap. II]). According to (G2) we now see that the translations form a group generated by the straight translations. This group operates transitively on the points. On the other hand it is easy to verify that any translation is uniquely determined by its action on one point. Hence the transitivity is sharp.

It remains to prove that the translation-group is abelian. To do so we first show by straightforward methods using (Pa) and (G1) that two straight translations commute if they have different "directions". Using this result one verifies this commutativity also for straight translations with the same direction. Due to the fact that the straight translations generate the whole group the commutativity is now straightforward.

By standard methods we can now shift the structure of the translation-group to the points. In this way X becomes an abelian group whose operation and neutral element are, as usual, denoted by + and 0 resp. The translations are the mappings of the type

$$(1.1) \qquad x \mapsto x+v, \qquad (x,v \in X).$$

PROPOSITION 1.2. *A subspace* U *containing* 0 *is a subgroup of* (X,+).

PROOF. Let $u \in U$ and define

$$(1.2) \qquad S-u := \{x-u \mid x \in S\}$$

for any subset S of X. By the definition of a translation (chapter I, definition 2.1) we have

$$(1.3) \qquad L \parallel L-u$$

for all lines L. Hence $L \subseteq U$ implies $L-u \subseteq U$ due to $u \in U$ and chapter II, proposition 1.2. But

$$U = \bigcup \{L \mid 0 \in L \subseteq U, \ L \text{ line}\}$$

*) A detailed proof of this theorem will be given in ANDRÉ [5].

yields now $U-u \subseteq U$, hence U is a subgroup of X. □

LEMMA 1.2. $U \in \mathcal{U}_0$ *and* $G \in \mathcal{G}_0$ *imply* $U+G = [U,G] \in \mathcal{U}_0$.

PROOF. First U, G and [U,G] are subgroups of X. Hence $U+G \subseteq [U,G] = \bigcup_{u \in U} (u \parallel G)$ (cf. chapter II,(1.3)). On the other hand let $v \in [U,G]$ and $\bar{v}$ its projection on U in the direction G (cf. chapter II,(1.4)). Then $x \mapsto x + (v-\bar{v})$ is a translation with direction G and $v = \bar{v} + (v-\bar{v}) \in U+G$. □

By induction we easily see using this lemma

PROPOSITION 1.3. $U,V \in \mathcal{U}_0$ *imply* $[U,V] = U+V \in \mathcal{U}_0$.

We conclude this section with a description of the lattice $\mathcal{U}_0$ of all subspaces through 0.

THEOREM 1.3. *The lattice* $(\mathcal{U}_0,\cap,[\,])$ (cf. also chapter II, corollary of theorem 5.1) *is complete, of finite length, modular and complementary.* *)

PROOF. The lattice $\mathcal{U}_0$ is complete and of finite length because it is finite. The modularity follows from proposition 1.3 and the fact that the lattice of all subgroups of an abelian group is modular. Due to theorem 3.3 of chapter II any element $\neq \{0\}$ of $\mathcal{U}_0$ can be represented as a join of atoms (i.e. straight lines in our case); this yields that $\mathcal{U}_0$ is complementary.

2. COMPATIBILITY OF STRAIGHT LINES

Let $F = (X,L,\sqcup,\parallel)$ be a finite nearaffine space. We fix a point in X, say 0. Remember that $\mathcal{G}_0$ and $\mathcal{U}_0$ are the sets of all straight lines and all subspaces resp. going through 0.

DEFINITION 2.1. Two straight lines $G,G' \in \mathcal{G}_0$ are called *compatible (verträglich)*, symbolically

(2.1) $\quad G' \text{ cp } G,$

*) For these and other lattice-theoretical concepts see e.g. BIRKHOFF [9].

if either $G = G'$ or there exist $x \in G\backslash\{0\}$ and $x' \in G'\backslash\{0\}$ such that $x \sqcup x'$ is straight.

Trivially cp is a reflexive and symmetric relation on $\mathcal{G}_0$. We shall see later that cp is also transitive.

DEFINITION 2.2. A subspace $U \in \mathcal{U}_0$ is called *compatible* if any two straight lines $G, G' \leq U, \in \mathcal{G}_0$ are compatible in the sense of definition 2.1. This obviously means that either U itself is a straight line or the triangle condition (G3) (cf. chapter I, section 3) holds in $\mathcal{F}_U$ (cf. chapter II, (1,2)).

PROPOSITION 2.1. *A plane* $E \in \mathcal{U}_0$ *is compatible iff any point of* E *contains at least three different lines contained in* E. *)

This is an easy consequence of (P1), (P2) and proposition 4.5 of chapter II.

COROLLARY. G cp G' *implies* $[G, G']$ *is compatible.*

The following proposition is similar to the diagonal condition (Di) of quadrilaterals (cf. chapter I, section 4).

PROPOSITION 2.2. *Let* x, y, z, w *be pairwise different points not all lying in one plane and such that* $x \sqcup y$, $y \sqcup z$, $z \sqcup w$ *and* $w \sqcup x$ *are straight. Then the diagonals* $x \sqcup z$ *and* $y \sqcup w$ *are straight too.*

PROOF. The intersections $[x \sqcup y, x \sqcup w] \cap [z \sqcup y, z \sqcup w]$ and $[y \sqcup x, y \sqcup z] \cap [w \sqcup x, w \sqcup z]$ are straight lines because x, y, z, w do not belong to the same plane; these straight lines coincide with $x \sqcup z$ and $y \sqcup w$ resp. $\square$

PROPOSITION 2.3. *The diagonal condition* (Di) *holds in a compatible subspace of dimension at least three.*

PROOF. Assume that x, y, z, w are the four corners of the quadrilateral under consideration. If they do not belong to the same plane (Di) holds because of proposition 2.2. Assume that x, y, z, w lie in the same plane $E \leq U$. Due to

---

*) This means that E considered as a nearaffine plane has the type (n,s) with $s \geq 2$ in the sense of ANDRÉ [2, II, §1]; see also section 3 of this chapter.

$\operatorname{Dim} U \geq 3$ there exists a straight line $G \not\leq E$ with $x \in G \leq U$. Select a $u \in G \setminus \{x\}$ in such a way that $u \sqcup z$ is straight; this is possible because $x \sqcup z$ is straight by hypothesis of (Di) and U is compatible. By proposition 2.2 both $y \sqcup u$ and $w \sqcup u$ are straight. Further application of proposition 2.2 now to the quadrilateral x,y,u,w yields that $y \sqcup w$ is straight. □

THEOREM 2.1. *The compatibility relation in the sense of definition 2.1 is transitive. The equivalence classes with respect to this relation are subspaces of* $\mathcal{U}_0$, *hence maximal compatible subspaces, having pairwise only* {0} *as intersection.*

PROOF. Fix a $G \in \mathcal{G}_0$ and consider

$$\text{(2.2)} \qquad \langle G \rangle := \{G' \in \mathcal{G}_0 \mid G' \text{ cp } G\}.$$

If $\langle G \rangle = \{G\}$ then all is proved. Let $V \in \mathcal{U}_0$ be a subspace with $G \leq V$ and in which any straight line through 0 and contained in V is compatible to the fixed G. Assume it has already been proved that V is compatible. If V contains all straight lines compatible to G the proof is finished. Otherwise let G' be compatible to G and $G' \not\leq V$. We shall enlarge the subspace V to another compatible subspace. If $V = G$ then $[G,G']$ is compatible by proposition 2.1. Assume now $\operatorname{Dim} V \geq 2$ and $0 \in G_1 \leq V$, $G_1 \neq G$. Using proposition 2.2 we easily see $G' \text{ cp } G_1$. Hence G' is compatible to all lines of $\mathcal{G}_0$ lying in V. Now select a $G'' \leq [V,G']$, $\not\leq V$, belonging to $\mathcal{G}_0$ and different from G'. Then $[G',G'']$ is a plane. Using chapter II,(1.3) we see that its intersection with V is a straight line G_2 different from G' and G". Proposition 2.1 yields that G" is compatible to G' and a further use of proposition 2.2 gives the compatibility of G" to all lines of $\mathcal{G}_0$ contained in V. This yields that $[V,G'] > V$ is also compatible.

Repeating this method we finally get a maximal compatible subspace U and the lines of $\mathcal{G}_0$ compatible with the given G are exactly those straight lines which are subspaces of U. Hence cp is an equivalence relation and the lines of $\mathcal{G}_0$ contained in U form the equivalence class generated by G. Of course two different maximal compatible subspaces have only 0 as common element. This proves the theorem. □

THEOREM 2.2. *Let* $\{X_1, \ldots, X_t\} \subseteq \mathcal{U}_0$ *be the set of all maximal compatible subspaces of the finite nearaffine space* $\mathcal{F} = (X, \mathcal{L}, \sqcup, \|)$ *satisfying* (D1). *Then the group* (X,+) (cf. section 1, esp. theorem 1.2) *is a direct sum of the subgroups* X_i.

PROOF. As subspaces going through 0 the sets X_i are subgroups of X due to proposition 1.2. Moreover,

$$(2.3) \qquad X = \sum_{i=1}^{t} X_i$$

because any $G \in G_0$ is contained in (exactly) one X_i, and any $x \in X$ is a sum of elements y such that $0 \sqcup y$ is straight (i.e. the mapping $z \mapsto z+y$ is a straight translation); this is a consequence of theorem 1.2. It remains to prove that the sum (2.3) is direct, i.e.

$$(2.4) \qquad X_j \cap \sum_{i=1}^{j-1} X_i = \{0\}$$

holds for all $j \in \{2,\ldots,t\}$. Assume the contrary of (2.4). Then there exists a j and an x in the intersection stated on the left hand side of (2.4) such that $0 \sqcup x$ is straight. But theorem 2.1 yields that every straight line lying in X_j not compatible with any straight line on $\sum_{i=1}^{j-1} X_i$ is not contained in this sum. This yields (2.4). □

Let $U_0(X_i)$ be the lattice of all subspaces of X_i ($i \in \{1,\ldots,t\}$) through 0 and $U_0 = U_0(X)$, as before, the lattice of all subspaces through 0.

Theorem 1.3 stated that U_0 forms a complete complementary modular lattice of finite length. By simple lattice-theoretical arguments we are now able to sharpen this theorem.

THEOREM 2.3. *The lattice* U_0 *is the direct product of the lattices* $U_0(X_i)$ *of all subspaces of* X_i *going through* 0 *where the* X_i ($i \in \{1,\ldots,t\}$) *are the maximal compatible subspaces of* X. *The lattices* $U_0(X_i)$ *are irreducible complementary modular lattices of finite length. Especially* U_0 *itself is irreducible iff* X *is compatible, i.e.* (G3) *holds in* X.

We conclude this section with a sufficient condition for the validity of (Di) in a finite nearaffine space.

THEOREM 2.4. *Let* $F = (X, L, \sqcup, \|)$ *be a finite nearaffine space in which all* *maximal compatible subspaces* X_i *are at least three-dimensional. Then the diagonal condition* (Di) *holds* *) *in* F.

PROOF. This is an obvious consequence of proposition 2.3 and theorem 2.2. □

It is unknown whether the condition on the X_i can be weakened.

3. THE TYPE OF A NEARAFFINE SPACE

Let F be a finite nearaffine space of order n. As we have seen in chapter II, proposition 2.3 the class of F, i.e. the number γ of straight lines going through a given point in F does not depend on the special choice of this point. If the space is a plane then define s by $\gamma =: s+1$ and call (n,s) the *type* of this plane (see also ANDRÉ [2,II,§1]). We shall generalize this concept for arbitrary finite nearaffine spaces. The fundamental idea for this is given by the following theorem.

THEOREM 3.1. *All subplanes of a finite compatible nearaffine space (i.e. a space with* (G3); cf. section 2*) have the same type.* **)

PROOF. We may assume that the compatible space under consideration is at least three-dimensional. Let E and E' be two different subplanes; they are also compatible. We have to consider two cases.

(1) $E \cap E' =: G$ is a straight line. Select $x \in G$ and two straight lines $H, H' \neq G$ with $x \in H \leq E$ and $x' \in H' \leq E'$. Fix $y \in H\setminus\{x\}$ and $y' \in H'\setminus\{x\}$ such that $y \sqcup y'$ is straight; this is possible because the space under consideration is compatible. Let K be any straight line through y and contained in E. Define a mapping f by

$$f(K) = \begin{cases} y' \sqcup (K \cap G) \;^{***)} & \text{if} \quad K \nparallel G, \\ (y' \parallel K) & \text{if} \quad K \parallel G. \end{cases}$$

---

*) Compatible spaces with (Di) are also called *regular* (cf. ANDRÉ [2,I,§3]).

**) For this theorem no results of section 1 are used.

***) Here we identify the set $K \cap G =: \{z\}$ with the point z.

Due to proposition 2.3 the mapping f is a bijection from the pencil of all straigt lines in E through y onto the respective pencil of the straigt lines in E' through y'. Hence the plane E and E' have the same type.

(2) E,E' are in an arbitrary situation. Then one can find a chain $E = E_0, E_1, \ldots, E_t = E'$ of planes such that $E_i \cap E_{i+1}$ $(i \in \{0,\ldots,t-1\})$ are straight lines and hence E and E' are of the same type. In order to find such a chain we consider the following possibilities.

(a) $E \cap E' = \{x\}$. Select different straight lines G, G' with $x \in G \leq E$ and $x' \in G' \leq E'$. Then $E = E_0$, $E_1 = [G,G']$, $E_2 = E'$ is a chain with the desired properties.

(b) *There are* $x \in E$, $x' \in E'$ *such that* $x \sqcup x'$ *is straight.* Then choose a straight line H such that $(x \sqcup x') \cap H$ is a point. We see $E'' := [x \sqcup x', H]$ is a plane related to E and E' by (a) or (1).

(c) *Arbitrary case.* Select $x \in E$ and $x' \in E'$. By (G2) these points can be connected by a chain of straight lines $x_i \sqcup x_{i+1}$. Choose a plane $E_i \ni x_i$. Using induction on the length of the chain of straight lines we can reduce this case to (a). □

DEFINITION 3.1. A compatible finite d-dimensional $(d \geq 2)$ nearaffine space is said to be of *type* (n,s,d) if all subplanes are of type (n,s). A straight line with n points is said to be of *type* (n,1,1).

PROPOSITION 3.1. *In a nearaffine space of type* (n,s,d) *with* $d \geq 2$ *the number of all straight lines through a fixed point is*

$$\gamma = 1 + s + \ldots + s^{d-1} = \frac{s^d - 1}{s-1}. \tag{3.1}$$

PROOF. We prove this proposition by induction on d. It is true for d = 2 due to the definition of s. Let us consider the general case and assume d > 2. Select a hyperplane H, a point $x \in H$ and a straight line $G \not\leq H$ through x. Obviously H is compatible. Fix a $y \in G \setminus \{x\}$. By our induction hypothesis the number of all straight lines in H through x is

$$\gamma' := 1 + s + \ldots + s^{d-2}.$$

Given any straight line G' with $x \in G' \leq H$, then [G,G'] is a plane containing y and $x \in G'$, $G'' \leq H$, $G' \neq G''$ imply $[G,G'] \cap [G,G''] = G$. There exist s straight lines $\neq G$ through y on any such [G,G']. Hence the number of all straight lines through y is

$$\gamma = 1 + s\gamma' = 1 + s \ldots + s^{d-1}. \quad \square$$

Using well-known structure theorems on irreducible complementary modular lattices (see e.g. BIRKHOFF [9]) we get the following properties of compatible spaces.

THEOREM 3.2. *If the compatible nearaffine space* $F = (X, L, \sqcup, ||)$ *of type* (n,s,d) *is three-dimensional the irreducible complementary modular lattice* U_0 *is a projective plane of order* s. *If* F *is at least four-dimensional then* U_0 *forms an at least three-dimensional, hence desarguesian projective space. In this case* s *must be a power of a prime.*

It is an open question whether s is a divisor of n in the general case.

Select now a point 0 of the finite d-dimensional nearaffine space $F = (X, L, \sqcup, ||)$ and consider the maximal compatible subspaces $X_1, \ldots, X_t$ through 0. Assume that X_i is of type (n, s_i, d_i). Then we say that F is of *type*

$$(3.2) \qquad (n, \{(s_1, d_1), \ldots, (s_t, d_t)\}).$$

(The curved brackets { } notify of course that the order of the (s_i, d_i) is inessential.) Using theorem 2.2 we can easily prove the following theorem.

THEOREM 3.3. *A finite* d*-dimensional nearaffine space of a type given by* (3.2) *has the class* (cf. the remark after proposition 2.2 from chapter II)

$$(3.3) \qquad \gamma = \sum_{i=1}^{t} \frac{s_i^{d_i} - 1}{s_i - 1}$$

and its dimension is

$$(3.4) \qquad d = \sum_{i=1}^{t} d_i .$$

4. THE STRUCTURE OF DESARGUESIAN NEARAFFINE SPACES

In this section we add to the conditions (D1) and (Di) the Desargues configuration (D2) (cf. chapter I, section 4). But while we have reasonable sufficient conditions for the validity of (D1) and (Di) (cf. sections 1 and 2 resp.) no such condition for (D2) is known to the author. But if we assume (D1), (D2) and (Di) and another condition stated below for the nearaffine space F then a complete survey can be given even in the infinite case; this will establish an analogue of Hilbert's characterization of desarguesian affine spaces. Let us now state the additional condition which is a weakened form of (G3):

(G3') Weakened triangle condition. *For any straight line* G *and* $x \in G$ *there exists a straight line* $G' \neq G$ *with* $x \in G'$ *such that* G *and* G' *are compatible* (cf. section 2), *i.e. there exist* $y \in G\setminus\{x\}$ *and* $y' \in G'\setminus\{x\}$ *such that* $y \sqcup y'$ *is straight.*

REMARK. This condition is e.g. true if the hypotheses of theorem 2.4 hold.

THEOREM 4.1. (Structure theorem for desarguesian nearaffine spaces). *Let* $F = (X, L, \sqcup, \|)$ *be a nearaffine space with the additional conditions* (D1), (D2), (Di) *and* (G3'). *Then* F *is isomorphic to a nearfield space in the sense of definition 3.3 of chapter I.*

SKETCH OF PROOF.*) Using (D1) and theorem 1.2 we get a group-theoretical structure on X in a natural way. For the sake of brevity we call the elements of $(X,+)$ *vectors*. Define the set Q by

(4.1) $$Q := \{x \in X \mid 0 \sqcup x \text{ is straight}\} \cup \{0\}.$$

Introduce now *scalars* as mappings α from Q into itself such that

(4.2) $$\alpha 0 = 0$$

and

(4.2') $$u, v \in Q \text{ implies } \alpha u = \alpha v \text{ or } u \sqcup v \parallel \alpha u \sqcup \alpha v .$$

*) A detailed proof of this theorem will be given in ANDRÉ [5]. The proof goes essentially the same way as ARTIN's proof (cf.[7]) of the structure theorem for desarguesian affine spaces.

We denote the set of all scalars by F. It is rather difficult to prove that for all $\alpha \in F$ either $\alpha = 0$ or α is a bijection on Q and

(4.3) $\alpha(u+v) = \alpha u + \alpha v$

provided that $u,v,u+v \in Q$. For this proof (Di) must be used. If $\alpha \in F\backslash\{0\}$ then also $\alpha^{-1} \in F$. Moreover, the product of scalars is again a scalar. Using (D2) one can prove that there is an $\alpha \in F$ mapping $u \neq 0$ onto v iff $v \in 0 \sqcup u$. In this case α is uniquely determined by u and v. A $u \in Q$ is called *dependent* of $U \subseteq Q$ if u can be represented as a linear combination of finitely many elements of U whose coefficients belong to F. For this relation all usual dependence conditions hold (see e.g. VAN DER WAERDEN [24,§20], also chapter II, theorem 3.1) so that Q is a dependence space. Let B be a basis of Q, i.e. an independent set generating Q; such a basis exists because of Zorn's lemma. Let $\langle B\rangle$ be the set of all those $x \in X$ which can be represented as a linear combination of B with coefficients in F. Using (G2) one can prove $\langle B\rangle = X$ and extend the multiplication of the elements of F with those of Q to a general multiplication of scalars with vectors. In this way every $\alpha \in F\backslash\{0\}$ becomes an automorphism of (X,+) thus generalizing (4.3). Moreover, the following cancellation law holds:

(4.4) $\alpha x = \beta x$ implies $x = 0$ or $\alpha = \beta$.

By heavily using (D2) and also (G3'), (Di) (at one step of the proof) one can show that the lines are just the sets of the form $Fx+y$ with $x \in X\backslash\{0\}$, $y \in X$, and two lines $Fx+y$ and $Fx'+y'$ are parallel iff $Fx = Fx'$. Moreover, $u \in Q\backslash\{0\}$ iff for all $\alpha,\beta \in F$ there exists a $\gamma \in F$ such that

(4.5) $\alpha u + \beta u = \gamma u$.

Using (4.4) we can define an addition $\underset{u}{+}$ on F by $\alpha \underset{u}{+} \beta := \gamma$. It is easy to see that $(F,\underset{u}{+},\cdot)$ becomes a nearfield and that (X,+) together with F will be a structure defined in chapter I, (3.2) to (3.4). Hence $\mathcal{F}$ is isomorphic to a nearfield space. □

REMARK. One can further prove that this algebraic structure is uniquely determined by $\mathcal{F}$ up to isomorphism and that (D3) holds in $\mathcal{F}$.

If the desarguesian nearaffine space is finite, d-dimensional and compatible then $X = F^d$ for a suitable nearfield F. If the space is of type (n,s,d) (cf. section 3) it is easy to see that $|F| = n$ and one can prove $s = |K|$ where K is the *nucleus* of F defined by

$$K := \{\alpha \in F \mid (\xi+\eta)\alpha = \xi\alpha + \eta\alpha \quad \text{for all } \xi,\eta \in F\}.$$

Hence in this case s is a divisor of n.

APPENDIX: UNSOLVED PROBLEMS

(1) State a complete set of geometric axioms characterizing the spaces of the type I_Γ (cf. section 2 of chapter I)! This has only been done if Γ is a group as defined in chapter I, (3.5), i.e. essentially a Frobenius group; the essential result in this case is given by the hypotheses of chapter III, theorem 4.1.

(2) Search for further relations between Γ and I_Γ. Especially it might be possible to describe the full automorphism group of I_Γ (which has of course Γ as a subgroup) and the translations T of I_Γ (cf. chapter I, definition 2.1). Under what conditions is T a transitive subgroup of Γ?

(3) This problem stated in the 1st ed. has been solved: If in a group space I_Γ all points are joinable (cf. chapter I, def. 1.2) then Γ is either imprimitive or double transitive (cf. chapter I, Theorem 2.2 of this edition).

(4) Do (V) and (Pa) imply (T) (cf. sections 1 and 3 of chapter I) at least in the finite case? (chapter I, section 1, consequence (f) together with theorem 3.1 imply the converse.)

REMARK. The Veblen-condition (V) is independent of the axioms (L), (R), (P) and (G) of a nearaffine space given in the plane case; this has been proved by J. VAN DE SCHOOT & H. WILBRINK: *Nearaffine planes I*, to appear in Proceedings of the Koninklijke Nederlandse Akademie van Wetenschappen.

(5) What further reasonable geometrical configurations may hold in quasi-affine or nearaffine spaces? Are there generalizations of the Reidemeister-, Thomsen-, Klingenberg- and other configurations (cf. also KLINGENBERG [17])? State relations among them! State connections between transitive permutation groups Γ and configurations in I_Γ!

(6) What further properties hold in quasi-affine, nearaffine and especially nearfield spaces if suitable topological properties of $\sqcup$ and $\|$ are

required? Characterize the spaces over the Kalscheuer nearfields (cf. KALSCHEUER [16]).

(7) Let F be a Kalscheuer nearfield and consider

$$X := \{(\xi_i)_{i\in\mathbb{N}} \mid \xi_i \in F, \sum_{i=1}^{\infty} |\xi_i|^2 < \infty\}$$

(for the absolute value in a Kalscheuer nearfield see KALSCHEUER [16]), leading to generalizations of Hilbert spaces. Search for properties of such spaces, especially with respect to functional analysis! Are there reasonable relations to quantum mechanics?

(8) Let $S = (X, L, \sqcup, \parallel)$ be a quasi-affine (or more specially a nearaffine) space and let G be the set of all straight lines in S. Then the incidence structure $\bar{S} := (X, G, \in)$ is a partial plane (cf. DEMBOWSKI [10,p.9]), i.e. any two different points are incident with at most one line of G. Under what conditions for $\bar{S}$ is it possible to reproduce S from $\bar{S}$ in a unique way?

(9) This problem stated in the first ed. has been solved by O. BACHMANN; the answer is affirmative (cf. chapter II, theorem 5.1, including the footnote, in this edition).

(10) Let F be a finite d-dimensional nearaffine space. It is easy to see that two of its points can be connected by a chain of at most d straight lines. Define A_i ($i \in \{1,\dots,d\}$) as the set of all those 2-subsets $\{x,y\}$ of different points which can be connected by i but no fewer straight lines. Is X together with these A_i an association scheme of class number d (DEMBOWSKI [10,p.281])? If the answer is affirmative then F would be a partial design (l.c.,p.282). Compute the parameters p_{ij}^h and n_i in this case!

Remark. If d = 2 then X with A_1, A_2 is an association scheme of class number 2 and thus a strongly regular graph (cf. ANDRÉ [2,II,§2]; for strongly regular graphs see e.g. GOETHALS & SEIDEL [11]).

(11) For any finite nearaffine space $F = (X, L, \sqcup, \parallel)$ there exists a commutative join ∇ defined by (6.5), chapter II. Some of its properties are given in chapter II, theorem 6.2. State further conditions on ∇, if possible in such a way that the original space F can uniquely be reproduced from $(X, L, \nabla, \parallel)$! Give a complete survey of all possible situations of the nodes of $x\nabla y$!

(12) Let $\mathcal{F}$ be a compatible nearaffine space of type (n,s,d) (cf. chapter III, section 3). Under what conditions does $s|n$ hold?

<u>Remark</u>. This is true if $\mathcal{F}$ is desarguesian (cf. chapter III, section 4).

(13) State sufficient conditions for the validity of (D2) (cf. chapter I, section 4)! Does $\mathrm{Dim}\, \mathcal{F} \geq 3$ suffice in the finite case? Is the following *anti-hyperplane-condition* (H) sufficient?

(H) *If* x *and* y *are different points then* [x,y] *is neither the whole space* X *nor a hyperplane in it.**)

(14) Give examples of *proper* finite nearaffine spaces, esp. planes, being no nearfield spaces, hence not dearguesian, or prove that there cannot exist such spaces!

(15) Let $\mathcal{F} = (X, L, \sqcup, ||)$ be a finite nearaffine space such that in X is defined an addition + such that (X,+) becomes an abelian automorphism-group of $\mathcal{F}$ generated by straight translations with neutral element 0 (cf. chapter III, esp. theorem 1.2 and (1.1) which show that such group exists if $\mathrm{Dim}\, \mathcal{F} \geq 3$). Define kx $(k \in z,\ x \in X)$ as usual. Prove or disprove $kx \in 0 \sqcup x$! (Trivially this is true if $0 \sqcup x$ is straight.)

<u>REMARK</u>. If $kx \in 0 \sqcup x$ holds for all $k \in Z$ then all $x \in X \setminus \{0\}$ have the same order in (X,+) and hence this group is elementary abelian: Assume first $x, y \in X \setminus \{0\}$ such that $0 \sqcup x \neq 0 \sqcup y$, and $0 \sqcup x$ is straight. If $kx \neq 0$, $ky = 0$ for a $k \in Z$ then $k(x-y) = kx \in 0 \sqcup x$. If $x-y \notin 0 \sqcup x$ then $0 \sqcup x$ and $0 \sqcup (x-y)$ would have the two different intersection points 0 and $k(x-y)$ contradicting (G1). Hence $x-y \in 0 \sqcup x$, $y \in 0 \sqcup x$ because $0 \sqcup x$ is straight, thus $0 \sqcup x = 0 \sqcup y$ contradicting the hypothesis $0 \sqcup x \neq 0 \sqcup y$. The case $0 \sqcup x = 0 \sqcup y$ straight can be reduced to the previous case in the usual manner. All straight and thus all translations have the same order. (The proof of theorem 1.3 in chapter III of the 1st ed. is so incorrect.)

ADDED IN PROOF

O. BACHMANN (Bern) informed me that II, theorem 5.1 can be proved without condition (2) of weak subspaces (cf. II, def. 5.1) provided that the order n of the space is ≥ 3. Thus problem (9) is solved affirmatively. In this proof the steps (1), (2), (4), (5) and (6) go as in the proof of II, theorem 5.4, noted in this paper. The proof of the other steps as it differs will be stated subsequently.

(3) We have to show $G' \subseteq S$. If $x \in G$ we have $G' = G \subseteq S$. Assume, therefore, $x \in G$. Since $n \geq 3$ we can choose three pairwise different points $r,s,t \in G$. Now (Pa) implies the existence of a $p \in (t \| s \sqcup x) \cap (x \| G)$. By I, proposition 3.3 we have $p \neq x$. Now (V) implies $q \in (r \sqcup x) \cap (t \| s \sqcup x)$ exists. Now $q \in S$ by $r \sqcup x \leq S$. Moreover, $q = t$ would imply $x \in G$, hence $q \neq t$ and $(t \| s \sqcup x) = t \sqcup q \subseteq S$, whence $p \in S$ and thus $G' = x \sqcup p \subseteq S$.

(7) We additionally assume that the base point b of L is in U and have to prove $L \| W$. For this we have to show $z \in S$. Select $y' \in U$ such that $y' \sqcup b$ is straight; this is possible by I, §3, (G2). Choose $h \in (y' \sqcup b) \setminus \{y', b\}$ and $b' \in L \setminus \{b\}$. Then (V) implies the existence of a $d \in (y' \| L) \cap (h \sqcup b')$ with $d \neq y'$. From $h, b' \in S$ it follows $d \in S$, hence $(y' \| L) = y' \sqcup d \subseteq S$. Applying (G2) on U we conclude $(y \| L) \subseteq S$ for all $y \in U$, hence $z \in S$.

(8) In this step the conclusion $z \sqcup y = (z \| L') \subseteq W$ is a consequence of (P1) holding on the hyperplane W.

REFERENCES

[1] ANDRÉ, J., *On flats and dual-flats in incidence structures*, in: Atti del Convegno di Geometria Combinatoria e sue Applicazioni, Perugia, 1971, pp.7-15.

[2] ANDRÉ, J., *Some new results on incidence structures*, to appear in Proceedings Academia Lincei, Roma.

[3] ANDRÉ, J., *Eine Kennzeichnung der Dilatationsgruppen desarguesscher affiner Räume als Permutationsgruppen*, to appear in Arch. Math. (Basel), 25.

[4] ANDRÉ, J., *Lineare Algebra über Fastkörpern*, Math.Z. 136 (1974), 295-313.

[5] ANDRÉ, J., *Affine Geometrien über Fastkörpern*, to be submitted to Mitt. Math. Sem. Giessen.

[6] ARNOLD, H.J., *Die Geometrie der Ringe im Rahmen allgemeiner affiner Strukturen*, Vandenhoeck & Ruprecht, Göttingen, 1971.

[7] ARTIN, E., *Geometric algebra*, Interscience, New York etc., 1957.

[8] BACHMANN, O., *Ueber eine Klasse verallgemeinerter affiner Räume*, to appear in Monatshefte für Math.

[9] BIRKHOFF, G., *Lattice theory*, third ed., Amer. Math. Soc., Coll. Publ., New York, 1967.

[10] DEMBOWSKI, P., *Finite geometries*, Springer-Verlag, Berlin etc., 1968.

[11] GOETHALS, J.M. & J.J. SEIDEL, *Strongly regular graphs derived from combinatorial designs*, to appear in Canad. Math. J.

[12] HUGHES, D. & F. PIPER, *Projective planes*, Springer-Verlag, Berlin etc., 1973.

[13] JORDAN, P., *Algebraische Betrachtungen zur Theorie des Wirkungsquantums und der Elementarlänge*, Abh. Math. Sem. Univ. Hamburg, 18 (1952) 99-119.

[14] JORDAN, P., *Halbgruppen von Idempotenten und nicht-kommutative Verbände*, J. Reine Angew. Math., 211 (1962) 136-161.

[15] JORDAN, P., J. VON NEUMANN & E. WIGNER, *On an algebraic generalization of the quantum mechanical formalism*, Ann. of Math., 35 (1934) 29-64.

[16] KALSCHEUER, F., *Die Bestimmung aller stetigen Fastkörper*, Abh. Math. Sem. Univ. Hamburg, 13 (1940) 413-435.

[17] KLINGENBERG, W., *Beziehungen zwischen einigen affinen Schliessungssätzen*, Abh. Math. Sem. Univ. Hamburg, 18 (1952) 120-143.

[18] PICKERT, G., *Projektive Ebenen*, Springer-Verlag, Berlin etc., 1955.

[19] SPERNER, E., *Affine Räume mit schwacher Inzidenz und zugehörige algebraische Strukturen*, J. Reine Angew. Math., 204 (1960) 205-215.

[20] TAMASCHKE, O., *Projektive Geometrie I*, BI, Mannheim, 1969.

[21] TAMASCHKE, O., *Projektive Geometrie II*, BI, Mannheim, 1972.

[22] VARADARAJAN, V.S., *Geometry of quantum theory I*, van Nostrand, Princeton, 1968.

[23] VARADARAJAN, V.S., *Geometry of quantum theory II*, van Nostrand, Princeton, 1970.

[24] WAERDEN, B.L. VAN DER, *Algebra I*, 7. Aufl., Springer-Verlag, Berlin etc., 1966.

[25] WIELANDT, H., *Finite permutation groups*, Academic Press, New York etc., 1964.

[26] WILLE, R., *Kongruenzklassengeometrien*, Springer-Verlag, Berlin etc., 1970.

WEIGHT ENUMERATORS OF CODES

N.J.A. SLOANE

Bell Laboratories, Murray Hill, New Jersey 07974, USA

ABSTRACT

A tutorial paper dealing with the weight enumerators of codes, especially of self-dual codes. We prove MACWILLIAMS' theorem on the weight distribution of the dual code, GLEASON's theorem on the weight distribution of a self-dual code, some generalizations of this theorem, and then use GLEASON's theorem to show that very good self-dual codes do not exist.

1. INTRODUCTION

We shall mostly consider codes which are *binary* (have symbols from F_2, the field with two elements) or *ternary* (have symbols from F_3, the field with 3 elements). Let F_q^n denote the vector space of all vectors of length n, i.e., having n components, from F_q.

An $[n,k]$ *code* C over F_q is a subspace of F_q^n of dimension k. The vectors of C are called *codewords*. So a binary code is a set of vectors which is closed under addition. A ternary code is closed under addition and under multiplication by -1.

The (*Hamming*) *weight* of a vector $x = (x_1,\ldots,x_n) \in F_q^n$, denoted by $wt(x)$, is the number of non-zero x_i; and the (*Hamming*) *distance* between vectors $x,y \in F_q^n$ is $dist(x,y) = wt(x-y)$.

If every non-zero codeword in C has weight $\geq d$, the code is said to have *minimum weight* d, and is called an $[n,k,d]$ code; n,k,d are the basic parameters of the code. The codewords contain n symbols, and so the *rate* or *efficiency* of the code is $\frac{k}{n}$. Furthermore the code can correct $[\frac{d-1}{2}]$ errors.

M. Hall, Jr. and J. H. van Lint (eds.), Combinatorics, 115-142.

The *dual code* $C^\perp$ is the orthogonal subspace to C:

$$C^\perp = \{u \mid u \cdot v = \sum_{i=1}^{n} u_i v_i = 0, \text{ for all } v \in C\}.$$

$C^\perp$ is an $[n,n-k]$ code.

If $C \subset C^\perp$, C is called *self-orthogonal*, while if $C = C^\perp$ it is called *self-dual*. (See the examples below.)

Let A_i be the number of codewords in C with weight i. Then the set $\{A_0,\ldots,A_n\}$ is called the *weight distribution* of C. It is more convenient to make a polynomial out of the A_i's. The *weight enumerator* of C is

$$W_C(x,y) = A_0x^n + A_1x^{n-1}y + \ldots + A_ny^n =$$

$$= \sum_{i=0}^{n} A_i x^{n-i}y^i = \sum_{u \in C} x^{n-wt(u)}y^{wt(u)}.$$

This is a homogeneous polynomial of degree n in the indeterminates x and y. We could get rid of x by setting $x = 1$, but the theorems are simpler if W is homogeneous.

The weight enumerator gives a good deal of information about the code (see for example [1, §16.1] for some things you can do with the weight enumerator). But is has been calculated for only a few families of codes (e.g. Hamming codes [34], second order Reed Muller codes [41]).

OPEN PROBLEM 1. Find the weight enumerators of all Reed Muller codes (cf. [15]).

We mention in passing a related problem. The distribution of coset leaders by weight is also important for finding the error probability of a code, and for other reasons. But almost nothing is known about calculating it ([4],[12],[42]).

OPEN PROBLEM 2. Find the weight distribution of the coset leaders of the first order Reed Muller codes.

A code is *maximal self-orthogonal* if it is self-orthogonal and is not contained in any larger self-orthogonal code. For binary codes, a maximal self-orthogonal code has dimension $k = \frac{n-1}{2}$ if n is odd, or $k = \frac{n}{2}$ (and is self-dual) if n is even. This paper is concerned with weight enumerators of maximal self-orthogonal codes. First we give some examples.

EXAMPLES. These are binary codes

	[n,k,d]	Code	Weight enumerator
1.	[2,1,2]	{00,11}	$\phi_2 = x^2+y^2$.
2.	[3,1,3]	{000,111}	$\phi_3 = x^3+y^3$.
3.	[3,2,2]	{000,011,101,110}	x^3+3xy^2.
4.	[7,3,4]	Even weight Hamming code	$\phi_7 = x^7+7x^3y^4$.
5.	[7,4,3]	Hamming code	$x^7+7x^4y^3+7x^3y^4+y^7$.
6.	[8,4,4]	Extended Hamming code	$\phi_8 = x^8+14x^4y^4+y^8$.
7.	[17,8,4]	$I_{17}^{(3)}$ of [37]	$\phi_{17} = x^{17}+17x^{13}y^4+187x^9y^8+51x^5y^{12}$.
8.	[23,11,8]	Even weight Golay code	$\phi_{23} = x^{23}+506x^{15}y^8+1288x^{11}y^{12}+ +253x^7y^{16}$.
9.	[24,12,8]	Extended Golay code	$\phi_{24} = x^{24}+759(x^{16}y^8+x^8y^{16})+ +2576x^{12}y^{12}+y^{24}$.

(The subscript of a polynomial almost always gives its degree.)

All except examples 3 and 5 are maximal self-orthogonal. Examples 1, 6, 9 are self-dual.

Observe that examples 4, 6, 7, 8, 9 have the property that every codeword has weight divisible by 4 (because only powers of y^4 appear in the weight enumerators). Codes with this property are important because they have connections with block designs [1], sphere packings [17], lattices and finite groups [6-8], and projective planes [26] (see also [22]).

For non-binary codes it is often useful to have more detailed information than is given by the Hamming weight enumerator. Let C be a code over F_q, where the elements of F_q are labeled $\omega_0=0,\omega_1,\ldots,\omega_{q-1}$ in some fixed order. Then the *composition* of a vector $v \in F_q^n$ is defined to be $s(v) = (s_0(v),s_1(v),\ldots,s_{q-1}(v))$, where $s_i(v)$ denotes the number of coordinates of v that are equal to ω_i. Clearly $\sum_i s_i(v) = n$. Let $A(s)$ be the number of codewords $v \in C$ such that $s(v) = s$. Then the complete weight enumerator of C is the polynomial

$$V_C(z_0,\ldots,z_{q-1}) = \sum_s A(s) z_0^{s_0} \ldots z_{q-1}^{s_{q-1}}.$$

This is a homogeneous polynomial of degree n in the q indeterminates $z_0,\ldots,z_{q-1}$.

The next two examples are of self-dual codes over $F_3 = \{0,1,2\}$, having all weights divisible by 3. The exponents of x,y,z give the number of 0's,

1's, 2's respectively.

[n,k,d]	Code	Complete (V) and Hamming (W) weight enumerators
10. [4,2,3]	{0000,1110,0121, 2220,0212,1201, 1022,2011,2102}	$V = \bar{\Psi}_4 = x^4+x(y+z)^3$, $W = \psi_4 = x^4+8xy^3$.
11. [12,6,6]	Extended Golay code over GF(3), containing the vector 11...1	$V = \bar{\Psi}_{12} = x^{12}+y^{12}+z^{12}+22(x^6y^6+x^6z^6+y^6z^6)+ +220(x^6y^3z^3+x^3y^6z^3+x^3y^3z^6)$, $W = \psi_{12} = x^{12}+264x^6y^6+440x^3y^9+24y^{12}$.

2. MACWILLIAMS' THEOREM

This theorem, due to Mrs. F.J. MACWILLIAMS [20,21], is one of the most remarkable results in coding theory. It says that the weight enumerator of the dual code $C^\perp$ is completely determined by the weight enumerator of C.

We shall prove the binary case. The proof depends on the following lemma, which can be considered as a version of the Poisson summation formula [9,p.220]. Here $F = F_2$.

LEMMA 2.1. (cf. [19]). *Let* $f\colon F^n \to A$ *be any mapping from* F^n *into a vector space* A *over the complex numbers. Define the Fourier transform* $\hat{f}\colon F^n \to A$ *by*

$$\hat{f}(u) = \sum_{v\in F^n} f(v)(-1)^{u\cdot v}.$$

Then for any linear code $C \subset F^n$ *we have*

$$\sum_{v\in C^\perp} f(v) = \frac{1}{|C|}\sum_{u\in C}\hat{f}(u).$$

PROOF.

$$\sum_{u\in C}\hat{f}(u) = \sum_{u\in C}\sum_{v\in F^n} f(v)(-1)^{u\cdot v} = \sum_{v\in F^n} f(v)\sum_{u\in C}(-1)^{u\cdot v}.$$

If $v \in C^\perp$, the inner sum is equal to $|C|$. But if $v \notin C^\perp$, $u\cdot v = 0$ and 1 equally often and the inner sum is zero. □

THEOREM 2.1. (MACWILLIAMS' theorem, binary case). *Let C be an $[n,k]$ binary code and $C^{\perp}$ its dual code. Then the weight enumerator of $C^{\perp}$ is given by*

$$W_{C^{\perp}}(x,y) = \frac{1}{2^k} W_C(x+y,x-y).$$

PROOF. In the lemma, let A be the set of polynomials in x,y with complex coefficients, and $f(v) = x^{n-wt(v)}y^{wt(v)}$. Then

$$\hat{f}(u) = \sum_{v\in F^n} x^{n-wt(v)}y^{wt(v)}(-1)^{u\cdot v} =$$
$$= \sum_{v_1=0}^{1}\sum_{v_2=0}^{1}\cdots\sum_{v_n=0}^{1}\prod_{i=1}^{n} x^{1-v_i}y^{v_i}(-1)^{u_iv_i} =$$
$$= \prod_{i=1}^{n}\sum_{v=0}^{1} x^{1-v}y^{v}(-1)^{u_iv}.$$

If $u_i = 0$ the inner sum is x+y; if $u_i = 1$ the inner sum is x-y. Therefore

$$\hat{f}(u) = (x+y)^{n-wt(u)}(x-y)^{wt(u)}. \quad \square$$

Examples of MACWILLIAMS' theorem

(i) $C = \underline{0}$, $W_C = x^n$; $C^{\perp} = F^n$, $W_{C^{\perp}} = (x+y)^n$.

(ii) $C = \{\underline{0},\underline{1}\}$, $W_C = x^n+y^n$; $C^{\perp}$ = {even weight vectors of length n}, $W_{C^{\perp}} = \frac{1}{2}\{(x+y)^n+(x-y)^n\}$.

(iii) Example 1 of section 1: $C = \{00,11\} = C^{\perp}$, $W_C = x^2+y^2 = \phi_2$.

(iv) Verify the theorem for the pairs of examples 2 and 3, and 4 and 5.

We state without proof two more general versions.

THEOREM 2.2. (MACWILLIAMS' theorem for Hamming weight enumerators). *Let C be an $[n,k]$ code over F_q. Then*

$$W_{C^{\perp}}(x,y) = \frac{1}{q^k} W_C(x+(q-1)y,x-y).$$

For the next theorem we need a little more notation. Let $q = p^a$ where p is prime. Let f(x) be a primitive irreducible polynomial of degree a over F_p and let α be a root of f(x). Any element λ of F_q can be written uniquely as

$$\lambda = \lambda_0 + \lambda_1\alpha + \ldots + \lambda_{a-1}\alpha^{a-1}, \qquad 0 \le \lambda_i < p.$$

Let $\xi = e^{2\pi i/p}$, a complex p-th root of unity. Then the mapping $\chi: \lambda \to \xi^{\lambda_0}$ is a *character* of F_q, i.e., a homomorphism from the additive group of F_q to the multiplicative group of the complex numbers. E.g., if $q = p = 2$, χ maps $x \in F_2$ onto $(-1)^x$.

THEOREM 2.3. (MACWILLIAMS' theorem for complete weight enumerators.) *Let C be an $[n,k]$ code over F_q. Then the complete weight enumerator of $C^\perp$ is given by*

$$V_{C^\perp}(z_0,\ldots,z_{q-1}) = \frac{1}{q^k} V_C\left(\sum_{j=0}^{q-1} \chi(\omega_0\omega_j)z_j,\ldots,\sum_{j=0}^{q-1} \chi(\omega_{q-1}\omega_q)z_j\right).$$

EXAMPLE. $q = 3$, $\xi = \omega = e^{2\pi i/3}$.

$$V_{C^\perp}(x,y,z) = \frac{1}{3^k} V_C(x+y+z, x+\omega y+\omega^2 z, x+\omega^2 y+\omega z).$$

Verify that theorems 2.2 and 2.3 hold for the code of example 10. For the proofs of theorems 2.2 and 2.3 see [20],[21],[19], and for other generalizations see [23],[24].

3. GLEASON'S THEOREM

If the code is self-dual, $C = C^\perp$, then the MACWILLIAMS' theorems 2.1, 2.2, 2.3 give identities which the weight enumerators must satisfy. For example theorem 2.1 states that the weight enumerator of a binary self-dual code must satisfy

$$W(x,y) = \frac{1}{2^{n/2}} W(x+y, x-y),$$

or since $W(x,y)$ is homogeneous of degree n,

$$W(x,y) = W\left(\frac{x+y}{\sqrt{2}}, \frac{x-y}{\sqrt{2}}\right). \tag{3.1}$$

From this GLEASON [11] was able to prove the following remarkable theorems.

THEOREM 3.1. (GLEASON's theorem I). *Let C be a binary self-dual code. Then the weight enumerator* $W_C(x,y)$ *of C is a polynomial in the weight enumerators*

$$\phi_2 = x^2 + y^2 \quad \text{and} \quad \phi_8 = x^8 + 14x^4y^4 + y^8$$

of section 1. Equivalently, W *is a polynomial in* ϕ_2 *and*

$$\theta_8 = x^2y^2(x^2-y^2)^2 = \tfrac{1}{4}(\phi_2^4-\phi_8).$$

THEOREM 3.2. (GLEASON's theorem II). *Let C be a binary self-dual code in which every codeword has weight divisible by* 4. *Then the weight enumerator* $W_C(x,y)$ *of C is a polynomial in the weight enumerators* ϕ_8 *and*

$$\phi_{24} = x^{24} + 759x^{16}y^8 + 2576x^{12}y^{12} + 759x^8y^{16} + y^{24}$$

of section 1; or equivalently in ϕ_8 *and*

$$\theta_{24} = x^4y^4(X^4-y^4)^4 = \frac{1}{42}(\phi_8^3-\phi_{24}).$$

THEOREM 3.3. (GLEASON's theorem III). *Let C be a self-dual code over* F_3. *Then the Hamming weight enumerator* $W_C(x,y)$ *of C is a polynomial in the weight enumerators*

$$\psi_4 = x^4 + 8xy^3 \quad \text{and} \quad \psi_{12} = x^{12} + 264x^6y^6 + 440x^3y^9 + 24y^{12}$$

of section 1; or equivalently in ψ_4 *and*

$$\theta_{12} = y^3(x^3-y^3)^3 = \frac{1}{24}(\psi_4^3-\psi_{12}).$$

Applications of theorem 3.1.

(i) A self-dual code of length 12 contains no codewords of weight 2. What is its weight enumerator W? Answer: By theorem 3.1 W has the form

$$W = a_1\phi_2^6 + a_2\phi_2^2\theta_8 =$$
$$= a_1(x^{12}+6x^{10}y^2+\ldots) + a_2(x^4+2x^2y^2+y^4)x^2y^2(x^2-y^2)^2.$$

But since there are no words of weight 2, this is also

$$= x^{12} + 0x^{10}y^2 + \dots .$$

Therefore $a_1 = 1$, $a_2 = -6$, and

$$W = x^{12} + 15x^8y^4 + 32x^6y^6 + 15x^4y^8 + y^{12}.$$

(ii) Is there a self-dual code of length 32 with minimum distance 10?
Answer: By theorem 3.1 its weight enumerator W has the form

$$W = a_1\phi_2^{16} + a_2\phi_2^{12}\theta_8 + a_3\phi_2^8\theta_8^2 + a_4\phi_2^4\theta_8^3 + a_5\theta_8^4 =$$

$$= x^{32} + 0x^{30}y^2 + 0x^{28}y^4 + 0x^{26}y^6 + 0x^{24}y^8 + A_{10}x^{22}y^{10} + \dots .$$

Equating coefficients we find that $a_1,\dots,a_5$ are uniquely determined and that

$$W = x^{32} + 4960x^{22}y^{10} - 3472x^{20}y^{12} + \dots .$$

Since a weight enumerator cannot have a negative coefficient, no such code exists.

(iii) The extended Golay code (example 9) has

$$W = \phi_{24} = \phi_2^{12} - 12\phi_2^8\theta_8 + 6\phi_2^4\theta_8^2 - 64\theta_8^3.$$

(iv) Exercise: Take all the codewords in the extended Golay code which begin either with 00... or 11..., and delete the first two coordinates. Use theorem 3.1 to obtain the weight distribution of this code. (Answer: $x^{22} + y^{22} + 77(x^{16}y^6+x^6y^{16}) + 330(x^{14}y^8+x^8y^{14}) +$
$+ 616(x^{12}y^{10}+x^{10}y^{12})$).

Application of theorem 3.2.

The extended quadratic residue [48,24,12] code ([1,p.433]) has weights divisible by 4, so its weight enumerator has the form (from theorem 3.2)

(3.2) $$W = a_0\phi_8^6 + a_1\phi_8^3\theta_{24} + a_2\theta_{24}^2 =$$

$$= a_0(x^8+14x^4y^4+y^8)^6 + a_1x^4y^4(x^4-y^4)^4(x^8+14x^4y^4+y^8)^3 +$$

$$+ a_2x^8y^8(x^4-y^4)^8.$$

But since the minimum weight of this code is 12, W is also equal to

(3.3) $$x^{48} + 0x^{44}y^4 + 0x^{40}y^8 + \dots .$$

Equating coefficients in (3.2), (3.3) we find a_0, a_1, a_2 are uniquely determined: $a_0 = 1$, $a_1 = -84$, $a_2 = 246$, and

(3.4) $$W = x^{48} + 17296x^{36}y^{12} + 535095x^{32}y^{16} + 3995376x^{28}y^{20} +$$

$$+ 7681680x^{24}y^{24} + 3995376x^{20}y^{28} + \dots .$$

This example shows how powerful theorem 3.2 can be in obtaining weight enumerators: the fact that the minimum weight was 12 was enough to determine the full weight distribution!

We return to this example, and give further consequences of theorem 3.2, in section 6.

4. INVARIANT THEORY, AND PROOF OF GLEASON'S THEOREMS

Introduction

Other methods of proof are possible (see [3]) but the following proof from invariant theory is the simplest, once the necessary machinery has been developed, and is the easiest to generalize.

Suppose $\mathcal{C}$ is a binary self-dual code with weight enumerator W(x,y). We have already seen in (3.1) that W(x,y) must satisfy

(4.1) $$W(x,y) = W\left(\frac{x+y}{\sqrt{2}}, \frac{x-y}{\sqrt{2}}\right) .$$

Since $\mathcal{C}$ is self-dual, for any $x \in \mathcal{C}$, $x \cdot x = 0$, so x has even weight, and only even powers of y appear in W(x,y). Therefore

(4.2) $W(x,y) = W(x,-y)$.

For an n×n matrix $A = (a_{ij})$ and a polynomial $f(\underline{x}) = f(x_1,\ldots,x_n)$, the result of transforming the variables of f by A is denoted $A \circ f(\underline{x}) = f(\sum a_{1j}x_j,\ldots,\sum a_{nj}x_j)$. Note that $B\circ(A\circ f(\underline{x})) = (AB)\circ f(\underline{x})$.

So (4.1), (4.2) state that $T_1 \circ W = W$, $T_2 \circ W = W$, where

$$T_1 = \frac{1}{\sqrt{2}}\begin{pmatrix}1 & 1\\ 1 & -1\end{pmatrix}, \qquad T_2 = \begin{pmatrix}1 & 0\\ 0 & -1\end{pmatrix}.$$

This implies $T\circ W = W$, where T is any matrix in the group generated by T_1 and T_2. We denote this group by $G_1 = \langle T_1, T_2\rangle$. It is not difficult to check that G_1 is isomorphic to the dihedral group of order 16.

Invariants

Let G be any finite group of n×n complex matrices, and let χ be a homomorphism from G into the multiplicative group of the complex numbers. (I.e., χ is a character of G.) Then $f(\underline{x})$ is called a *relative invariant of G with respect to* χ if

$$A\circ f(\underline{x}) = \chi(A)f(\underline{x}) \qquad \text{for all } A \in G.$$

In particular, if χ is identically 1, and

$$A\circ f(\underline{x}) = f(\underline{x}) \qquad \text{for all } A \in G,$$

then $f(\underline{x})$ is called an *(absolute) invariant* of G.

Clearly if f, g are absolute invariants of G so are f+g and fg, so the absolute invariants form a ring. If f is an absolute invariant and g,h are relative invariants with respect to χ, then g+h and fg are also relative invariants with respect to χ.

(4.1), (4.2) state that W(x,y) is invariant under G_1. To prove theorem 3.1, it will be sufficient to specify the ring of invariants of G_1.

For any finite group G let $\mathcal{J}(G)$ denote the ring of absolute invariants. The problem of characterizing $\mathcal{J}(G)$ is very old, and there are many classical results (see [5,Chapter 17],[32,Part II], [44]). It is enough to characterize the invariants which are homogeneous polynomials, since any invariant is a sum of homogeneous invariants.

Existence of a basic set of invariants for finite groups

DEFINITION 4.1. Polynomials $f_1(\underline{x}),\ldots,f_m(\underline{x})$ are *algebraically dependent* if there is a polynomial p with complex coefficients, not all zero, such that $p(f_1(\underline{x}),\ldots,f_m(\underline{x}))$ is identically zero. Otherwise $f_1(\underline{x}),\ldots,f_m(\underline{x})$ are *algebraically independent*.

THEOREM 4.1. ([13,p.154]). *Any* n+1 *polynomials in* n *variables are algebraically dependent.*

By far the most convenient description of $\mathcal{J}(G)$ is a set of invariants $f_1,\ldots,f_m$ such that any invariant is a *polynomial* in $f_1,\ldots,f_m$. Then $f_1,\ldots,f_m$ is called a *polynomial basis* for $\mathcal{J}(G)$. By theorem 4.1 if $m > n$ there will be polynomial equations, which are called *syzygies*, relating $f_1,\ldots,f_m$.

E. NOETHER'S THEOREM 4.2. (cf. [44,pp.275-276]). $\mathcal{J}(G)$ *has a polynomial basis consisting of not more than* $\binom{g+n}{n}$ *invariants, of degree not exceeding* g, *where* g *is the order of* G.

Theorem 4.2 says that a polynomial basis for $\mathcal{J}(G)$ can always be found. Finding invariants is fairly easy using:

THEOREM 4.3. *If* f(x) *is any polynomial then*

$$h(\underline{x}) = \sum_{A\in G} A\circ f(x)$$

is an invariant of G.

PROOF. For any $A' \in G$,

$$A'\circ h(\underline{x}) = \sum_{A\in G} A'\circ(A\circ f(\underline{x})) = \sum_{A\in G} (AA')\circ f(\underline{x}) =$$

$$= \sum_{B\in G} B\circ f(\underline{x}) = h(\underline{x}). \quad \square$$

Furthermore, it is clear that all invariants of G can be obtained in this way. In fact the proof of theorem 4.2 shows that a polynomial basis for the invariants of G can be obtained by averaging over G all monomials

$$x_1^{b_1} x_2^{b_2} \ldots x_n^{b_n}$$

of total degree $\sum b_i \le g$.

More generally, any symmetric function of the g polynomials $\{A\circ f(\underline{x}) \mid A \in G\}$ is an invariant of G.

MOLIEN's theorem

The next three theorems enable one to determine when enough invariants have been found to make a basis:

THEOREM 4.4. (cf. [32,p.258]). *The number of linearly independent invariants of* G *of the first degree is*

$$\frac{1}{|G|} \sum_{A\in G} \mathrm{trace}(A).$$

THEOREM 4.5. (MOLIEN [33],[32,p.259]). *The number of linearly independent invariants of* G *of degree* ν *is the coefficient of* λ^ν *in the expansion of*

$$\Phi(\lambda) = \frac{1}{|G|} \sum_{A\in G} \frac{1}{\det|I-\lambda A|}. \tag{4.3}$$

$\Phi(\lambda)$ is called the *Molien series* of G.

A similar result holds for relative invariants:

THEOREM 4.6. (MOLIEN [33],[32,p.259]). *The number of linearly independent relative invariants with respect to* χ *of degree* ν *is the coefficient of* λ^ν *in the expansion of*

$$\frac{1}{|G|} \sum_{A\in G} \frac{\bar{\chi}(A)}{\det|I-\lambda A|}, \tag{4.4}$$

where the bar denotes the complex conjugate.

A simple example

Let $\mathcal{C}$ be a self-dual code over GF(q) with Hamming weight enumerator W(x,y). By theorem 2.2, W(x,y) is invariant under the transformation

$$T_3 = \frac{1}{\sqrt{q}}\begin{pmatrix} 1 & q-1 \\ 1 & -1 \end{pmatrix}.$$

Now $T_3^2 = I$, so T_3 generates the group $G_2 = \{I, T_3\}$ of order 2. We shall find the invariants of G_2.

By averaging x over the group, using theorem 4.3, we obtain the invariant $x + \frac{1}{\sqrt{q}}(x+(q-1)y)$, or equivalently $\sigma_1 = x + (\sqrt{q}-1)y$. By averaging x^2 we obtain the invariant $x^2 + \frac{1}{q}(x+(q-1)y)^2$, or equivalently, subtracting $(1+1/q)\sigma_1^2$, $\sigma_2 = y(x-y)$.

Any polynomial in σ_1, σ_2 is of course an invariant of G_2, and the number of products $\sigma_1^i\sigma_2^j$ of degree ν is equal to the number of solutions of $i+2j = \nu$, which is the coefficient of λ^ν in

$$(1+\lambda+\lambda^2+\ldots)(1+\lambda^2+\lambda^4+\ldots) = \frac{1}{(1-\lambda)(1-\lambda^2)} . \tag{4.5}$$

To see if this includes all the invariants of G_2 we compute the Molien series (4.3). This is

$$\Phi(\lambda) = \tfrac{1}{2}\left(\frac{1}{(1-\lambda)^2} + \frac{1}{1-\lambda^2}\right) = \frac{1}{(1-\lambda)(1-\lambda^2)}$$

which agrees with (4.5)! We conclude that we have found all the invariants, i.e., that σ_1, σ_2 are a polynomial basis for the invariants of G_2.

For coding applications we are interested in invariants of even degree. This corresponds to extending G_2 by adding the matrix $-I$, and the Molien series becomes

$$\Phi_e(\lambda) = \tfrac{1}{2}\,(\Phi(\lambda)+\Phi(-\lambda)) = \frac{1}{(1-\lambda^2)^2} ,$$

and as a basis for the new invariants we may take σ_1^2, σ_2^2 or equivalently $\sigma_3 = x^2+(q-1)xy$, $\sigma_4 = xy-y^2$. Thus we have shown that the Hamming weight enumerator of any self-dual code over GF(q) is a polynomial in σ_3 and σ_4.

For example, the code generated by {11} (which is self-dual if q is even) has weight enumerator $\sigma_3-(q-1)\sigma_4$.

The preceding argument enables us to give a short proof of a recent result of LEONT'EV.

THEOREM 4.7. (LEONT'EV [18]). *For a linear code C over* GF(q)

$$W_C(x,y)W_C\left(\frac{x+(q-1)y}{\sqrt{q}}, \frac{x-y}{\sqrt{q}}\right)$$

is a polynomial in σ_3 *and* σ_4.

PROOF. This product is clearly invariant under T_3 and $-I$, and so the result follows from what we have just proved. □

Notation. The following notation is convenient for describing invariants. $\mathbb{C}$ denotes the complex numbers. If $f,g,h,\ldots$ are polynomials $\mathbb{C}[f,g,h,\ldots]$ denotes the ring of polynomials in $f,g,h,\ldots$ with complex coefficients. If R and S are rings, $R \oplus S$ denotes their *direct sum*.

Using this notation, we see that the following result implies theorem 3.1.

THEOREM 3.1*. *The ring of invariants of* $G_1 = \langle T_1, T_2 \rangle$ *is* $\mathbb{C}[\phi_2, \phi_8]$.

PROOF. Let $\mathcal{J}(G_1)$ denote the ring of invariants of G_1. We know from coding theory that ϕ_2 and ϕ_8 are in $\mathcal{J}(G_1)$, and so $\mathcal{J}(G_1) \supseteq M = \mathbb{C}[\phi_2, \phi_8]$.

To show $\mathcal{J}(G_1) = M$, let a_d (or b_d) be the number of linearly independent polynomials of degree d in $\mathcal{J}(G_1)$ (or M). Clearly

$$\sum_{d=0}^{\infty} b_d \lambda^d = \frac{1}{(1-\lambda^2)(1-\lambda^8)} .$$

But from MOLIEN's theorem 4.5,

$$\Phi(\lambda) = \sum_{d=0}^{\infty} a_d \lambda^d = \frac{1}{16} \sum_{A \in G_1} \frac{1}{|I-\lambda A|} = \frac{1}{(1-\lambda^2)(1-\lambda^8)} \tag{4.6}$$

after a straightforward calculation. Therefore $a_d = b_d$ for all d, and so $\mathcal{J}(G_1) = M$. □

In a similar manner we deduce theorems 3.2 and 3.3 from:

THEOREM 3.2*. *The ring of invariants of the group* $G_3 = \langle T_1, T_4 \rangle$, *where*

$$T_1 = \frac{1}{\sqrt{2}} \begin{pmatrix} 1 & 1 \\ 1 & -1 \end{pmatrix}, \quad T_4 = \begin{pmatrix} 1 & 0 \\ 0 & i \end{pmatrix},$$

is $\mathbb{C}[\phi_8, \phi_{24}]$.

G_3 has order 192, and the Molien series (4.3) is

$$\Phi(\lambda) = \frac{1}{(1-\lambda^8)(1-\lambda^{24})} . \tag{4.7}$$

THEOREM 3.3*. *The ring of invariants of* $G_4 = \langle T_5, T_6 \rangle$, *where*

$$T_5 = \frac{1}{\sqrt{3}}\begin{pmatrix} 1 & 2 \\ 1 & -1 \end{pmatrix}, \quad T_6 = \begin{pmatrix} 1 & 0 \\ 0 & e^{2\pi i/3} \end{pmatrix}.$$

is $\mathbb{C}[\psi_4, \psi_{12}]$.

G_4 has order 48, and the Molien series (4.3) is

$$\Phi(\lambda) = \frac{1}{(1-\lambda^4)(1-\lambda^{12})} . \tag{4.8}$$

Theorems 3.1*, 3.2*, 3.3* must have been known to KLEIN and BURNSIDE (see [16],[5,p.362],[39]).

Note that in all these examples the degrees of the basic invariants can be read off the Molien series (4.5)-(4.8) (cf. [28]).

5. GENERALIZATIONS OF GLEASON'S THEOREMS

In this section we give three generalizations of GLEASON's theorems. Other generalizations will be found in [23],[29]. The proofs always follow the same procedure:

Step 1. Translate assumptions about the code into algebraic constraints on the weight enumerator.

Step 2. Use invariant theory to find all possible polynomials satisfying these constraints.

But each of the three examples given has special features of its own. Theorem 5.1 is unusual in that it is rather difficult to find the basic invariants. (Usually one quickly finds what one thinks is a basis for the invariants and the difficulty lies in proving that it *is* a basis.) Theorems 5.2 and 5.3 use a group whose order becomes arbitrarily large, and theorem 5.4 also requires the introduction of new indeterminates and the use of relative rather than absolute invariants.

(I) COMPLETE WEIGHT ENUMERATOR OF A TERNARY SELF-DUAL CODE

Let C be an $[n,\frac{1}{2}n]$ self-dual code over GF(3) which contains the codeword $\underline{1} = 11\ldots1$. Let the complete weight enumerator of C be

$$V(x,y,z) = \sum_{u \in C} x^{s_0(u)} y^{s_1(u)} z^{s_2(u)},$$

where $s_i(u)$ is the number of components of u which are equal to i (as in section 1).

THEOREM 5.1. (cf. [30]).

$$V(x,y,z) \in \mathbb{C}[\alpha_{12},\beta_{12},\delta_{36}] \oplus \gamma_{24}\mathbb{C}[\alpha_{12},\beta_{12},\delta_{36}]$$

(i.e., V(x,y,z) can be written uniquely as a polynomial in α_{12}, β_{12}, δ_{36} plus γ_{24} times another such polynomial), where

$$a = x^3 + y^3 + z^3,$$

$$p = 3xyz,$$

$$b = x^3y^3 + x^3z^3 + y^3z^3,$$

$$\alpha_{12} = a(a^3+8p^3),$$

$$\beta_{12} = (a^2-12b)^2,$$

$$\gamma_{24} = b[(9b-a^2)^3 - 3ap^3(9b-a^2) - a^3p^3 - p^6],$$

$$\delta_{36} = p^3(a^3-p^3)^3.$$

(As usual, the subscript of a polynomial gives its degree.)

PROOF. We carry out the two steps just mentioned.

Step 1. Let a typical codeword $u \in C$ contain a 0's, b 1's, and c 2's. Then since C is self-dual and contains $\underline{1}$

$$u \cdot u = 0 \pmod 3 \Rightarrow 3 \mid b + c,$$

$$u \cdot \underline{1} = 0 \pmod 3 \Rightarrow 3 \mid b - c \Rightarrow 3 \mid b \text{ and } 3 \mid c,$$

$$\underline{1} \cdot \underline{1} = 0 \pmod 3 \Rightarrow 3 \mid a + b + c \Rightarrow 3 \mid a.$$

Therefore V(x,y,z) is invariant under the transformations

$$\begin{pmatrix} \omega & & \\ & 1 & \\ & & 1 \end{pmatrix}, \; J = \begin{pmatrix} 1 & & \\ & \omega & \\ & & 1 \end{pmatrix}, \; \begin{pmatrix} 1 & & \\ & 1 & \\ & & \omega \end{pmatrix}, \quad \omega = e^{2\pi i/3}.$$

Also -u contains a 0's, c 1's, b 2's, and $\underline{1}$ + u contains c 0's, a 1's, b 2's. Therefore V(x,y,z) is invariant under

$$\begin{pmatrix}1&0&0\\0&0&1\\0&1&0\end{pmatrix},\begin{pmatrix}0&1&0\\0&0&1\\1&0&0\end{pmatrix},$$

i.e., under any permutation of its arguments.

Finally from the MACWILLIAMS' theorem (the example after theorem 2.3), $V(x,y,z)$ is invariant under

$$M = \frac{1}{\sqrt{3}}\begin{pmatrix}1&1&1\\1&\omega&\omega^2\\1&\omega^2&\omega\end{pmatrix}.$$

These 6 matrices generate a group G_5, of order 2592, consisting of 1944 matrices of the type

$$s^{\nu}\begin{pmatrix}1&&\\&\omega^a&\\&&\omega^b\end{pmatrix} M^e\begin{pmatrix}1&&\\&\omega^c&\\&&\omega^d\end{pmatrix}, \qquad s = e^{2\pi i/12},$$

and 648 matrices of the type

$$s^{\nu}\begin{pmatrix}1&&\\&\omega^a&\\&&\omega^b\end{pmatrix} P,$$

where $0 \le \nu \le 11$, $0 \le a,b,c,d \le 2$, $e = 1$ or 3, and P is any 3×3 permutation matrix.

Step 2. This step consists of showing that the ring of invariants of G_5 is equal to $\mathbb{C}[\alpha_{12},\beta_{12},\delta_{36}] \oplus \gamma_{24}\mathbb{C}[\alpha_{12},\beta_{12},\delta_{36}]$. First, it is straightforward to show that the Molien series (4.3) of G_5 is

$$\Phi(\lambda) = \frac{1+\lambda^{24}}{(1-\lambda^{12})^2(1-\lambda^{36})},$$

which suggests the degrees of the basic invariants that we should look for.

Next, G_5 is generated by J, M, and all permutation matrices P. Obviously the invariants must be symmetric functions of x,y,z. So we take the symmetric functions a,p,b, and find functions of them which are invariant under J and M. For example,

$$a \overset{M}{\longleftrightarrow} \frac{1}{\sqrt{3}}(a+2p) \overset{J}{\longrightarrow} \frac{1}{\sqrt{3}}(a+2\omega p) \overset{M}{\longleftrightarrow} \frac{i}{\sqrt{3}}(a+2\omega^2 p),$$

so one invariant is

$$\alpha_{12} = a(a+2p)(a+2\omega p)(a+2\omega^2 p) = a(a^3+8p^3).$$

Again

$$a^2-12b \overset{M}{\longleftrightarrow} -(a^2-12b),$$

so another invariant is $\beta_{12} = (a^2-12b)^2$. Again

$$b \overset{M}{\longleftrightarrow} \frac{1}{9}(a^2+ap+p^2) - b \overset{J}{\longrightarrow} \frac{1}{9}(a^2+\omega ap+\omega^2p^2) - b \overset{M}{\longleftrightarrow}$$
$$- \frac{1}{9}[(a^2+\omega^2ap+\omega p^2)-b]$$

gives the invariant

$$\gamma_{24} = b(9b-a^2-ap-p^2)(9b-a^2-\omega ap-\omega^2p^2)(9b-a^2-\omega^2ap-\omega p^2) =$$
$$= b[(9b-a^2)^3-a^3p^3-p^6-3ap^3(9b-a^2)].$$

Finally

$$p \overset{M}{\longleftrightarrow} \frac{1}{\sqrt{3}}(a-p) \overset{J}{\longrightarrow} \frac{1}{\sqrt{3}}(a-\omega p) \overset{M}{\longleftrightarrow} \frac{s}{\sqrt{3}}(a-\omega^2p)$$

gives the invariant

$$\delta_{36} = p^3(a-p)^3(a-\omega p)^3(a-\omega^2p)^3 = p^3(a^3-p^3)^3.$$

One can show that $\alpha_{12},\beta_{12},\delta_{36}$ are algebraically independent and that there is a syzygy of degree 48:

$$(768\gamma_{24}+\alpha_{12}^2+18\alpha_{12}\beta_{12}-27\beta_{12}^2)^2 = 64\beta_{12}(\alpha_{12}^3-64\delta_{36}). \quad \square$$

<u>Remark</u>. Without the assumption that the code contains the all-ones vector the theorem (due to R.J. McELIECE [23,§4.7]) becomes much more complicated.

<u>Applications of theorem 5.1.</u>

For the ternary Golay code (example 11 of section 1), $V = \alpha_{12} + \frac{4}{3}\beta_{12}$. For PLESS's [24,12,9] symmetry code ([35],[36]),

$$V = \frac{179}{432}\alpha_{12}^2 - \frac{19}{24}\alpha_{12}\beta_{12} + \frac{595}{432}\beta_{12}^2 - \frac{352}{9}\gamma_{24}.$$

We have also found the complete weight enumerator of the symmetry codes of lengths 36 and 48 ([30]).

(II) SPLIT WEIGHT ENUMERATORS

We define the *left* and *right weight* of a vector v = $= (v_1,\dots,v_m,v_{m+1},\dots,v_{2m})$ to be respectively $w_L = \mathrm{wt}(v_1,\dots,v_m)$ and $w_R = \mathrm{wt}(v_{m+1},\dots,v_{2m})$. The *split weight enumerator* of a [2m,k] binary code C is

$$\mathcal{W}_C(x,y,X,Y) = \sum_{v\in C} x^{m-w_L(v)} y^{w_L(v)} X^{m-w_R(v)} Y^{w_R(v)}.$$

THEOREM 5.2. (cf. [29]). *Let C be a* [2m,m] *self-dual binary code satisfying:*

(B1) *C contains the vectors* $0^m 1^m = 0\dots01\dots1$ *and* $\underline{1}$; *and*

(B2) *the number of codewords with* $(w_L,w_R) = (j,k)$ *is equal to the number with* $(w_L,w_R) = (k,j)$.

Then

(i) $\mathcal{W}_C(x,y,X,Y)$ *is an element of* $\mathbb{C}[\rho_4,\eta_8,\theta_{16}]$, *where*

$$\rho_4 = (x^2+y^2)(X^2+Y^2),$$

$$\eta_8 = x^4X^4 + x^4Y^4 + y^4X^4 + y^4Y^4 + 12x^2y^2X^2Y^2,$$

$$\theta_{16} = (x^2X^2-y^2Y^2)^2(x^2Y^2-y^2X^2)^2.$$

(ii) *Furthermore, if all weights in C are multiples of* 4, *then* $\mathcal{W}_C(x,y,X,Y)$ *is an element of* $\mathbb{C}[\eta_8,\theta_{16},\gamma_{24}]$, where

$$\gamma_{24} = x^2y^2X^2Y^2(x^4-y^4)^2(X^4-Y^4)^2.$$

A code satisfying (B1), (B2) *is "balanced" about its midpoint, and the division into two halves is a natural one.*

Applications of theorem 5.2.

If $u = (u_1,\dots,u_n)$ and $v = (v_1,\dots,v_n)$ let $u|v = (u_1,\dots,u_n,v_1,\dots,v_n)$. For j = 1,2 let C_j be a code of length n with ordinary weight enumerator $W_j(x,y)$ and split weight enumerator $\mathcal{W}_j(x,y,X,Y)$. The code $C_1|C_2$ = $= \{u|v\colon u \in C_1,\ v \in C_2\}$ has ordinary and split weight enumerators $W_1(x,y)W_2(x,y)$ and $W_1(x,y)W_2(X,Y)$. The equivalent code $C_1 \| C_2 = \{u'|v'|u''|v''\colon$ $: u = u'|u'' \in C_1;\ v = v'|v'' \in C_2\}$, where u and v are broken in half, has ordinary and split weight enumerators $W_1(x,y)W_2(x,y)$ and $\mathcal{W}_1(x,y,X,Y)\mathcal{W}_2(x,y,X,Y)$.

There is a MACWILLIAMS theorem for split weight enumerators (easily obtained from lemma 2.1):

$$(5.1) \qquad \mathcal{W}_{C^{\perp}}(x,y,X,Y) = \frac{1}{|C|}\mathcal{W}_C(x+y,x-y,X+Y,X-Y).$$

We use a detached-coefficient notation for $\mathcal{W}$, and instead of the terms

$$\alpha(x^a y^b X^c Y^d + x^a y^b X^d Y^c + x^b y^a X^c Y^d + x^b y^a X^d Y^c)$$

we write a row of a table:

c/0	x	y	X	Y	#
α	a	b	c	d	4

giving respectively the coefficient, the exponents, and the number of terms of this type. The sum of the products of the first and last columns is the total number of codewords.

A quadratic residue code of length 81 = q+1, where q is a prime, with

Table I. Split weight enumerators

Code	$\mathcal{W}$	c/0	x	y	X	Y	#
H_8	η_8	1	4	0	4	0	4
		12	2	2	2	2	1
	θ_{16}	1	8	0	4	4	4
		-2	6	2	6	2	4
		4	4	4	4	4	1
	γ_{24}	1	10	2	10	2	4
		-2	10	2	6	6	4
		4	6	6	6	6	1
G_{24}		1	12	0	12	0	4
		132	10	2	6	6	4
		495	8	4	8	4	4
		1584	6	6	6	6	1
Z_{48}		1	24	0	24	0	4
		276	22	2	14	10	8
		3864	20	4	16	8	8
		13524	20	4	12	12	4
		9016	18	6	18	6	4
		125580	18	6	14	10	8
		256335	16	8	16	8	4
		950544	16	8	12	12	4
		1835400	14	10	14	10	4
		3480176	12	12	12	12	1

generator matrix, in the canonical form of figures 1-7 of [14], satisfies the hypotheses of theorem 5.2(ii). Table I shows 3 such codes, the [8,4,4] Hamming code H_8 (example 6 of section 1), the [24,12,8] extended Golay code G_{24} (example 9 of section 1) for which $W = \eta_8^3 - 3\eta_8\theta_{16} - 42\gamma_{24}$, and the [48,24,12] quadratic residue code Z_{48}. Also if $S_2 = \{00,11\}$, $S_2|S_2$ has $W = \rho_4$. $H_8|H_8$ has $W = \eta_8^2 + 12\theta_{16}$. Let $R(r,m)$ denote an r-th order Reed-Muller (RM) code of length 2^m. Then RM codes can be constructed recursively from $R(r+1,m) * R(r,m) = R(r+1,m+1)$, where $C_1 * C_2 = \{u|(u+v): u \in C_1, v \in C\}$ [43]. The first order RM code of length n obtained in this way has

$$W = (x^{\frac{1}{2}n}+y^{\frac{1}{2}n})(X^{\frac{1}{2}n}+Y^{\frac{1}{2}n}) + (2n-4)(xyXY)^{\frac{1}{4}n}.$$

PROOF OF THEOREM 5.2(ii). (Part (i) is similar.) Let C satisfy the hypotheses of theorem 5.2(ii) and have split weight enumerator $W = W(x,y,X,Y)$. From the hypotheses, equation (5.1), and the fact that in each term $x^j y^k X^l Y^m$ of W, j+k = l+m, it follows that W is invariant under

$$\frac{1}{\sqrt{2}}\begin{pmatrix}1&1&0&0\\1&-1&0&0\\0&0&1&1\\0&0&1&-1\end{pmatrix}, \begin{pmatrix}1&&&\\&i&&\\&&1&\\&&&i\end{pmatrix}, \begin{pmatrix}1&&&\\&1&&\\&&&1\\&&1&\end{pmatrix}, \begin{pmatrix}&&1&\\&&&1\\1&&&\\&1&&\end{pmatrix} \text{ and } \begin{pmatrix}\alpha&&&\\&\alpha&&\\&&\alpha^{-1}&\\&&&\alpha^{-1}\end{pmatrix},$$

for any complex number α. Let us choose α to be a primitive complex p-th root of unity, where p is a prime greater than deg W = length of C = 2m.

Then one can show that these matrices generate a group G_6 of order 6144p, and that the Molien series for G_6 is

$$\frac{1}{(1-\lambda^8)(1-\lambda^{16})(1-\lambda^{24})} + O(\lambda^p). \tag{5.2}$$

(See [29] for details.) On the other hand, we know from coding theory that $J(G_6)$ contains $M = \mathbb{C}[\eta_8,\theta_{16},\gamma_{24}]$. But the number of linearly independent polynomials of degree 2m in M is the coefficient of λ^{2m} in

$$\frac{1}{(1-\lambda^8)(1-\lambda^{16})(1-\lambda^{24})}. \tag{5.3}$$

Because p > 2m, the coefficients of λ^{2m} in (5.2) and (5.3) agree, and so $J(G_6) = M$. □

(III) WEIGHT ENUMERATORS OF MAXIMAL BINARY SELF-ORTHOGONAL CODES OF ODD LENGTH

THEOREM 5.3.

(A) (cf. [29]). *For* n *odd, let* C *be an* $[n,\frac{1}{2}(n-1)]$ *binary self-orthogonal code. Thus* $C^{\perp} = C \cup (\underline{1}+C)$. *Then*

(i) $W_C(x,y)$ *is an element of the direct sum* $x\mathbb{C}[\phi_2,\phi_8] \oplus \phi_7\mathbb{C}[\phi_2,\phi_8]$, *where* $\phi_2 = x^2+y^2$, $\phi_7 = x^7+7x^3y^4$, $\phi_8 = x^8+14x^4y^4+y^8$. *In words:* $W_C(x,y)$ *can be written in a unique way as* x *times a polynomial in* ϕ_2 *and* ϕ_8, *plus* ϕ_7 *times another such polynomial.*

(B) *Suppose in addition that all weights in* C *are multiples of* 4. *Then*

(ii) n *must be of the form* $8m\pm1$.

(iii) *If* $n = 8m-1$, *then* $W_C(x,y)$ *is an element of* $\phi_7\mathbb{C}[\phi_8,\phi_{24}] \oplus \phi_{23}\mathbb{C}[\phi_8,\phi_{24}]$, *where* $\phi_{23} = x^{23} + 506x^{15}y^8 + 1288x^{11}y^{12} + 253x^7y^{16}$, $\phi_{24} = x^4y^4(x^4-y^4)^4$.

(iv) *If* $n = 8m+1$, *then* $W_C(x,y)$ *is an element of* $x\mathbb{C}[\phi_8,\phi_{24}] \oplus \phi_{17}\mathbb{C}[\phi_8,\phi_{24}]$, *where* $\phi_{17} = x^{17} + 17x^{13}y^4 + 187x^9y^8 + 51x^5y^{12}$.

See the examples in section 1. Some other examples: The [31,15,8] quadratic residue code: $W = -14\phi_7\phi_{24} + \phi_{23}\phi_8$. The [47,23,12] QR code: $W = \frac{1}{7}\{-253\phi_7\phi_8^2\phi_{24} + \phi_{23}(7\phi_8^3-41\phi_{24})\}$. See [37] for additional examples.

It is not presently known if a projective plane of order 10 exists. If it does exist, then from [26] the rows of its incidence matrix generate a [111,55,12] code with

$$W = \frac{1}{7}\{\phi_7(-253\phi_8^{10}\phi_{24} + 24123\phi_8^7\phi_{24}^2 - 430551\phi_8^4\phi_{24}^3 + c_1\phi_8\phi_{24}^4) + \phi_{23}(7\phi_8^{11} - 825\phi_8^8\phi_{24} + 22077\phi_8^5\phi_{24}^2 + c_2\phi_8^2\phi_{24}^3)\},$$

where c_1, c_2 are constants, at present unknown.

PROOF OF THEOREM 5.3. Let C be a code of length $4m-1$ satisfying the hypotheses A,B of theorem 5.3, with weight enumerator $W(\underline{x}) = W(x,y)$. Let

$$M = \frac{1}{\sqrt{2}}\begin{pmatrix}1 & 1\\ 1 & -1\end{pmatrix}, \quad J = \begin{pmatrix}1 & 0\\ 0 & i\end{pmatrix}, \quad R = \begin{pmatrix}0 & 1\\ 1 & 0\end{pmatrix} = MJ^2M.$$

By the MACWILLIAMS theorem 2.1, $M\circ W(\underline{x}) = 2^{-1/2}(W(\underline{x}) + R\circ W(\underline{x}))$. Also $J\circ W(\underline{x}) = W(\underline{x})$. Let M be the set of all polynomials satisfying these two equations.

From coding theory M contains $N = \phi_7\mathbb{C}[\phi_8,\phi_{24}] \oplus \phi_{23}\mathbb{C}[\phi_8,\phi_{24}]$. To show $M = N$, let a_d (or b_d) be the number of linearly independent polynomials of degree d in M (or N). Clearly $\sum_{d=0}^{\infty} b_d\lambda^d = (\lambda^7+\lambda^{23})/(1-\lambda^8)(1-\lambda^{24})$. We show $M = N$ by showing $a_d = b_d$ for all d.

The key device is to consider not $W(x,y)$ but $f(u,v,x,y) = uW(x,y) + vW(y,x)$. Then $f(u,v,x,y)$ is invariant under

$$M^+ = \begin{pmatrix} M & 0 \\ 0 & M \end{pmatrix} \quad \text{and} \quad J^+ = \begin{pmatrix} J & 0 \\ 0 & J \end{pmatrix} \quad \text{acting on} \begin{pmatrix} u \\ v \\ x \\ y \end{pmatrix}.$$

As in theorem 5.2, let ω be a primitive complex p-th root of unity, where p is a prime greater than deg W = length of C. Then $f(u,v,x,y)$ is a relative invariant under $P = \mathrm{diag}(\omega,\omega,1,1)$ with respect to $\chi(P) = \omega$.

Now M,J generate a group G_{192} of order 192, consisting of the matrices

$$(5.4) \qquad r^\nu\begin{pmatrix} 1 & 0 \\ 0 & \alpha \end{pmatrix},\ r^\nu\begin{pmatrix} 0 & 1 \\ \alpha & 0 \end{pmatrix},\ \frac{r^\nu}{\sqrt{2}}\begin{pmatrix} 1 & \beta \\ \alpha & -\alpha\beta \end{pmatrix},$$

where $r = (1+i)/\sqrt{2}$, $0 \le \nu \le 7$, $\alpha,\beta \in \{1,i,-1,-i\}$ (cf. [23]). So M^+, J^+, P generate a group G of order 192p consisting of the matrices $\begin{pmatrix} \omega^\nu A & \\ & A \end{pmatrix}$, $0 \le \nu \le p-1$, $A \in G_{192}$. Then the set M^+ of the relative invariants of G with respect to $\chi(M^+) = \chi(J^+) = 1$, $\chi(P) = \omega$ is in 1-1-correspondence with M up to degree p-1. Therefore from theorem 5.1, for all $p > d$, a_d is the coefficient of λ^{d+1} in

$$\frac{1}{192p}\sum_{B\in G}\frac{\bar{\chi}(B)}{|I-\lambda B|} = \frac{1}{192}\sum_{A\in G_{192}}\frac{1}{p}\sum_{\nu=0}^{p-1}\frac{\omega^{-\nu}}{|I-\lambda A||I-\lambda\omega^\nu A|} \rightarrow$$

$$(\text{as } p\to\infty, |\lambda|<1) \quad \rightarrow \frac{1}{192}\sum_{A\in G_{192}}\frac{1}{|I-\lambda A|}\frac{1}{2\pi}\int_0^{2\pi}\frac{e^{-i\theta}d\theta}{|I-\lambda e^{i\theta}A|} =$$

$$= \frac{\lambda}{192}\sum_{A\in G_{192}}\frac{\mathrm{trace}(A)}{|I-\lambda A|} \overset{(5.4)}{=} \lambda\,\frac{\lambda^7+\lambda^{23}}{(1-\lambda^8)(1-\lambda^{24})}.$$

This proves (iii) and half of (ii). The case n = 4m+1 is treated similarly, taking $M^+ = \begin{pmatrix} \bar{M} & 0 \\ 0 & M \end{pmatrix}$, $J^+ = \begin{pmatrix} \bar{J} & 0 \\ 0 & J \end{pmatrix}$. For part (i) we take $J = \begin{pmatrix} 1 & \\ & -1 \end{pmatrix}$, obtaining a group of order 16p. □

6. VERY GOOD SELF-DUAL CODES DO NOT EXIST

Let C be a binary self-dual code with all weights divisible by 4 and of length $n = 8j = 24\mu+8\nu$, $\nu=0,1$ or 2. From GLEASON's theorem 3.1, the weight enumerator of C has the form

$$(6.1) \qquad W = \sum_{i=0}^{\mu} a_i f^{j-3i} g^i ,$$

where $f = 1+14y+y^2$, $g = y(1-y)^4$. (We have replaced x by 1 and y^4 by y.)

Suppose the $\mu+1$ coefficients a_i are chosen so that

$$(6.2) \qquad W = W^* = 1 + A_{4\mu+4} y^{\mu+1} + A_{4\mu+8} y^{\mu+2} + \ldots .$$

This determines the a_i and A_i uniquely. The resulting W^* is the weight enumerator of that self-dual code with the greatest minimum weight we could hope to attain, and is called an *extremal* weight enumerator.

If a code exists with weight enumerator W^*, it has minimum weight $d^* = 4\mu+4$, unless $A_{4\mu+4}$ is accidentally zero, in which case $d^* \geq 4\mu+8$.

But it can be shown (cf. [27]) that $A_{4\mu+4}$, the number of codewords of minimum non-zero weight enumerator, is equal to:

$$\binom{n}{5}\binom{5\mu-2}{\mu-1} \Big/ \binom{4\mu+4}{5}, \qquad \text{if } n = 24\mu,$$

$$\frac{1}{4} n(n-1)(n-2)(n-4)\frac{(5\mu)!}{\mu!(4\mu+4)!}, \qquad \text{if } n = 24\mu+8,$$

$$\frac{3}{2} n(n-2)\frac{(5\mu+2)!}{\mu!(4\mu+4)!}, \qquad \text{if } n = 24\mu+16,$$

and is never zero. This proves

THEOREM 6.1. (cf. [27]). *The minimum weight of a binary self-dual code of length* n *with all weights divisible by* 4 *is* $\leq 4[\frac{n}{24}] + 4$.

However, the next coefficient, $A_{4\mu+8}$, turns out to be negative if n is large (above about 3712), and so a self-dual code with weight enumerator W^* does not exist for large n. In fact one can show that no self-dual code can even have minimum distance within a constant of $\frac{n}{6}$, if n is sufficiently large:

THEOREM 6.2. (cf. [31]). *Let* b *be any constant. Suppose the* a_i *in* (6.1) *are chosen so that*

$$W = 1 + A_{4d}y^{d} + A_{4d+4}y^{d+1} + \dots ,$$

where $d \geq \frac{n}{6} - b$. *Then one of the coefficients* A_i *is negative, for all sufficiently large* n. *So a binary self-dual code of length* n, *weights divisible by* 4, *and minimum weight* d *does not exist for all sufficiently large* n.

On the other hand it is known that self-dual codes exist which meet the Gilbert bound ([25],[38]).

Similar results hold for self-dual codes over GF(3) (see [31]).

OPEN PROBLEM 3. What is the greatest n for which equality holds in theorem 6.1? (cf. [40],[10]).

ACKNOWLEDGEMENTS

This paper is based on joint work of F.J. MACWILLIAMS, C.L. MALLOWS, A.M. ODLYZKO, and the author; see [23],[27],[29] and [31].

REFERENCES

[1] ASMUS Jr., A.F. & H.F. MATTSON Jr., *New 5-designs*, J. Combinatorial Theory, 6 (1969) 122-151.

[2] BERLEKAMP, E.R., *Algebraic coding theory*, McGraw-Hill, New York, 1968.

[3] BERLEKAMP, E.R., F.J. MACWILLIAMS & N.J.A. SLOANE, *Gleason's theorem on self-dual codes*, IEEE Trans. Information Theory, IT-18 (1972) 409-414.

[4] BERLEKAMP, E.R. & L.R. WELCH, *Weight distributions of the cosets of the (32,6) Reed-Muller code*, IEEE Trans. Information Theory, IT-18 (1972) 203-207.

[5] BURNSIDE, W., *Theory of groups of finite order*, 2nd edition, 1911; reprinted by Dover, New York, 1955.

[6] CONWAY, J.H., *A group of order 8, 315, 553, 613, 086, 720, 000*, Bull. London Math. Soc., 1 (1969) 79-88.

[7] CONWAY, J.H., *A characterization of Leech's lattice*, Invent. Math., 7 (1969) 137-142.

[8] CONWAY, J.H., *Groups, lattices and quadratic forms*, in: *Computers in algebra and number theory*, SIAM-AMS Proceedings IV, Amer. Math. Soc., Providence, R.I., 1971, pp.135-139.

[9] DYM, H. & H.P. MCKEAN, *Fourier series and integrals*, Acad. Press, New York, 1972.

[10] FEIT, W., *A self-dual even (96,48,16) code*, IEEE Trans. Information Theory, IT-20 (1974) 136-138.

[11] GLEASON, A.M., *Weight polynomials of self-dual codes and the MacWilliams identities*, in: Actes Congrès Internat. Math. 1970, vol. 3, Gauthiers-Villars, Paris, 1971, pp. 211-215.

[12] HOBBS, C.F., *Approximating the performance of a binary group code*, IEEE Trans. Information Theory, IT-11 (1965) 142-144.

[13] JACOBSON, N., *Lectures in abstract algebra, vol. 3*, Van Nostrand, Princeton, N.J., 1964.

[14] KARLIN, M., *New binary coding results by circulants*, IEEE Trans. Information Theory, IT-15 (1969) 81-92.

[15] KASAMI, T. & N. TOKURA, *On the weight structure of Reed-Muller codes*, IEEE Trans. Information Theory, IT-16 (1970) 752-759.

[16] KLEIN, F., *Lectures on the icosahedron and the solution of equations of the fifth degree*, 2nd revised edition, 1913; reprinted by Dover, New York, 1956.

[17] LEECH, J. & N.J.A. SLOANE, *Sphere packings and error-correcting codes*, Canad. J. Math., 23 (1971) 718-745.

[18] LEONT'EV, V.K., *Spectra of linear codes*, in: Third International Symp. on Information Theory, Tallinn, Estonia, June 1973, Abstracts of papers, part II, pp. 102-106.

[19] LINT, J.H. VAN, *Coding theory*, Lecture Notes in Mathematics 201, Springer-Verlag, Berlin, 1971.

[20] MACWILLIAMS, F.J., *Combinatorial problems of elementary abelian groups*, thesis, Dept. of Math., Harvard University, May 1962.

[21] MACWILLIAMS, F.J., *A theorem on the distribution of weights in a systematic code*, Bell System Tech. J., 42 (1963) 79-84.

[22] MACWILLIAMS, F.J. & N.J.A. SLOANE, *Combinatorial coding theory*, to appear.

[23] MACWILLIAMS, F.J., C.L. MALLOWS & N.J.A. SLOANE, *Generalizations of Gleason's theorem on weight enumerators of self-dual codes*, IEEE Trans. Information Theory, IT-18 (1972) 794-805.

[24] MACWILLIAMS, F.J., N.J.A. SLOANE & J.M. GOETHALS, *The MacWilliams identities for nonlinear codes*, Bell System Tech. J., 51 (1972) 803-819.

[25] MACWILLIAMS, F.J., N.J.A. SLOANE & J.G. THOMPSON, *Good self-dual codes exist*, Discrete Math., 3 (1972) 153-162.

[26] MACWILLIAMS, F.J., N.J.A. SLOANE & J.G. THOMPSON, *On the existence of a projective plane of order 10*, J. Combinatorial Theory A, 14 (1973) 66-78.

[27] MALLOWS, C.L. & N.J.A. SLOANE, *An upper bound for self-dual codes*, Information and Control, 22 (1973) 188-200.

[28] MALLOWS, C.L. & N.J.A. SLOANE, *On the invariants of a linear group of order 336*, Proc. Cambridge Philos. Soc., 74 (1973) 435-440.

[29] MALLOWS, C.L. & N.J.A. SLOANE, *Weight enumerators of self-orthogonal codes*, Discrete Math., to appear

[30] MALLOWS, C.L. & N.J.A. SLOANE, unpublished.

[31] MALLOWS, C.L., A.M. ODLYZKO & N.J.A. SLOANE, *Upper bounds for modular forms, lattices and codes*, J. Algebra, to appear.

[32] MILLER, G.A., H.F. BLICHFELDT & L.E. DICKSON, *Theory and applications of finite groups*, 1916; reprinted by Dover, New York, 1961.

[33] MOLIEN, T., *Über die Invarianten der lineare Substitutionsgruppe*, Sitzungsber. König. Preuss. Akad. Wiss., 1897, pp.1152-1156.

[34] PETERSON, W.W. & E.J. WELDON Jr., *Error-correcting codes*, 2nd edition, MIT Press, Cambridge, Mass., 1972.

[35] PLESS, V., *On a new family of symmetry codes and related new five-designs*, Bull. Amer. Math. Soc., 75 (1969) 1339-1342.

[36] PLESS, V., *Symmetry codes over GF(3) and new five-designs*, J. Combinatorial Theory, 12 (1972) 119-142.

[37] PLESS, V., *A classification of self-orthogonal codes over GF(2)*, Discrete Math., 3 (1972) 209-246.

[38] PLESS, V. & J.N. PIERCE, *Self-dual codes over GF(q) satisfy a modified Varshamov bound*, Information and Control, 23 (1973) 35-40.

[39] SHEPHARD, G.C. & J.A. TODD, *Finite unitary reflection groups*, Canad. J. Math., 6 (1954) 274-304.

[40] SLOANE, N.J.A., *Is there a (72,36) d = 16 self-dual code?*, IEEE Trans. Information Theory, IT-19 (1973) 251.

[41] SLOANE, N.J.A. & E.R. BERLEKAMP, *Weight enumerators for second order Reed-Muller codes*, IEEE Trans. Information Theory, IT-16 (1970) 745-751.

[42] SLOANE, N.J.A. & R.J. DICK, *On the enumeration of cosets of first order Reed-Muller codes*, IEEE Internat. Conf. on Commun., Montreal 1971, 7: pp.36-2 to 36-6.

[43] SLOANE, N.J.A. & D.S. WHITEHEAD, *New family of single-error correcting codes*, IEEE Trans. Information Theory, IT-16 (1970) 717-719.

[44] WEYL, H., *The classical groups*, Princeton University Press, Princeton, N.J., 1946.

THE ASSOCIATION SCHEMES OF CODING THEORY

P. DELSARTE *)

MBLE Research Laboratory, Brussels, Belgium

1. INTRODUCTION

This paper contains the bases of an algebraic theory of certain association schemes, called *polynomial schemes*. Special emphasis is put on concepts arising from the theories of error correcting codes and of combinatorial designs. The main goal is to provide a general framework in which various applications can be treated by similar methods. In this respect, an interesting formal duality is exhibited between non-constructive coding and design theory.

First, in section 2, some general definitions and preliminary results are given about an *association scheme* (for short, a scheme), especially from the point of view of its *Bose-Mesner algebra* [3]. The natural schemes of coding theory are defined and some polynomial properties of their Bose-Mesner algebras are emphasized.

Section 3 is devoted to the concept of *inner* and *outer distribution* of a subset in an association scheme. Essentially, it is shown that the inner distribution of any subset satisfies certain well-defined inequalities. This result leads to *linear programming problems* having interesting applications in classical theory of codes and designs. Useful relations between the inner and outer distributions are also obtained.

In section 4, we give an axiomatic definition of *polynomial schemes*, which generalize the "coding schemes". In this context, several results about generalized codes and designs are presented. To be more specific, let us mention the questions of *perfect codes* and *tight designs*, among others.

Finally, in section 5, as an application, we briefly consider the *linear codes*, for which certain aspects of the general theory have simple interpretations.

The present paper essentially constitutes part of the author's recent

---

*) The author's participation in this meeting was not supported by NATO.

M. Hall, Jr. and J. H. van Lint (eds.), Combinatorics, 143-161. *All Rights Reserved.*

work [8], where proofs of all theorems given below can be found. A few proofs are incorporated herein, mainly to illustrate the methods.

2. DEFINITIONS AND PRELIMINARIES

Before examining more general notions, we shall briefly describe a suitable framework for classical coding theory. Let F be a finite alphabet, of cardinality $q \geq 2$. Then, for a given $n \geq 1$, the set $X = F^n$ of all n-tuples over F is made a metric space (X,d_H) by definition of the *Hamming distance*:

$$d_H(x,y) = |\{\nu \mid 1 \leq \nu \leq n,\ x_\nu \neq y_\nu\}|, \quad x = (x_\nu)_{\nu=1}^n,\ y = (y_\nu)_{\nu=1}^n.$$

(The distance between two n-tuples is the number of coordinate positions in which they differ.) A *q-ary code of length* n then simply is any non-empty subset of X, endowed with the Hamming metric.

Let us now define the set $R = \{R_0,R_1,\ldots,R_n\}$ of distance relations on X:

$$R_i = \{(x,y) \in X^2 \mid d_H(x,y) = i\}, \quad i=0,1,\ldots,n.$$

We shall call the pair (X,R) a *Hamming scheme*, using the notation H(n,q) for it. (Another terminology is "hypercubic type association schemes"; we refer to YAMAMOTO, FUJII & HAMADA [28].) The following properties of H(n,q) are easily checked:

A1. *The set* R *is a partition of* X^2, *the* R_i *are symmetric (i.e.* $R_i^{-1} = R_i$ *for all* i*) and* R_0 *is the diagonal* $(= \{(x,x) \mid x \in X\})$.

A2. *Let* $(x,y) \in R_k$. *The integer* $p_{i,j,k} = |\{z \in X \mid (x,z) \in R_i,\ (y,z) \in R_j\}|$ *is a constant; it only depends on* (i,j,k).

We now turn to a more general situation. Let X be a finite set of cardinality ≥2, and, for a given integer $n \geq 1$, let $R = \{R_0,R_1,\ldots,R_n\}$ be a set of n+1 relations R_i on X satisfying the axioms A1 and A2. Then (X,R) is called an *association scheme with* n *classes*. This is the concept introduced by BOSE & SHIMAMOTO [4]. (According to HIGMAN's terminology, it is a homogeneous coherent configuration with trivial pairing [13].) Two points $x,y \in X$ are said to be *i-th associates* whenever $(x,y) \in R_i$ holds. The $p_{i,j,k}$ are the *intersection numbers* of the scheme. They satisfy the symmetry relations $p_{i,j,k} = p_{j,i,k}$; for less trivial identities, the reader is referred to BOSE & MESNER [3].

The case $n = 2$ corresponds to the *strongly regular graphs* introduced by BOSE [2]. In general, an association scheme (X,R) may be viewed as a complete graph on the vertex set X, with n distinct colours $c_1,\ldots,c_n$ for the edges: the edge (x,y) is coloured in c_i if and only if x and y are i-th associates, for $i=1,2,\ldots,n$.

REMARKS

(i) Let G be a transitive permutation group on a set X, with $2 \leq |X| < \infty$, and let $R = \{R_0,R_1,\ldots,R_n\}$ denote the set of all G-orbits of X^2. Then, provided the R_i are symmetric, it is well-known that (X,R) is an association scheme. This is called the *group case* (cf. HIGMAN [13]). For instance, the Hamming schemes belong to the group case.

(ii) Assume that, for a given scheme (X,R), two points $x,y \in X$ are i-th associates if and only if they are at distance $\rho(x,y) = i$ in the graph (X,R_1). Then (X,R_1) is called a *perfectly regular graph* (cf. HIGMAN [13]) or a *metrically regular graph* (cf. DOOB [9]). It can easily be seen that an association scheme has such a property of being "generated by a graph" if and only if the intersection numbers satisfy $p_{i,j,k} \neq 0$ whenever $k = i+j$ and

$$(p_{i,j,k} \neq 0) \Rightarrow (|i-j| \leq k \leq i+j).$$

In this case, we call (X,R) a *metric scheme*. The Hamming schemes clearly are metric, with $\rho = d_H$.

Let R_i be any relation on X. We shall denote by D_i the *adjacency matrix* of (X,R_i), i.e. the square matrix of order $|X|$ over $\mathbb{R}$, having X as row and column labeling set, whose (x,y)-entry is

$$D_i(x,y) = \begin{cases} 1 & \text{if } (x,y) \in R_i, \\ 0 & \text{otherwise,} \end{cases}$$

for all $x,y \in X$. The axioms A1, A2 for an association scheme (X,R) can now be expressed in matrix form as follows:

A'1. $\sum_i D_i = J$ (= *all-one matrix*), $D_i^T = D_i$ *for all* i, $D_0 = I$.

A'2. $D_iD_j = D_jD_i = \sum_k p_{i,j,k}D_k$ *for all* i,j.

As an immediate consequence, we have the following result, due to BOSE & MESNER [3]: Let R be a set of n+1 relations R_i on X, satisfying A1. Then (X,R) is an association scheme, with n classes, if and only if the adjacency matrices D_i generate an (n+1)-dimensional commutative algebra over $\mathbb{R}$.

We shall now examine in some detail the properties of the algebra $A = \langle D_0, D_1, \ldots, D_n \rangle$ of an association scheme (X,R), which we call the *Bose-Mesner algebra*. It is easy to show that A is a semisimple algebra, isomorphic to $\mathbb{R}^{n+1}$. In other words, A admits a unique basis $(J_0, J_1, \ldots, J_n)$ of *minimal idempotents* J_k, being mutually orthogonal: $J_i J_k = \delta_{i,k} J_k$, and satisfying $\sum J_k = I$. Moreover, for a suitable numbering, we have $J_0 = |X|^{-1} J$. This structure has been described first by OGAWA [22].

Let us write the expansion of the adjacency matrices D_k with respect to the basis of minimal idempotents J_i:

(2.1) $$D_k = \sum_{i=0}^{n} P_k(i) J_i, \quad k=0,1,\ldots,n,$$

for some real numbers $P_k(i)$, uniquely defined. These are the eigenvalues of D_k. Indeed (2.1) yields $D_k J_i = P_k(i) J_i$. This shows, in addition, that the column spaces of the J_i are the common eigenspaces of all matrices belonging to A. We shall denote by μ_i, and call i-th *multiplicity*, the dimension of the i-th eigenspace, i.e. $\mu_i = \mathrm{rank}(J_i)$, i=0,1,...,n. We also point out that the eigenvalue $v_k = P_k(0)$ has an obvious combinatorial meaning, namely
v_k = *valency* of R_k = number of k-th associates of a fixed point in X
(= number of ones in each row of D_k).

It is often more interesting to characterize an association scheme by the eigenvalues $P_k(i)$ rather than by the intersection numbers $p_{i,j,k}$, although either set of parameters can be derived from the other. In fact we shall make use of the *eigenmatrices* P and Q: these are the non-singular square matrices of order n+1 defined to be

$$P = [P_k(i) : i,k \in \{0,1,\ldots,n\}], \qquad Q = |X| P^{-1}.$$

So $P_k(i)$ is the (i,k)-entry of P. Similarly, $Q_k(i)$ will denote the (i,k)-entry of Q, for i,k=0,1,...,n. Thus, according to (2.1), we have

(2.2) $$|X| J_k = \sum_{i=0}^{n} Q_k(i) D_i, \quad k=0,1,\ldots,n.$$

It is easily seen that $Q_0(i) = 1$ and $Q_k(0) = \mu_k$ hold, for all i,k.

Given an (n+1)-tuple $\underline{c} = (c_0, c_1, \ldots, c_n)$ of real numbers c_i, we shall denote by $\Delta_{\underline{c}}$ the diagonal matrix $\text{diag}(c_0, c_1, \ldots, c_n)$. The following theorem, due to YAMAMOTO, FUJII & HAMADA [28], exhibits interesting *orthogonality relations* on the eigenmatrices.

THEOREM 1. $Q^T \Delta_{\underline{v}} Q = |X| \Delta_{\underline{\mu}}, \quad P^T \Delta_{\underline{\mu}} P = |X| \Delta_{\underline{v}}.$

PROOF. One easily obtains the first identity by expressing $J_i J_k = \delta_{i,k} J_k$ in the basis of matrices D_j. The details will not be given. Then the second identity follows from the first one, by use of $P = |X| Q^{-1}$. □

When the eigenmatrices P and Q are interchanged, the relations of theorem 1 are transformed into each other. This is the first occurrence of a nice *formal duality* appearing at several places in the theory. In this respect, one may ask the question whether there exists a *dual scheme* of (X,R), that is, an association scheme (X',R') whose eigenmatrices are P' = Q and Q' = P. Such an "actual duality" can indeed by defined for certain association schemes, such as those admitting an Abelian regular automorphism group. (In this case the Bose-Mesner algebras are in fact *Schur rings*. The duality introduced by TAMASCHKE [25] for Schur rings can be used to define dual association schemes.) For instance, the Hamming schemes admit a dual. More precisely, they are self-dual so that one has P = Q, ie. $P^2 = q^n I$.

For the Hamming schemes, it turns out that the eigenvalue $P_k(i)$ can be expressed as a polynomial of degree k with respect to i:

$$P_k(i) = \sum_{j=0}^{k} (-q)^j (q-1)^{k-j} \binom{n-j}{k-j} \binom{i}{j}, \qquad k=0,1,\ldots,n.$$

In fact, the $P_k(z)$ constitute a well-known class of orthogonal polynomials, first introduced by KRAWTCHOUCK [16]. Theorem 1, with $v_k = \mu_k = \binom{n}{k}(q-1)^k$, precisely contains the orthogonality relations on the *Krawtchouck polynomials*:

$$\sum_{k=0}^{n} P_i(k) P_j(k) \binom{n}{k} (q-1)^k = q^n \binom{n}{i} (q-1)^i \delta_{i,j}.$$

It can be shown that the association schemes whose eigenmatrix P has such "polynomial properties" are exactly the metric schemes. They will be studied in section 4.1.

We conclude the present section with a definition of a class of metric schemes that constitute a natural framework for the theory of *constant weight binary codes*, first considered by JOHNSON [15]: for F = {0,1} and an integer $v \geq 2$, the weight of an element $x \in F^v$ is the number of coordinates x_ν being equal to 1. Let X denote the set of elements of a given weight n in F^v, with $1 \le n \le v/2$. For k=0,1,...,n, we define the distance relations

$$R_k = \{(x,y) \in X \mid d_H(x,y) = 2k\}.$$

Then it is easily verified that (X,R) is an association scheme, with n classes, for $R = \{R_0,\ldots,R_n\}$. We shall call it a *Johnson scheme*, using the notation J(n,v). (The classical terminology is "triangular type association scheme".)

It turns out that, for the Johnson scheme J(n,v), the eigenvalue $P_k(i)$ can be expressed as a polynomial of degree k with respect to the variable $z_i = i(v+1-i)$. Explicit formulae for $P_k(i)$ have been discovered by OGASAWARA [21] and by YAMAMOTO, FUJII & HAMADA [28]:

$$P_k(i) = \sum_{j=0}^{k} (-1)^{k-j}\binom{n-j}{k-j}\binom{n-i}{j}\binom{v-n+j-i}{j}.$$

On the other hand, it can be shown that the elements of the second eigenmatrix have similar properties: $Q_k(i)$ is a polynomial of degree k in the variable $z_i = i$, for all k. Notice that, according to theorem 1, the P- and Q-polynomials of the scheme J(n,v) form two families of orthogonal polynomials, with $v_k = \binom{n}{k}\binom{v-n}{k}$ and $\mu_k = \binom{v}{k} - \binom{v}{k-1}$.

3. DISTRIBUTION OF A SUBSET IN AN ASSOCIATION SCHEME

Let Y be a non-empty subset of X for any given association scheme (X,R). We define the *inner distribution* of Y to be the (n+1)-tuple $\underline{a} = (a_0,a_1,\ldots,a_n)$ of rational numbers a_i given by

$$|Y|a_i = |R_i \cap Y^2|, \quad i=0,1,\ldots,n.$$

Thus, a_i is the average number of points of Y being i-th associates of a fixed point of Y. Clearly, $a_0 = 1$ and $\sum a_i = |Y|$ hold. For a metric scheme, Y is called a *code* in (X,R) and $\underline{a}$ is the *distance distribution* of the code.

A central question in the rest of this paper will be the following. What can be said about a subset (a "code", a "design") when its inner distribution is given? First, however, we wish to characterize those (n+1)-tuples that are inner distributions of subsets. We shall now establish a very useful result in this direction.

THEOREM 2. *The inner distribution* $\underline{a}$ *of any subset* $Y \subseteq X$ *satisfies* $\underline{a}Q_k \geq 0$, *for all* k, *where* $Q = [Q_0, Q_1, \ldots, Q_n]$ *is the second eigenmatrix of the association scheme* (X,R).

PROOF. We shall make use of the vector $\underline{u} \in \mathbb{R}(X)$, characterizing Y as a subset of X, defined by u(x) = 1 or 0 according to whether $x \in Y$ or $x \in X-Y$. Then we clearly have $|Y|a_i = \underline{u}^T D_i \underline{u}$. Hence, using (2.2), we deduce

$$|Y|\underline{a}Q_k = \underline{u}^T\left(\sum_{i=0}^{n} Q_k(i)D_i\right)\underline{u} = |X|\underline{u}^T J_k \underline{u}.$$

Now the idempotent matrix J_k is positive semi-definite. Therefore, the right member is ≥0 and the theorem is proved. □

REMARK. Let us briefly indicate an interpretation of theorem 2 in the theory of linear codes. When the alphabet F is a field GF(q), the set $X = F^n$ has the structure of a vector space over F. Then a q-ary code of length n is called *linear* whenever it is a subspace of X. The distance distribution $\underline{a}$ of a linear code with respect to the Hamming scheme simply is the classical *weight distribution*: a_i is the number of codevectors having weight i. (The Hamming weight of $x \in X$ is the number of non-zero components x_ν.) The *dual* Y' of a linear code Y is the set of vectors $x \in X$ such that $x_1y_1 + \ldots + x_ny_n = 0$ holds for every $y \in Y$. Clearly, Y' is itself linear, with dim(Y') = n - dim(Y). Then the weight distributions $\underline{a}$ and $\underline{a}'$ of Y and Y' are related by $|Y|\underline{a}' = \underline{a}Q$, where Q = P is the matrix of Krawtchouck polynomials $P_k(z)$. This is a version of the celebrated *MacWilliams identities* [20] on the weight distributions of dual linear codes. Thus, in the linear case, theorem 2 reduces to the trivial property $a'_k \geq 0$. We have seen that this remains valid for unrestricted codes in Hamming schemes when $\underline{a}'$ is defined as the formal "Krawtchouck-MacWilliams transform" of the distance distribution $\underline{a}$.

Theorem 2 leads to *linear programming problems* in the theory of subsets $Y \subseteq X$ whose specific properties can be expressed in terms of linear relations on the inner distribution with respect to a given association scheme. One is interested in upper or lower bounds on the cardinality of subsets satisfying these conditions. The linear programming problems in the (n+1)-tuple $\underline{a} = (a_0, a_1, \ldots, a_n)$ of real variables a_i, have the following form:

$$\begin{cases} \sum_{i=0}^{n} f_{i,j} a_i = 0, \quad j=1,2,\ldots,m, \\ a_0 = 1, \quad a_i \geq 0, \quad \underline{a}Q_k \geq 0, \quad \forall\, i,k, \\ \text{maximize (or minimize) } g = \sum_{i=0}^{n} a_i . \end{cases}$$

The first line contains the specifications of the problem, whereas the second line contains general necessary conditions (cf. theorem 2). So, since $g = |Y|$ holds, each subset Y under consideration satisfies the *linear programming bounds*

$$\min(g) \leq |Y| \leq \max(g).$$

In fact, we are mainly interested in two types of applications (i.e. in two types of matrices $[f_{i,j}]$), that are "dual to each other":

(i) *The codes with specified distance* δ, characterized by $a_1 = a_2 = \ldots = a_{\delta-1} = 0$, for a given $\delta \geq 1$.

(ii) *The designs with specified strength* τ, characterized by $\underline{a}Q_1 = \underline{a}Q_2 = \ldots = \underline{a}Q_\tau = 0$, for a given $\tau \geq 0$.

The significance of problem (i) of δ-*codes* in metric schemes is obvious and needs no comment. As for the problem (ii) of τ-*designs* (cf. section 4.2), we can give no general "combinatorial interpretation" of it. However, for the Hamming and Johnson schemes, we have the following result.

THEOREM 3. *A* τ-*design* Y *in* H(n,q) *is equivalent to an orthogonal array* $[\lambda q^\tau, n, q, \tau]$, *without repeated columns, of strength* τ, *having* n *constraints, index* λ, *in* q *symbols, with* $|Y| = \lambda q^\tau$. (This is the concept introduced by RAO [23].) *A* τ-*design* Y *in* J(n,v) *is equivalent to a classical* τ-*design* $S_\lambda(\tau,n,v)$, *without repeated blocks, on* v *points, block size* n, *with* $|Y| = \lambda\binom{v}{\tau}/\binom{n}{\tau}$. (This is the concept introduced by HANANI [12] and HUGHES [14].)

PROOF. We shall only consider the case of Johnson schemes: $(X,R) = J(n,v)$. Let X_i denote the subset of $\{0,1\}^v$ formed by all elements of weight i, for $i=0,1,\ldots,n$. Next, let Y be a non-empty subset of X ($= X_n$). Given $z \in X_i$, we shall denote by $\lambda_i(z)$ the number of "blocks" $y \in Y$ such that $d_H(y,z) = n-i$. Then the average value of $\lambda_i(z)$ over X_i is equal to

$$\lambda_i = |Y|\binom{n}{i}/\binom{v}{i}, \quad i=0,1,\ldots,n.$$

By definition, Y forms a τ-design $S_{\lambda}(\tau,n,v)$, with $\lambda = \lambda_{\tau}$, if and only if $\lambda_i(z)$ is a constant ($= \lambda_i$), for all $z \in X_i$, whenever $i=0,1,\ldots,\tau$.

On the other hand, it is easy to show, by a counting argument, that

$$\sum_{z \in X_i} (\lambda_i(z) - \lambda_i)^2 = |Y| \left\{ \sum_{j=0}^{n} a_j \binom{n-j}{i} - \lambda_i \binom{n}{i} \right\}$$

holds, for all $i \le n$, where $\underline{a}$ is the distance distribution of Y. Therefore, Y forms a τ-design if and only if the right member vanishes, i.e.

$$\sum_{j=0}^{n} a_j \binom{n-j}{i} = \lambda_i \binom{n}{i},$$

for $i=0,1,\ldots,\tau$. Observing that X itself forms a trivial τ-design, and that the distance distribution $\underline{v}$ of X is given by the valencies v_j, we can also write this as follows

$$\sum_{j=0}^{n} \{|X| a_j - |Y| v_j\} \binom{n-j}{i} = 0.$$

Since the polynomials $\binom{n-z}{i} \in \mathbb{R}[z]$, with $i=0,1,\ldots,\tau$, form a basis of the vector space of polynomials of degree $\le\tau$, the above system is equivalent to

$$\sum_{j=0}^{n} \{|X| a_j - |Y| v_j\} Q_k(j) = 0,$$

for $k=0,1,\ldots,\tau$. (We have used the property $\deg(Q_k(z)) = k$ of the eigenmatrix Q.) From the orthogonality relations on Q we finally obtain the desired characterization of a τ-design, namely $\sum_j a_j Q_k(j) = |Y| \delta_{0,k}$ for all $k \le \tau$. □

According to theorem 3, we may use the linear programming method in order to obtain a lower bound on the index λ of orthogonal arrays with given τ,n,q and of ordinary τ-designs with given τ,n,v. It can be shown that the linear programming bound improves the RAO inequality [23] for orthogonal arrays as well as the FISHER-PETRENJUK-WILSON inequality [27] for τ-designs (cf. section 4.2).

Similarly, the linear programming method implies several classical bounds for the problem of δ-codes in the Hamming and Johnson schemes, such as the elementary sphere packing bound, the Singleton bound and the Plotkin bound (for references, cf. [8]). A major interest of the method lies in the fact that, by use of duality in linear programming, it yields strong characterizations for codes achieving the bounds. In particular, the Lloyd theorem for perfect codes can be obtained in this manner (see section 4.1).

Let us finally introduce the *outer distribution* of a non-empty subset $Y \subseteq X$ with respect to a given association scheme (X,R): it is the integer matrix B, having X and $\{0,1,\ldots,n\}$ as row and column labeling sets, respectively, the (x,i)-entry being defined as

$$B(x,i) = |R_i \cap (\{x\} \times Y)|,$$

for $x \in X$ and $i=0,1,\ldots,n$. Thus, $B(x,i)$ is the number of points in Y that are i-th associates of the fixed point x. We shall now establish a useful relation between the inner distribution $\underline{a}$ and the outer distribution B of any $Y \subseteq X$. Like before, P and Q denote the eigenmatrices and $\Delta_{\underline{c}}$ stands for $\mathrm{diag}(c_0,c_1,\ldots,c_n)$.

THEOREM 4. $B^TB = |X|^{-1}|Y|P^T\Delta_{\underline{a}Q}P.$

PROOF. Counting in two different ways the number of triples $(x,y,y') \in X \times Y \times Y$ such that $(x,y) \in R_i$ and $(x,y') \in R_j$ we obtain the identity

$$(B^TB)(i,j) = |Y| \sum_{k=0}^{n} p_{i,j,k}a_k.$$

Define $\underline{b} = \underline{a}Q$, whence $|X|\underline{a} = \underline{b}P$. Using the formulas $P_i(u)P_j(u) = \sum_k p_{i,j,k}P_k(u)$ deduced from axiom A'2 we readily deduce

$$(B^TB)(i,j) = |X|^{-1}|Y| \sum_{u=0}^{n} b_uP_i(u)P_j(u),$$

which is the desired result. □

It follows from theorem 4 that the rank of the matrix B is equal to the number of non-zero components of the vector $\underline{a}Q$. Notice also that, since B^TB is positive semi-definite, we have obtained a new proof of theorem 2: all components of $\underline{a}Q$ are non-negative.

Let $Q_0,Q_1,\ldots,Q_n$ be the columns of Q. Then theorem 4 clearly implies the identity

$$\|BQ_k\|^2 = |X|\,|Y|\,\underline{a}Q_k, \quad 0 \le k \le n,$$

where $\|\cdot\|$ denotes the euclidean norm. This not only shows that $\underline{a}Q_k \ge 0$ holds, but also that the vector BQ_k is zero if and only if $\underline{a}Q_k$ is zero. This property is very useful for actual computation of the outer distribution B. In certain cases, it allows to determine B from the inner distribution $\underline{a}$. More details about this matter are given below, in the context of metric schemes. The reader shall have noticed the significance of the outer distribution in the classical theory of linear codes: the rows of B are the weight distributions of the cosets of the given code (cf. section 5).

4. POLYNOMIAL SCHEMES

We have mentioned the polynomial properties of the eigenmatrices P and Q for the Hamming and Johnson schemes occurring in the classical theory of codes and designs. In the present section, we shall take these properties as axioms and set up the bases of a general theory of "codes and designs in polynomial schemes". The main idea consists in trying to derive as much information as possible about a code (or a design) from its inner distribution, and, more precisely, from certain fundamental parameters depending on the inner distribution.

DEFINITION. Let $z_0, z_1, \ldots z_n$ be distinct non-negative real numbers, with $z_0 = 0$. Assume that the entries of the eigenmatrix P can be written as

$$P_k(i) = \Phi_k(z_i), \quad i,k=0,1,\ldots,n,$$

where $\Phi_k(z) \in \mathbb{R}[z]$ is a polynomial of degree k. Then the given association scheme is said to be P-*polynomial* with respect to the z_i. A Q-polynomial scheme is defined analogously from the properties of the eigenmatrix Q.

We recall that the Hamming and Johnson schemes are P-polynomial for $z_i = i$ and $z_i = i(v+1-i)$, respectively. They both are Q-polynomial for $z_i = i$. In fact, there exist several other families of association schemes having the P- and Q-polynomial properties. Some of them also have interesting applications in coding theory.

The orthogonality relations of theorem 1 can be interpreted as follows. For a P-polynomial scheme, $\Phi_0(z), \Phi_1(z), \ldots, \Phi_n(z)$ form a class of orthogonal polynomials over the set $\{z_0, z_1, \ldots, z_n\}$, the weight function w being given by $w(z_i) = \mu_i$. Moreover, it can be shown that the *sum polynomials*

$$\Psi_k(z) = \Phi_0(z) + \Phi_1(z) + \ldots + \Phi_k(z),$$

with $k=0,1,\ldots,n-1$, form a class of orthogonal polynomials over the set

$\{z_1, z_2, \ldots, z_n\}$ for the weight function $w'(z) = zw(z)$. The dual results hold in the theory of Q-polynomial schemes, the weights being $w(z_i) = v_i$ and $w'(z_i) = z_i v_i$.

4.1. P-polynomial schemes

It turns out that a given association scheme is P-polynomial if and only if it is metric, as defined in section 2. Thus the mapping ρ, of X^2 into $\{0,1,\ldots,n\}$, given by $\rho(x,y) = i$ whenever $(x,y) \in R_i$, is a *distance function* on X. (It is the natural distance in the graph (X,R_1).)

A *code* Y, that is, a non-empty subset $Y \subseteq X$ in a metric scheme (X,R), will be characterized by two parameters, d and r, both deduced from the distance distribution $\underline{a}$:

(i) The *minimum distance* d is the largest integer ≥ 1 such that $a_i = \delta_{0,i}$ for $i=0,1,\ldots,d-1$.

(ii) The *external distance* r is the number of non-zero components of the n-tuple $(\underline{a}Q_1, \underline{a}Q_2, \ldots, \underline{a}Q_n)$.

We shall only consider *non-trivial codes*, i.e. proper subsets of X containing at least two elements. Then the parameters satisfy $1 \leq r, d \leq n$. The significance of d is obvious: it is the smallest positive value assumed by the distance function $\rho(x,y)$ for $x,y \in Y$. As for the external distance r, our terminology is based on the following result.

THEOREM 5. *Each point of* X *is at distance* $\leq r$ *from at least one point belonging to the code* Y*:*

$$\min_{y \in Y} \rho(x,y) \leq r, \quad \forall x \in X.$$

PROOF. Let us first define the following two subsets of $\{0,1,\ldots,n\}$, with equal cardinalities: $K = \{0,1,\ldots,r\}$ and $L = \{k \mid \underline{a}Q_k \neq 0\}$. We shall denote by $\bar{B}$, $\bar{P}$ and $\bar{\Delta}$ the restrictions of the matrices B, P and $\Delta_{\underline{a}Q}$ to the sets $X \times K$, $L \times K$ and $L \times L$, respectively. Then the equation of theorem 4 implies

$$\bar{B}^T\bar{B} = |X|^{-1}|Y|\bar{P}^T\bar{\Delta}\bar{P}.$$

By definition, $\bar{\Delta}$ is a non-singular diagonal matrix. On the other hand, from the P-polynomial property it clearly follows that $\bar{P}$ also is non-singular. Hence $\bar{B}^T\bar{B}$ is non-singular and we deduce $\text{rank}(\bar{B}) = \text{rank}(B) = r+1$. In other words, the columns $B_0, B_1, \ldots, B_r$ of $\bar{B}$ form a basis for the column space of B. So, since a row B(x) of B cannot be identically zero, at least one of the integers $B_0(x), B_1(x), \ldots, B_r(x)$ must be non-zero, for every given $x \in X$. Remembering that $B_i(x)$ is the number of points $y \in Y$ such that $\rho(x,y) = i$, we obtain the desired result. □

It must be observed that, in general, r is not the "true external distance": there may exist no point $x \in X$ at minimum distance r from the code Y. As an easy consequence of theorem 5 we deduce an interesting inequality, first discovered by MACWILLIAMS [19] for linear codes.

THEOREM 6. *For any code,* $\lfloor (d-1)/2 \rfloor \leq r$ *holds.*

PROOF. Let us define the spheres $S_e(y) = \{x \in X \mid 0 \leq \rho(x,y) \leq e\}$, of radius $e = \lfloor (d-1)/2 \rfloor$, centred at the points $y \in Y$. By definition, these spheres are mutually disjoint. So, any point $x_0 \in X$ at distance e from some $y \in Y$ is at minimum distance e from Y. According to theorem 5, this implies $e \leq r$. □

Moreover, it can be shown that the equality $e = r$ holds if and only if the spheres $S_e(y)$ form a partition of X, for y running through Y. In this case, by extension of the classical notion, Y is called a *perfect code of order* e. This concept has also been independently introduced and investigated by BIGGS [1] in the context of distance transitive graphs (which corresponds to the group case for metric schemes).

In connection with theorem 6, let us mention the following two bounds on the cardinality of any code with given parameters d and r:

(4.1) $$\sum_{i=0}^{e} v_i \leq |Y|^{-1}|X| \leq \sum_{i=0}^{r} v_i,$$

with $e = \lfloor (d-1)/2 \rfloor$. The left member is the obvious sphere packing bound. The right member is the "covering bound", which easily follows from theorem 5. It turns out that, if one of the bounds is achieved, then so is the other. Therefore, we may take either equality in (4.1) as a definition for perfect codes.

Let us now mention, without proof, a generalized version of the *Lloyd theorem* [1,7,8,17] which yields a strong necessary condition for perfect codes.

THEOREM 7. *Let there exist a perfect code of order* e *in a* P-*polynomial (= metric) scheme. Then the sum polynomial* $\Psi_e(z)$ *admits* e *distinct zeros in the set* $\{z_1, z_2, \ldots, z_n\}$, *namely those* z_k *such that* $\underline{a}Q_k \neq 0$.

From the fact that $\Psi_0(z), \Psi_1(z), \ldots, \Psi_{n-1}(z)$ form a class of orthogonal polynomials, it is possible to derive explicit expressions for the distance distribution of a perfect code. In fact, it turns out that the full outer distribution of perfect codes of a given order only depends on the parameters of the scheme.

Unfortunately, there are "very few" perfect codes in the classical coding schemes. (We refer to VAN LINT [18].) Let us now define a weaker property, which however seems interesting: the complete regularity. A code Y is called *completely regular* if the row B(x) of its outer distribution only depends on the minimum distance between x and Y, for all $x \in X$. The following theorem contains a sufficient condition for this property, in terms of the fundamental parameters.

THEOREM 8. *If the minimum distance* d *and the external distance* r *of a code satisfy* $d \geq 2r-1$, *then the code is completely regular.*

This result not only applies to perfect codes ($d = 2r+1$), but also to the nearly perfect codes defined by GOETHALS & SNOVER [11] and to the uniformly packed codes introduced by SEMAKOV, ZINOV'EV & ZAITZEV [24].

4.2. Q-polynomial schemes

The theory is formally similar to that of the preceding section. One would of course like to know an intrinsic interpretation of the concept of Q-polynomial schemes (i.e. the "dual" of the metric property). Unfortunately, this is an unsolved question, although some useful algebraic criteria are known for the Q-polynomial property.

We shall now investigate the dual notion of a code in a metric scheme. A *design* Y is a non-empty subset $Y \subseteq X$ for a Q-polynomial scheme (X,R); it will be characterized by two fundamental parameters, t and s, deduced from the inner distribution $\underline{a}$:

(i) The *maximum strength* t is the largest integer ≥0 such that $\underline{a}Q_k = |Y|\delta_{0,k}$ for $k=0,1,\ldots,t$.

(ii) The *degree* s is the number of non-zero components of the n-tuple $(a_1, a_2, \ldots, a_n)$.

We shall only consider non-trivial designs (i.e. assume $1 < |Y| < |X|$). Then $1 \le s$, $t+1 \le n$ holds. The interpretation of the degree is clear: s is the number of distinct colours appearing in the subgraph of (X,R) whose vertex set is Y. The meaning of the maximum strength is less obvious; it must be discovered in each particular case. For the Hamming and Johnson schemes, the significance of t-designs of strength t has been emphasized in theorem 3. This motivates a study of t-designs in general Q-polynomial schemes.

The following result is dual to theorem 6 about codes in metric schemes (for the correspondence $t \leftrightarrow d-1$, $s \leftrightarrow r$).

THEOREM 9. *For any design,* $\lfloor t/2 \rfloor \le s$ *holds.*

PROOF. Let $\underline{a}$ be the inner distribution of Y. Assume $s < e = \lfloor t/2 \rfloor$. Then there exists a polynomial $f(z) \in \mathbb{R}[z]$, of degree e, vanishing at each point z_i such that $a_i \neq 0$, for $i=0,1,\ldots,n$. Consider the expansion of $(f(z))^2$, which has degree $\le t$, in the basis of polynomials $\Phi_k(z)$ associated to the eigenmatrix Q:

$$(f(z))^2 = \sum_{k=0}^{t} b_k \Phi_k(z).$$

(The real numbers b_k are uniquely derived from the values $f(z_i)$ by the formulas $|X| b_k = \sum_i P_i(k)(f(z_i))^2$.) Using $\Phi_k(z_i) = Q_k(i)$, we obtain

$$\sum_{i=0}^{n} a_i (f(z_i))^2 = \sum_{k=0}^{t} b_k (\underline{a}Q_k) = b_0 |Y|.$$

Now, by definition of f(z), the left member is zero, whereas the right member, being equal to $|X|^{-1}|Y| \sum_i v_i (f(z_i))^2$, is strictly positive. The desired inequality $e \le s$ follows from this contradiction. □

We have seen that $\lfloor (d-1)/2 \rfloor = r$ can be taken for the definition of a perfect code of order e in a metric scheme. Analogously, we take the equality $\lfloor t/2 \rfloor = s$ as a definition of a *tight design* of degree s in a Q-polynomial scheme. It turns out that this coincides with the concept introduced by WILSON [26] for classical t-designs (in the Johnson schemes). In the case of Hamming schemes, a tight design is equivalent to a *generalized Hadamard code* [7].

The result of theorem 9 can also be viewed as a direct consequence of the following two inequalities (dual to those of (4.1)):

$$\sum_{i=0}^{e} \mu_i \le |Y| \le \sum_{i=0}^{s} \mu_i, \tag{4.2}$$

with $e = \lfloor t/2 \rfloor$. The left bound is due to RAO [23] for the Hamming schemes and to WILSON & RAY-CHAUDHURI [27] for the Johnson schemes (where it takes the simple form $|Y| \ge \binom{v}{e}$). As for the right bound, it has been first discovered by WILSON [26] and by the author [7] in the case of Johnson and Hamming schemes, respectively.

It turns out that, if one of the bounds (4.2) is achieved, then so is the other. Therefore, we may take either equality in (4.2) to define tight designs. Let us also mention a dual of the Lloyd theorem, extending known results on classical tight designs [26] and generalized Hadamard codes [7].

THEOREM 10. *Let there exist a tight design of degree* s *in a* Q*-polynomial scheme. Then the sum polynomial* $\Psi_s(z)$ *admits* s *distinct zeros in the set* $\{z_1, z_2, \ldots, z_n\}$, *namely those* z_k *such that* $a_k \neq 0$.

The above result implies that the s colours of the subgraph of (X,R) having Y as vertex set are determined from the parameters of the scheme, when Y is a tight design. Moreover, it can be shown that this subgraph itself is a Q-polynomial association scheme, with s classes. This is a particular case of a more general result (to be compared with theorem 8):

THEOREM 11. *If the maximum strength* t *and the degree* s *of a design satisfy* $t \ge 2s-2$, *then the design carries a* Q*-polynomial scheme, with* s *classes.*

For s = 2, this theorem has been discovered first by GOETHALS & SEIDEL [10] in the case of Johnson schemes and by the author [6] in the case of Hamming schemes (at least for linear codes). This led to some interesting constructions of strongly regular graphs. Recently, CAMERON [5] also obtained theorem 11 for designs of any degree s in a Johnson scheme.

5. APPLICATION. LINEAR CODES IN HAMMING SCHEMES

In the preceding section we have exhibited a formal duality between the theories of P- and Q-polynomial schemes, between the concepts of codes and designs, both characterized by a pair of fundamental parameters. In

Hamming schemes this duality becomes actual for the class of *linear codes* (or designs). We recall that the eigenmatrices P and Q of a Hamming scheme H(n,q) are equal, and they are determined from the Krawtchouk polynomials.

For a given linear code Y of length n over GF(q), the *Hamming weight* of an n-tuple over F = GF(q) is defined to be the number of its non-zero components. Let $w_1, w_2, \ldots, w_s$ be the values assumed by the Hamming weight over non-zero elements of Y. The w_i are called the *weights* of Y. The degree of Y clearly is equal to the number of distinct weights, and the minimum distance is equal to the minimum weight.

We shall denote by Y' the dual of the linear code Y. Then the respective weight distributions $\underline{a}$ and $\underline{a}'$ of Y and Y' are related by the *MacWilliams identities* $|Y|\underline{a}' = \underline{a}Q$. Let d,r,t,s be the four fundamental parameters of Y and d',r',t',s' the corresponding parameters of Y'. Then it immediately follows from the MacWilliams identities that we have

$$d' = t+1 \quad , \quad r' = s \quad , \quad t' = d-1 \quad , \quad s' = r \quad .$$

Consequently, a linear tight design (= generalized Hadamard code) is nothing but the dual of a linear perfect code (that is, a Hamming code, a Golay code, or a binary repetition code of odd length).

Finally, let us mention a criterion for a linear code Y having s weights to carry an association scheme with s classes. By definition, the row B'(x) of the outer distribution matrix B' of Y' is the weight distribution of the coset code x+Y', for any given $x \in X = F^n$. We shall denote by s^* the number of distinct weight distributions B'(x), with $x \notin Y'$. It follows from theorem 4 that the degree s of Y (i.e. the external distance r' of Y') is equal to rank(B')−1. Hence, $s \le s^*$ holds. It turns out that we have $s = s^*$ if and only if the code Y carries a subscheme, with s classes, of the Hamming scheme H(n,q). Moreover, this subscheme admits a dual (cf. section 1), which has a natural representation on the cosets of Y' in X: two cosets are associated according to the weight distribution of their difference.

REFERENCES

[1] BIGGS, N.L., *Perfect codes in graphs*, J. Combinatorial Theory B, 15 (1973) 289-296.

[2] BOSE, R.C., *Strongly regular graphs, partial geometries and partially balanced designs*, Pacific J. Math., 13 (1963) 389-419.

[3] BOSE, R.C. & D.M. MESNER, *On linear associative algebras corresponding to association schemes of partially balanced designs*, Ann. Math. Statist., 30 (1959) 21-38.

[4] BOSE R.C. & T. SHIMAMOTO, *Classification and analysis of partially balanced incomplete block designs with two associate classes*, J. Amer. Statist. Assoc., 47 (1952) 151-184.

[5] CAMERON, P.J., *Near-regularity conditions for designs*, Geometriae Dedicata (to appear).

[6] DELSARTE, P., *Weights of linear codes and strongly regular normed spaces*, Discrete Math., 3 (1972) 47-64.

[7] DELSARTE, P., *Four fundamental parameters of a code and their combinatorial significance*, Information and Control, 23 (1973) 407-438.

[8] DELSARTE, P., *An algebraic approach to the association schemes of coding theory*, Philips Res. Repts. Suppl., 10 (1973).

[9] DOOB, M., *On graph products and association schemes*, Utilitas Math., 1 (1972) 291-302.

[10] GOETHALS, J.M. & J.J. SEIDEL, *Strongly regular graphs derived from combinatorial designs*, Canad. J. Math., 22 (1970) 597-614.

[11] GOETHALS, J.M. & S.L. SNOVER, *Nearly perfect binary codes*, Discrete Math., 3 (1972) 65-88.

[12] HANANI, H., *The existence and construction of balanced imcomplete block designs*, Ann. Math. Statist., 32 (1961) 361-386.

[13] HIGMAN, D.G., *Combinatorial considerations about permutation groups*, Lecture Notes, Mathematical Institute, Oxford, 1972.

[14] HUGHES, D.R., *Combinatorial analysis. t-designs and permutation groups*, Amer. Math. Soc., Proc. Symp. Pure Math., 6 (1962) 39-41.

[15] JOHNSON, S.M., *A new upper bound for error-correcting codes*, IEEE Trans. Information Theory, IT-8 (1962) 203-207.

[16] KRAWTCHOUK, M., *Sur une généralisation des polynômes d'Hermite*, Comptes Rendus de L'Académie des Sciences, Paris, 189 (1929) 620-622.

[17] LENSTRA, Jr., H.W., *Two theorems on perfect codes*, Discrete Math., 3 (1972) 125-132.

[18] LINT, J.H. VAN, *A survey of perfect codes*, Rocky Mountain J. Math. (to appear).

[19] MACWILLIAMS, F.J., Doctoral Dissertation, Harvard University, 1961 (unpublished).

[20] MACWILLIAMS, F.J., *A theorem on the distribution of weights in a systematic code*, Bell Syst. Tech. J., 42 (1963) 79-94.

[21] OGASAWARA, M., *A necessary condition for the existence of regular and symmetrical PBIB designs of* T_m *type*, Inst. Statist. mimeo series 418, Chapel Hill, N.C., 1965.

[22] OGAWA, J., *The theory of the association algebra and the relationship algebra of a partially balanced incomplete block design*, Inst. Statist. mimeo series 224, Chapel Hill, N.C., 1959.

[23] RAO, C.R., *Factorial experiments derivable from combinatorial arrangements of arrays*, J. Roy. Statist. Soc., 9 (1947) 128-139.

[24] SEMAKOV, N.V., V.A. ZINOV'EV & G.V. ZAITZEV, *Uniformly packed codes*, Problemy Peredači Informacii, 7 (1971) 38-50, (in Russian).

[25] TAMASCHKE, O., *Zur Theorie der Permutationsgruppen mit regulärer Untergruppe, I and II*, Math. Z., 80 (1963) 328-352 and 443-465.

[26] WILSON, R.M., *Lectures on t-designs*, Ohio State University, 1971, communicated by J. DOYEN.

[27] WILSON, R.M. & D.K. RAY-CHAUDHURI, *Generalization of Fisher's inequality to t-designs*, Amer. Math. Soc. Notices, 18 (1971) 805.

[28] YAMAMOTO, S., Y. FUJII & N. HAMADA, *Composition of some series of association algebras*, J. Sci. Hiroshima Univ., (A-I) 29 (1965) 181-215.

RECENT RESULTS ON PERFECT CODES AND RELATED TOPICS

J.H. VAN LINT

Technological University, Eindhoven, The Netherlands

1. INTRODUCTION

In this paper we shall use the framework and terminology explained by DELSARTE in the previous paper [3]. We consider perfect codes in the Hamming and Johnson schemes and in association schemes corresponding to distance-transitive graphs. For these schemes we illustrate some recent examples and concentrate on the recent developments concerning non-existence proofs. Here the main tools are the *sphere packing bound* (cf. [3, § 4.1]) and *Lloyd's theorem* [3, Theorem 7].

The completely regular codes briefly mentioned by DELSARTE contain two classes which have properties similar to the perfect codes. The most interesting of these properties is that such codes can be used to construct t-designs. The search for such codes started a few years ago. Again we shall illustrate some examples and prove some non-existence theorems.

2. HAMMING SCHEMES H(n,q) WITH q A PRIME POWER

Let $q = p^r$ (p a prime). We consider a perfect e-error-correcting code Y, i.e. a perfect code of order e. The minimum distance of Y is $d = 2e+1$. For $e = 1$ there are many known examples of perfect codes (cf. [7]). For $e > 1$ one always has the trivial example $e = n$ and $|Y| = 1$. For $q = 2$ and $n = 2e+1$ the repetition code $Y := \{(0,0,\ldots,0),(1,1,\ldots,1)\}$ provides an example. Besides these there are 2 non-trivial perfect codes known as the Golay codes (cf. [7]). The parameters of these codes are $n = 23$, $q = 2$,

M. Hall, Jr. and J. H. van Lint (eds.), Combinatorics, 163-183.

$e = 3$ respectively $n = 11$, $q = 3$, $e = 2$. For both of these codes it was shown recently [4], [12] that they are unique (up to translations and permutations of coordinate places). In 1970 VAN LINT [8] proved that if there are any other perfect codes then they have $e > 3$ and $p|e$. A year later TIETÄVÄINEN [13] proved that if $e \geq 3$ and $p \leq e$ then there is no perfect code of order e in $H(n,q)$, thus completely settling the problem for the case where q is a prime power. The same result was obtained independently by ZINOV'EV & LEONT'EV [15]. All these theorems had quite complicated proofs. A few months ago TIETÄVÄINEN [14] succeeded in shortening the proof considerably for $q > 2$. It turns out that very little more is needed if $q = 2$. We shall present the complete proof below.

We remind the reader of the definition of the *Krawtchouk polynomial* $K_k(n,q;u)$ of degree k:

$$K_k(n,q;u) := \sum_{j=0}^{k} (-1)^j (q-1)^{k-j} \binom{u}{j}\binom{n-u}{k-j} . \tag{2.1}$$

The sum polynomial Ψ_e occurring in Lloyd's theorem which we quote below is

$$\Psi_e(x) := K_e(n-1,q;x-1) = \sum_{i=0}^{e} (-1)^i \binom{n-x}{e-i}\binom{x-1}{i}(q-1)^{e-i} . \tag{2.2}$$

The two necessary conditions for the existence of a perfect code of order e in $H(n,q)$ mentioned in [3] are

$$\sum_{i=0}^{e} \binom{n}{i}(q-1)^i = q^k \tag{2.3}$$

for some integer k (cf. [7]) and Theorem 7 of [3] which states

$$\left\{ \begin{array}{l} \Psi_e \text{ has } e \text{ distinct zeros } x_1 < x_2 < \dots < x_e \text{ which are integers} \\ \text{in } [1,n]. \end{array} \right. \tag{2.4}$$

The following properties of Ψ_e and its zeros are easily obtained by substitution or by calculating suitable coefficients of Ψ_e.

$$\Psi_e(0) = \sum_{i=0}^{e} \binom{n}{i}(q-1)^i , \tag{2.5}$$

$$\Psi_e(1) = \binom{n-1}{e}(q-1)^e , \tag{2.6}$$

(2.7) $$\sum_{i=1}^{e} x_i = \frac{e(n-e)(q-1)}{q} + \frac{e(e+1)}{2} ,$$

(2.8) $$\sum_{i=1}^{e} x_i \le \frac{ne(q-1)}{q} \qquad \text{(for } q > 2) ,$$

(2.9) $$\prod_{i=1}^{e} x_i = e!q^{-e}\Psi_e(0) ,$$

(2.10) $$\prod_{i=1}^{e} (x_i-1) = e!q^{-e}\Psi_e(1) .$$

For integral values of x between 1 and n the terms of the sum in (2.2) alternate in sign. Since the terms decrease in absolute value, with increasing i, for $x < \frac{(n-e+1)(q-1)+e}{q-1+e}$ we have (if Ψ_e has integral zeros) :

(2.11) $$x_1 \ge \frac{(n-e+1)(q-1)+e}{q-1+e} .$$

It was suggested by D.H. SMITH that the following lemma could prove to be useful in non-existence proofs of perfect codes.

LEMMA 1. *If a non-trivial perfect code of order* e *in* H(n,q) *exists then*

(i) $n \ge \frac{1}{2}e^2 + \frac{5}{2}e + 1$ *if* $q > 2$,

(ii) $n \ge e^2 + 4e + 2$ *if* $q = 2$.

PROOF. Let Y be a non-trivial perfect code of order e and let $\underline{a} = (a_0,a_1,\ldots,a_n)$ be the distance distribution of Y. Then we have

$$a_{2e+1} = e!n!(q-1)^{e+1} \{(n-e-1)!(2e+1)!\}^{-1} ,$$

$$a_{2e+2} = \tfrac{1}{2}a_{2e+1} \{\frac{q-1}{e+1} (n-e^2-3e-1) + e\} ,$$

and

$$a_{2e+3} = a_{2e+1} (n-2e-1)(n-e^2-4e-2) \{(2e+2)(2e+3)\}^{-1} \text{ if } q = 2.$$

Then (i) and (ii) follow from the observation that $a_{2e+2} \ge 0$ and $a_{2e+3} \ge 0$. □

We need one more concept which will play an important role in the non-existence proof. We define for $n \in \mathbb{N}$

(2.12) $$a_p(n) := \max\{m \in \mathbb{N} \mid m|n,\ p\nmid m\}\ ,$$

i.e. $a_p(n)$ is the largest divisor of n which is not divisible by p. We call n_1 and n_2 *p-equivalent* if $a_p(n_1) = a_p(n_2)$.

Since one can explicitly determine the zeros of Ψ_2 it is easy to show that the ternary Golay code is the only non-trivial perfect code of order 2 (cf. [7]). In the following theorem we therefore take $e > 2$.

THEOREM 1. *If* $q = p^r$, $e > 2$ *then there is no perfect code* Y *of order* e, *with* $|Y| > 2$, *in* H(n,q).

PROOF. Assume Y is a perfect code of order $e < (n-1)/2$ in H(n,q). For the zeros of Ψ_e we find from (2.3), (2.5) and (2.9)

(2.13) $$\prod_{i=1}^{e} x_i = e!q^{k-e}\ ,$$

and hence

$$a_p(x_1)a_p(x_2)\ \ldots\ a_p(x_e) = a_p(e!) \le e!\ .$$

It follows that there are zeros x_i, x_j which are p-equivalent or $\{a_p(x_1), a_p(x_2), \ldots, a_p(x_e)\} = \{1,2,\ldots,e\}$, i.e. $p > e \ge 3$. In the latter case there is a zero $x_i = p^\alpha$ and a zero $x_j = 2p^\beta$ and then either $x_i \ge 2x_j$ or $x_j \ge 2x_i$. Hence we always have

(2.14) $$2x_1 \le x_e \text{ and hence } x_1x_e \le \frac{2}{9}(x_1+x_e)^2\ .$$

Now by (2.5), (2,8), (2.14) and the arithmetic-geometric mean inequality we find

(2.15) $$(q-1)^e q^{-e} n(n-1)\ldots(n-e+1) < e!q^{-e}\Psi_e(0) = \prod_{i=1}^{e} x_i \le$$

$$\le \frac{8}{9}\left(\frac{x_1+x_e}{2}\right)^2 \left(\frac{x_2+x_3+\ldots+x_{e-1}}{e-2}\right)^{e-2} \le$$

$$\le \frac{8}{9}\left(\frac{x_1+\ldots+x_e}{e}\right)^e \le \frac{8}{9}(q-1)^e q^{-e} n^e$$

(for $q > 2$).

If $q = 2$ the final expression is $\frac{8}{9}(\frac{n+1}{2})^e$.

Now let $q > 2$. From (2.6) and (2.10) we find $q^e \mid (n-1)\ldots(n-e)$ and therefore

$$n > p^{re-[e/p]-[e/p^2]-\ldots} > p^{e(r-\frac{1}{p-1})} \geq q^{\frac{1}{2}e} . \tag{2.16}$$

By (2.15) we have

$$1 - \frac{e(e-1)}{2n} < \frac{n(n-1)\ldots(n-e+1)}{n^e} \leq \frac{8}{9} ,$$

i.e.

$$n < \frac{9}{2}e(e-1)$$

and hence (2.16) implies

$$3^{\frac{1}{2}e} \leq q^{\frac{1}{2}e} < \frac{9}{2}e(e-1) , \text{ i.e. } e \leq 11.$$

It then follows that $q \leq 8$ and $n \leq 495$. In 1967 (cf. [9]) a computer search had found all solutions of (2.5) for $n \leq 1000$, $q \leq 100$, $e \leq 1000$. This yielded no new codes. Hence for $q > 2$ the proof is finished.

For the case $q = 2$ ($n \neq 2e+1$) we do not have an inequality of type (2.16). We now have to use lemma 1 to get a lower bound on n and generalize the method used above to obtain an upper bound. Starting from (2.14) one shows by induction that

$$\prod_{i=1}^{s} \xi_i \leq \left(\frac{8}{9}\right)^{s-1} \left(\frac{1}{s} \sum_{i=1}^{s} \xi_i\right)^s$$

if $\xi_1, \xi_2, \ldots, \xi_s$ are 2-equivalent. Then in the same way as above the arithmetic-geometric mean inequality yields

$$\prod_{i=1}^{e} x_i \leq \left(\frac{8}{9}\right)^{e-m} \left(\frac{x_1+\ldots+x_e}{e}\right)^e \tag{2.17}$$

if $x_1, x_2, \ldots, x_e$ are divided over m equivalence classes under 2-equivalence. This means that if we can prove that $m \leq e-6$ then the analogue of (2.15) extended by (2.17) gives us

$$n(n-1)\ldots(n-e+1) < \left(\frac{8}{9}\right)^6 (n+1)^e ,$$

i.e. $n < e^2+e$, which contradicts inequality (ii) of lemma 1.

In the proof of the remaining step we let $p(x)$ denote the product of the odd integers $\leq x$. Then

$$a_2(x_1x_2\ldots x_e) = a_2(e!) \le p(e)[\tfrac{e}{2}]!2^{-[\frac{e}{4}]} <$$

$$< p(e)e^{[\frac{e}{2}]-5} \qquad (\text{for } e \ge 16)\ .$$

Furthermore,

$$p(e) \le p(2m)e^{[\frac{e+1}{2}]-m}$$

and

$$a_2(x_1x_2\ldots x_e) \ge 1.3.5\cdots(2m-1) = p(2m)\ .$$

Combining these inequalities we find

$$1 < e^{[\frac{e}{2}]+[\frac{e+1}{2}]-m-5}\ ,$$

i.e. $m \le e-6$ (for $e \ge 16$). Hence the proof is complete for $e \ge 16$.

This leaves $e \le 15$ and then by (2.15) $n < 1000$ and we again refer to the computer search. □

Admittedly the case $q = 2$ is still rather messy. However, it seems likely that further simplifications of the proof are possible. We advise the reader to study the proof given here carefully in order to appreciate the great difficulties that arise when one tries to generalize to values of q which are not prime powers. In the next section we shall see that even the case $e = 2$, where one can explicitly determine the zeros of Ψ_e, is difficult.

3. HAMMING SCHEMES H(n,q) WITH q NOT A PRIME POWER

If q is not a prime power the sphere packing condition no longer has the form (2.3) which is replaced by

$$(3.1) \qquad \sum_{i=0}^{e} \binom{n}{i}(q-1)^i \mid q^n\ .$$

The other necessary condition for the existence of a perfect code still has the form (2.4).

Condition (3.1) is satisfied for e = 1 and n = q+1. It has been shown that a perfect code of order 1 in H(7,6) does not exist. As far as we know this is the only case where non-existence has been proved (for q not a prime power). Attempts to generalize the non-existence proofs for $e \geq 2$ have failed up to now because (3.1) is so much weaker than (2.3) that as a consequence (2.13) is replaced by a weaker statement. But even if (2.13) remained true the idea of splitting the zeros of Ψ_e into equivalence classes, which was the essential step in section 2, cannot be generalized.

As a small step on the road to complete understanding of perfect codes we shall completely treat the case q = 10, e = 2 since the alphabet of 10 symbols is of practical interest and this case illustrates how some of the ideas of section 2 can still be used.

We assume that a perfect code of order e in H(n,q) exists. From now on we take q = 10 but continue to use the symbol q in view of application to other examples. We shall keep e arbitrary as long as possible and then specialize to e = 2. The sphere packing bound now reads

$$\sum_{i=0}^{e} \binom{n}{i}(q-1)^i = q^k p^\alpha , \tag{3.2}$$

where p = 2 or p = 5. We define $\bar{p}$ by $q = p\bar{p}$. Since $\sum_{i=0}^{n} \binom{n}{i}(q-1)^i = q^n$ we find by subtraction that

$$q^k p^\alpha (q^{n-k-\alpha}\, \bar{p}^\alpha - 1) \equiv 0 \pmod{(q-1)^e} .$$

For $e \geq 2$ this implies that $\alpha \equiv 0 \pmod 6$. For the zeros of Ψ_e we again have (2.8) and (2.11). Instead of (2.9) we have

$$\prod_{i=1}^{e} x_i = e!\,q^{k-e} p^\alpha . \tag{3.3}$$

Furthermore we find from (2.6) and (2.10) in precisely the same way as (2.16) the inequality

$$n > 5^{\frac{3}{4}e} \tag{3.4}$$

(from the fact that 5^e divides $(n-1)\ldots(n-e)$).

Next we remark that the argument of (2.15) still holds if $2x_1 \leq x_e$. This would again yield the inequality $n < \frac{9}{2}e(e-1)$, which contradicts (3.4). Hence we now have

(3.5) $$2x_1 > x_e .$$

THEOREM 2. *There is no perfect code of order* 2 *in* H(n,10) *for* n > 2.

PROOF. Assume on the contrary that such a code exists. By the sphere packing bound we have

(3.6) $$(18n-7)^2 - 41 = 8q^k p^\alpha .$$

Since $29^2 - 41 = 8q^2$ we find by subtraction

(3.7) $$36(9n+11)(n-2) = 8(q^k p^\alpha - q^2) .$$

Lloyd's theorem states that the zeros x_1, x_2 of

(3.8) $$x^2 - \{\tfrac{9}{5}(n-2)+3\}x + 2q^{k-2}p^\alpha$$

are integers between 1 and n.

We can already draw the following conclusions:

(3.9) $$n \equiv 2 \pmod 5 ,$$

(3.10) $$k \geq 2 ,$$

the latter because x_1x_2 is an integer only if $k \geq 2$ or $k = 1$ and $p = 5$ but in that case $x_1x_2 = 5^{\alpha-1}$ which contradicts (3.5). From (3.9), (3.10) and (3.7) we find that $n \equiv 2 \pmod{25}$ and using this we see from (3.8) that $x_1 + x_2 \not\equiv 0 \pmod 5$. It follows that one of the zeros is a power of 2. Let

$$x_1 = 2^\nu , \ x_2 = 2^\mu 5^\sigma$$

(where we no longer require x_1 to be the smaller of the two zeros). We consider the equation (3.7) mod 32. The right-hand side is 0 and therefore we have two possibilities to consider, namely $n \equiv 2 \pmod 8$ and $n \equiv 5 \pmod 8$.

Case i : $n \equiv 2 \pmod 8$. In (3.8) one of the zeros is odd and since $x_1 \neq 1$ we must have $\mu = 0$. Furthermore, $x_1 + x_2$ is divisible by 3 which implies that $\nu + \sigma$ is odd. We now return to the sphere packing bound (3.6) with the knowledge that

$$\tfrac{9}{5}(n-2) + 3 = 2^\nu + 5^\sigma \quad (\nu+\sigma \text{ odd}) .$$

Substitution yields

$$(5\cdot 2^{\nu+1}+2\cdot 5^{\sigma+1}-1)^2 - 41 = 8\cdot 10^k\cdot p^\alpha ,$$

i.e.

$$2^2\cdot 5^{2\sigma+2} - 2^2\cdot 5^{\sigma+1} \equiv 40 \pmod{32},$$

which is a contradiction.

Case ii : $n \equiv 5 \pmod 8$. We now have $x_1 + x_2 \equiv 2 \pmod 8$ and since (2.11) implies $x_1 \neq 2$ we must have $x_2 = 2\cdot 5^\sigma$. Again we substitute in (3.6). We find

$$-5\cdot 2^{\nu+2} \equiv 40 \pmod{25}, \text{ i.e. } \nu \equiv 3 \pmod 4$$

and

$$2^4\cdot 5^{2\sigma+2} - 2^3\cdot 5^{\sigma+1} \equiv 40 \pmod{64} \text{ unless } k = 2 \text{ and } p = 5 .$$

If $k = 2$ and $p = 5$ then $x_1 = 16$ and hence by (2.11) $n \leq 20$ contradicting $n \equiv 2 \pmod{25}$. So we must have σ even. Then

$$x_1 + x_2 = 2^\nu + 2\cdot 5^\sigma \equiv 1 \pmod 3$$

which contradicts (3.8).

This completes the proof. □

Of course this is an isolated example of a non-existence proof. In order to generalize this, e.g. to all q which are twice a prime, more ideas are necessary. We hope that some of the ideas used above will prove fruitful in future research on perfect codes.

4. JOHNSON SCHEMES J(n,v)

Let $n = 2e+1$ and $v = 2n$. We consider the two word code $Y := \{(1,1,\dots,1,0,0,\dots,0),(0,0,\dots,0,1,1,\dots,1)\}$ which we can interpret as an analogue of the repetition code in a Hamming scheme. Clearly every element of $J(n,v)$ has distance $\leq e$ to exactly one of the elements of Y, i.e. Y is a perfect code of order e. These examples and the perfect codes with $|Y| = 1$ and $e = v$ we again consider trivial. No example of a non-trivial perfect code in $J(n,v)$ is known. In a search for such codes one quickly sees that it is again the sphere packing bound which is difficult to exploit. We shall briefly illustrate the case $e = 2$. We then have the two necessary conditions for the existence of a perfect code of order 2 in $J(n,v)$:

$$\left\{1 + \binom{n}{1}\binom{v-n}{1} + \binom{n}{2}\binom{v-n}{2}\right\} \mid \binom{v}{n}, \tag{4.1}$$

$$\begin{aligned} 4\Psi_2(x) &= x^2 + \{2n^2 - 2vn + v - 6\}x + \\ &\quad + \{n^4 - 2vn^3 + (v^2+v-5)n^2 + (-v^2+5v)n + 4\} \end{aligned} \tag{4.2}$$

has two zeros which are both integers of the form $i(v+1-i)$ with $0 \le i \le n$. So far the only solutions to these two conditions which we have been able to find correspond to trivial codes.

5. OTHER METRIC SCHEMES; GRAPHS

Let X be the set of rowvectors with 7 coordinates, three of which are 1 and the others 0. For $\underline{x},\underline{y} \in X$ we define

$$\rho(\underline{x},\underline{y}) = 1 \text{ if } (\underline{x},\underline{y}) = 0\ ,$$

$$\rho(\underline{x},\underline{y}) = 2 \text{ if } (\underline{x},\underline{y}) = 2\ ,$$

$$\rho(\underline{x},\underline{y}) = 3 \text{ if } (\underline{x},\underline{y}) = 1\ ,$$

where $(\underline{x},\underline{y})$ is the inner product over $\mathbb{R}$. It is easy to check that this is a distance function for X. With each vector in X we associate a vertex $v(\underline{x})$ of a graph G and we join the vertices $v(\underline{x})$ and $v(\underline{y})$ by an edge iff $\rho(\underline{x},\underline{y}) = 1$. It turns out that $\rho(\underline{x},\underline{y})$ is the distance of $v(\underline{x})$ and $v(\underline{y})$ in the graph G. It is straightforward to check that the distance ρ defines a metric scheme in the sense of Remark (ii) of [3, §2].

G is a perfectly regular graph with valency $v_1 = 4$. Let Y be the set of 7 rowvectors of the incidence matrix of PG(2,2). If $\underline{x}$ and $\underline{y}$ are two distinct rows of Y then clearly $\rho(\underline{x},\underline{y}) = 3$. Since $|X| = 35$ we have

$$(1+v_1)\ |Y| = |X|\ ,$$

i.e. equality in the sphere packing bound [3, § 4.1, formula (3)]. Hence Y is a perfect code of order 1 in the metric scheme. This is the first example given by BIGGS [1] in his paper on perfect codes in distance-transitive graphs. In [2] a number of other examples is given, all with $e = 1$. We have illustrated the example here in the setting of [3]. In both points of view it is clear that proving that Y is a perfect code is easy compared to showing that we have a metric scheme to start with. In the terminology of graph theory the difficult problem is to show that G is a distance-transitive graph and not to find the perfect code. Theorems of the type we discussed for the Hamming schemes were possible because we had an infinite class of schemes in which we could search for perfect codes. It does not seem likely that this will be the case for distance-transitive graphs. So even though we still have Lloyd's theorem as a tool we do not know where to use it. It would be extremely interesting if a perfect code of order $e > 1$ would be found in a scheme of the type considered here. Of course the Golay code is such a code and there is a code derived from the Golay code which is also of order 3 (O. HEDEN, private communication) but this code is essentially the same as the Golay code. An example not corresponding to a Hamming scheme is not known.

6. NEARLY PERFECT CODES

In this section we discuss a class of completely regular codes namely the nearly perfect codes introduced by GOETHALS & SNOVER [5].

JOHNSON [6] proved the following extension of the sphere packing bound (see also section 7):

LEMMA 2. *If* Y *is a code with minimum distance* $d = 2e+1$ *in* $X := H(n,2)$ *then*

$$(6.1)\qquad |Y| \left\{ \sum_{i=0}^{e} \binom{n}{i} + \frac{1}{[n/(e+1)]} \binom{n}{e} \left(\frac{n-e}{e+1} - \left[\frac{n-e}{e+1}\right]\right) \right\} \le 2^n = |X| \,.$$

If $e+1$ divides $n+1$ then this reduces to the sphere packing bound. It is well known that $(e+1)|(n+1)$ is a necessary condition for the existence of a perfect code in $H(n,2)$. The code Y is called *nearly perfect* if equality holds in (6.1). From the proof of (6.1) it immediately follows that if Y is a nearly perfect code then for every $\underline{x} \in X$ with $\rho(\underline{x},Y) > e$ there are exactly $[n/(e+1)]$ points $\underline{y} \in Y$ with $\rho(\underline{x},\underline{y}) = e+1$. Furthermore, it follows that if $\rho(\underline{x},\underline{y}) = e$ for some $\underline{y} \in Y$, then there are exactly $[(n-e)/(e+1)]$ points $\underline{z} \in Y$ with $\rho(\underline{x},\underline{z}) = e+1$. In fact, such a code Y is completely regular. The distance distribution of Y is determined in [5].

The following theorem is an example of the theorems given in [5] showing the importance of nearly perfect codes in the theory of designs. We shall interpret a point of Y as the incidence vector of a subset of a set S of n points.

THEOREM 3. *If* Y *is a nearly perfect code with minimum distance* $d = 2e+1$ *in* $X = H(n,2)$ *and* $\underline{0} \in Y$ *then the words of weight* d *in* Y *form an* e-*design with* $\lambda = [(n-e)/(e+1)]$.

PROOF. Any e-subset D of S corresponds to a point $\underline{x} \in X$ with weight e, i.e. distance e to $\underline{0}$. We mentioned above that $\underline{x}$ then has distance e+1 to exactly λ points of Y each of which therefore has weight 2e+1. Hence D is a subset of exactly λ sets corresponding to code words of weight 2e+1. □

It is also shown in [5] that such designs can be extended to (e+1)-designs.

The non-linear codes known as the Preparata codes (cf. [10]) have $n = 4^m-1$, $|Y| = 2^{n-r}$ where $r = 4m-1$ $(m \geq 2)$, and $d = 5$. By substitution we see that these codes satisfy (6.1) with equality, i.e. they are nearly perfect. We thus obtain an infinite class of 3-designs.

Of course the question now rises whether nearly perfect codes with $e > 2$ can be found. (From now on we exclude perfect codes.) The definition alone is enough to show that the answer is negative for $e = 3$ and $e = 4$. GOETHALS & SNOVER mention this in their paper without giving the proof. We present their proof here.

THEOREM 4. *Except for the known perfect codes there are no nearly perfect codes with minimum distance* 7 *or* 9.

PROOF.

(i) Suppose $n \not\equiv 3 \pmod 4$. Let $n+1 \equiv s \pmod 4$ where $s = 1,2$ or 3. Then substituting this and $e = 3$ in (6.1) yields

$$|Y|\{1 + n + \binom{n}{2} + \frac{n+1}{n+1-s}\binom{n}{3}\} = 2^n . \tag{6.2}$$

This can be written as

$$|Y|\{6 + (n+1)(n^2+(s-1)n+(s-2)(s-3))\} = 3\cdot 2^{n+1} . \tag{6.3}$$

If $s = 1$, i.e. $n \equiv 0 \pmod 4$, we find from (6.3)

$$|Y|(n+2)(n^2-n+4) = 3\cdot 2^{n+1}$$

and hence $(n+2)|6$ which gives us $n = 4$, which does not correspond to a nearly perfect code. If $s = 2$ or 3 we obtain from (6.3)

$$|Y|(n+1)n(n+s-1) = 6\{2^n - |Y|\} ,$$

where $|Y| = 2^k$ or $3\cdot 2^k$ $(k < n)$. This implies that

$$(n+1)n(n+s-1) = 6(2^{n-k}-1) \text{ or } 2(2^{n-k}-3) .$$

Here the left-hand side is $\equiv 0 \pmod 4$ and the right-hand side is $\equiv 2 \pmod 4$, a contradiction.

(ii) We now consider the case $e = 4$. Let $n+1 \equiv s \pmod 5$, where $s = 1,2,3$ or 4. Again we substitute in (6.1) and replace $n+1$ by m. We find

$$3\cdot 2^{n+3}/|Y| = \begin{cases} m\ (m^3-5m^2+14m+8) & \text{if } s = 1 , \\ m\ (m^3-4m^2+7m+20) & \text{if } s = 2 , \\ m\ (m^3-3m^2+2m+24) & \text{if } s = 3 , \\ m\ (m^3-2m^2-m+26) & \text{if } s = 4 . \end{cases}$$

Clearly $16 \nmid m$ and therefore $m \mid 24$ which leaves only a finite number of cases which are all easily ruled out. □

Of course the first really interesting case is $e = 5$ since it could lead to a 6-design. In this case we were not able to do anything with (6.1) alone. However, GOETHALS & SNOVER also proved that there is an analogue of Lloyd's theorem for nearly perfect codes. We quote the theorem.

THEOREM 5. *Let there exist a nearly perfect code in* $H(n,2)$ *with minimum distance* $2e+1$ *and let* $n+1 \not\equiv 0 \pmod{e+1}$. *Then the polynomial*

(6.4) $$Q(x) := \Psi_{e-1}(x) + \frac{1}{[(n+1)/(e+1)]}(\Psi_{e+1}(x) - \Psi_{e-1}(x))$$

(where $\Psi_e(x)$ *is the polynomial of* (2.2) *with* $q = 2$*) has* $e+1$ *distinct integral zeros between* 1 *and* n.

As was to be expected the case $e = 5$ does not yield any solutions either.

THEOREM 6. *There is no nearly perfect code with minimum distance* 11 *in* H(n,2) *for* n > 11.

PROOF. Assume that Y is such a nearly perfect code. We write $n = 6v+\ell$ where $\ell = 0,1,2,3$ or 4 (since $\ell = 5$ is excluded by the theorem on perfect codes). By substitution in (6.1), taking e = 5, we see that $|Y|$ is either a power of 2 or 5 times a power of 2. Hence we have

$$(6.5) \qquad \sum_{i=0}^{5} \binom{n}{i} + \frac{1}{[n/6]} \binom{n}{5} \left\{ \frac{n-5}{6} - \left[\frac{n-5}{6}\right] \right\} = 2^r/a ,$$

where a = 1 or 5. As in the case of perfect codes the left-hand side of (6.5) is Q(0). By substitution we see that Q(1) > 0.

In the same way as (2.7) and (2.9) we find the sum and product of the zeros of Q:

$$(6.6) \qquad \sum_{i=1}^{6} x_i = 3(n+1) ,$$

$$(6.7) \qquad \prod_{i=1}^{6} x_i = \left[\frac{n+1}{6}\right] 6!\, 2^{r-6}/a .$$

We observe that Q(n+1-x) = Q(x), i.e.

$$x_{7-i} = n+1 - x_i \qquad (i=1,2,3).$$

We now introduce the variable $z := (2x-n-1)^2$. On substitution in (6.4) we then find that

$$(6.8) \qquad Q^*(z) := z^3 + 5(-2n+5-\ell)z^2 + \{15n^2 + 10(3\ell-5)n - (70\ell-19)\}z + \\ - 15(\ell+1)(n-1)(n-3)$$

has three integral zeros $z_i := (2x_i-n-1)^2$, i=1,2,3. Again a simple computation shows that $Q^*(0) < 0$ and $Q^*(8) > 0$. Hence by theorem 5 either n is even and $Q^*(1) = 0$ or n is odd and $Q^*(4) = 0$. However $Q^*(4) < 0$ for $\ell > 0$ and $Q^*(1) = -15\ell(n-2)(n-4) = 0$ only if $\ell = 0$. Hence we now know that n = 6v and that $x_3 = 3v$ and $x_4 = 3v+1$. Furthermore, z_1 and z_2 satisfy

$$z^2 + (26-10n)z + 15(n-1)(n-3) = 0 ,$$

i.e.

$$z_{1,2} = 5n - 13 \pm (10n^2-70n+124)^{\frac{1}{2}} = 5n - 13 \pm x ,$$

where

$$(6.9) \qquad 5(2n-7)^2 + 3 = 2x^2 .$$

Substituting $x_3 = 3v$ and $x_4 = 3v+1$ in (6.7) we find

$$(3v+1)x_1x_2x_5x_6 = 3\cdot 5\cdot 2^{r-2}/a ,$$

so either $3v+1 = 2^\sigma$ or $3v+1 = 5\cdot 2^\sigma$, i.e. $n = 2(2^\sigma-1)$ or $n = 2(5\cdot 2^\sigma-1)$. First substitute $n = 2(2^\sigma-1)$ in (6.9). This yields

$$16(5\cdot 2^{2\sigma-1}-55\cdot 2^{\sigma-2}+19) = x^2 .$$

The expression in brackets is $\equiv 3 \pmod 4$ if $\sigma \geq 4$. Hence only $n = 14$ is a possibility, but this does not yield a solution. Substitution of $n = 2(5\cdot 2^\sigma-1)$ yields a contradiction in exactly the same way. □

7. UNIFORMLY PACKED CODES

The uniformly packed codes were introduced by SEMAKOV, ZINOV'EV & ZAITSEV [11]. Once again the codes are in $H(n,2)$. The definition generalizes the idea of perfect and nearly perfect codes (in fact these are uniformly packed).

Let Y be a binary code of length n and minimum distance d. Let $e := [(d-1)/2]$. The set of all words of length n is again denoted by X. We define

$$(7.1) \qquad Y_e := \{\underline{x} \in X \mid \rho(\underline{x},Y) \geq e\} ,$$

and

$$(7.2) \qquad \forall \underline{z} \in Y_e, [r(\underline{z}) := |Y \cap S_{e+1}(\underline{z})|] ,$$

i.e. $r(\underline{z})$ is the number of points $\underline{y} \in Y$ with $e \leq \rho(\underline{y},\underline{z}) \leq e+1$. Clearly we have

$$(7.3) \qquad \forall \underline{z} \in Y_e, [r(\underline{z}) \leq [(n+1)/(e+1)]] .$$

Since

$$\sum_{\underline{z}\in Y_e} r(\underline{z}) = \sum_{\underline{y}\in Y} |S_{e+1}(\underline{y}) \setminus S_{e-1}(\underline{y})| = |Y|(\binom{n}{e} + \binom{n}{e+1})$$

and

$$|Y_e| = 2^n - |Y| \sum_{i=0}^{e-1} \binom{n}{i}$$

we find that the average value r of $r(\underline{z})$ is

$$\text{(7.4)} \qquad r = \frac{|Y| \binom{n+1}{e+1}}{2^n - |Y| \sum_{i=0}^{e-1} \binom{n}{i}} .$$

Observe that (7.3) and (7.4) together yield a proof of the Johnson bound (6.1).

If $\forall \underline{z} \in Y_e$, $[r(\underline{z}) = r]$ the code Y is called *uniformly packed*. Clearly a uniformly packed code with $r = [(n+1)/(e+1)]$ is nearly perfect (and of course perfect if $r = (n+1)/(e+1)$ and also if $r = 1$). From now on we only consider uniformly packed codes with $1 < r < [(n+1)/(e+1)]$. In [11] the distance distribution of a uniformly packed code is determined and it turns out that these codes are also completely regular. Furthermore, theorem 3 also generalizes. In fact, in the extended code of a uniformly packed code the words of a given weight w form an (e+1)-design. So once again the search starts!

We use the following notation

$$\text{(7.5)} \qquad r = (n-s)/(e+1), \text{ where } s \ge e.$$

We restrict the search to $e = 1$, $s \le 15$ and $e = 2$, $s \le 5$.

From (7.4) we find

$$\text{(7.6)} \qquad 2^n(n-s) = |Y| \, (n^2+2n-s) .$$

Since

$$n^2 + 2n - s > (n-s)(n+s+2)$$

and

$$n + s + 2 \ge 2s + 6 \ge 8,$$

we see that $16 \mid (n^2+2n-s)$, i.e. $s \equiv 0,3,8$ or $15 \pmod{16}$.

(a) <u>$e = 1$, $s = 3$</u>. There are two cases to consider. If $3 \nmid n$ we have

$$(n+3)(n-1) = n^2 + 2n - 3 = 2^k ,$$

i.e. $n = 5$, contradicting $r > 1$. If $n = 3m$ we find

$$(3m-1)(m+1) = 2^k ,$$

i.e. $m = 1$ or $m = 3$. So we have a possible solution with $n = 9$ and $r = 3$. Then $|Y| = 2^5$. There is indeed a uniformly packed code with these parameters.

Let $C :=$ circulant(0,1,1,0). Then the rows of

$$G := \left(I_5 \;\middle|\; \begin{matrix} 1\ 1\ 1\ 1 \\ C \end{matrix} \right)$$

generate a 5-dimensional linear subspace Y of X. The rows of G have weight ≥ 3 (equality occurring) and no two rows of G coincide in the final 4 positions. Hence the non-zero elements of Y have weight ≥ 3. We know that the maximal value of $r(\underline{z})$ for $\underline{z} \in Y_1$ is at most 5 and the average value of $r(\underline{z})$ is 3. Suppose there is a $\underline{z} \in Y_1$ with $r(\underline{z}) = 5$. Then without loss of generality we may assume $w(\underline{z}) = 1$ and then the sum of the 4 points of Y at distance 2 from $\underline{z}$ would have to be $(1,1,\ldots,1) - \underline{z}$. Since $(1,1,\ldots,1) \in Y$ this would imply $\underline{z} \in Y$, a contradiction. If for some $\underline{z}$ we would have $r(\underline{z}) = 4$ then again without loss of generality we may take $w(\underline{z}) = 1$ or 2. If $w(\underline{z}) = 1$ then the sum of the 3 points of Y at distance 2 from $\underline{z}$ would have weight 7, again a contradiction. If $w(\underline{z}) = 2$ then the sum of the 4 points of Y at distance 2 from $\underline{z}$ would have weight 8, once more a contradiction. This proves that Y is uniformly packed. The 2-design formed by the words of weight 4 of the extended code $\bar{Y}$ is the residual of the symmetric (16,6,2)-design.

(b) $\underline{e = 1,\ s = 8}$. Now (7.6) reads

$$2^n(n-8) = |Y|(n^2+2n-8) = |Y|(n+4)(n-2) .$$

If $n \equiv 2 \pmod 3$ then $n = 9m+8$ and we find

$$m \cdot 2^n = |Y|(3m+4)(3m+2)$$

which implies $m = 0$. If $n \not\equiv 2 \pmod 3$ then $n-2 = 2^k$, $n+4 = 2^\ell$, i.e. $n = 4$, a contradiction.

(c) $\underline{e = 1,\ s = 15}$. In this case we find from (7.6)

$$(n-15)2^n = |Y|(n+5)(n-3) .$$

As before we distinguish 4 cases depending on the g.c.d. of n and 15. In the same way as above we then find two possible parameter sets for uniformly packed codes:

(7.7) $$n = 27,\ r = 6,\ |Y| = 2^{21},$$

(7.8) $$n = 35,\ r = 10,\ |Y| = 2^{29}.$$

At present we do not know whether such codes exist.

We turn to the case e = 2. Then $n \equiv s \pmod 3$ and $n \geq s+6$. We find from (7.4) the equation

(7.9) $$(n-s)2^{n+1} = |Y|(n+1)(n^2+n-2s)$$

and we observe that

(7.10) $$(n-s,n+1) = (s+1,n+1),$$

(7.11) $$(n-s,\ n^2+n-2s) = (n-s,\ s(s-1)).$$

(d) $e = 2,\ s = 2$. We then have from (7.9), (7.10) and (7.11)

$$n + 1 = 3\cdot 2^k,$$

$$n^2 + n - 4 = 2^m.$$

If $k > 2$ then $m = 2$ which is impossible. If $k = 2$ we find $n = 11$ which yields a possible set of parameters: $n = 11$, $r = 3$, $|Y| = 24$. We now demonstrate a uniformly packed code with these parameters. Consider a Hadamard matrix H_{12} of order 12. From

$$A := \begin{pmatrix} \frac{1}{2}(H_{12} + J) \\ \frac{1}{2}(-H_{12} + J) \end{pmatrix}$$

we leave out the first column. The rows of the remaining matrix are the 24 words of the punctured Hadamard code Y. From the properties of the Hadamard matrix it follows that Y has minimum distance 5. For any $\underline{z} \in Y_2$ we know that $r(\underline{z}) \leq 4$ and the average value of $r(\underline{z})$ is 3. Suppose $r(\underline{z}) = 4$ for some $\underline{z}$. This implies that after a suitable multiplication of rows and columns by -1 and a permutation of columns there are 4 rows of H_{12} which have the form

$$
\begin{array}{llll}
x_1{+}{+} & {+}{+}{+} & {+}{+}{+} & {+}{+}{+} \\
x_2{-}{-} & {-}{-}{-} & {+}{+}{+} & {+}{+}{+} \\
x_3{-}{-} & {+}{+}{+} & {-}{-}{-} & {+}{+}{+} \\
x_4{-}{-} & {+}{+}{+} & {+}{+}{+} & {-}{-}{-}
\end{array}
$$

(where in this notation $\underline{z}$ corresponds to (x-- +++ +++ +++)). Taking $x_1 = +1$ we must have $x_2 = x_3 = x_4 = -1$ and then there is no other row of 12 +1's and -1's which is orthogonal to these 4 rows, a contradiction. Hence there is no $\underline{z}$ with $r(\underline{z}) = 4$, and therefore the code Y is uniformly packed. If we extend Y we find A. The words of weight 6 are obtained by leaving out the first and thirteenth row of A (if A has standard form). This yields a well-known 3-design.

(e) $e = 2,\ s = 3$. From (7.9), (7.10), (7.11) we find $n+1 = 2^k$. Since $n \equiv 0 \pmod 3$ k must be even and $k \geq 4$. Then

$$n^2 + n - 2s = 2^{2k} - 2^k - 6 = 3\cdot 2^m ,$$

i.e. $m = 1$, a contradiction.

(f) $e = 2,\ s = 4$ and $e = 2,\ s = 5$ are treated in exactly the same way. We omit the details. No possible parameter sets come up.

The equation (7.4) has a number of infinite families of solutions. We mention one below, the others are still being investigated. Without going into details we mention that there is also a generalization of Lloyd's theorem for uniformly packed codes. In some cases the infinite families of solutions of (7.4) also satisfy the conditions of this theorem.

Let $k \geq 2$, $n = 2^{2k-1}-1$, $e = 2$, $r = \frac{1}{3}(4^{k-1}-1)$, $d = n-2(2k-1)$ and $|Y| = 2^d$. Then these numbers satisfy (7.4). For $k = 2$ these are the parameters of the repetition code (which is perfect). For $k \geq 2$ the parameters are those of the 2-error-correcting primitive binary BCH-codes of length n and dimension d (cf. [7]). These codes are indeed uniformly packed and therefore we find from these codes several infinite sequences of 3-designs. We leave the details for a later paper.

REFERENCES

[1] BIGGS, N.L., *Perfect codes in graphs*, J. Combinatorial Theory B, 15 (1973) 289-296.

[2] BIGGS, N.L., *Perfect codes and distance-transitive graphs*, in: *Combinatorics*, T.P. MCDONOUGH & V.C. MAVRON (eds.), Proc. of the Third British Comb. Conference, Aberystwyth 1973, London Math. Soc. Lecture Notes (1974).

[3] DELSARTE, P., *The association schemes of coding theory*, this volume, pp. 143-161.

[4] DELSARTE, P. & J.M. GOETHALS, *Unrestricted codes with the Golay parameters are unique*, Report R 238, M.B.L.E. Research Laboratory, Brussels, 1973.

[5] GOETHALS, J.M. & S.L. SNOVER, *Nearly perfect binary codes*, Discrete Math., 3 (1972) 65-88.

[6] JOHNSON, S.M., *A new upper bound for error-correcting codes*, IEEE Trans. Information Theory, IT-8 (1962) 203-207.

[7] LINT, J.H. VAN, *Coding theory*, Lecture Notes in Mathematics 201, Springer-Verlag, Berlin, 1971.

[8] LINT, J.H. VAN, *Nonexistence theorems for perfect error correcting codes*, in: *Computers in algebra and number theory*, SIAM-AMS Proceedings IV, Amer. Math. Soc., Providence, R.I., 1971, pp. 89-95.

[9] LINT, J.H. VAN, *1967-1969 Report of the discrete mathematics group*, Report 69-WSK-04 of the Technological University, Eindhoven, 1969.

[10] PREPARATA, F.P., *A class of optimum nonlinear double-error-correcting codes*, Information and Control, 13 (1968) 378-400.

[11] SEMAKOV, N.V., V.A. ZINOV'EV & G.V. ZAITSEV, *Uniformly packed codes*, Problems of Information Transmission, 7 (1971) 30-39 (translated from Problemy Peredači Informacii, 7 (1971) 38-50).

[12] SNOVER, S.L., *The uniqueness of the Nordstrom-Robinson and the Golay binary codes*, thesis, Michigan State Univ., 1973.

[13] TIETÄVÄINEN, A., *On the nonexistence of perfect codes over finite fields*, SIAM J. Appl. Math., 24 (1973) 88-96.

[14] TIETÄVÄINEN, A., *A short proof for the nonexistence of unknown perfect codes*, Ann. Acad. S . Fenn. A580, 1974.

[15] ZINOV'EV, V.A. & V.K. LEONT'EV, *A theorem on the nonexistence of perfect codes over finite fields*, Problemy Peredači Informacii, to appear (in Russian).

IRREDUCIBLE CYCLIC CODES AND GAUSS SUMS *)

R.J. McELIECE

Jet Propulsion Labaratory, California Institute of Technology, Pasadena, Cal. 91109, USA

1. INTRODUCTION

In this paper we wish to point out the existence of a close connection between *irreducible cyclic codes* and *Gauss sums* over finite fields, and then to apply the well-developed theory of Gauss sums to the much less well-developed theory of irreducible cyclic codes.

We begin by giving two equivalent definitions of an irreducible cyclic code.

Let p be a prime, and let $q = p^e$ be a power of p. Denote by F_e the finite Galois field GF(q). Let n be a positive integer not divisible by p, and let $h(x) = h_0+h_1x+\ldots+h_kx^k$ be an F_e-irreducible divisor of $f_n(x)$, the n-th cyclotomic polynomial. It follows from the theory of finite fields that k, the degree of h(x), is the order of q mod n, i.e., the least positive integer such that $q^k \equiv 1 \pmod n$. The set of n-tuples $(c_0,c_1,\ldots,c_{n-1})$ from F_e such that

$$(1.1) \qquad \sum_{i=0}^{k} h_i c_{i+t} = 0 , \qquad (t=0,1,\ldots,n-1),$$

(subscripts are to be reduced mod n if necessary) is called an *(n,k) irreducible cyclic code* over F_e; h(x) is called the *parity-check polynomial* of the code.

Alternatively, if θ is a zero of h(x) in the field $F_{ek} = GF(q^k)$, the code can be characterized as the set of n-tuples $c(x) = (c_0(x),c_1(x),\ldots,c_{n-1}(x))$ from F_e of the form

---

*) This paper presents the results of one phase of research carried out at the Jet Propulsion Laboratory, California Institute of Technology, under Contract No. NAS 7-100, sponsored by the National Aeronautics and Space Administration.

M. Hall, Jr. and J. H. van Lint (eds.), Combinatorics, 185-202.

$$(1.2)\qquad c_i(x) = T_e^{ek}(x\,\theta^i)\ , \qquad (i=0,1,\ldots,n-1),$$

for some $x \in F_{ek}$, where $T_e^{ek}(\cdot)$ is the trace of F_{ek} over F_e. VAN LINT [10,Ch.3] gives a proof of the equivalence of these two formulations but for our purposes we shall take (1.2) as the definition of an irreducible cyclic code.

Now for $x \in F_{ek}$, $a \in F_e$, we denote by $\nu(x;a)$ the number of integers $i \in \{0,1,\ldots,n-1\}$ such that $c_i(x) = a$, i.e.,

$$(1.3)\qquad \nu(x;a) = |\{i: T_e^{ek}(x\,\theta^i) = a,\ 0 \le i \le n-1\}|^{*)}\ .$$

It is the study of the numbers $\nu(x;a)$ that interests us, but after section 2 we will consider only the numbers $\nu(x;0)$. Since $n-\nu(x;0)$ is the weight of $c(x)$, i.e., the number of non-zero components of $c(x)$, this restriction is equivalent to the study of the *weight distributions* of irreducible cyclic codes. It is probable, however, that many of the techniques developed in this paper can be extended to the numbers $\nu(x;a)$ for $a \neq 0$.

We will show in section 2 that the numbers $\nu(x;a)$ are intimately related to certain Gauss sums in the field F_{ek}. If μ is a character (a homomorphism into the complex numbers) of the multiplicative group $F_{ek}^{(\cdot)}$ of F_{ek} and λ is a character of the additive group $F_{ek}^{(+)}$ of F_{ek}, the Gauss sum $G(\mu,\lambda)$ is defined by

$$(1.4)\qquad G(\mu,\lambda) = \sum_{x\in F_{ek}^*} \mu(x)\lambda(x)\ .$$

We shall see in section 2 that the numbers $\nu(x;a)$ can be expressed in terms of the Gauss sums $G(\mu,\lambda)$ for characters μ of order $N = (q^k-1)/n$, i.e., characters satisfying $\mu^N(x) = 1$ for all $x \neq 0$. The values $\nu(x;0)$ actually only depend on the sums $G(\mu,\lambda)$ for which $\mu^{N_1} = 1$, where $N_1 =$ $=$ g.c.d.$(N,(q^k-1)/(q-1))$. By invoking known theorems on Gauss sums we shall succeed in computing the numbers $\nu(x;0)$ for all irreducible cyclic codes for

---

*) Although the definition (1.2) depends upon the particular n-th root of unity θ, the replacement of θ by another such primitive n-th root of unity θ^j with $(j,n) = 1$ will only permute the coordinates of $c(x)$ and so will not affect the numbers $\nu(x;a)$.

which: $N_1 = 1$ (section 2), $N_1 = 2$ (section 5), $N_1 = 3$ (section 6), $N_1 = 4$ (section 7), $p^l \equiv -1 \pmod{N_1}$ for some l (section 3), $\mathrm{ord}_p(N_1) = (N_1-1)/2$ and N_1 is prime (section 4). We have collected the necessary facts about Gauss sums in the appendix.

Some of the results in this paper are already known, at least for $e = 1$. In particular, MCELIECE & RUMSEY [11] first noticed that the Davenport-Hasse theorem (G5) about Gauss sums could be applied to the study of irreducible cyclic codes. BAUMERT & MCELIECE [1] calculated $\nu(x;a)$ for all a if $p^l \equiv -1 \pmod{N}$ for some l, or if $N = 2$, and for $q = 2$ and all $N < 100$. BAUMERT & MYKKELTVEIT [2] calculated $\nu(x;a)$ for all a when N is a prime for which p generates the quadratic residues. Later, in unpublished manuscripts, MYKKELTVEIT settled the cases $N = 3$ and $N = 4$. The main contributions of this paper are the extension of previous results to GF(q), and the observation that if one is only interested in the values $\nu(x;0)$, it is the number N_1 rather than N which is important.

2. GENERAL RESULTS

Let β be a complex primitive N-th root of unity, and ψ a primitive root in F_{ek} such that $\psi^N = \theta$. Then the function μ defined by

$$\mu(\psi^i) = \beta^i \tag{2.1}$$

is a character of order N of the multiplicative group $F_{ek}^{(\cdot)}$. Similarly if ζ is a complex primitive p-th root of unity the function λ defined by

$$\lambda(x) = \zeta^{T_1^{ek}(x)} \tag{2.2}$$

is a character of order p of the additive group $F_{ek}^{(+)}$. Now it is easily shown that an element $y \in F_{ek}^*$ is of the form $x\theta^j$ for some j if and only if $\mu^i(x) = \mu^i(y)$ for $i=0,1,\ldots,N-1$. Hence $\nu(x;a)$ is equal to the number of elements $y \in F_{ek}^*$ such that

$$\mu^i(yx^{-1}) = 1 , \qquad (i=0,1,\ldots,N), \tag{2.3}$$

$$T_e^{ek}(y) = a. \tag{2.4}$$

If ξ is a fixed element of F_{ek} with $T_e^{ek}(\xi) = 1$, condition (2.4) becomes $T_e^{ek}(y-a\xi) = 0$. If we define the character λ_b for $b \in F_{ek}$ by

$$(2.5) \qquad \lambda_b(x) = \lambda(bx) ,$$

it follows that (2.4) is equivalent to

$$(2.4') \qquad \lambda_b(y-\xi a) = 1 \qquad \text{for all } b \in F_e .$$

Thus it follows from the fact

$$\sum_{i=0}^{N-1} \mu^i(x) = \begin{cases} N & \text{if } \mu(x) = 1, \\ 0 & \text{if } \mu(x) \neq 1, \end{cases} \qquad \sum_{b \in F_e} \lambda_b(x) = \begin{cases} q & \text{if } \lambda_b(x) = 1 \ \forall_b \in F_e, \\ 0 & \text{otherwise,} \end{cases}$$

that

$$(2.6) \qquad \begin{aligned} qN\nu(x;a) &= \sum_{y \in F_{ek}^*} \sum_{i=0}^{N-1} \sum_{b \in F_e} \mu^i(x^{-1}y)\lambda_b(y-a\xi) = \\ &= \sum_{i=0}^{N-1} \mu^{-i}(x) \sum_{b \in F_e} \lambda_b^{-1}(a\xi) \sum_{y \in F_{ek}^*} \mu^i(y)\lambda_b(y) . \end{aligned}$$

Now the sum $\sum \mu^i(y)\lambda_b(y)$ which appears in (2.6) is just the *Gauss sum* $G(\mu^i,\lambda_b)$ over the field F_{ek}. We begin our simplification of (2.6) by separating out the sums $G(\mu^i,\lambda_b)$ for which either $i = 0$ or $b = 0$, using property (G1). The result is

$$qN\nu(x;a) = \begin{cases} q^k - q + \displaystyle\sum_{i=1}^{N-1} \sum_{b \in F_e^*} \mu^{-i}(x)\lambda_b^{-1}(a\xi)\, G(\mu^i,\lambda_b) & \text{if } a = 0 , \\ q^k + \displaystyle\sum_{i=1}^{N-1} \sum_{b \in F_e^*} \mu^{-i}(x)\lambda_b^{-1}(a\xi)\, G(\mu^i,\lambda_b) & \text{if } a \neq 0 . \end{cases}$$

By (G2), if $b \neq 0$, $G(\mu^i,\lambda_b) = \mu(b^{-1})G(\mu^i,\lambda)$. Thus if we denote $G(\mu^i,\lambda)$ by G_i and $\mu(x)$ by β^j,

$$(2.7)\qquad qN\nu(x;a) = \begin{cases} q^k - q + \sum_{i=1}^{N-1} \beta^{-ij} G_i \sum_{b\in F_e^*} \mu(b)^{-i} \lambda_{a\xi}(b)^{-1} & \text{if } a=0, \\ q^k + \sum_{i=1}^{N-1} \beta^{-ij} G_i \sum_{b\in F_e^*} \mu(b)^{-i} \lambda_{a\xi}(b)^{-1} & \text{if } a\neq 0. \end{cases}$$

The inner sum in (2.7) is the complex conjugate $\overline{g(\mu^i,\lambda_{a\xi})}$ of the Gauss sum of μ^i and $\lambda_{a\xi}$ over the subfield F_e of F_{ek}. Now the character μ^i will be identically equal to 1 on F_e^* if and only if it is equal to 1 on a primitive root of F_e^*. But $N_e^{ek}(\psi) = \psi^{(q^k-1)/(q-1)} = g$ is such a primitive root, and so $\mu^i(g) = \beta^{i(q^k-1)/(q-1)}$. It follows that the character μ^i acts trivially on F_e^* if and only if $i \equiv 0 \pmod{N_2}$, where

$$(2.8)\qquad \begin{cases} N_1 = (N,(q^k-1)/(q-1)) \\ N_2 = N/N_1 . \end{cases}$$

We now distinguish two cases, $a=0$ and $a\neq 0$. If $a=0$, the inner sum in (2.7) is simply $\sum_{b\in F_e^*} \mu(b)^{-i}$, which is by our above remarks $q-1$ if $i \equiv 0 \pmod{N_2}$ and 0 otherwise. Hence (2.7) becomes in this case

$$(2.9)\qquad \nu(x;0) = \frac{q^{k-1}-1}{N} + \frac{(q-1)}{qN} \sum_{i=1}^{N_1-1} G_{N_2 i}\beta^{-N_2 ij} .$$

Using property (G3), we obtain the estimate

$$(2.10)\qquad \left|\nu(x;0) - \frac{q^{k-1}-1}{N}\right| \leq \frac{(q-1)(N_1-1)}{N} q^{k/2-1} .$$

In particular if $N_1 = 1$ (i.e., if $q \equiv 1 \pmod N$ and $(N,k) = 1$) it follows that

$$\nu(x;0) = \frac{q^{k-1}-1}{N} \qquad \text{for all } x \in F_{ek}^* ,$$

a result recently proved for $e = 1$ by OGANESIAN & YADZJAN [12].

Next consider the case $a \neq 0$. According to (G2), $g(\mu^i, \lambda_{a\xi}) = \mu(a\xi)^{-i} \cdot g(\mu^i, \lambda)$. Hence if we denote the sum $g(\mu^i, \lambda)$ by g_i and the root of unity $\mu(x^{-1}a\xi)$ by γ, (2.7) becomes

$$(2.11) \qquad \nu(x;a) = \frac{q^{k-1}}{N} + \frac{1}{qN}\sum_{i=1}^{N-1} G_i \overline{g_i} \gamma^i, \qquad a \neq 0.$$

We saw above that μ^i acts trivially on F_e^* if and only if $i \equiv 0 \pmod{N_2}$; hence (G1) and (G3) yield the estimate

$$(2.12) \qquad \left|\nu(x;a) - \frac{q^{k-1}}{N}\right| \leq \frac{(N_1-1)+(N-N_1)q^{\frac{1}{2}}}{N} q^{k/2-1} .$$

Because of the relative simplicity of the formula (2.9), for the rest of the paper we shall restrict our attention to the case $a = 0$; that is, we shall content ourselves with a study of the weights of the codewords $c(x)$. (It is probable, however, that the formula (2.11) can be used to extend our results to the case $a \neq 0$ as well.) The heart of the formula (2.9) is the sum

$$\sum_{i=1}^{N_1-1} G_{N_2 i} \beta^{-N_2 ij} .$$

But $G_{N_2 i} = G(\mu^{N_2 i}, \lambda)$ is just a Gauss sum for a character $\mu^{N_2 i}$ of order N_1, and β^{-N_2} is an N_1-st root of unity. Thus we change our notation and for the remainder of the paper let β denote an N_1-st root of unity, and μ a character of F_{ek}^* such that $\mu(\psi) = \beta$. We denote the Gauss sum $G(\mu^i, \lambda)$ by G_i. Our goal is thus the calculation of the sum

$$(2.13) \qquad \sum_{i=1}^{N_1-1} G_i \beta^{-ij} .$$

We shall succeed in evaluating the sum (2.13) if $p^l \equiv -1 \pmod{N_1}$ for some l; if N_1 is prime and p generates the quadratic residues mod N_1; if $N_1 = 2$; $N_1 = 3$; or $N_1 = 4$.

3. THE SEMIPRIMITIVE CASE

If $N_1 > 2$ and if there exists an integer l such that $p^l \equiv -1 \pmod{N_1}$ *) we say that p is *semiprimitive* mod N_1. In this case it is possible to determine the Gauss sums of order N_1 in the field F_{2l} explicitly. The result is (G6) that for $i=1,2,\dots,N_1-1$,

$$G(\mu^i,\lambda) = (-1)^i p^l, \quad (N_1 \text{ even and } \frac{p^l+1}{N_1} \text{ odd}),$$

$$G(\mu^i,\lambda) = p^l, \quad (N_1 \text{ odd or } \frac{p^l+1}{N_1} \text{ even}).$$

We are however interested in the Gauss sums of order N_1 in the field F_{ek}. Since $N_1 > 2$ it follows that 2l is a divisor of ek, say 2lm = ek. Thus by the theorem of DAVENPORT & HASSE (G5),

$$(3.1)\quad \begin{cases} G_i = (-1)^{m+1+im}\, q^{k/2}, & (N_1 \text{ even and } \frac{p^l+1}{N_1} \text{ odd}), \\[2ex] G_i = (-1)^{m+1}\, q^{k/2}, & (N_1 \text{ odd or } \frac{p^l+1}{N_1} \text{ even}). \end{cases}$$

From (3.1) it is a simple matter to compute the sums (2.13), and thus also to compute $\nu(x;0)$ from (2.9). The result is

$$(3.2)\qquad \nu(x;0) = \begin{cases} \dfrac{q^{k-1}-1}{N} + \dfrac{(-1)^{m+1}(q-1)(N_1-1)}{N}\, q^{k/2-1} \\[3ex] \dfrac{q^{k-1}-1}{N} + \dfrac{(-1)^{m}(q-1)}{N}\, q^{k/2-1}. \end{cases}$$

The first alternative holds when $j \equiv 0 \pmod{N_1}$, unless N_1 is even, $(p^l+1)/N_1$ is odd, and m is odd, in which case it holds when $j \equiv N_1/2 \pmod{N_1}$. The second alternative holds in all other cases. Incidentally it is not necessary to know m in order to use the formulas (3.2) since if the wrong

---

*) We assume that l is in fact the least positive integer with this property.

sign is used the resulting numbers will not be integers.

As an example, consider the (91,6) irreducible cyclic code over GF(3). Then $N = 8$ and $N_1 = 4$. Since $3 \equiv -1 \pmod 4$ the results of this section apply. From (3.2) it thus follows that

$$\nu(x;0) = 37 \quad \text{if} \quad \text{ind}(x) \equiv 2 \pmod 4,$$

$$\nu(x;0) = 28 \quad \text{if} \quad \text{ind}(x) \not\equiv 2 \pmod 4 .$$

4. THE QUADRATIC RESIDUE CASE

Suppose that N_1 is an odd prime and that p generates the quadratic residues mod N_1, i.e., $p^{(N_1-1)/2}$ is the least power of p congruent to 1 (mod N_1). If $N_1 \equiv 1 \pmod 4$, -1 is a quadratic residue mod N_1 and the results of section 3 apply. If $N_1 \equiv 3 \pmod 4$ the calculation of a Gauss sum of order N_1 in $F_{(N_1-1)/2}$ is not so easy, but BAUMERT & MYKKELTVEIT [2] have shown (cf. G10) that $G(\mu,\lambda) = p^{(N_1-1)/4} e^{\pm i\theta}$, where θ is as described in the appendix. We are interested in the Gauss sums of order N_1 in the field F_{ek}, however. Since $p^{ek} \equiv 1 \pmod{N_1}$ it follows that ek is divisible by $(N_1-1)/2$; we denote the quotient $2ek/(N_1-1)$ by m. Then by the Davenport-Hasse result, by choosing the primitive root ψ of F_{ek} properly,

$$(4.1) \qquad G_1 = (-1)^{m+1} q^{k/2} e^{-im\theta} = -q^{k/2} e^{im\alpha}, \qquad \alpha = \pi - \theta .$$

By (G4), $G_i = G_1$ if i is a quadratic residue mod N_1 and $G_i = \overline{G_1}$ if i is a non-residue. Hence if we denote the sum $\sum\{\beta^i : (\frac{i}{p}) = +1\}$ by η, the sum (2.13) assumes one of the three values $(N_1-1)(G_1+\overline{G}_1)$, $(G_1\eta+\overline{G}_1\overline{\eta})$, $(G_1\overline{\eta}+\overline{G}_1\eta)$. GAUSS [5] article 356 showed that η is a solution to the quadratic equation $x^2+x+(N_1+1)/4 = 0$, i.e., $\eta = (-1\pm\sqrt{-N})/2$. The ambiguity in η is immaterial for our purposes since the mapping $\eta \to \overline{\eta}$ merely interchanges $G_1\eta+\overline{G}_1\overline{\eta}$ and $G_1\overline{\eta}+\overline{G}_1\eta$. Hence without loss of generality we may take

$$(4.2) \qquad \eta = -\frac{(N_1+1)^{\frac{1}{2}}}{2} e^{i\rho}, \quad \tan\rho = N_1^{\frac{1}{2}}, \qquad 0 < \rho < \pi/2 .$$

It follows from (4.1) and (4.2) that the sum (2.13) assumes one of the three values $-(N_1-1)q^{k/2} \cos m\alpha$, $(N_1+1)^{\frac{1}{2}} q^{k/2} \cos(m\alpha\pm\rho)$. Thus by (2.9)

$$(4.3)\qquad \nu(x;0) = \begin{cases} \dfrac{q^{k-1}-1}{N} - \dfrac{(q-1)(N_1-1)}{N}\, q^{k/2-1} \cos m\alpha\,, \\[2ex] \dfrac{q^{k-1}-1}{N} + \dfrac{(q-1)(N_1+1)^{\frac{1}{2}}}{N}\, q^{k/2-1} \cos(m\alpha\pm\rho)\,. \end{cases}$$

The first alternative in (4.3) holds when $j \equiv \mathrm{ind}(x) \equiv 0 \pmod{N_1}$. If $\mathrm{ind}(x) \not\equiv 0 \pmod{N_1}$ the quadratic character of ind(x) determines whether the second or third alternative holds. Finally, the angles α and ρ are given by

$$(4.3')\qquad \begin{cases} \tan\rho = N_1^{\frac{1}{2}},\ 0 < \rho < \pi/2\,, \\ \alpha = \pi-\theta,\ \tan\theta = bN_1^{\frac{1}{2}}/a,\ 0 < \theta < \pi\,, \\ a^2+N_1b^2 = 4p^{s-2t},\ a \equiv -2p^{s-t} \pmod{N_1}\,, \\ s = (N_1-1)/2,\ t = w_p(n_1)/(p-1),\ n_1 = (p^s-1)/N_1\,. \end{cases}$$

As an example consider the (71,5) code over GF(5). Here $N = 44$ and $N_1 = 11$. Since $11 \equiv 3 \pmod 4$ and $5^6 \equiv 1 \pmod{11}$ the results of this section apply. From (4.3') we compute $\rho = \tan^{-1}\sqrt{11} = 73.221345^\circ$, $\alpha = \pi-\tan^{-1}\sqrt{11}/3 = 132.130415^\circ$. From (4.3) it then follows that

$$\nu(x;0) = \begin{cases} 21 & \mathrm{ind}(x) \equiv 0 \pmod{11}\,, \\ 16 & \mathrm{ind}(x) \equiv 1,3,4,5,9 \pmod{11}\,, \\ 11 & \mathrm{ind}(x) \equiv 2,6,7,8,10 \pmod{11}\,, \end{cases}$$

provided the primitive root ψ of $GF(5^5)$ is properly chosen. (Otherwise the second two values would be interchanged.)

5. $N_1 = 2$

We now suppose that p is odd and that $N_1 = 2$. Then since $(q^k-1)/(q-1) = 1+q+\ldots+q^{k-1} \equiv k \pmod 2$, it follows that k must be even. In the field F_1, the Gauss sum of order 2 is given by (cf. G7)

$$G(\mu,\lambda) = \begin{cases} \sqrt{p} & \text{if } p \equiv 1 \pmod 4, \\ \sqrt{-p} & \text{if } p \equiv 3 \pmod 4. \end{cases}$$

Hence by the Davenport-Hasse theorem, in the field F_{ek}

$$(5.1) \qquad G_1 = \begin{cases} q^{k/2} & \text{if } p \equiv 3 \pmod 4,\ ek \equiv 2 \pmod 4, \\ -q^{k/2} & \text{otherwise}. \end{cases}$$

Thus by (2.9),

$$\nu(x;0) = \frac{q^{k-1}-1}{N} \pm \frac{(q-1)}{N} q^{k/2-1}.$$

If $p \equiv 3 \pmod 4$ and $ek \equiv 2 \pmod 4$, the + sign applies when $j \equiv \text{ind}(x) \equiv 0 \pmod 2$. In all other cases the + sign applies when ind(x) is odd.

As an example consider the (410,4) code over GF(9). Then $N = 16$, $N_1 = 2$. Thus by (5.2)

$$\nu(x;0) = \begin{cases} 41 & \text{ind}(x) \equiv 0 \pmod 2, \\ 50 & \text{ind}(x) \equiv 1 \pmod 2. \end{cases}$$

6. $N_1 = 3$

If $N_1 = 3$ and $p \equiv 2 \pmod 3$ the results of section 3 will apply. Thus suppose $p \equiv 1 \pmod 3$. Then $(q^k-1)/(q-1) \equiv 1+q+\ldots+q^{k-1} \equiv k \pmod 3$. Hence $k \equiv 0 \pmod 3$. According to (G8), in the field F_1, the Gauss sum of order 3 is given by

$$G(\mu,\lambda) = p^{\frac{1}{2}} e^{\pm i\theta},$$

where θ is as given in the appendix. Thus if we denote ek by m, the cubic Gauss sum G_1 in the field F_{ek} can be assumed to be

$$(6.1) \qquad G_1 = (-1)^{m+1} q^{k/2} e^{-im\theta} = -q^{k/2} e^{im\alpha}, \qquad \alpha = \pi - \theta.$$

With reasoning similar to that of section 4, it follows that the sum (2.13) assumes one of the three values $-2q^{k/2}\cos m\alpha$, $-2q^{k/2}\cos(m\alpha\pm 2\pi/3)$. Hence

$$(6.2)\qquad \nu(x;0) = \begin{cases} \dfrac{q^{k-1}-1}{N} - \dfrac{2(q-1)}{N} q^{k/2-1} \cos m\alpha , \\[2ex] \dfrac{q^{k-1}-1}{N} - \dfrac{2(q-1)}{N} q^{k/2-1} \cos(m\alpha\pm 2\pi/3) , \end{cases}$$

where α is determined by

$$(6.2')\qquad \begin{cases} \alpha = \pi-\theta,\ \tan 3\theta = 3b\sqrt{3}/a,\ 0 < \theta < \pi/3 \\[2ex] a^2+27b^2 = 4p,\ a \equiv 1 \pmod 3) . \end{cases}$$

As an example consider the (57,3) code over GF(7). Then $N = 6$, $N_1 = 3$. From (6.2') we calculate $a = 1$, $b = \pm 1$, $\theta = 26.368868^\circ$, $\alpha = 153.631132^\circ$. It follows from (6.2) that

$$\nu(x;0) = \begin{cases} 9 & \operatorname{ind}(x) \equiv 0 \pmod 3 , \\ 12 & \operatorname{ind}(x) \equiv 1 \pmod 3 , \\ 3 & \operatorname{ind}(x) \equiv 2 \pmod 3 , \end{cases}$$

provided the primitive root ψ of $GF(7^3)$ has been chosen properly. If the wrong primitive root is selected the second and third values of $\nu(x;0)$ would be interchanged.

7. $N_1 = 4$

If $N_1 = 4$ and $p \equiv 3 \pmod 4$ the results of section 3 apply. Hence we assume that $p \equiv 1 \pmod 4$. Also, since $(q^k-1)/(q-1) = 1+q+\ldots+q^{k-1} \equiv k \pmod 4$, it follows that $k \equiv 0 \pmod 4$. By (G9) the Gauss sum of order 4 in F_1 is $p^{\frac{1}{2}} e^{\pm i\theta}$, where θ is given in the appendix. Hence in the field F_{ek}

$$(7.1)\qquad G_1 = -q^{k/2} e^{im\theta} .$$

By (5.1), $G_2 = -q^{k/2}$. Furthermore $G_3 = \overline{G_1}$, and so the sum (2.13) is

$$q^{k/2}((-1)^{j+1} - 2\cos(m\theta-j\pi/2)).$$

Thus $\nu(x;0)$ is given by

$$(7.2)\qquad \nu(x;0) = \begin{cases} \dfrac{q^{k-1}-1}{N} - \dfrac{(q-1)(1\pm 2\cos m\theta)}{N}\, q^{k/2-1}, & j\equiv 0,2 \pmod 4. \\ \\ \dfrac{q^{k-1}-1}{N} + \dfrac{(q-1)(1\pm 2\sin m\theta)}{N}\, q^{k/2-1}, & j\equiv 1,3 \pmod 4. \end{cases}$$

The angle θ is determined by

$$(7.2')\qquad \begin{cases} \tan 4\theta = 4ab/(4b^2-a^2),\ 0 < \theta < \pi/4\ , \\ \\ a^2+4b^2 = p,\ a \equiv 1 \pmod 4)\ . \end{cases}$$

As an example consider the (39,4) code over GF(5). Then $N = 16$, $N_1 = 4$. The angle $\theta = \frac{1}{4}\tan^{-1}\frac{4}{3} = 13.282526^\circ$. Thus the values of $\nu(x;0)$ are

$$\nu(x;0) = \begin{cases} 5 & j \equiv 0 \pmod 4)\ , \\ 11 & j \equiv 1 \pmod 4)\ , \\ 8 & j \equiv 2 \pmod 4)\ , \\ 7 & j \equiv 3 \pmod 4)\ . \end{cases}$$

APPENDIX: SOME PROPERTIES OF GAUSS SUMS

Let $F_k = GF(p^k)$ and let N be an integer dividing p^k-1. If ψ is a primitive root of F_k and if $x \neq 0$ is an element of F_k, we define the index of x (with respect to ψ) as $\mathrm{ind}(x) = i$, where $\psi^i = x$ and $i \in \{0,1,\ldots,p^k-2\}$. Let ζ be a complex p-th root of unity, i.e., $\zeta = \exp(2\pi i h/p)$ for some $h \in \{0,1,\ldots,p-1\}$. Then for any $b \in F_k$ we may define a character of the additive group $F_k^{(+)}$ of F_k by

$$(A1)\qquad \lambda(x) = \zeta^{T_1^k(bx)}\ .$$

Similarly if β is any complex N-th root of unity we define a character of the multiplicative group $F_k^{(\cdot)}$ of non-zero elements of F_k by

$$(A2)\qquad \mu(x) = \beta^{\mathrm{ind}(x)}\ .$$

It turns out that every character of $F_k^{(+)}$ has the form (A1), and every character of $F_k^{(\cdot)}$ of order N has the form (A2).

The *Gauss sum* of the characters μ,λ in F_k is now defined by

$$\text{(A3)} \qquad G(\mu,\lambda) = \sum_{\substack{x\in F_k \\ x\neq 0}} \mu(x)\,\lambda(x)\ .$$

This is also called a Gauss sum of order N. Such sums (sometimes they are called *Lagrange resolvents*) have been intensively studied since GAUSS considered the special case N = 2, f = 1 in [5, article 356], and much is known about them. The article by IWASAWA [8] is a very good introduction to the subject, but is difficult to obtain. It contains an especially good treatment of Stickleberger's theorem. The recent book by IRELAND & ROSEN [7] is also a good source of information. The last chapter of HASSE's book [6] is not as elementary as the other two accounts but is more complete. In this appendix we shall list, but not prove, the results needed for this paper.

The first result deals with the sums $G(\mu,\lambda)$ where either μ or λ is trivial.

$$\text{(G1)} \qquad G(\mu,\lambda) = \begin{cases} p^k-1 & \text{if } \mu = 1,\ \lambda = 1, \\ 0 & \text{if } \mu \neq 1,\ \lambda = 1, \\ -1 & \text{if } \mu = 1,\ \lambda \neq 1. \end{cases}$$

Property (G1) is quite easy to prove from the definition (A3).

The next property shows how changing the character λ affects the value of $G(\mu,\lambda)$. For this purpose we denote the character in (A1) by λ_b.

$$\text{(G2)} \qquad G(\mu,\lambda_b) = \mu(b)^{-1}G(\mu,\lambda_1) \quad \text{if } b \neq 0.$$

Property (G2) also follows easily from the definition.

The next property is the first really interesting property of Gauss sums, but it does not lie very deep. A proof may be found on p.92 of LANG [9].

$$\text{(G3)} \qquad \begin{cases} G(\mu,\lambda)G(\mu^{-1},\lambda) = \mu(-1)\ p^k, \quad \text{and so} \\ |G(\mu,\lambda)| = p^{k/2}, \ \text{if } \mu \neq 1,\ \lambda \neq 1. \end{cases}$$

The next property shows that certain automorphisms of $Q(\beta)$ leave $G(\mu,\lambda)$ invariant.

(G4) $$G(\mu^{p^i},\lambda) = G(\mu,\lambda) \qquad \text{for all } i=0,1,2,\ldots ,$$

Property (G4) follows directly from (A3) and the fact that $T_1^e(x^{p^i}) = T_1^e(x)$.

We now come to the remarkable theorem of DAVENPORT & HASSE [4], the result which is easily the most important for the applications to irreducible cyclic codes. Let k' be the order of p (mod N), i.e., the smallest positive integer such that $p^{k'} \equiv 1$ (mod N). Since also $p^k \equiv 1$ (mod N), k is divisible by k', say $k = k'm$, and so $F_{k'}$ is a subfield of F_k. If λ' is a non-trivial character on $F_{k'}^{(+)}$ and μ' is a character on $F_{k'}^{(\cdot)}$ of order N, we may form the sum $G(\mu',\lambda')$ in $F_{k'}$. We now "lift" the characters μ',λ' from $F_{k'}$ to F_k by defining

$$\lambda(x) = \lambda'(T_{k'}^k(x))$$

$$\mu(x) = \mu'(N_{k'}^k(x)) .$$

The character λ is non-trivial on F_k since the trace $T_{k'}^k$ is onto, and the character μ is of order N since the norm $N_{k'}^k$ is onto. The theorem of Davenport and Hasse shows that there is a simple relationship between the sum $G(\lambda,\mu)$ in the field F_k and the sum $G(\mu',\lambda')$ in the smaller field $F_{k'}$:

(G5) $$G(\mu,\lambda) = (-1)^{m+1}G(\mu',\lambda')^m.$$

Since every character of order N in F_k can be obtained by lifting a character of order N from $F_{k'}$, and since the value of $G(\mu,\lambda)$ is not materially dependent upon which character $\lambda \neq 1$ is chosen, (G5) allows us to compute *any* Gauss sum of order N in F_k in terms of a Gauss sum in a smaller field. A proof of the Davenport-Hasse theorem was given by McELIECE & RUMSEY in [11].

The remaining results concern the explicit calculation of certain Gauss sums. The first is what BAUMERT & McELIECE [1] called the *semiprimitive case*.

Here it is assumed that there exists an integer l such that $p^l \equiv -1$ (mod N). It is further assumed that β in (A2) is a primitive N-th root of unity. Then STICKLEBERGER [13, §3.6 and 3.10] showed that for any $\lambda \neq 1$, in F_{2l}

(G6)
$$\begin{cases} G(\mu^i,\lambda) = (-1)^i p^l, & (N_1 \text{ even and } \frac{p^l+1}{N_1} \text{ odd}), \\ G(\mu^i,\lambda) = p^l, & (N_1 \text{ odd or } \frac{p^l+1}{N_1} \text{ even}). \end{cases}$$

This result also appears as a lemma (p.168) in BAUMERT & MCELIECE [1].

The remaining results of this section concern the explicit determination of the sum $G(\mu,\lambda)$ in a variety of other special cases. However, the determination is not as explicit as (G6) in general for the following reasons. First, we did not specify exactly either the N-th root of unity β or the primitive root ψ of F_k. This uncertainty will in general cause an ambiguity in the determination of $G(\mu,\lambda)$ of an automorphism of the field $\mathbb{Q}(\beta)$. Second, we did not specify either the p-th root of unity ζ or the choice of b in the definition (A1) of the character λ. Property (G2) shows that this uncertainty will in general cause an ambiguity of a multiplicative factor of an N-th root of unity. The first ambiguity is inevitable because there is no "canonical" way to choose a primitive root of a finite field. However, the second ambiguity is in principle resolvable if we take $\zeta = \exp(2\pi i/p)$, $b = 1$. Unfortunately even if this is done the problem of resolving the ambiguity is in general intractable. GAUSS spent a year on the case $N = 2$, and the case $N = 3$ has never been resolved (CASSELS [3]). Thus we shall not specify the p-th root of unity ζ exactly, and accept this nagging but essentially harmless ambiguity.

We now come to the earliest result about Gauss sums. It is due to GAUSS himself [5, art.356]. We assume p is odd, $N = 2$, $f = 1$. In this case

(G7)
$$G(\mu,\lambda) = \begin{cases} \pm\sqrt{p}, & p \equiv 1 \pmod 4, \\ \pm\sqrt{-p}, & \text{otherwise}. \end{cases}$$

Incidentally, GAUSS succeeded in determining the doubtful sign in (G7) as +, if $\zeta = \exp(2\pi i/p)$.

Next we assume that $p \equiv 1 \pmod 3$ and that μ is a non-trivial character of order 3 in F_1. Then with a suitable choice of ζ,

(G8)
$$\begin{cases} G(\mu,\lambda) = \sqrt{p}\, e^{\pm i\theta}, \\ \tan 3\theta = 3b\sqrt{3}/a,\ 0 < \theta < \pi/3, \\ a^2+27b^2 = 4p,\ a \equiv 1 \pmod 3,\ b > 0. \end{cases}$$

These results appear implicity in GAUSS [5, art.358] but a clearer proof is given by HASSE [6, §20.4]. It turns out that the representation $4p = a^2+27b^2$ in integers a,b is unique except for the signs of a and b. The sign of a is determined by the congruence $a \equiv 1 \pmod 3$. The sign of b cannot be determined, since it reflects the uncertainty in the choice of ψ.

Next we consider primes $\equiv 1 \pmod 4$ and consider Gauss sums of order 4 in F_1. Here the result is that with a suitable choice of ζ,

$$\text{(G9)} \qquad \begin{cases} G(\mu,\lambda) = \sqrt{p}\, e^{\pm i\theta}, \\ \tan 4\theta = 4ab/(4b^2-a^2),\ 0 < \theta < \pi/4, \\ a^2+4b^2 = p,\ a \equiv 1 \pmod 4,\ b > 0. \end{cases}$$

Once again this result is essentially due to GAUSS, see HASSE [6, §20.4].

We now come to a result about Gauss sums which is apparently new, and which has arisen from the study of irreducible cyclic codes. The assumption is that N is an odd prime $\equiv 3 \pmod 4$ and that p generates the quadratic residues mod N, i.e., that the order of p (mod N) is $s = (N-1)/2$. In this case BAUMERT & MYKKELTVEIT [2] have shown that in the field F_s

$$\text{(G10)} \qquad \begin{cases} G(\mu,\lambda) = p^{s/2}\, e^{\pm i\theta}, \\ \tan \theta = b\sqrt{N}/a,\ 0 < \theta < \pi, \\ a^2+Nb^2 = 4p^{s-2t},\ a \equiv -2p^{s-t} \pmod N. \end{cases}$$

In (G10) the integer t is determined as follows. Let $n = (p^s-1)/N$, and let $n = \sum_{j=0}^{s-1} n_j p^j$ be the expansion of n in the base p. Then t is the p-weight of n divided by p-1, i.e.,

$$\text{(G10')} \qquad (p-1)t = w_p(n) = n_0+n_1+\ldots+n_{s-1}.$$

Alternatively BAUMERT & MYKKELTVEIT show that if $r_1,r_2,\ldots,r_s$ are the quadratic residues mod N reduced mod N, i.e., $0 < r_1 < r_2 < \ldots < r_s < N$, then

$$\text{(G10")} \qquad Nt = r_1+r_2+\ldots+r_s.$$

The key to the proof of (G10) is the determination of the highest power of p which devides $G(\mu,\lambda)$. It turns out that a famous theorem of STICKLEBERGER [13, §6] shows that this highest power is $\min(w_p(n), w_p((N-1)n))$. The fact that $w_p(n) < w_p((N-1)n)$ follows easily from the famous theorem of Gauss that for primes N of the form 4k+3 there are more quadratic residues in the range (0,N/2) than in the range (-N/2,0). (For a proof of this result see WEYL [14,ch.IV].)

LIST OF SYMBOLS

p, a prime
q, a power p^e of p
n, an integer not divisible by p
k, the least positive integer such that $q^k \equiv 1 \pmod n$
$F^{(+)}$, the additive group of the field F
$F^{(\cdot)}$, the multiplicative group of non-zero elements of F
F^*, the set of non-zero elements of F
F_i, the field $GF(p^i)$
T_i^{ij}, the trace of F_{ij} over F_i
N_i^{ij}, the norm of F_{ij} over F_i
$N = (q^k-1)/n$
$N_1 = \text{g.c.d.}(N, 1+q+\ldots+q^{k-1})$
$N_2 = N/N_1$
β, a complex primitive N or N_1-st root of unity
ζ, a complex primitive p-th root of unity
ξ, an element of F_{ek} with $T_e^{ek}(\xi) = 1$
ψ, a primitive root in F_{ek}
θ, a primitive n-th root of unity in F_{ek}, usually $\theta = \psi^N$
ind(x), the least positive integer j such that $\psi^j = x$ in F_{ek}
c(x), a codeword (vector) whose i-th component is $T_e^{ek}(x\theta^i)$, $x \in F_{ek}$
(i=0,1,...,n-1)
$\nu(x;a)$, the number of components of c(x) equal to a
μ, a character of $F_{ek}^{(\cdot)}$, usually of order N or N_1
λ, a character of $F_{ek}^{(+)}$, usually $\lambda(x) = \zeta^{T_1^{ek}(x)}$
λ_b, a character defined by $\lambda_b(x) = \lambda(bx)$
$G(\mu,\lambda)$, the Gauss sum $\sum\{\mu(x)\lambda(x) : x \in F_{ek}^*\}$
G_i, the Gauss sum $G(\mu^i,\lambda)$
$\mathfrak{g}_i$, the Gauss sum $\sum\{\mu^i(x)\lambda(x) : x \in F_e^*\}$

REFERENCES

[1] BAUMERT, L.D. & R.J. MCELIECE, *Weights of irreducible cyclic codes*, Information and Control, 20 (1972) 158-175.

[2] BAUMERT, L.D. & J. MYKKELTVEIT, *Weight distributions of some irreducible cyclic codes*, DSN Progress Report 16 (1973) 128-131 (published by Jet Propulsion Laboratory, Pasadena, California).

[3] CASSELS, J.W.S., *On Kummer sums*, Proc. London Math. Soc. (3), 21 (1970) 19-27.

[4] DAVENPORT, H. & H. HASSE, *Die Nullstellen der Kongruenzzetafunktionen in gewissen zyklischen Fallen*, J. Reine Angew. Math., 172 (1935) 151-182.

[5] GAUSS, C.F., *Disquisitiones Arithmeticae*, English translation published by Yale University Press, New Haven, 1966.

[6] HASSE, H., *Vorlesungen über Zahlentheorie*, Springer-Verlag, Berlin, 1964.

[7] IRELAND, K. & M.I. ROSEN, *Elements of number theory*, Bogden and Quigley, Tarrytown-on-Hudson, 1972.

[8] IWASAWA, K., *Stickelberger's theorem on Gauss sums*, Notes taken by J. SMITH at the National Science Foundation Advanced Science Seminar, Bowdoin College, 1966.

[9] LANG, S., *Algebraic number theory*, Addison Wesley, Reading, 1970.

[10] LINT, J.H. VAN, *Coding theory*, Lecture Notes in Mathematics 201, Springer-Verlag, Berlin etc., 1971.

[11] MCELIECE, R.J. & H. RUMSEY, Jr., *Euler products, cyclotomy and coding*, J. Number Theory, 4 (1972) 302-311.

[12] OGANESIAN, S.S. & V.G. YAGDZIAN, *A class of optimal cyclic codes with base* p, Problemy Peredači Informacii, 8 (1972), vyp. 2, 109-111 (in Russian).

[13] STICKELBERGER, L., *Ueber eine Verallgemeinerung der Kreisteilung*, Math. Ann., 37 (1890) 321-367.

[14] WEYL, H., *Algebraic theory of numbers*, Annals of Mathematics Studies 1, Princeton University Press, Princeton, 1940.

P A R T 2

GRAPH THEORY

FOUNDATIONS, PARTITIONS AND COMBINATORIAL GEOMETRY

ISOMORPHISM PROBLEMS FOR HYPERGRAPHS

C. BERGE

Université de Paris, Paris 5e, France

1. INTRODUCTION

A *hypergraph* $H = (X, \mathcal{E}) = (E_1, E_2, \ldots, E_m) = (E_i : i \in M)$ is a family $\mathcal{E}$ of subsets E_i of a set $X = \{x_j : j \in N\}$ of *vertices*. The sets E_i are called *edges*.

The *rank* $r(H)$ of a hypergraph H is the maximum cardinality of the edges. If all edges have the same cardinality, the hypergraph is said to be *uniform*. The *subhypergraph* induced by a subset A of X is the hypergraph $H_A = (E_i \cap A : i \in M, E_i \cap A \neq \emptyset)$.

If $I \subseteq M$, the *partial hypergraph* generated by I is the hypergraph $(E_i : i \in I)$. The *section hypergraph* is the partial hypergraph $H \times A = (E_i : i \in M, E_i \subseteq A \subseteq X)$.

The *dual* H^* of H is a hypergraph with vertex set $E = \{e_1, \ldots, e_m\}$, and having edges which are certain subsets of E, namely edges X_j where $X_j = \{e_i : i \in M, x_j \in E_i\}$.

Consider two hypergraphs $H = (E_1, \ldots, E_m)$ and $H' = (F_1, \ldots, F_m)$. H is *equivalent* to H' ($H \equiv H'$) if the mapping $\phi: X \to Y$, $\phi(x_i) = y_i$, satisfies $\phi(E_i) = F_{\pi i}$ $(i \in M)$ for some permutation π of M.

H is *equal* to H' (or $H = H'$) if the permutation π in the above definition can be the identity.

H is *isomorphic* to H' (or $H \cong H'$) if there is a bijection $\phi: X \to Y$ and if there is a permutation π of M such that $\phi(E_i) = F_{\pi i}$ $(i \in M)$. The bijection ϕ is called an *isomorphism*.

H is *strongly isomorphic* to H' (or $H \underline{\cong} H'$) if there is a bijection $\phi: X \to Y$ for which $\phi(E_i) = F_i$ for all $i \in M$.

M. Hall, Jr. and J. H. van Lint (eds.), Combinatorics, 205-214.

Observe that equality implies the other three relations and any of the relations imply isomorphism.

We give some examples:

EXAMPLE. Consider the following:

$H =$ 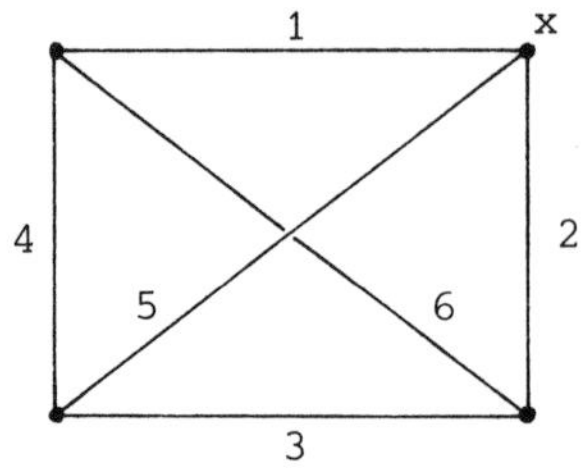

$H' =$ 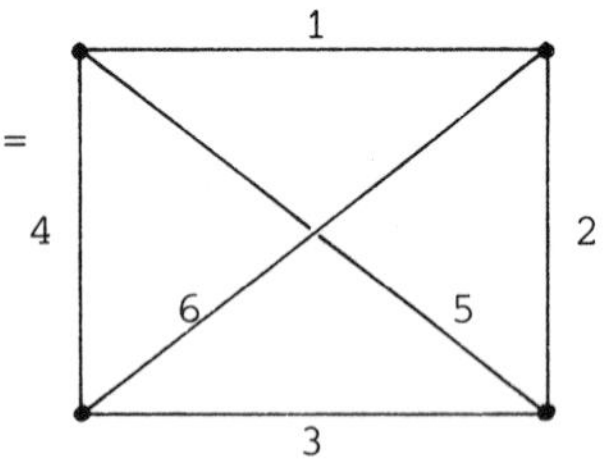

Observe that $H \backsimeq H'$, but $H \not\stackrel{\frown}{=} H'$, since if $H \stackrel{\frown}{=} H'$, the vertex x would map to the non-existent vertex meeting edges 1, 2 and 5 in H'.

EXAMPLE. Consider the line graph L(H) of the graph H above:

$L(H) =$ 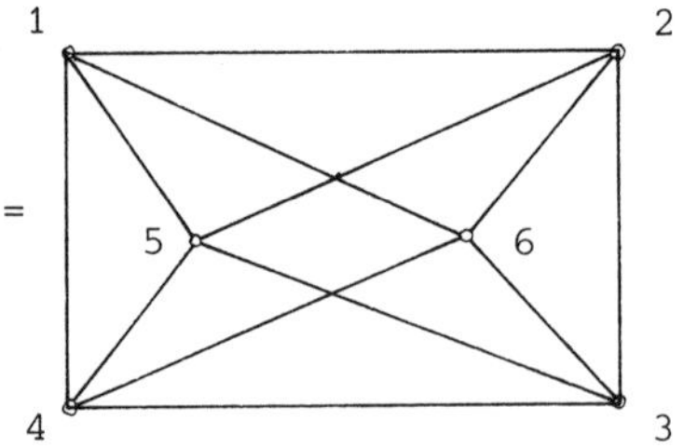

Observe that $L(H) = L(H')$, but since the edges are unlabeled here, equality is meaningless.

Our purpose in this paper is to present some general results concerning isomorphisms and other relations among hypergraphs.

A *multigraph* is a hypergraph with $|E_i| \leq 2$ for all $i \in M$.

PROPOSITION 1. *If* $H = (X, (E_i)_{i \in M})$ *and* $H' = (Y, (F_i)_{i \in M})$ *are multigraphs, and if* $\phi: X \to Y$ *is a bijection, then the following are equivalent:*

(i) ϕ *is an isomorphism;*

(ii) $m_H(x,y) = m_{H'}(\phi(x), \phi(y))$ *for all* $x, y \in X$, *where* $m_H(x,y)$ *is the number of edges joining* x *and* x' *in* H.

PROPOSITION 2. *If* H *is a hypergraph, then* $(H^*)^* = H$.

PROPOSITION 3. *If* H, H' *are hypergraphs, then* $H \simeq H'$ *if and only if* $H^* \simeq H'^*$.

PROPOSITION 4. *If* H, H' *are hypergraphs, then* $H \equiv H'$ *if and only if* $H^* \cong H'^*$.

PROPOSITION 5. *The dual of the partial hypergraph of* H *generated by* $(E_i : i \in I)$ *equals the subhypergraph of* H^* *induced by* $\{e_i : i \in I\}$.

PROPOSITION 5*. *The dual of the subhypergraph* $H_A = (E_i \cap A : i \in M,\ E_i \cap A \neq \emptyset)$ *with* $A = \{x_j : j \in J\}$ *equals the partial hypergraph of* H^* *generated by* $(X_j : j \in J)$.

PROPOSITION 6. *The dual of the section hypergraph* $H \times A$, $A = \{x_j : j \in J\}$, *equals the subhypergraph of* H^* *induced by* $\bigcup_{j\in N} X_j - \bigcup_{j\in N-J} X_j$.

2. TRANSITIVE HYPERGRAPHS

Let $H = (X,E)$ be a hypergraph. Two vertices x and y of H are *symmetric* if there exists an automorphism ϕ of H such that $\phi(x) = y$. Two edges E_i and E_j are *symmetric* if there exists an automorphism ϕ of H such that $\phi(E_i) = E_j$.

H is said to be *vertex-transitive* (resp. *edge-transitive*) if any two vertices (resp. edges) are symmetric. A hypergraph that is both vertex-transitive and edge-symmetric is said to be *transitive*. *Because of the duality principle for hypergraphs, the study of vertex-transitive hypergraphs reduces to the study of edge-transitive hypergraphs.*

The following result is a generalization of a theorem for graphs due to E. DAUBER [3].

THEOREM 1. *For an edge-transitive hypergraph* $H = (X,E)$, *there exists a partition* $(X_1,X_2,\ldots,X_k)$ *of* X *such that*

(i) $\sum_\lambda r(X_\lambda) = r(X)$, *where* $r(A)$ *denotes the rank of* H_A,

(ii) H_{X_λ} *is transitive for all* λ.

Since H is edge-transitive, $|E_i| = h$ for all i. Let $E_1 = \{x_1,x_2,\ldots,x_h\}$. For $i \in M$, let ϕ_i be an automorphism such that $\phi_i(E_1) = E_i$.

Let $Y_p = \{\phi_i(x_p) : i \in M\}$, $(p=1,2,\ldots,h)$. Then, $\overline{H} = (Y_1,\ldots,Y_h)$ is a hypergraph on X, because

$$\bigcup_p Y_p = \bigcup_{i\in M} \phi_i(E_1) = \bigcup_{i\in M} E_i = X .$$

Let $X_1, X_2, \ldots, X_k$ be the connected components of $\overline{H}$.

PROOF OF (i). Let $E_1^\lambda = \{x_p : p \le h,\ Y_p \subseteq X_\lambda\}$ for $\lambda = 1,2,\ldots,k$. For $x_p \in E_1^\lambda$,

$$\phi_i(x_p) \in E_i \cap Y_p \subseteq E_i \cap X_\lambda .$$

Hence

$$\phi_i(E_1^\lambda) \subseteq E_i \cap X_\lambda .$$

Thus,

$$h = \sum_\lambda |E_1^\lambda| = \sum_\lambda |\phi_i(E_1^\lambda)| \le \sum_\lambda |E_i \cap X_\lambda| = h .$$

Hence the equality holds, and

$$E_i \cap X_\lambda = \phi_i(E_1^\lambda) .$$

Hence

$$|E_i \cap X_\lambda| = |E_1^\lambda| , \qquad (i \in M).$$

This shows that H_{X_λ} is uniform with rank $|E_1^\lambda|$, and furthermore,

$$\sum_\lambda r(X_\lambda) = \sum_\lambda |E_1^\lambda| = \sum_\lambda |E_i \cap X_\lambda| = h .$$

PROOF OF (ii). In H_{X_λ} the edges $E_i \cap X_\lambda$ and $E_j \cap X_\lambda$ are symmetric, since

$$\phi_j\phi_i^{-1}(E_i \cap X_\lambda) = \phi_j(E_1^\lambda) = E_j \cap X_\lambda .$$

Hence H_{X_λ} is edge-transitive.

Furthermore, two vertices $x,y \in Y_p$ are symmetric, since

$$\left.\begin{array}{l} x = \phi_i(x_p) \\ y = \phi_j(x_p) \end{array}\right\} \text{ implies } y = \phi_j\phi_i^{-1}(x) .$$

Now consider two vertices x, x' in X_λ with $x \in Y_p$, $x' \in Y_q$. There exists a sequence $(Y_p, Y_{p_1}, Y_{p_2}, \ldots, Y_q)$ such that any two consecutive sets of the sequence intersect. Let $x_k \in Y_{p_{k-1}} \cap Y_{p_k}$. In the sequence $(x, x_1, x_2, \ldots, x_q = x')$, any two consecutive vertices are symmetric. Therefore x and x' are symmetric.

Thus, H_{X_λ} is both edge-transitive and vertex-transitive. □

COROLLARY 1. *If* H *is an edge-transitive hypergraph that is not vertex-transitive, then* H *is bicolorable.*

PROOF. This follows, since the partition of X has at least two classes, and they are both transversal sets of H. □

COROLLARY 2. (DAUBER). *If* H *is an edge-transitive graph that is not vertex-transitive, then* H *is bipartite.*

PROOF. This follows from corollary 1. □

3. EXTENSIONS OF THE WHITNEY THEOREM

Let $G = (E_i : i \in M)$ and $G' = (F_i : i \in M)$ be two connected simple graphs with $|M| = m > 2$. H. WHITNEY [6] has shown that $|E_i \cap E_j| = |F_i \cap F_j|$ for all i,j implies that $G \cong G'$, unless $G = K_3$ and $G' = K_{1,3}$, or vice versa.

An easy corollary of WHITNEY's theorem states that if G and G' are two simple graphs different from K_3 and $K_{1,3}$, then $G - E_i \cong G' - F_i$ for all i implies that $G \cong G'$. (The weak reconstruction conjecture states only that $n(G) > 4$, $G - E_i \cong G' - F_i$ for all i, implies that $G \cong G'$).

In [2], BERGE & RADO have proved several extensions of these theorems for hypergraphs.

Denote by $\mathcal{P}(M)$ the set of all subsets of $M = \{1,2,\ldots,m\}$, by $\mathcal{P}_1(M)$ the set of all subsets $I \subseteq M$ such that $|I| \equiv 1$ modulo 2, and by $\mathcal{P}_0(M)$ the set of all subsets $J \subseteq M$ such that $|J| \equiv 0$ modulo 2 and $J \neq \emptyset$. Clearly, $|\mathcal{P}_1(M)| = 2^{m-1}$ and $|\mathcal{P}_0(M)| = 2^{m-1} - 1$ (because the regular bipartite graph whose vertex-sets are $\mathcal{P}_1(M)$ and $\mathcal{P}_0(M) \cup \{\emptyset\}$ and where (S,T) is an edge iff $-1 \le |S|-|T| \le 1$, has a perfect matching).

The two *Whitney hypergraphs* $W_1(M)$ and $W_0(M)$ are defined as follows: The vertex set of $W_1(M)$ is $\mathcal{P}_1(M)$, and its edges are

$$A_i = \{I : I \in \mathcal{P}_1(M),\ I \ni i\}\ , \qquad (i \in M).$$

The vertex set of $W_0(M)$ is $\mathcal{P}_0(M)$, and its edges are

$$B_i = \{J : J \in \mathcal{P}_0(M),\ J \ni i\}\ , \qquad (i \in M).$$

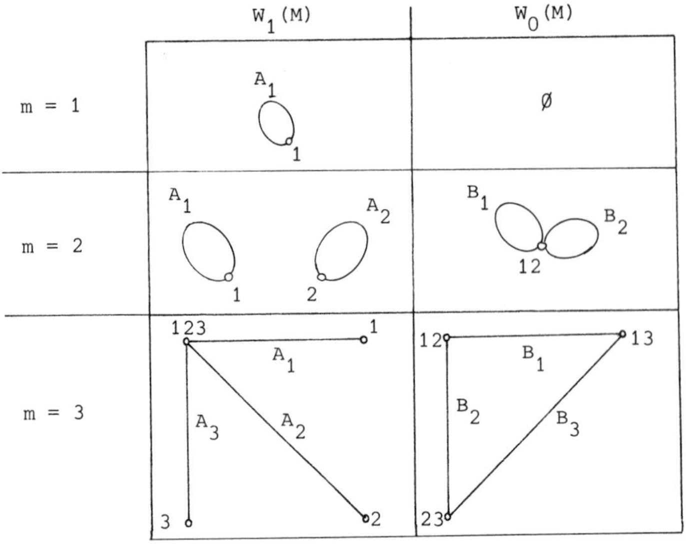

PROPOSITION 7. *For* $m \geq 2$, *the Whitney hypergraphs* $W_1(M) = (A_i : i \in M)$ *and* $W_0(M) = (B_i : i \in M)$ *are two uniform hypergraphs of rank* 2^{m-2}; *their boolean atoms have cardinality one, and they are not isomorphic. However, they satisfy*

$$W_1(M) - A_i \cong W_0(M) - B_i , \qquad (i \in M).$$

PROOF. For $K \subseteq M$, $K \neq \emptyset$, put

$$A_K = \bigcup_{i \in K} A_i ,$$

$$A_{[K]} = \bigcap_{i \in K} A_i ,$$

$$A_\emptyset = \emptyset .$$

Clearly, $W_1(M)$ and $W_0(M)$ are not isomorphic, since $|A_M| = 2^{m-1}$ and $|B_M| = 2^{m-1} - 1$.

For $K \subseteq M$, we have

$$A_{[K]} - A_{M-K} = \{I : I \in \mathcal{P}_1(M),\ I \supseteq K\} - \{I : I \in \mathcal{P}_1(M),\ I \cap (M-K) \neq \emptyset\} =$$

$$= \{K\} \text{ or } \emptyset .$$

If this set of vertices is not empty, it has cardinality one, and it is a boolean atom of $W_1(M)$. Therefore, all the boolean atoms have cardinality one.

Now, let $N = \{2,3,\dots,m\}$, and let us show that $(A_i : i \in N)$ and $(B_i : i \in N)$ are strongly isomorphic.

If $K \subseteq N$, $|K| \equiv 1$ modulo 2, we have

$$A_{[K]} - A_{N-K} = \{K\} ,$$

$$B_{[K]} - B_{N-K} = \{K \cup \{1\}\}$$

(and vice versa if $|K| \equiv 0$ modulo 2). Hence, for all $K \subseteq N$, $K \neq \emptyset$, we have

$$|A_{[K]} - A_{N-K}| = |B_{[K]} - B_{N-K}| = 1 .$$

This shows that $(A_i : i \in N) \cong (B_i : i \in N)$. □

A converse of this proposition is:

THEOREM 2. *Let* $H = (E_i : i \in M)$ *and* $H' = (F_i : i \in M)$ *be two families of subsets* $E_i \subseteq X$ *and* $F_i \subseteq Y$ *(with possibly empty edges or infinite edges), with at least one finite edge, such that*

$$(3.1) \qquad \begin{cases} \text{for all } k, \text{ there exists a bijection } \phi_k : X \to Y \text{ such that} \\ \phi_k(E_i) = F_i \quad (i \in M,\ i \neq k). \end{cases}$$

Then $H \cong H'$, *unless there exist two sets* $A \subseteq X$ *and* $B \subseteq Y$ *with* $|A| = |B|$ *such that* $(A \cap E_i : i \in M) \cong W_\varepsilon(M)$ *and* $(B \cap F_i : i \in M) \cong W_{1-\varepsilon}(M)$.

PROOF. The proof, by induction on m, is the same as in [2, theorem 2]. For the finite case, a direct proof, shorter than our original one, was found recently by LOVÁSZ [5]. □

Note that the statement of theorem 2 would not be true if there is no finite edge: take four infinite sets D_0, D_1, D_2, D_3 of the same cardinality, and put $X = Y = D_0 \cup D_1 \cup D_2 \cup D_3$, $E_1 = D_1 \cup D_2 \cup D_3$, $F_1 = D_2 \cup D_3$, $E_2 = F_2 = D_2$, $E_3 = F_3 = D_3$. $H = (E_1, E_2, E_3)$ and $H' = (F_1, F_2, F_3)$ satisfy (1), and there is no $A \subseteq X$ such that $(A \cap E_i) \cong W_1(1,2,3)$ and no B such that $(B \cap F_i) \cong W_1(1,2,3)$. However, $H \not\cong H'$.

COROLLARY 1. *Let* $|M| \geq p \geq 2$, *and let* $H = (E_i : i \in M)$ *and* $H' = (F_i : i \in M)$ *be two hypergraphs such that*

$$(3.2) \qquad (E_i : i \in I) \cong (F_i : i \in I) , \qquad (I \subseteq M,\ |I| = p-1).$$

Then $H \quad H'$, *unless there exist sets* $A \subseteq E_M$, $B \subseteq F_M$ *and* $P \subseteq M$ *such that* $|P| = p$ *and*

$$(3.3) \qquad \begin{cases} (A \cap E_i : i \in P) \cong W_\varepsilon(P) , \\ (B \cap F_i : i \in P) \cong W_{1-\varepsilon}(P) . \end{cases}$$

<u>PROOF</u>. For $m = p$, consider two hypergraphs H and H' with m edges which satisfy (3.2) and not (3.3).

Let us show first that $|E_M| = |F_M|$; if we have for instance $|E_M| < |F_M|$, consider a set X obtained from E_M by adding $|F_M| - |E_M|$ additional points, and put $Y = F_M$. By theorem 2, there exists a bijection $\phi: X \to Y$ such that $\phi(E_i) = F_i$, and therefore

$$\phi(E_M) = \cup F_i = F_M .$$

This shows that $|E_M| = |F_M|$ which is a contradiction. Thus, $|E_M| = |F_M|$, and theorem 2, applied with $X = E_M$, $Y = F_M$ shows that $H \cong H'$.

Now, let $m = p+t$, $t \geq 1$, and assume that the statement of this corollary is true for hypergraphs with $p+t-1$ edges. Consider two hypergraphs $H = (E_i : i \in M)$, $H' = (F_i : i \in M)$, with $M = \{1,2,\dots,m\}$, satisfying (3.2) but not (3.3). By the induction hypothesis, we have, for $k \in M$,

$$(E_i : i \in M-\{k\}) \cong (F_i : i \in M-\{k\}) .$$

On the other hand, there exist no sets $A_0 \subseteq E_M$, $B_0 \subseteq F_M$, such that

$$(A_0 \cap E_i : i \in M) \cong W_\varepsilon(M) , \quad (B_0 \cap F_i : i \in M) \cong W_{1-\varepsilon}(M) ,$$

because this would imply the existence of two sets A and B satisfying (3.3). Since the theorem is true for $p = m$, we have also $H \cong H'$. □

<u>COROLLARY 2</u>. *Let* $H = (E_i : i \in M)$ *and* $H' = (F_i : i \in M)$ *be two hypergraphs of rank* $h < 2^{p-2}$. *If, for every* $J \subseteq M$ *with* $|J| = p-1$, *we have* $(E_i : i \in J) \cong (F_i : i \in J)$, *then* $H \cong H'$.

<u>PROOF</u>. The proof follows immediately from corollary 1. □

<u>COROLLARY 3</u>. *Let* $H = (E_i : i \in M)$ *and* $H' = (F_i : i \in M)$ *be two multigraphs such that*

(3.4) $|E_i \cap E_j| = |F_i \cap F_j|$ *for all* $i,j \in M$,

(3.5) $\begin{cases} H, H' \text{ do not contain as partial graphs} \\ W_\varepsilon(i,j,k) \text{ and } W_{1-\varepsilon}(i,j,k), \text{ respectively.} \end{cases}$

Then $H \cong H'$.

PROOF. This follows from corollary 1 with $p = 3$. □

The Whitney theorem follows easily from corollary 3 because, if H and H' are connected, of order >4, and if, say, $(E_i,E_j,E_k) \cong W_1(i,j,k)$ and $(F_i,F_j,F_k) \cong W_0(i,j,k)$, then there exists an edge E_q which has exactly one endpoint in $W_1(i,j,k)$; hence

$$|F_i \cap F_q| + |F_j \cap F_q| + |F_k \cap F_q| =$$

$$= |E_i \cap E_q| + |E_j \cap E_q| + |E_k \cap E_q| = 0 \text{ or } 3 ,$$

which is impossible. If H is of order 4 with more than three edges, it is easy to check that $H \cong H'$.

The following result is in fact due to LOVÁSZ [4], who stated it only for graphs.

THEOREM 3. *Let* $H = \{E_i : i \in M\}$ *and* $H' = \{F_i : i \in M\}$ *be* r*-uniform simple hypergraphs of order* n *with* $|M| = m > \frac{1}{2}\binom{n}{r}$, *such that*

$$H - E_i \cong H' - F_i , \qquad (i \in M).$$

Then $H \cong H'$.

PROOF. Denote by $\overline{H} = P_r(X)-H$ the complement hypergraph of H, whose number of edges is

$$m(\overline{H}) = \binom{n}{r} - m < m .$$

We may assume that $X = Y$. If $S \subseteq P_r(X)$, denote by $\alpha(S,H')$ the number of isomorphisms $\pi: X \to Y$ such that $\{\pi S : S \in S\} \subseteq \{F_i : i \in M\}$. By the sieve formula,

$$\alpha(H,H') = \sum_{k=0}^{m} (-1)^k \sum_{\substack{I \subseteq M \\ |I|=k}} \alpha(\{E_i : i \in I\},\overline{H}') .$$

Since the terms with $|I| > m(\overline{H})$ are null,

$$\text{(3.6)} \qquad \alpha(H,H') = \sum_{k=0}^{m-1} (-1)^k \sum_{\substack{I \subseteq M \\ |I|=k}} \alpha(\{E_i : i \in I\}, \overline{H}')$$

and

$$\text{(3.7)} \qquad \alpha(H',H') = \sum_{k=0}^{m-1} (-1)^k \sum_{\substack{J \subseteq M \\ |J|=h}} \alpha(\{F_j : j \in J\}, \overline{H}') \ .$$

Since, by hypothesis, H and H' have the same proper partial hypergraphs, the terms in (3.6) and in (3.7) are equal, hence:

$$\alpha(H,H') = \alpha(H',H') \geq 1 \ . \ \square$$

REFERENCES

[1] BERGE, C., *Graphs and hypergraphs*, North Holland Publ. Coy., Amsterdam, 1973.

[2] BERGE, C. & R. RADO, *On isomorphic hypergraphs, and some extensions of Whitney's theorem to families of sets*, J. Combinatorial Theory B, 13 (1972) 226-241.

[3] DAUBER, E., in: HARARY, F., *Graph theory*, Addison-Wesley, Reading, Mass., 1969, p. 172.

[4] LOVÁSZ, L., *A note on the line reconstruction problem*, J. Combinatorial Theory B, 13 (1972) 309-310.

[5] LOVÁSZ, L., Private communication.

[6] WHITNEY, H., *Congruent graphs and the connectivity of graphs*, Amer. J. Math., 54 (1932) 160-168.

EXTREMAL PROBLEMS FOR HYPERGRAPHS

G.O.H. KATONA

Hungarian Academy of Sciences, 1053 Budapest, Hungary

By a *hypergraph* we mean a pair $(V,\mathcal{A})$, where V is a finite set, and $\mathcal{A} = \{A_1,\ldots,A_m\}$ is a family of its different subsets. $|V|$ means the number of elements of V; this is usually denoted simply by n. Similarly, $|\mathcal{A}| = m$. The elements of V are called *vertices*, the elements of $\mathcal{A}$ are the *edges*.

We use the term hypergraph, because it becomes more and more familiar, but the questions concerned here did not develop directly from the theory of graphs (with some exceptions); the particular cases of these theorems give usually trivialities for graphs.

A hypergraph is a k-*graph* if $|A| = k$ holds for all $A \in \mathcal{A}$. $(V,\mathcal{A})$ is a *complete* k-*graph* if $\mathcal{A}$ consists of all the k-tuples of V.

In this paper we try to give a survey of some extremal problems of hypergraphs, namely, the problems developed from SPERNER's [74] theorem. We shall mention briefly some other areas, too. On the other hand we give some remarks on the possible generalizations for more general structures.

We have the feeling, that the classification of the problems in this paper is not good. However, the various questions are connected in many ways, thus the only proper way of classification would be a graph whose vertices are the problems and the "connected" problems are connected. (The most interesting question concerning this graph would be "how to get nice new vertices?")

For the interested readers it is suggested to read the survey paper of ERDÖS & KLEITMAN [21] on this subject, since our paper contains it only partly.

1. $|V|$ IS FIXED, MAXIMIZE $|\mathcal{A}|$

The typical problem of this type: A set of conditions is given on $\mathcal{A}$, and we are interested in determining the maximum (minimum) of $m = |\mathcal{A}|$ if

M. Hall, Jr. and J. H. van Lint (eds.), Combinatorics, 215-244. All Rights Reserved.

$n = |V|$ is fixed and $(V,\mathcal{A})$ runs over all the possible $\mathcal{A}$'s satisfying the given conditions.

The origin of these theorems is the well-known theorem of SPERNER [74].

THEOREM 1. *If* $(V,\mathcal{A})$ *satisfies* $A_i \not\subset A_j$ $(i \neq j)$, *then*

(1) $$m \leq \binom{n}{[n/2]},$$

where equality holds for the complete $[n/2]$*-graph.*

The following beautiful proof is due to LUBELL.

PROOF. $\mathcal{C} = \{C_0,\ldots,C_n\}$ is called a *complete chain*, if $C_0 \subset C_1 \subset \ldots \subset C_n$. ($\subset$ denotes inclusion without $=$); ($|C_i| = i$ follows). Let us count in two different ways the number of pairs $(\mathcal{C},A_i)$, where $A_i \in \mathcal{A}$ and $A_i = C_j \in \mathcal{C}$ for some j. For a given A_i, C_j must be equal to $C_{|A_i|}$, we have $|A_i|!$ possibilities in choosing $C_0,C_1,\ldots,C_{|A_i|-1}$, and $(n-|A_i|)!$ possibilities for $C_{|A_i|+1},\ldots,C_n$. The number of possible $\mathcal{C}$'s is $|A_i|!\,(n-|A_i|)!$, and the total number of pairs $(\mathcal{C},A_i)$ is $\sum_{i=1}^m |A_i|!\,(n-|A_i|)!$. On the other hand, fixing $\mathcal{C}$, there is at most one A_i since $A_i = C_j \subset C_k = A_l$ would contradict the condition given on $\mathcal{A}$. Thus, the number of pairs $(\mathcal{C},A_i)$ is at most $n!$, the total number of $\mathcal{C}$'s. We obtain the inequality

$$\sum_{i=1}^{m} |A_i|!\,(n-|A_i|)! \leq n!$$

or

(2) $$\sum_{i=1}^{m} \frac{1}{\binom{n}{|A_i|}} \leq 1 .$$

(1) follows from (2) easily, using

$$\binom{n}{|A_i|} \leq \binom{n}{[n/2]} .$$

The proof is completed. □

Equation (2) (which was discovered by LUBELL [67], MESHALKIN [68] and YAMAMOTO [77]) is perhaps more important than (1) itself. If $\sum_{i=1}^m f(|A_i|)$, where f is an arbitrary function, is maximized, then the maximum is attained by the complete k-graph, where k is defined by

$$f(k)\binom{n}{k} = \max_{0\le i\le n} f(i)\binom{n}{i}.$$

The proof of this statement (cf. [58,45]) easily follows from (2)

$$1 \ge \sum_{i=1}^{m} \frac{1}{\binom{n}{|A_i|}} = \sum_{i=1}^{m} \frac{f(|A_i|)}{f(|A_i|)\binom{n}{|A_i|}} \ge \sum_{i=1}^{m} \frac{f(|A_i|)}{f(k)\binom{n}{k}},$$

that is,

$$\sum_{i=1}^{m} f(|A_i|) \le \binom{n}{k} f(k).$$

In some other cases LUBELL's method works again. In order to show, what properties of $\mathcal{C}$ are used in general, (may be) it is worthwhile to formulate the method as a separate lemma. $(W',\mathcal{B}')$ is called a *sub-hypergraph* of $(W,\mathcal{B})$ if $W'\subseteq W$ and $\mathcal{B}'\subseteq\mathcal{B}$. $(W',\mathcal{B})$ is a *spanned sub-hypergraph* if $\mathcal{B} = \{B\colon B\subseteq W',\ B\in\mathcal{B}\}$. We say that U is an *independent set* in $(W,\mathcal{B})$ if $U\subseteq W$, and there is no $B\in\mathcal{B}$ such that $B\subseteq U$.

LEMMA 1. *Let* $(W_1,\mathcal{B}_1),\dots,(W_z,\mathcal{B}_z)$ *be spanned sub-hypergraphs of* $(W,\mathcal{B})$, *the maximal number of independent elements being* $f_1,\dots,f_z$ *and* f, *respectively. Then*

$$f \le \frac{\sum_{i=1}^{z} f_i}{\min_{w\in W} |\{i\colon w\in W_i\}|}. \tag{3}$$

If, additionally, $|W_1| = \dots = |W_z|$, $(W_1,\mathcal{B}_1),\dots,(W_z,\mathcal{B}_z)$ *are isomorphic, and* $|\{i\colon w\in W_i\}|$ *does not depend on* w, *then*

$$\frac{f}{|W|} \le \frac{f_i}{|W_i|}. \tag{4}$$

PROOF. Let $F\subseteq W$ ($|F| = f$) be an independent set in $(W,\mathcal{B})$. Let us count in two different ways the number of pairs $((W_i,\mathcal{B}_i),w)$ where $w\in F$ and $w\in W_i$. For a given $w\in F$ there are $|\{i\colon w\in W_i\}|$ sub-hypergraphs, thus the total number is $\sum_{w\in F} |\{i\colon w\in W_i\}|$. On the other hand, fixing a sub-hypergraph $(W_i,\mathcal{B}_i)$, the maximal number of w's satisfying $w \in W_i$ can be f_i. Thus the number of pairs is at most $\sum_{i=1}^{z} f_i$. The resulting inequality

$$(5) \qquad \sum_{w \in F} |\{i: w \in W_i\}| \le \sum_{i=1}^{z} f_i .$$

However, since

$$f \min_{w \in W} |\{i: w \in W_i\}| \le \sum_{w \in F} |\{i: w \in W_i\}| ,$$

the inequality (3) follows from (5).

Using the additional suppositions

$$\sum_{w \in W} |\{i: w \in W_i\}| = \sum_{i=1}^{z} |W_i| = z|W_i| ,$$

and

$$|\{i: w \in W_i\}| = \frac{z|W_i|}{|W|} .$$

On the other hand $\sum_{i=1}^{z} f_i = zf_i$. Substituting this result into (3) the inequality (4) is obtained, which completes the proof. □

How to apply this lemma to our problems? W equals 2^V (the power set of V) and $\mathcal{B}$ consists of the subsets of 2^V which are excluded by the given condition. If the conditions exclude only elements and pairs of elements of 2^V, then $(W, \mathcal{B})$ is a simple graph. For instance, in the case of SPERNER's theorem: two vertices $A_i, A_j \in W$ are connected iff $A_i \subset A_j$ or $A_j \subset A_i$. For $W_1, \ldots, W_z$ we choose all possible chains $\mathcal{C}$ given in LUBELL's proof. In this case (5) leads to (2), and (3) leads to (1).

The next natural condition (see [19]) for $\mathcal{A}$ is

$$(6) \qquad A_i \cap A_j \ne \emptyset .$$

This question is, however, trivial: $\mathcal{A}$ can contain at most one of the sets A, V-A, thus $|\mathcal{A}| \le$ half the number of all subsets of V:

$$(7) \qquad |\mathcal{A}| \le \frac{2^n}{2} = 2^{n-1} .$$

(The application of lemma 1 gives the same if we take $W_i = \{A_i, V-A_i\}$ for all $A_i \in 2^V$; then (4) gives (7).) This is the best possible bound: $\mathcal{A} = \{A: v\in A, v\in V, v \text{ fixed}\}$ gives equality in (7).

The other classical theorem (ERDÖS, KO & RADO [19]) solves the problem for a combined condition, with a small modification.

THEOREM 2. *If* $(V,\mathcal{A})$ *is a hypergraph satisfying the condition*

(8) $$A_i \not\subset A_j, A_i \cap A_j \neq \emptyset, |A_i| \leq k \text{ if } A_i, A_j \in \mathcal{A} \ (i\neq j),$$

where $k \leq \frac{n}{2}$, *then*

(9) $$m \leq \binom{n-1}{k-1},$$

and this is the best possible bound.

PROOF. First the constructions concerning (9):

$$\mathcal{A} = \{A: |A|=k, v\in A, v\in V, v \text{ fixed}\}.$$

In the proof lemma 1 is used again. W consists of all elements of 2^V having at least k elements. $(W,\mathcal{B})$ is a simple graph. Two different vertices A,A' are connected iff $A \subset A'$, $A \supset A'$ or $A \cap A' = \emptyset$. W_i's are defined in the following way. Let us consider all possible cyclic orderings of V. W_i consists of all subsets of V with size $\leq k$, and with consecutive elements according to the i-th ordering. The $(W_i,\mathcal{B}_i)$'s are isomorphic, f_i does not depend on i.

We shall show that $f_i \leq k$ if $k \leq n/2$. Fix the i-th cyclic ordering $v_1,\dots,v_n$ (the indices are mod n), and suppose $w_1,\dots,w_{f_i}$ are independent vertices in $(W_i,\mathcal{B}_i)$. By the symmetry we can suppose $w_1 = \{v_1,\dots,v_n\}$. If the first and last elements of a w_j are outside w_1 then either $w_j \supset w_1$, or $w_j \cap w_1 = \emptyset$ holds. Then the first or last element of each w_j is in w_1. Fix an l, $(1\leq l<r)$, and consider all sets $A \in W_i$, the last element of which is v_l or the first element of which is v_{l+1}. These vertices are all connected in $(W,\mathcal{B})$ (or in $(W_i,\mathcal{B}_i)$), thus there is at most one w_j among them. Altogether, we have at most (r-1) w_j's with last element from $v_1,\dots,v_{r-1}$ or with first element from $v_2,\dots,v_r$. v_1 can be the first element of w_1, only. (Other $A \in W_i$ with this property either contain or are contained in w_1.)

The same holds for the w_j's having v_1 as a last element. We obtained $f_i \le r \le k$.

We need $|\{i: A \in W_i\}| = |A|!\,(n-|A|)!$. This is simply the number of cyclic orderings in which A has consecutive members. (5) gives the following inequality:

$$\sum_{i=1}^{m} \frac{1}{\binom{n}{|A_i|}} \le \frac{k}{n} \qquad \text{if } k \le \frac{n}{2} \tag{10}$$

and hence, using that in the case $|A_i| \le k \le n/2$,

$$\binom{n}{|A_i|} \le \binom{n}{k}$$

holds, we obtain (9), and the proof is completed. □

This proof is a stronger version of the proof given in [42]. By (10) it is also easy to determine max $\sum_{i=1}^{m} f(A_i)$ under (8).

An obvious question: what happens if the condition $|A_i| \le k$ is omitted (or more generally, $n/2 < k \le n$). If n is odd, then theorem 1 gives the estimation $\binom{n}{(n+1)/2}$, and the complete $\frac{n+1}{2}$-graph satisfies the conditions. The case of even n is solved by BRACE & DAYKIN [2].

Another type of conditions is $A_i \cup A_j \ne V$. This does not seem to be a new condition, since it is equivalent to $(V-A_i) \cap (V-A_j) \ne \emptyset$. However, in some combinations of conditions we can not use the complement sets. For instance if

$$A_i \cap A_j \ne \emptyset \quad \text{and} \quad A_i \cup A_j \ne V,$$

this is the case. Under this condition $m \le 2^{n-2}$, as DAYKIN & LOVÁSZ [12] proved; equality holds with $\mathcal{A} = \{A: v \in A,\ w \notin A$, where $v \ne w$ are fixed elements of $V\}$.

The next type of conditions is the constraint on the sizes of $A_i \cap A_j$ or $A_i \cup A_j$ $(i \ne j)$ (perhaps of $A_i \cap A_j \cap A_l$, and so on). An example: in [19] the following condition is considered

$$|A_i| = k, \quad |A_i \cap A_j| \ge l, \quad (k \ge l). \tag{11}$$

The result [19]: if n is large enough (relatively to k and l), then

(12) $m \le \binom{n-l}{k-l}$,

where equality holds for $\mathcal{A} = \{A\colon L \subset A$ where $|L| = l$, L a fixed subset of $V\}$. The result does not hold for small n, as the following example shows (given by MIN): $n = 8$, $k = 4$, $l = 2$, $\mathcal{A} = \{A\colon |A| = 4, |A \cap \{1,2,3,4\}| = 3\}$, $m = 16 > \binom{6}{2}$. This result gives a good example for the case, that sometimes the exact formulas are valid only for large values.

There is a large class of problems, where the solution (the extremal hypergraph) can be constructed by finite geometries or block designs. We shall not consider these problems, because their methods are completely different from the problems treated here. Thus, we do not investigate (with some exceptions) the conditions of such type, where $|A_i \cap A_j|$ has to be small, or $|A_i - A_j|$ has to be large. However, the questions (11)-(12) give an opportunity for a glimpse at the connections between the two areas. Consider the case $k = 3$, $l = 2$ (in this simple case (12) holds if $n \ge 6$ [39]). A Steiner triple system is a 3-graph $(V, \mathcal{C})$ with the property, that each pair $v,w \in V$ $(v \ne w)$ is contained by exactly one $C \in \mathcal{C}$. It is well known [71], that such a system exists iff $n \equiv 1$ or 3 (mod 6). Use lemma 1; W consists of all the triples of V; w_1 and w_2 are connected in $\mathcal{B}$ iff $|w_1 \cap w_2| < 2$. W_i consists of the triples arising from a fixed Steiner triple system by the i-th permutation of V. It is easy to see, that $(W_i, \mathcal{B}_i)$ is a complete graph, so $f_i = 1$. Trivially, $|W_i| = \binom{n}{2}/3$, $|W| = \binom{n}{3}$, thus (4) gives $f \le n-2$, and this is (12) for $k = 3$, $l = 2$.

By the combinations of the above conditions we obtain a lot of problems. We try to list some of them.

If

(13) $A_i \not\subset A_j$, $A_i \cap A_j \ne \emptyset$, $A_i \cup A_j \ne V$,

then [2] (see also [45,59]) gives

(14) $m \le \binom{n-1}{[(n-2)/2]}$.

If

(15) $|A_i \cap A_j| \ge 1$,

then [39] gives

(16) $$m \le \sum_{i=\frac{l+n}{2}}^{n} \binom{n}{i} \quad \text{if } n+l \text{ is even}$$

and

(17) $$m \le \binom{n-1}{\frac{n+l-1}{2}} + \sum_{i=\frac{n+l+1}{2}}^{n} \binom{n}{i} \quad \text{if } n+l \text{ is odd.}$$

If

(18) $$A_i \not\subset A_j, \ |A_i \cap A_j| \ge l ,$$

then [69] gives

(19) $$m \le \binom{n}{[(n+l+1)/2]} .$$

Let $1 \le k \le n$ and $1 \le h \le \min(k,n-k)$, and suppose

(20) $$A_i \cap A_j \ne \emptyset, \quad h \le |A_i| \le k,$$

then [36] gives

(21) $$m \le \sum_{i=h}^{k} \binom{m-1}{i-1}.$$

If $1 \le k \le n$, and there is no pair $i \ne j$ such that

(22) $$A_i \supset A_j \quad \text{and} \quad |A_i - A_j| \ge k,$$

then [17] gives

(23) $m \le$ (the sum of k largest binomial coefficients of order n).

Conversely, if there is no pair satisfying

(24) $$A_i \supset A_j \quad \text{and} \quad |A_i - A_j| < k ,$$

then [43] gives

(25) $$m \le \sum_{i \equiv [n/2] \pmod k} \binom{n}{i} .$$

Concerning the combinations which are missing, three cases can happen.

1) It is an easy consequence of another one.
2) The author of this paper does not know the result.
3) It is a nice open problem.

An example for case 3):

If $|A_i \cap A_j| \ge 1$ but there is no pair with $A_i \cup A_j = V$, then probably the inequalities (16) and (17) hold with $n-1$ rather than n. (We can not give examples for case 2).)

2. CONDITIONS VARYING ON A WIDER SCALE

In this section we consider the same kind of problems as in section 1, but the *conditions vary on a wider scale.*

The most general form of theorem 2 (and (11)-(12)) is the following theorem of HAJNAL & ROTHSCHILD [29].

If

(26) $$\begin{cases} |A_i| = k, \text{ and for any } i_1, \ldots, i_{r+1} \\ \text{there are } i_j \text{ and } i_h \text{ with } |A_{i_j} \cap A_{i_h}| \ge l, \end{cases}$$

then

(27) $$m \le \sum_{i=1}^{r} (-1)^{i+1} \binom{r}{i} \binom{n-il}{k-il},$$

provided n *is large enough* ($n \ge n(k,r,l)$).

What are the best values for $n(k,r,l)$? By theorem 2, $n(k,1,1) = 2k$. For the cases of $n(k,1,l)$ we can not expect a nice smallest value. The estimations of [19] are improved in [37]. The hopeful case is $n(k,r,1)$. For instance, $n(k,2,1) = 3k+1$ might be true.

The same question without $|A_i| = k$, and only for $l = 1$ is solved by KLEITMAN [55]. So, if for any $i_1, \ldots, i_{r+1}$ there is a pair i_j, i_h such that

$$(28) \qquad A_{i_j} \cap A_{i_h} \neq \emptyset$$

and $n = (r+1)q$, then

$$(29) \qquad m \leq \sum_{i=q+1}^{(r+1)q} \binom{(r+1)q}{i} + \binom{(r+1)q}{q} \frac{r}{r+1}.$$

If $n = (r+1)q-1$, another exact estimation is given. For other n's there is a small gap between the estimations and the constructions [55].

An obvious open question is the case $l > 1$ ($|A_{i_j} \cap A_{i_h}| \geq l$). This is solved only for $r = 1$ (see (15)-(17)).

A third variant of these questions was posed by D. PETZ and solved by P. FRANKL [27] (students in Budapest):

If

$$(30) \qquad A_i \not\supset A_j \quad \text{and} \quad |A_{i_1} \cup \ldots \cup A_{i_r}| \leq qr+s$$

where $0 \leq s < r$, then

$$(31) \qquad m \leq \sum_{i=0}^{\min(q,s/2)} \binom{n-s}{q-i} \binom{s}{[s/2]+i},$$

provided n is large enough depending on r and qr+s. The construction: let $C \subset V$, $|C| = s$, then $\mathcal{A} = \{A\colon |A| = q+[s/2], |A \cap (V-C)| \leq q\}$. The cases $s = 0$ and $s = 1$ are solved independently by E. BOROS. Observe that (30)-(31) is a generalization of (18)-(19) using the complement set. In (18)-(19) $r = 2$, $s = 0$ or 1.

It seems that in (9) equality can hold ($k < n/2$) only for the given extremal hypergraph (all the A's containing a given $v \in V$). In [19] it is asked, what happens, if we exclude this extremal hypergraph, or suppose $\bigcap_{i=1}^{m} A_i = \emptyset$. HILTON & MILNER [33] have given the answer:

$$(32) \qquad m \leq 1 + \binom{n-1}{k-1} - \binom{m-k-1}{k-1}.$$

They have more general theorems, too: If $1 \leq \min(3,s+1) \leq k \leq n/2$, and $|A_i| \leq k$, $A_i \not\subset A_j$ $(i \neq j)$, $A_i \cap A_j \neq \emptyset$,

$$(33) \qquad A_{i_1} \cap \ldots \cap A_{i_{m-s+1}} = \emptyset$$

for any different indices $i_1,\dots,i_{m-s+1}$, then

(34)
$$m \le \begin{cases} \binom{n-1}{k-1} - \binom{n-k}{k-1} + n-k & \text{if } 2 < k \le s+2, \\ s + \binom{n-1}{k-1} - \binom{n-k}{k-1} + \binom{n-k-s}{k-s-1} & \text{if } k \le 2 \text{ or } k \ge s+2. \end{cases}$$

A combination of (32) and (20)-(21) is given in [36]:

Let $1 \le k \le n-1$, $1 \le h \le \min(k,n-k)$. If

(35)
$$A_i \cap A_j \ne \emptyset,\ h \le |A_i| \le k \quad \text{and} \quad \bigcap_{i=1}^{m} A_i = \emptyset,$$

then

(36)
$$m \le 1 + \sum_{i=h}^{k} \left[\binom{n-1}{i-1} + \binom{n-k-1}{i-1} \right].$$

The following three results [36] are modifications of the above ones, when besides $A_1,\dots,A_m$ there is an additional edge B of our hypergraph with slightly different conditions. Let h and k satisfy $1 \le k \le n/2$, $1 \le h \le n-1$. If $A_i \cap A_j \ne \emptyset$, $A_i \cap B \ne \emptyset$, $|A_i| \le k$, $|B| = h$, $A_i \not\subset A_j$, $A_i \not\subseteq B$ ($B \subset A_i$ is not excluded), then

$$m \le \begin{cases} \binom{n-1}{k-1} - \binom{n-h-1}{k-1} & \text{if } h \ge n/2, \\ \binom{n-1}{k-1} - \binom{h-1}{k-1} & \text{if } h < n/2. \end{cases}$$

If $A_i \cap A_j \ne \emptyset$, $A_i \cap B \ne \emptyset$, $|A_i| \le k$, $|B| = h$, $A_i \not\subset A_j$, $B \cap A_1 \cap \dots \cap A_m = \emptyset$, then

$$m \le \begin{cases} \binom{n-1}{k-1} - \binom{n-h-1}{k-1} & \text{if } k \le h, \\ 1 + \binom{n-1}{k-1} - \binom{n-k-1}{k-1} & \text{if } h < k. \end{cases}$$

Finally, if $A_i \cap A_j \ne \emptyset$, $A_i \cap B \ne \emptyset$, $g \le |A_i| \le k$, $|B| = h$, $B \cap A_1 \cap \dots \cap A_n = \emptyset$, then

$$m \le \begin{cases} \sum_{i=g}^{k} \left[\binom{n-1}{i-1} - \binom{n-h-1}{i-1} \right] & \text{if } k \le h, \\ 1 + \sum_{i=g}^{k} \left[\binom{m-1}{i-1} - \binom{m-k-1}{i-1} \right] & \text{if } h < k. \end{cases}$$

In a paper of HILTON [35] the concept of the *simultaneously disjoint pairs of edges* is defined by

$$A_{i_1} \cap A_{j_1} = \ldots = A_{i_s} \cap A_{j_s} = \emptyset .$$

Let $2 \le 2k \le m$ and $s \le \binom{n-k-1}{k-1} - 1$. If

$$(37) \qquad \begin{cases} |A_i| \le k,\ A_i \not\subset A_j \text{ and there are no } s+1 \\ \text{simultaneously disjoint pairs of edges,} \end{cases}$$

then (cf. [35])

$$(38) \qquad m \le \binom{n-1}{k-1} + s.$$

If $A_i \subset A_j$ is allowed, then we obtain (cf.[35])

$$m \le \sum_{i=1}^{k} \binom{n-1}{i-1} + s.$$

Or more generally [32], if $h \le |A_i| \le k$, then

$$m \le \sum_{i=h}^{k} \binom{n-1}{i-1} + s.$$

(For a recent result of this type see [11].)

Similar results are obtained in [32] in case we exclude the existence of s+1 simultaneously disjoint r-tuples of edges. For s = 0 it was solved earlier by ERDÖS & GALLAI [18] for 2-graphs and later by ERDÖS [20] for k-graphs. ERDÖS' case is also included by (26)-(27), but the common generalization of HAJNAL & ROTHSCHILD [29] and HILTON [32] is still open.

3. WEAKENING THE CONDITIONS

Could we weaken the conditions of our theorems with the same conclusions? In this section we give examples for that.

First KLEITMAN [49] and KATONA [40] independently observed, that if we fix a partition $V_0 \cup V_1 = V$, $(V_0 \cap V_1 = \emptyset)$ of V and we exclude the edges satisfying

(39) $$A_i \cap V_\delta = A_j \cap V_\delta \quad \text{and} \quad A_i \cap V_{1-\delta} \subset A_j \cap V_{1-\delta}$$

(instead of $A_i \subset A_j$), then under this weaker condition the conclusion

(40) $$m \le \binom{n}{[n/2]}$$

remains the same.

A natural question: what happens for the partition $V_0 \cup V_1 \cup V_2 = V$ (V_0, V_1, V_2 are pairwise disjoint), if we exclude edges equal in two V_i's and containing each other in the third? The answer is disappointing: m can be larger than $\binom{n}{[n/2]}$. In [47] an additional condition is given, under which (40) remains true. This additional condition is rather complicated. It excludes some 4-tuples of edges of the hypergraph. Recently, GREENE & KLEITMAN [28] determined weak conditions from the symmetric chain method (see [3]).

A combination of (39) and (22) is given in [44], and a combination of (39) and (24) in [43]. Recent generalizations of this type can be found in [60].

A question: how could we weaken the conditions of theorem 2 with the same conclusion?

4. ONE CONDITION CONTAINING MORE OPERATIONS OR RELATIONS

In this section we treat the problems where *one condition contains more operations or relations*.

Probably the oldest result of this type is due to KLEITMAN [56]. If there is no triple satisfying

$$A_i \cap A_j = \emptyset \quad \text{and} \quad A_i \cup A_j = A_h$$

simultaneously, then

$$m \le \sum_{i=r+1}^{2r+1} \binom{n}{i},$$

provided $n = 3r+1$, and this is the best estimation. For $n = 3r$ and $n = 3r+2$ the results are near best possible.

Another problem: there are no 4 different edges in the hypergraph satisfying both

$$A_i \cup A_j = A_k \quad \text{and} \quad A_i \cap A_j = A_l.$$

ERDÖS & KLEITMAN [24] have constructed $c_1 \frac{2^n}{n^{\frac{1}{4}}}$ edges with this condition and they proved that

$$m \le c_2 \frac{2^n}{n^{\frac{1}{4}}}$$

but $c_1 < c_2$.

Many obvious general questions can be asked.

In the next problem $|V|$ is not fixed, but we list it here, because its character is similar to the other problems treated here. Now $|A| = m$ is fixed and $f(m)$ is the largest number such that there are always $f(m)$ edges in the hypergraph no three different ones of them having the property

$$A_i \cup B_i = C_i.$$

The first result is given by KLEITMAN [50]:

$$f(m) \le cm\sqrt{\log m}\ .$$

J. RIDDEL proved $\sqrt{m} < f(m)$, and finally ERDŐS & KOMLÓS [22] determined

$$f(m) < 2\sqrt{2n}+4.$$

BOLLOBÁS proved for 3-graphs that if

(41) there are no three different edges $A_h \supseteq (A_i - A_j) \cup (A_j - A_i)$,

then

$$m \leq \left(\frac{n}{3}\right)^3$$

if $3|n$. The hypergraph with equality: V is divided into 3 equal parts, and we choose the edges having exactly one vertex from each part.

It is conjectured also by BOLLOBÁS, that a similar theory might be true for k-graphs. For 2-graphs it is a particular case of TURÁN's graph theorem [76]. A conjecture of ERDÖS & KATONA: Under the condition (41) (without size restrictions) the best hypergraph can be constructed in the following way. Divide V into $[\frac{n}{3}]$ classes of 3 and 2 elements, and choose those edges which contain exactly one vertex from each class.

5. MISCELLANY

We will treat three further problems which do not really fit into any of these sections. The first question was proposed by RÉNYI [70]. The edges of the hypergraph are called *qualitatively independent* if

(42) $A_i \cap A_j,\quad A_i \cap \bar{A}_j,\quad \bar{A}_i \cap A_j,\quad \bar{A}_i \cap \bar{A}_j$

are all non-empty. What is the maximum of m under this condition? The answer is

$$m \leq \binom{n-1}{[(n-2)/2]}.$$

This is an easy consequence of theorems 1 and 2, as it is pointed out by KLEITMAN & SPENCER [59] and independently in [45]. (Observe, that (13) and (42) are equivalent, thus [2] also gives the solution.) In [59] a harder problem is also considered. We say, the edges are k-*qualitatively independent* if

$$A_{i_1}^{\delta_1} \cap \ldots \cap A_{i_k}^{\delta_k} \neq \emptyset$$

for any different $i_1,\dots,i_k$, where A^δ is either A or $\bar{A} = V-A$. Under this condition

$$m \le 2^{c\frac{n}{2^k}}$$

and a hypergraph is constructed with

$$2^{d\frac{n}{k2^k}}$$

edges, where c and d are constants, k fixed and $n \to \infty$.

An unsolved question: maximize m under the condition that any of (42) has a size $\ge r$.

The density of a hypergraph was defined by ERDÖS. It is the largest s such that there is a $U \subseteq V$ such that $|U| = s$ and $|A \cap U| = 2^s$. SAUER [72] proved, that supposing

$$s \le k,$$

we obtain

$$m \le \sum_{i=0}^{k} \binom{n}{i}.$$

A similar problem of ERDÖS & KATONA: what is the maximum of m under the condition that $|A_i \cap A_j|$ are all different $(1 \le i < j \le m)$?

A new area of problems is considered in [2]. The valency $v = v((V,A))$ of a hypergraph is the minimal valency of its vertices. In [2] the maximum of v is asked for under several conditions.

If $A_i \cap A_j \neq 0$, then

$$v \le \begin{cases} 2^{n-2} + \frac{1}{2}\binom{n-1}{(n-1)/2} & \text{if n is odd,} \\ 2^{n-2} + [\frac{1}{4}\binom{n}{n/2}] & \text{if n is even.} \end{cases}$$

If $A_i \not\subseteq A_j$, then

$$v \le \binom{n-1}{[(n-1)/2]}$$

and if $A_i \not\subset A_j$, $A_i \cap A_j = \emptyset$, then the same holds.

6. THE PROBLEMS WE SHALL NOT CONSIDER HERE

These problems -although they have many points in common with our subject- require different methods, and are approached from various points of view. These problems are also extremal problems for hypergraphs, but this concept is too wide.

1) If $A_i \cap A_j$ is small, $A_i - A_j$ or $(A_i - A_j) \cup (A_j - A_i)$ are large, the problems are usually *coding problems*. Their methods are closer to block designs and finite geometries.

2) *Covering problems*. Usually a smallest family of edges is sought under some conditions, covering all the edges of a given hypergraph. In 1) and 2) the solutions give hypergraphs where the edges are "far" from each other, in our cases they are "close".

3) *Ramsey type theorems*. See the paper of GRAHAM & ROTHSCHILD in this tract (pp. 61-76).

4) *Turán type theorems*. Certain generalizations are very near (see [46]).

5) *Combinatorial search problems*. They are closely related to the coding problems (see [46]).

6) We did not touch the question of the number of optimal hypergraphs. In many cases there is only one. In some other cases it is an open problem how many of them exist. A closely related problem: how many hypergraphs do we have under several conditions? For these questions see [21].

7) *Δ-systems and* B-*property*. A hypergraph is a *strong Δ-system* if $A_i \cap A_j$ $(i \neq j)$ does not depend on i and j. In the case of a *weak Δ-system* $|A_i \cap A_j|$ $(i \neq j)$ is independent of i and j. $f_s(k,l)$ denotes the minimum of $|A|$ with the property that in the case $|A_i| = k$, $(1 \leq i \leq m)$, there are always l A_i's forming a strong Δ-system. $f_w(k,l)$ denotes the same for weak Δ-systems. There are lower and upper estimations for $f_s(k,l)$ and $f_w(k,l)$. We say that (V,A) has *property* B, if there is a set $B \subset V$ such that $|B \cap A_i| \geq 1$ but $B \not\supset A_i$ $(1 \leq i \leq m)$. The questions concerning Δ-systems and the B-property are closely related to our problems; however, ERDÖS [25] has recently published a survey paper on this subject.

7. $|\mathcal{A}|$ IS FIXED

Perhaps, the main feature of the problems in this section is not $|\mathcal{A}|$ being fixed, because in many cases we obtain an inequality and in an inequality usually it is not important, which variable is fixed and which one is not. However, the problems treated here -as we shall see- have a definitely different character.

SPERNER's theorem says, if we have $\binom{n}{[n/2]}+1$ edges in a hypergraph (with $|V|=n$), then there is a pair of different edges $A_i \subset A_j$. Observe, however, that adding one edge to the complete $[n/2]$-graph there are always more pairs with $A_i \subset A_j$. What is the minimum? More generally, if m and n are fixed, what is the minimal number of pairs $A_i \subset A_j$? The solution is given by KLEITMAN [51]. The optimal hypergraph is constructed easily. Order all subsets of V, first take all $[n/2]$-tuples, then all $[n/2]+1$-tuples, all $[n/2]-1$-tuples, all $[n/2]+2$-tuples, and so on. The edges of the optimal hypergraph are the first m subsets according to this order.

The corresponding question is not solved yet, not even for the case of (15)-(16). This latter one can not be too hard for $l=1$. The optimal hypergraph could be constructed by taking the subsets of V according to their sizes, starting from n. (For the case of theorem 2 see later in this section.)

Let $(V,\mathcal{A})$ now be a k-graph, and let $\mathcal{C}(\mathcal{A})$ denote the family of subsets $C: C\subset A$ for an $A\in\mathcal{A}$ and $|C|=k-1$. SPERNER [74] used in his proof the easy fact

$$|\mathcal{C}(\mathcal{A})| \geq \frac{|\mathcal{A}|\cdot k}{n-k+1}\ .$$

The question arises, what is $\min|\mathcal{C}(\mathcal{A})|$ if n,k,m are fixed ($m \leq \binom{n}{k}$). The construction of the optimal k-graph is as follows. Fix an order $v_1,\dots,v_n$ of the vertices in V. Form a sequence of 0's and 1's in the usual way from each k-set of V. The first m sequences in the lexicographic order give the optimal k-graph. A formula can also be given for $\min|\mathcal{C}(\mathcal{A})|$. There is a unique expression of the form

$$m = \binom{a_k}{k}+\binom{a_{k-1}}{k-1}+\dots+\binom{a_t}{t}\ , \tag{43}$$

where $t\geq 1$, $a_k > a_{k-1} > \dots > a_t$ and $a_i \geq i$. Then

(44) $$\min|\mathcal{C}(\mathcal{A})| = \binom{a_k}{k-1}+\ldots+\binom{a_t}{t-1} = f_k(m).$$

The result is more clear if m has the form $\binom{a_k}{k}$. Then we have for the optimum a complete k-graph in $V' \subset V$ where $|V'| = a_k$. An interesting thing: (44) does not depend on n. (44) was first proved by KRUSKAL [63]. Some years later it was rediscovered in [41]. Then CLEMENTS & LINDSTRÖM [4] proved a more general theorem by a different method. They also proved the theorem independently, but they found [41] and [63] before publishing it. HANSEL [30] also has a paper, and recently DAYKIN [13] found a relatively short proof.

A similar result was found earlier by KLEITMAN [52]: If $(V,\mathcal{A})$ is a hypergraph with $A_i \not\supseteq A_j$ and $|\mathcal{A}| = \binom{n}{k}$ then the number of different sets C for which there exists an $A_i \in \mathcal{A}$ with $C \subseteq A_i$ is at least

(45) $$\sum_{i=0}^{k} \binom{n}{i}.$$

This question was solved for any m by CLEMENTS [9], using (44). In this solution only an algorithm is given determining the optimal $\mathcal{A}$, no formula of type (45) is given for the minimum in general. This remains open. [9] also contains useful inequalities concerning (44).

There are a lot of other consequences of (44). E.g., recently DAYKIN [14] observed that theorem 2 (ERDÖS, KO & RADO) follows from (44). Now we give some examples, where (44) is used in the proof.

Let $(V,\mathcal{A})$ be a k-graph. A (k-1)-*representation* of $(V,\mathcal{A})$ is a set $\{B_1,\ldots,B_m\}$ of (k-1)-tuples such that $B_i \subset A_i$ $(1 \le i \le m)$. ERDÖS asked what is the maximal m for which any $(V,\mathcal{A})$ with $|\mathcal{A}| = m$ has a (k-1)-representation. The answer [41] is

$$m = \binom{2k-1}{k}+\binom{2(k-1)-1}{k-1}+\ldots+\binom{1}{1}.$$

From inequality (2) it is trivial that if we modify the conditions of theorem 1 in such a way that $|A_i| = n/2$ (let n be even) is excluded, then $m \le \binom{n}{(n/2)-1}$ and this is the best. However, if we describe the number of edges A_i with $|A_i| = n/2$ (and this number is > 0, but < $\binom{n}{n/2}$), then usually we do not obtain an exact estimation for m. This question was solved in

[10] more generally: $p_j = |\{A: A\in\mathcal{A}, |A|=j\}|$ are called the *parameters* of a hypergraph. Let $0 \le i_0 < n$ and the parameters $p_{i_0} \ne 0, p_{i_0+1},\dots,p_n$ be fixed. $\max|\mathcal{A}|$ is determined under this condition, provided $A_i \not\supseteq A_j$.

Another formulation is given independently by DAYKIN, GODFREY & HILTON [15]: If $p_0 = 0$, $p_1,\dots,p_g > 0$ are given integers, then the least integer n such that there exists a hypergraph $(V,\mathcal{A})$ with $|V| = n$, $A_i \not\supseteq A_j$ and with the parameters $p_0,p_1,\dots,p_g$ is

$$n = p_1+f_2(p_2+f_3(p_3+\dots+f_g(p_g)\dots)),$$

where f_i is defined in (44).

[15] solves a conjecture of KLEITMAN & MILNER, too: If $(V,\mathcal{A})$ satisfies $A_i \not\supseteq A_j$ and has the parameters $p_0,p_1,\dots,p_n$, then there is an other hypergraph $(V,\mathcal{A}')$ satisfying $A'_i \not\supseteq A'_j$ and with parameters $0,\dots,0,p_{n/2},p_{n/2+1}+p_{n/2-1}, \dots,p_n+p_0$ (if n is odd, then the middle is: $\dots,0,0,p_{\frac{1}{2}(n+1)}+p_{\frac{1}{2}(n-1)}, p_{\frac{1}{2}(n+3)}+ +p_{\frac{1}{2}(n-3)},\dots$).

Let the parameters $p_l,\dots,p_{l+r}$ be fixed. What is the minimal number of $(l-1)$-tuples contained in any edge $A \in \mathcal{A}$? This is answered in [8].

CLEMENTS [11] dealt with the problem what happens in theorem 2 if we take more edges than $\binom{n-1}{k-1}$. However, he did not minimize the number of disjoint pairs, but maximized the number of edges meeting all other edges of $(V,\mathcal{A})$.

As is clear from the examples, (44) is almost necessary if containment is involved and the optimal arrangement does not consist of complete i-graphs. We had to write "almost", because KLEITMAN's result in [51] is an exception.

Another type of problems where $|\mathcal{A}|$ is fixed: what is the maximal number of pairs $A_i \supset A_j$, $|A_i - A_j| = 1$? An extremal hypergraph can be constructed by choosing the first $m = |\mathcal{A}|$ edges according to the lexicographic order. This is proved in [1,31,65]. However, as CLEMENTS [6] pointed out it is an easy consequence of (44).

$\min|\mathcal{C}(\mathcal{A})|$ can be asked for under several conditions. For instance in [39] it is tried to do this supposing $|A_i \cap A_j| \ge l$ ($|A_i| = k$ remains true, $l < k$). However, only $\frac{|\mathcal{C}(\mathcal{A})|}{|\mathcal{A}|}$ is minimized. The optimal hypergraph is a complete k-graph on a $(2k-l)$-element subset of V. For fixed $|\mathcal{A}|$, the hypergraph minimizing $|\mathcal{C}(\mathcal{A})|$ seems rather complicated, but it is regular enough to have some hope for the solution.

P. FRANKL asked the following question of similar type. If $|\mathcal{A}|$ is fixed,

$|A| = k$ for $A \in \mathcal{A}$, what is the minimum of $(2k-1)$-tuples contained in a union $A_i \cup A_j$ $(A_i, A_j \in \mathcal{A})$?

8. MORE HYPERGRAPHS

In these problems we have more hypergraphs with the same vertex set. Usually it is supposed that the hypergraphs do not have common edges. The conditions and the questions are usually similar to those in the above sections.

The first result was achieved by ERDÖS [17]. If the hypergraphs $(V,\mathcal{A}_1),\dots,(V,\mathcal{A}_d)$ satisfy the condition

$$A_i \not\subseteq A_j, \quad A_i, A_j \in \mathcal{A}_h \quad (1 \le h \le d)$$

(and $\mathcal{A}_i \cap \mathcal{A}_j = \emptyset$ $(i \ne j)$), then

$$\sum_{i=1}^{d} |\mathcal{A}_i| \le \text{(the sum of the d largest binomial coefficients of order n)}. \tag{46}$$

By the same proof as in the case of theorem 1 we obtain the inequality

$$\sum_{i=1}^{d} \sum_{A \in \mathcal{A}_i} 1/\binom{n}{|A|} \le d, \tag{47}$$

where simply the hypergraph $(V, \bigcup_{i=1}^{d} \mathcal{A}_i)$ was considered; thus one chain $\mathcal{C}$ can contain at most d A's. (47) is equivalent to $\sum_{k=0}^{n} x_k/\binom{n}{k} \le d$, where x_k denotes the number of A's with $|A| = k$. It is clear, that under this inequality $\sum_{k=0}^{n} x_k$ is maximal if we take the maximal values of the x_k's with minimal coefficients, thus $x_k = \binom{n}{k}$ for the d middle k's and 0 otherwise.

The next question, what is max $\sum_{i=1}^{d} |\mathcal{A}_i|$, if the $\mathcal{A}_i$'s are disjoint and $A_j \cap A_h \ne \emptyset$, $A_j, A_h \in \mathcal{A}_i$ $(1 \le i \le d)$. The answer was found by KLEITMAN [53]:

$$\sum_{i=1}^{d} |\mathcal{A}_i| \le 2^n - 2^{n-d}.$$

The corresponding question for theorem 2 is unsolved. A problem of KNESER [62] is the following. If $(V,\mathcal{A}_1),\dots,(V,\mathcal{A}_d)$ are k-graphs $(k < n/2)$ $A_i \cap A_j \ne \emptyset$ for $A_i, A_j \in \mathcal{A}_h$, $\mathcal{A}_i \cap \mathcal{A}_j = \emptyset$ and $(V, \bigcup_{h=1}^{d} \mathcal{A}_h)$ is the complete k-graph, what is the minimum of d under these conditions?

Another line was started by HILTON & MILNER [33]. Let $(V,\mathcal{A})$ and $(V,\mathcal{B})$ be two hypergraphs such that

$$|A_i| \le k,\ |B_i| \le k,\ A_i \cap B_j \ne \emptyset,\ A_i \not\subset A_j,\ B_i \not\subset B_j;$$

then supposing $p \le |\mathcal{A}|, |\mathcal{B}|$ and $1 \le \min(2,p) \le k \le n/2$,

$$|\mathcal{A}|+|\mathcal{B}| \le \begin{cases} \binom{n}{k} - \binom{n-k+1}{k} + n-k+1 & \text{if } 1 < k \le p+1, \\ p + \binom{n}{k} - \binom{n-k+1}{k} + \binom{n-k-p+1}{k-p} & \text{otherwise,} \end{cases}$$

holds. HILTON [34] generalized, for the case $|B_i| \le l \ne k$, KLEITMAN's result [57] on the same subject: $(V,\mathcal{A})$ and $(V,\mathcal{B})$ are hypergraphs satisfying

$$|A_i| = k,\ |B_i| = l,\ k+l \le n,\ A_i \cap B_j \ne \emptyset$$

then either

$$|\mathcal{A}| \le \binom{n-1}{k-1}$$

or

$$|\mathcal{B}| \le \binom{n-1}{l-1} - \binom{n-1-k}{l-1}.$$

EHRENFEUCHT & MYCIELSKI [16] conjectured that if the hypergraphs satisfy

$$|A_i| = k\ (A_i \in \mathcal{A}),\ |B_i| = l\ (B_i \in \mathcal{B}),\ |\mathcal{A}| = |\mathcal{B}| = m$$

and

$$A_i \cap B_j \ne \emptyset \quad \text{iff } i \ne j$$

then

$$m \le \binom{k+l}{l}. \tag{48}$$

It is proved in [48]. T. TARJÁN [75] modified the proof yielding a stronger result:

Let $(V,\mathcal{A})$ and $(V,\mathcal{B})$ be two hypergraphs with $|\mathcal{A}| = |\mathcal{B}|$ and

$$A_i \cap B_j \neq \emptyset \quad \text{iff } i \neq j. \tag{49}$$

Lemma 1 will be applied for the following graph. W consists of all pairs (S,T) where $S,T \subseteq V$, $S \cap T = \emptyset$, and two distinct vertices $(S_1,T_1),(S_2,T_2)$ are connected iff one of the sets $S_1 \cap T_2, S_2 \cap T_1$ is empty. Fix an order on the elements in V. Let W_i consist of those vertices (S,T) in which all elements of S precede all elements of T according to the i-th permutation of the elements of V. Observe that W_i spans a complete graph. That means $f_i = 1$. We need the number

$$|\{i:\ (S,T)\in W_i\}| = \binom{n}{|S|+|T|}|S|!\,|T|!\,(n-|S|-|T|)! = \frac{n!\,|S|!\,|T|!}{(|S|+|T|)!}.$$

From inequality (5) we obtain

$$\sum_{i=1}^{m} \frac{1}{\binom{|A_i|+|B_i|}{|A_i|}} \leq 1. \tag{50}$$

If $|A_i| = k$, $|B_i| = l$, (48) trivially follows. Other variants follow, too. E.g. if $|A_i|+|B_i| \leq k$ then

$$m \leq \binom{k}{[k/2]}.$$

9. n-DIMENSIONAL LATTICE-POINTS

SPERNER's question can be formulated in the following way. A square-free integer $N = p_1p_2\ldots p_n$ is given; what is the maximal number of its divisors not dividing each other? After answering this question it is a must to answer the same for arbitrary $N = p_1^{\alpha_1} \ldots p_n^{\alpha_n}$, too. The divisors of N have the form $p_1^{x_1} \ldots p_n^{x_n}$, where $0 \leq x_i \leq \alpha_i$ $(1 \leq i \leq n)$. Thus, with the divisors we can associate the lattice-points of an $(\alpha_1+1)\times\ldots\times(\alpha_n+1)$ n-dimensional parallelotope. All questions can be extended to n-dimensional parallelotopes in this way. Some of these extensions are motivated by other applications.

If the character of the problem is such that in the parallelotopes there do not appear new phenomena (compared to the hypergraphs), then it is easier to start making a conjecture and proof for the 2- and 3-dimensional parallelotopes, since they are more graphic.

We briefly list the results which are generalizations of this type.

SPERNER's theorem was generalized in [3]. The bound for m is the maximal number of lattice-points with a fixed coordinate sum ($= [\sum \alpha_i/2]$).

SCHÖNHEIM [73] generalized (46) and (39)-(40). In [44] the common generalization is given. (25) is generalized in [43].

ERDÖS & SCHÖNHEIM [26], further ERDÖS, HERZOG & SCHÖNHEIM [23] have investigated the generalization of (6). The max of m is not equal to the minimal m for which there exists an m-element set of divisors such that any other divisor is coprime to one of them. Both values are determined.

An analogue of (15)-(16) is generalized in [54].

The analogue of (44) is proved in [4]. Of course, there are no formulas, but it is proved that one of the optimal sets of lattice-points gives the first m in the lexicographic order. Other results concerning this theorem can be found in [7]. [5] gives the generalization of ERDÖS' problem of (k-1)-representation of k-edges. [8] also concerns this generalization.

In [6] CLEMENTS shows, that the theorem of LINDSEY [65] (which maximizes the pairs of neighbouring lattice-points if their number is given) is an easy consequence of the generalized formula (44). Recently KLEITMAN, KRIEGER & ROTHSCHILD [61] determined the maximal number of such pairs which differ only in one coordinate.

LINDSTRÖM [66] solved an interesting question of KRUSKAL [64], which is an analogue of (44). A hypergraph can be imagined as a set of certain faces of an (n-1)-dimensional simplex. Thus, if we fix the number of (k-1)-dimensional faces, then (44) gives the minimal number of (k-2)-dimensional subfaces. LINDSTRÖM solved the same question for more-dimensional cubes.

10. FURTHER ANALOGUES AND GENERALIZATIONS

There is an attempt to put these combinatorial theorems in a more general -algebraic- form. Most results concern SPERNER's theorem and close modifications. All these papers state the theorems for certain partial orders. We do not even give the list of these papers because KLEITMAN's paper in this tract contains it. The results contain all important combinatorial analogues of SPERNER's theorem with one exception: the partitions of a finite set under refinement.

For generalizing other problems, there is only one result by HSIEH [38]. It solves an analogue of theorem 2: what is the maximal number of k-dimensional non-disjoint subspaces? And what is interesting, the harder problem, when the subspaces must have l-dimensional common subspaces, is also solved for small n's. Compare this with (11)-(12) which is true only for large n's. The reason for the difference is, that the middle levels of the partial order of the subspaces are much larger than those of the subsets of a set.

It would be nice to have an algebraic generalization of (44). However, it seems to be hard, because besides the partial order we need an ordering in the levels of the elements of the same rank.

REFERENCES

[1] BERNSTEIN, A.J., *Maximally connected arrays on the n-cube*, Siam J. Appl. Math., 15 (1967) 1485-1489.

[2] BRACE, B.A. & D.E. DAYKIN, *Sperner type theorems for finite sets*, Proc. Combinatorics Conf. Oxford, 1972.

[3] BRUIJN, N.G. DE, C.A. VAN EBBENHORST TENGBERGEN & D.K. KRUYSWIJK, *On the set of divisors of a number*, Nieuw Arch. Wisk. (2), 23 (1952) 191-193.

[4] CLEMENTS, G.F. & B. LINDSTRÖM, *A generalization of a combinatorial theorem of Macaulay*, J. Combinatorial Theory, 7 (1969) 230-238.

[5] CLEMENTS, G.F., *On the existence of distinct representative sets for subsets of a finite set*, Canad. J. Math., 22 (1970) 1284-1292.

[6] CLEMENTS, G.F., *Sets of lattice points which contain a maximal number of edges*, Proc. Amer. Math. Soc., 27 (1971) 13-15.

[7] CLEMENTS, G.F., *More on the generalized Macaulay theorem*, Discrete Math., 1 (1971) 247-255.

[8] CLEMENTS, G.F., *A minimization problem concerning subsets*, Discrete Math., 4 (1973) 123-128.

[9] CLEMENTS, G.F., *Inequalities concerning numbers of subsets of a finite set*, to appear.

[10] CLEMENTS, G.F., *Sperner's theorem with constraints*, to appear.

[11] CLEMENTS, G.F., *Intersection theorems for sets of subsets of a finite set*, to appear.

[12] DAYKIN, D.E. & L. LOVÁSZ, *On the number of values of a Boolean function*, to appear.

[13] DAYKIN, D.E., *A simple proof of Katona's theorem*, to appear (probably under a different title).

[14] DAYKIN, D.E., *Erdös-Ko-Rado from Kruskal-Katona*, to appear.

[15] DAYKIN, D.E., J. GODFREY & A.J.W. HILTON, *Existence theorems for Sperner families*, to appear.

[16] EHRENFEUCHT, A. & J. MYCIELSKI.

[17] ERDÖS, P., *On a lemma of Littlewood and Offord*, Bull. Amer. Math. Soc., 51 (1945) 898-902.

[18] ERDÖS, P. & T. GALLAI, *On the maximal paths and circuits of graphs*, Acta Math. Acad. Sci. Hung., 10 (1959) 337-357.

[19] ERDÖS, P., CHAO KO & R. RADO, *Intersection theorems for systems of finite sets*, Quart. J. Math. Oxford (2), 12 (1961) 313-318.

[20] ERDÖS, P., *A problem on independent r-tuples*, Ann. Univ. Sci. Budapest, 8 (1965) 93-95.

[21] ERDÖS, P. & D.J. KLEITMAN, *Extremal problems among subsets of a set*, in: Combinatorial Mathematics and Applications, Proc. Chapel Hill Conference 1970, pp.146-170.

[22] ERDÖS, P. & J. KOMLÓS, *On a problem of Moser*, in: *Combinatorial theory and its applications*, Proc. of the Coll. at Balatonfüred 1969, North-Holland Publ. Co. Amsterdam, 1970, pp.365-367.

[23] ERDÖS, P., M. HERZOG & J. SCHÖNHEIM, *An extremal problem on the set of noncoprime divisors of a number*, Israel J. Math., 8 (1970) 408-412.

[24] ERDÖS, P. & D.J. KLEITMAN, *On collections of subsets containing no 4-member Boolean algebra*, Proc. Amer. Math. Soc., 28 (1971) 507-510.

[25] ERDÖS, P., *Problems and results on finite and infinite combinatorial analysis*, in: Proc. of the Coll. at Keszthely, North-Holland Publ. Co., Amsterdam, to appear.

[26] ERDÖS, P. & J. SCHÖNHEIM, *On the set of non pairwise coprime divisors of a number*, to appear.

[27] FRANKL, P., *A theorem on Sperner-systems satisfying an additional condition*, to appear in Studia Sci. Math. Hungar.

[28] GREENE, C. & D.J. KLEITMAN, *Strong versions of Sperner's theorem*, to appear.

[29] HAJNAL, A. & B.L. ROTHSCHILD, *A generalization of the Erdös-Ko-Rado theorem on finite set systems*, J. Combinatorial Theory, 15 (1973) 359-362.

[30] HANSEL, G., *Complexes et décompositions binomiales*, J. Combinatorial Theory A, 12 (1972) 167-183.

[31] HARPER, L.H., *Optimal assignments of numbers to vertices*, J. Soc. Indust. Appl. Math., 12 (1964) 131-135.

[32] HILTON, A.J.W., *Simultaneously independent k-sets of edges of a hypergraph*, to appear.

[33] HILTON, A.J.W. & E.C. MILNER, *Some intersection theorems for systems of finite sets*, Quart. J. Math. Oxford (2), 18 (1967) 369-384.

[34] HILTON, A.J.W., *An intersection theorem for two families of finite sets*, in: *Combinatorial mathematics and its applications*, Proc. Conf. Oxford 1969, Academic Press, London, 1971, pp. 137-148.

[35] HILTON, A.J.W., *Simultaneously disjoint pairs of subsets of a finite set*, Quart. J. Math. Oxford (2), 24 (1973) 81-95.

[36] HILTON, A.J.W., *Analogues of a theorem of Erdös, Ko and Rado on a family of finite sets*, to appear in Quart. J. Math. Oxford (2).

[37], [38] HSIEH, W.N., *Intersection theorems for systems of finite vector spaces*, to appear in Discrete Math.

[39] KATONA, G.O.H., *Intersection theorems for systems of finite sets*, Acta Math. Acad. Sci. Hungar., 15 (1964) 329-337.

[40] KATONA, G.O.H., *On a conjecture of Erdös and a stronger form of Sperner's theorem*, Studia Sci. Math. Hungar., 1 (1966) 59-63.

[41] KATONA, G.O.H., *A theorem of finite sets*, in: *Theory of graphs*, Proc. Coll. held at Tihany, 1966, Akadémiai Kiadó, Budapest, 1968, pp. 187-207.

[42] KATONA, G.O.H., *A simple proof of Erdös-Chao Ko-Rado theorem*, J. Combinatorial Theory B, 13 (1972) 183-184.

[43] KATONA, G.O.H., *Families of subsets having no subset containing another with small difference*, Nieuw Arch. Wisk. (3), 20 (1972) 54-67.

[44] KATONA, G.O.H., *A generalization of some generalizations of Sperner's theorem*, J. Combinatorial Theory A, 12 (1972) 72-81.

[45] KATONA, G.O.H., *Two applications of Sperner type theorems (for search theory and truth functions)*, Period. Math. Hungar., 3 (1973) 19-26.

[46] KATONA, G.O.H., *Combinatorial search problems*, in: *A survey of combinatorial theory*, J.N. SRIVASTAVA (ed.), North-Holland Publ. Co., Amsterdam, 1973.

[47] KATONA, G.O.H., *A three-part Sperner-theorem*, to appear in Studia Sci. Math. Hungar.

[48] KATONA, G.O.H., *On a conjecture of Ehrenfeucht and Mycielski*, to appear in J. Combinatorial Theory A.

[49] KLEITMAN, D.J., *On a lemma of Littlewood and Offord on the distribution of certain sums*, Math. Z., 90 (1965) 251-259.

[50] KLEITMAN, D.J., *On a combinatorial problem of Erdös*, Proc. Amer. Math. Soc., 17 (1966) 139-141.

[51] KLEITMAN, D.J., *A conjecture of Erdös-Katona on commensurable pairs among subsets on n-set*, in: *Theory of graphs*, Proc. Coll. held at Tihany, 1966, Akadémiai Kiadó, Budapest, 1968, pp. 215-218.

[52] KLEITMAN, D.J., *On subsets containing a family of non-commensurable subsets of a finite set*, J. Combinatorial Theory, 1 (1966) 297-298.

[53] KLEITMAN, D.J., *Families of non-disjoint subsets*, J. Combinatorial Theory, 1 (1966) 153-155.

[54] KLEITMAN, D.J., *On a combinatorial conjecture of Erdös*, J. Combinatorial Theory, 1 (1966) 209-214.

[55] KLEITMAN, D.J., *Max number of subsets of a finite set no k of which are pairwise disjoint*, J. Combinatorial Theory, 5 (1968) 157-163.

[56] KLEITMAN, D.J., *On families of subsets of a finite set containing no two disjoint sets and their union*, J. Combinatorial Theory A, 5 (1968) 235-237.

[57] KLEITMAN, D.J., *On a conjecture of Milner on k-graphs with non-disjoint edges*, J. Combinatorial Theory, 5 (1968) 153-156.

[58] KLEITMAN, D.J., M. EDELBERG & D. LUBELL, *Maximal sized antichains in partial orders*, Discrete Math., 1 (1971) 47-53.

[59] KLEITMAN, D.J. & J. SPENCER, *Families of k-independent sets*, Discrete Math., 6 (1973) 255-262.

[60] KLEITMAN, D.J., *On an extremal property of antichains in partial orders. The LYM property and some of its implications and applications*, this volume, pp. 277-290.

[61] KLEITMAN, D.J., M.M. KRIEGER & B.L. ROTHSCHILD, *Configurations maximizing the number of pairs of Hamming-adjacent lattice points*, to appear.

[62] KNESER, R., *Aufgabe 360*, Jahresbericht d. Deutschen Math. Vereinigung, 58 (2) (1955).

[63] KRUSKAL, J.B., *The number of simplicies in a complex*, in: *Mathematical optimization techniques*, Univ. of California Press, Berkeley and Los Angeles, 1963, pp. 251-278.

[64] KRUSKAL, J.B., *The number of s-dimensional faces in a complex: An analogy between the simplex and the cube*, J. Combinatorial Theory, 6 (1969) 86-89.

[65] LINDSEY, J.H., *Assignment of numbers to vertices*, Amer. Math. Monthly, 71 (1964) 508-516.

[66] LINDSTRÖM, B., *Optimal number of faces in cubical complexes*, Arkiv för Mat., 8 (1971) 245-257.

[67] LUBELL, D., *A short proof of Sperner's lemma*, J. Combinatorial Theory, 1 (1966) 299.

[68] MESHALKIN, L.D., *A generalization of Sperner's theorem on the number of subsets of a finite set*, Teor. Verojatnost. i Primenen., 8 (1963) 219-220 (in Russian).

[69] MILNER, E.C., *A combinatorial theorem on systems of sets*, J. London Math. Soc., 43 (1968) 204-206.

[70] RÉNYI, A., *Foundations of probability*, North-Holland Publ. Co., Amsterdam, 1971.

[71] RIESS, M., *Über eine Steinerische combinatorische Aufgabe, welche im 45sten Bande dieses Journals, Seite 181, gestellt worden ist*, J. Reine Angew. Math., 56 (1859) 326-344.

[72] SAUER, N., *On the density of families of sets*, J. Combinatorial Theory A, 13 (1972) 145-147.

[73] SCHÖNHEIM, J., *A generalization of results of P. Erdös, G. Katona and D.J. Kleitman concerning Sperner's theorem*, J. Combinatorial Theory A, 11 (1971) 111-117.

[74] SPERNER, E., *Ein Satz über Untermenge einer endlichen Menge*, Math. Z., 27 (1928) 544-548.

[75] TARJÁN, T., Private communication.

[76] TURÁN, P., *An extremal problem in graph theory*, Mat. Fiz. Lapok, 48 (1941) 436-452 (in Hungarian).

[77] YAMAMOTO, K., *Logarithmic order of free distributive lattices*, J. Math. Soc. Japan, 6 (1954) 343-353.

APPLICATIONS OF RAMSEY STYLE THEOREMS TO EIGENVALUES OF GRAPHS *)

A.J. HOFFMAN

IBM Watson Research Center, Yorktown Heights, New York 10598, USA

1. INTRODUCTION

Let G be a graph, A(G) its adjacency matrix, i.e. $A = (a_{ij})$ is given by

$$a_{ij} = \begin{cases} 1 & \text{if } i \text{ and } j \text{ are adjacent vertices,} \\ 0 & \text{otherwise.} \end{cases}$$

Thus, A = A(G) is a symmetric matrix whose entries are 0 and 1, with every $a_{ii} = 0$. For any real symmetric A, we denote its eigenvalues by

$$\lambda_1(A) \geq \lambda_2(A) \geq \ldots$$

or

$$\lambda^1(A) \leq \lambda^2(A) \leq \ldots ,$$

as is convenient. For A = A(G), we sometimes write $\lambda_i(G)$ or $\lambda^i(G)$ for $\lambda_i(A(G))$ or $\lambda^i(A(G))$ respectively.

There have been many investigations in graph theory, experimental designs, group theory, etc. in which knowledge of properties of $\{\lambda_i(G)\}$ -which we shall henceforth call the *spectrum* of G- has been very useful even where the eigenvalues are not mentioned in either the hypotheses or conclusions of the theorems proved. These investigations furnish part of (the rest is natural curiosity) the motivation for study of questions where

*) The preparation of this manuscript was supported (in part) by US Army contract # DAHC04-72-C-0023.

M. Hall, Jr. and J. H. van Lint (eds.), Combinatorics, 245-259.

the eigenvalues play an explicit role. More specifically, we can ask (and sometimes answer) questions of the following type:

(1) What properties of a graph control the magnitude of $\lambda^i(G)$ or $\lambda_i(G)$? We can answer this in a limited way for λ_1, λ_2 and λ^1.

(2) If G is an induced subgraph of H, written $G \subset H$ then (as we know from matrix theory), $\lambda_i(G) \le \lambda_i(H)$, $\lambda^i(G) \ge \lambda^i(H)$. Suppose we specify that every vertex of H must have large valence. Then by how much must $\lambda_i(H)$ exceed $\lambda_i(G)$ or $\lambda^i(G)$ exceed $\lambda^i(H)$? We can answer this for λ_2 and λ^1.

(3) Define a relationship $\sim$ on V(G) to mean $i \sim j$ if for every $k \neq i,j$, $a_{ik}=a_{jk}$, and let e(G) be the number of equivalence classes so defined. The examples

e(G) = 2

and

e(G) = 3

.

show that e(G) is not uniquely determined by the spectrum of G (which is (2,0,0,0,-2) in both cases). But is the magnitude of e(G) roughly determined by the spectrum? The answer is yes in a sense to be made precise later (and this result will be completely proved in these notes, whereas other results will be sketched).

(4) What real numbers can be limit points of the $\{\lambda_1(G)\}$, as G ranges over all graphs, or $\{\lambda_2(G)\},\ldots,$ or $\{\lambda^1(G)\}$, etc.? We know the early limit points for $\{\lambda_1(G)\}$ and $\{\lambda^1(G)\}$.

(5) Suppose (as is sometimes done) we represent G by the matrix $B(G) = J-2A(G)$. What can be said about the spectrum of B(G)? To show that the study of B(G) has its own surprises, we mention (and will indicate the proof of): for $i > 1$, $|\lambda_i(B)| \le 2^{9(\lambda^1(B))^2}$.

Of course the study of these questions mixes ideas of graph theory and matrix theory, and the principal tools from graph theory are RAMSEY's theorem and some relatives.

2. THE RAMSEY STYLE THEOREMS

For theorems 2.1-2.3, S is a set of symbols, $|S| = s \geq 2$.

THEOREM 2.1. *There exists a function* R(n,s) *such that every symmetric matrix of order* R(n,s) *with entries in* S *contains a principal submatrix of order* n, *with all diagonal entries the same, all off diagonal entries the same.*

This is essentially RAMSEY's theorem, which needs no proof here.

THEOREM 2.2. *There exists a function* Z(n,s) *such that every square matrix with entries in* S *of order* Z(n,s) *contains a square submatrix of order* n, *every entry of which is the same.*

This is easy to prove directly (see, e.g., the special case s = 2 given in [1], which easily generalizes) but we will give another proof soon.

THEOREM 2.3. *There exists a function* H(n,s) *such that every matrix with entries in* S *containing* H(n,s) *rows, no two the same, contains a square submatrix* M *of order* n, *which (after permutations of rows and columns) has the appearance*

$$(2.1)\qquad \begin{bmatrix} a & & c \\ & \ddots & \\ b & & a \end{bmatrix}$$

(all diagonal entries a, *lower triangle entries* b, *upper triangle entries* c), a,b,c *not all the same.*

PROOF OF THEOREM 2.2. (assuming theorem 2.3). Let Z(n,s) = nH(2n,s), and let A be of order Z(n,s). If at least H(2n,s) rows are different, then the lower left (or upper right) part of (2.1) yields the desired submatrix of order n. Hence, we may assume that A has n rows the same. By symmetry, A has n columns the same. □

PROOF OF THEOREM 2.3. (cf. [3]). We first prove:

(2.2) If $f(r) = 2(s^r-1)/(s-1)$, if A has all entries in S and has f(r) different rows, then A contains a submatrix B of 2r rows and r columns which, after permutation of rows and columns has the form

$$b_{11} \neq b_{21}, b_{32} \neq b_{42}, \ldots, b_{2r-1,r} \neq b_{2r,r}$$

$$b_{2k-1,j} = b_{2k,j} \quad \text{for } j \neq k,\ k,j=1,\ldots,r.$$

When $r = 1$, $f(r) = 2$, verifying (2.2) in this case. Assume (2.2) has been shown for r-1, and let A have f(r) different rows. We may also assume that each column of A is essential for the statement that any two rows are different (otherwise, such an inessential column can be discarded). Now suppose the first column of A were discarded. Then two rows of A (say 1 and 2) would be the same. Hence we may assume $a_{11} \neq a_{21}$, but $a_{1j} = a_{2j}$ for $j > 1$. Of the f(r-2) remaining rows, at least $(f(r)-2)/s$ must be the same in column 1. Now

$$\frac{f(r)-2}{s} = \left(\frac{2(s^r-1)}{s-1} - 2\right)\Big/ s = f(r-1).$$

The induction hypothesis applied to these $(f(r)-2)/s$ rows and the remaining columns of A completes the proof of (2.2).

Consider the matrix B given by (2.2). We use it to define a symmetric matrix C of order r on $\binom{s}{2} + s^2$ symbols as follows: c_{ii} is the unordered pair $\{b_{2i-1,i}, b_{2i,i}\}$ of distinct symbols in S; c_{ij} = the ordered pair $(b_{2i,j}, b_{2j,i})$ if $i < j$, the reverse if $i < j$. Assume $r = R(n, \binom{s}{2}+s^2)$. By theorem 2.1, C contains a principal submatrix of order n in which all diagonal entries are the same, all off diagonal entries are the same. Referring back to B, this means B contains a matrix D with 2n rows and n columns such that (after possible row interchanges), we have

$$d_{11} = d_{32} = \ldots = d_{2n-1,n} = \text{(say) } a_1$$

$$d_{21} = d_{42} = \ldots = d_{2n,n} = \text{(say) } a_2$$

$$d_{2k-1,j} = d_{2k,j} = \text{(say) } b, \quad \text{for all } j > k$$

$$d_{2k-1,j} = d_{2k,j} = \text{(say) } c, \quad \text{for all } j < k.$$

Clearly the odd or even rows of D (either unless b = c, and one of a_1, a_2 is b=c) produce the desired submatrix. Thus we have shown

$$H(n,s) = 2\,\frac{s^{R(n,\binom{s}{2}+s^2)}-1}{s-1}\,.\ \square$$

The last Ramsey style theorem we will use is

THEOREM 2.4. *There exists a function* S(n) *such that, if* G *is a connected graph on* S(n) *vertices,* G *contains a vertex of valence at least* n, *or a path of length* n *as an induced subgraph.*

The proof of this theorem is too easy to give here.

3. QUESTION (1)

Define H_n to be the graph on 2n+1 vertices, in which one vertex is not adjacent to exactly n other vertices, but all other pairs of vertices are adjacent. For any graph G, let $\ell(G)$ be the smallest positive integer such that neither $K_{1,\ell}$ nor H_ℓ is an induced subgraph of G. The following theorem is proved in [2].

THEOREM 3.1. *The function* $\ell(G)$ *is bounded from above and below by a function of* $|\lambda^1(G)|$.

The proof of theorem 3.1 has three parts. The first part shows that $\ell(G)$ is bounded from above by a function of $|\lambda^1(G)|$. The second part proceeds as follows: The distance d(G,H) between two graphs G and H with V(G) = V(H) is defined by making d(G,H) the maximum valence of the vertices in the graph $(G-H) \cup (H-G)$. Then L(G) is defined to be the smallest integer L such that there exists a graph H with

$$d(G,H) \le L$$

and H having a distinguished family of cliques $K^1, K^2, \ldots$ satisfying

(i) every edge of H is in at least one K^i,

(ii) every vertex of H is in at most L of the K^i's,

(iii) $|V(K^i) \cap V(K^j)| \le L$ for $i \ne j$.

Finally, it is proved that L(G) is bounded from above by a function of $\ell(G)$. The third part of the proof shows that $|\lambda^1(G)|$ is bounded from above by a function of L(G).

The proofs in the first and third parts are arguments from matrix theory, and are in [2]. The proofs in the second part use RAMSEY's theorem (theorem 2.1 in these notes). For the remainder of this section, $S = \{0,1\}$, and we use $R(n)$ for $R(n,2)$.

The strategy is to look for "large" cliques in G ("large" depends on ℓ) and define an equivalence relation on large cliques. The equivalence classes of large cliques are themselves almost cliques, i.e., we can add a graph in which each vertex has valence bounded by a function of ℓ, so that they become cliques. These added edges are the edges of H-G. Further the edges of G not contained in any large clique form a graph in which each vertex has valence bounded by a function of ℓ, and these edges are the edges of G-H.

Let $N = N(\ell) = \ell^2+\ell+2$. Define W to be the set of all cliques $K \subset G$ such that $|V(K)| \geq N$. We shall prove the statements given in the preceding paragraph for the cliques in W (the full discussion, including (ii) and (iii), requires further conditions on N).

LEMMA 3.1. *If* $K, K' \in W$, *define*

$$K \sim K'$$

if each vertex of K *is adjacent to all but at most* $\ell-1$ *vertices of* K'. *Then* $\sim$ *is an equivalence relation.*

PROOF. Reflexivity is clear, since $|V(K)| \geq \ell$. To prove symmetry, assume there is a vertex ν in K' not adjacent to at least ℓ vertices in K, and let A denote that set of ℓ vertices in K. Each vertex in A is not adjacent to at most $\ell-2$ vertices in K' other than ν, since $K \sim K'$. Hence, the set of vertices in K' each not adjacent to at least one vertex in A consists of ν and at most $\ell(\ell-2)$ other vertices. Since $N > \ell+\ell(\ell-2)+1$, it follows that K' contains at least ℓ vertices each of which is adjacent to each vertex in A. Call that set of ℓ vertices B. Then ν,A,B generate an H_ℓ, contrary to the definition of ℓ. This contradiction proves that $\sim$ is symmetric.

To prove transitivity, assume $K^1 \sim K^2$, $K^2 \sim K^3$, $K^1 \not\sim K^3$. Then K^3 contains a vertex ν not adjacent to a set C of ℓ vertices in K^1. Since $N > 2\ell+\ell(\ell-\ell)-1$, and $K^1 \sim K^2$, it follows that K^2 contains a subset D of $2\ell-1$ vertices each of which is adjacent to all vertices in C. But since $K^3 \sim K^2$, D contains some subset F of ℓ vertices adjacent to ν. Then

C,F,ν generate an $H_\ell \subset G$, which is a contradiction. □

Henceforth, the letter E will denote any equivalence class of cliques in W, and V(E) will be the union of all vertices of all cliques in E.

LEMMA 3.2. *Let* E *be an equivalence class,* $\nu \in V(E)$. *Then* ν *is adjacent to all but at most* $R(\ell)-1$ *other vertices in* V(E).

PROOF. Let $K^\nu \in E$ be a clique containing ν. By RAMSEY's theorem, if $F \subset V(E)$, $|F| \geq R(\ell)$, and every vertex in F not adjacent to ν, then F contains a K_ℓ or its complement $\bar{K}_\ell$. If $\bar{K}_\ell \subset F$, then since $|V(K^\nu)| > \ell^2-2\ell+1$, there exists a vertex $w \in K^\nu$ adjacent to all vertices in $\bar{K}_\ell$. Thus $K_{1,\ell} \subset G$, a contradiction.

If $K_\ell \subset F$, then $|V(K^\nu)| > \ell+\ell(\ell-2)+1$ implies that K^ν contains a set of ℓ vertices each adjacent to all the vertices in K_ℓ, thus generating an H_ℓ. □

LEMMA 3.3. *Let* $\tilde{H}$ *be the graph formed by edges of* G *not in any clique in* W. *Then every vertex in* $\tilde{H}$ *has valence at most* R(N).

PROOF. If not, then by RAMSEY's theorem we would have $K_{1,\ell} \subset G$, or the vertices adjacent to ν in $\tilde{H}$ would contain a clique in W, contradicting the definition of $\tilde{H}$. □

Results analogous to theorem 3.1 have been established for $\lambda_2(G)$ by HOWES [6]. For $\lambda_1(G)$, it is easy to see that its size is controlled by the size of the smallest t such that $K_{1,t} \not\subset G$, $K_t \not\subset G$. Corresponding results for other $\lambda_i(G)$ or $\lambda^i(G)$ are not known.

4. QUESTION (2)

Let us define

$$\mu^i(G) \equiv \lim_{d\to\infty} \sup_{\substack{H\supset G \\ d(H)>d}} \lambda^i(H),$$

where d(G) is the minimum valence of the vertices of G. Also

$$\mu_i(G) \equiv \lim_{d\to\infty} \inf_{\substack{H\supset G \\ d(H)>d}} \lambda_i(H).$$

It is easy to see that $\mu_1(G) = \infty$. Further, we shall give formulas for μ_2 and μ^1 showing that they are finite. By matrix theory arguments, this will prove the existence and finiteness of all other μ^i and μ_i, although we have no formulas for them.

THEOREM 4.1. (cf. [4]). *Let* $|V(G)| = m$ *and let* $\mathcal{C}^1$ *be the class of all* (0,1) *matrices* C *with* m *rows such that every row sum of* C *is positive, but this property is lost if any column is deleted. Then*

$$\mu^1(G) = \max_{C\in\mathcal{C}^1} \lambda^1(A-CC^T).$$

THEOREM 4.2. *Let* $|V(G)| = m \geq 2$, *and let* $\mathcal{C}_2$ *be the class of all* (0,1) *matrices* C *with* m *rows and at least two columns such that every row sum of* C *is positive, and, if* C *has more than two columns, no column can be deleted without destroying the property that* C *has positive row sums. Then*

$$\mu_2(C) = \min_{C\in\mathcal{C}_2} \lambda_1(A-C(J-I)^{-1}C^T).$$

We shall not make any remarks about the proof of theorem 4.2, except to state that, as far as the theme of this lecture is concerned, the use of Ramsey style theorems is analogous to the uses in the proof of theorem 4.1.

To prove theorem 4.1, one proceeds as follows: Let any $C\in\mathcal{C}^1$ be given, and let C have k columns. Extend G to a graph $G_C(n)$ by adjoining k cliques $K^1,\ldots,K^k$, each with n vertices, such that every vertex of clique K^j is adjacent to vertex i of G when $c_{ij} = 1$, and not adjacent when $c_{ij} = 0$. Additionally, no vertices of K^j and K^ℓ, $j \neq \ell$, are adjacent. Then a little algebra shows that $\lim_{n\to\infty} \lambda^1(G(n)) = \lambda^1(A-CC^T)$, so

$$(4.1) \qquad \mu^1(G) \geq \max_{C\in\mathcal{C}^1} \lambda^1(A-CC^T).$$

Note that, if $d_1 > d_2$, $\sup_{\substack{H\supseteq G\\ d(H)>d_1}} \lambda_1(H) \leq \sup_{\substack{H\supseteq G\\ d(H)>d_2}} \lambda_1(H)$. Together with (4.1) this implies $\mu^1(G)$ exists and is finite. To prove that

$$(4.2) \qquad \mu^1(G) \leq \max_{C\in\mathcal{C}^1} \lambda^1(A-CC^T),$$

let $\varepsilon > 0$ be given. Choose an integer n such that, for all the finitely many graphs $G_C(n)$ discussed in the proof of inequality (4.1), we have

$$(4.3) \qquad \lambda^1(G_C(n)) \leq \lambda^1(A-CC^T) + \varepsilon.$$

From the definition of $\mu^1(G)$, we know that, for any d, there exists $H \supset G$ such that

$$(4.4) \qquad d(H) > d$$

and

$$(4.5) \qquad \mu^1(G) \leq \lambda^1(H).$$

We shall choose d (depending on n (hence on ε) and $\mu^1(G)$) such that, for any $H \supset G$ and satisfying (4.4) and (4.5), there is a $C \in \mathcal{C}^1$ such that

$$(4.6) \qquad G_C(n) \subset H,$$

which, by matrix arguments, shows

$$(4.7) \qquad \lambda^1(H) \leq \lambda^1(G_C(n)).$$

Combining (4.5), (4.7) and (4.3) yields (4.2).

So the critical thing is to prove (4.6). Besides RAMSEY's theorem, we need theorem 2.2, and we write Z(n) for Z(n,2) (our set S is $\{0,1\}$). $Z^{(k)}(n)$ will be the k-th iterate of Z(n).

By section 3 we know that there is a function ℓ such that

$$(4.8) \qquad K_{1,\ell(\mu^1)} \not\subset H$$

$$(4.9) \qquad H_{\ell(\mu^1)} \not\subset H.$$

Let D(G) be the maximum valence of the vertices of G, let

$$(4.10) \qquad N = R(\max\{\ell(\mu^1), Z^{(\binom{m}{2})}(\max\{n,\ell(\mu^1)\})\}),$$

$$(4.11) \qquad d \geq 2^{m-1}N+D(G).$$

Since every vertex of H has valence at least d, every vertex of $G \subset H$ is adjacent to at least $2^{m-1}N$ vertices in H-G. For each $S \subset \{1,...,m\}$, $S \neq \emptyset$, let H(S) be the set of vertices in H-G, each of which is adjacent to the vertices in S and to no other vertices in G. Hence, there exist sets $S_1, S_2, ..., S_t$ $(t \leq m)$ of subsets of $\{1,...,m\}$ such that every vertex in G is in at least one S_i, discarding any S_i loses this property, the sets $H(S_i)$ are disjoint, and each $|H(S_i)| \geq N$. Let

$$\bar{Z} = Z^{(\binom{m}{2})} (\max \{n, \ell(\mu^1)\}).$$

We contend $H(S_i)$ contains a clique $K_{\bar{Z}}$. For, from (4.10) and RAMSEY's theorem, the only other possibility is that $H \supset H(S_i)$ contains $K_{1,\ell(\mu^1)}$, (note that $H(S_i)$ is attached to at least one vertex of G), violating (4.8).

Next, consider each of the t disjoint cliques $K_{\bar{Z}^1}, ..., K_{\bar{Z}^t}$. Take any two of these cliques. By theorem 1.2, they either contain large subcliques (at least $\ell(u^1)$ in size) with all vertices adjacent (impossible by (4.9), since each is attached to a vertex of G not adjacent to the other), or large subcliques with no vertices of one adjacent to vertices of the other. Since $t \leq m$, we need only iterate this process at most $(m-1)+(m-2)+...+1 = \binom{m}{2}$ times. When we are done, we have the desired C and $G_C(n)$. (We have used here tacitly that $Z(a) \geq a$, of course).

5. QUESTION (3)

We shall show that the magnitude of e(G), defined in the introduction (3), is controlled by the spectrum of G. For $a \leq b$, define $\Lambda(a,b)(G)$ to be the number of eigenvalues of G each of which is at most a or at least b. For the function H(n,m) in theorem 2.3, write H(n) for H(n,2) $(S=\{0,1\})$.

THEOREM 5.1.

$$\Lambda((-\sqrt{5}-1)/2,1)(G) \leq e(G) < R(H(R([5\Lambda+1]))),$$

(where Λ on the right is the same as Λ on the left).

To prove the left inequality is easy. Suppose $i \sim j$ and i and j are adjacent. Then it is easy to see that if $j \sim k$, i and k, and j and k are adjacent. If i and j are not adjacent, and j ~ k, then i and k are not adjacent and j and k are not adjacent. Hence each equivalence class is a clique or an independent set. Now if there are e equivalence classes, the matrix A(G) can be partitioned

$$A(G) = \begin{bmatrix} A_{11} & A_{12} & \cdots & A_{1e} \\ A_{21} & A_{22} & \cdots & A_{2e} \\ \vdots & \vdots & \ddots & \\ A_{e1} & A_{e2} & & A_{ee} \end{bmatrix}$$

such that each A_{ij} $(i \neq j)$ is 0 or J, each A_{ii} is 0 or J-I. Thus the number of eigenvalues not 0 or -1 of A(G) is at most e. But every eigenvalue in Λ is not 0 or -1.

To prove the right inequality, assume it false and that $e(G) \geq R(H(R([5\Lambda+1])))$. It follows from theorem 2.1 that there exists a subset X of $H(R([5\Lambda+1]))$ vertices of G which are inequivalent, and form either a clique or an independent set. In any case, the rows of A(G) corresponding to these vertices, together with the columns of A(G) corresponding to the complementary set of vertices, form a submatrix of A(G) in which all rows are different. By theorem 2.3, there exists a square submatrix of this matrix of order $R([5\Lambda+1])$, which, after row and column permutations, has the form I,J-I, or triangular. Let S be the set of vertices corresponding to the columns of this submatrix. By theorem 1.1, since $|S| = R([5\Lambda+1])$, S contains a subset Y of vertices, with $|Y| = [5\Lambda+1]$ where Y is a clique or independent set. The corresponding $[5\Lambda+1]$ vertices of X form a subset W such that the incidence matrix of W versus Y is I,J-I, or triangular (say V). Thus, setting $m = [5\Lambda+1]$, A(G) contains a principal submatrix of order $(2m \times 2m)$ of one of the following forms

$$\begin{pmatrix} 0 & I \\ I & 0 \end{pmatrix}, \quad \begin{pmatrix} 0 & J-I \\ J-I & 0 \end{pmatrix}, \quad \begin{pmatrix} 0 & V \\ V^T & 0 \end{pmatrix}$$

$$\begin{pmatrix} J-I & I \\ I & 0 \end{pmatrix}, \quad \begin{pmatrix} J-I & J-I \\ J-I & 0 \end{pmatrix}, \quad \begin{pmatrix} J-I & V \\ V^T & 0 \end{pmatrix}$$

$$\begin{pmatrix} J-I & I \\ I & J-I \end{pmatrix}, \quad \begin{pmatrix} J-I & J-I \\ J-I & J-I \end{pmatrix}, \quad \begin{pmatrix} J-I & V \\ V^T & J-I \end{pmatrix}.$$

In each of these nine cases, the matrix has more than Λ eigenvalues $\le (-\sqrt{5}-1)/2$ or ≥ 1, which is impossible.

It is interesting that if 1 is replaced by $1 + \varepsilon$ or $(-\sqrt{5}-1)/2$ replaced by $((-\sqrt{5}-1)/2)-\varepsilon$, the right-hand inequality is no longer true.

The same arguments will establish a theorem similar to theorem 5.1 if 1 is replaced by $(\sqrt{5}-1)/2$ and $(-\sqrt{5}-1)/2$ is replaced by -2.

6. QUESTION (4)

We only know some small limit points of $\{\lambda_1(G)\}$ and of $\{\lambda^1(G)\}$ as G varies over all graphs. To be more specific, let Λ_1 be the set of distinct numbers each of which is $\lambda_1(G)$ for some G.

<u>THEOREM 6.1.</u> (cf. [5]). *Let* $P_n(x) = x^{n+1}-(1+x+\ldots+x^{n-1})$, $n=1,2,\ldots$. *Let* β_n *be the unique positive root of* P_n, *and* $\alpha_n=\beta_n^{\frac{1}{2}}+\beta_n^{-\frac{1}{2}}$. *Then the numbers* $2=\alpha_1<\alpha_2<\ldots$ *are all limit points of* Λ_1 *less than* $\tau^{\frac{1}{2}}+\tau^{-\frac{1}{2}}$, *where* $\tau=\frac{1}{2}(\sqrt{5}+1)$.

A very rough sketch of the proof proceeds as follows. If α is a limit point of Λ_1, then there must exist a sequence of connected graphs $G_1,G_2,\ldots$ such that $\lambda_1(G_i)$ are all different, and $\lambda_1(G_i) \to \alpha$. But this means that the graphs G_i are all different, hence $|V(G_i)| \to \infty$, hence (theorem 2.4) for sufficiently large i, G_i contains a vertex of large valence (impossible, for this implies $\lambda_1(G_i) \to \infty$) or a long path. For a path S_n of length n, $\lambda_1(S_n) \to 2$, which must be the smallest limit point. Next, assume $2 < \alpha < \tau^{\frac{1}{2}}+\tau^{-\frac{1}{2}}$. Then G_i cannot be a simple circuit infinitely often, for $\lambda_1(\text{circuit}) = 2$. One can also show that if G_i contains a circuit, and at least one additional edge, $\lambda_1(G_i) > \tau^{\frac{1}{2}}+\tau^{-\frac{1}{2}}$. Hence, this cannot occur infinitely often, so we may assume each G_i is a tree. One can also prove that, if a tree G_i contains at least three vertices of valence at least three, $\lambda_1(G_i) > \tau^{\frac{1}{2}}+\tau^{-\frac{1}{2}}$. Further, if (infinitely often), G_i contains a vertex of valence at least four, then $\lim \lambda_1(G_i) > \tau^{\frac{1}{2}}+\tau^{-\frac{1}{2}}$. Continuing arguments and calculations in this vein, one finds that the desired limit points must be limit points of the biggest eigenvalues of the sequence of graphs $\{G_i^k\}$, where k is fixed, $i\to\infty$, and G_i^k is

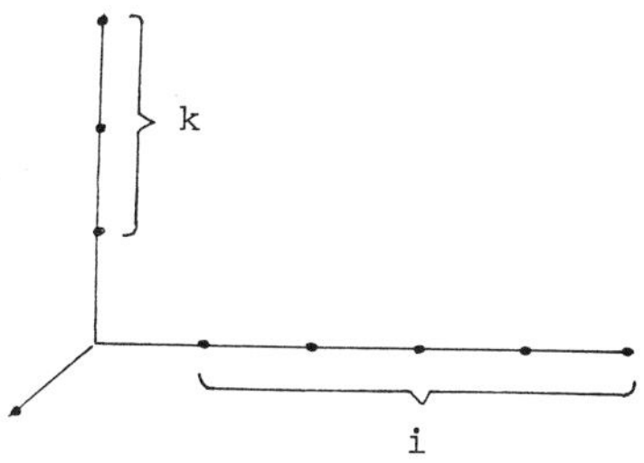

and $\lim_{i\to\infty} \lambda_1(G_i^k) = \alpha_k$.

If we define Λ^1 to be the set of distinct numbers each of which is $\lambda^1(G)$ for some G, then

THEOREM 6.2. *Let* T *be a tree on at least two vertices,* L(T) *its line graph,* e *an end of* T, $\tilde{A}(L(T,e))$ *be* A(L(T)) *modified so that there is* -1 *in the diagonal position corresponding to* e. *Then* $\lambda^1(\tilde{A}(L(T,e))) > -2$, *and every limit point of* $\Lambda^1 > -2$ *occurs in this way.*

Here again we use theorem 2.4. For if $\lim_{i\to\infty} \lambda^1(G_i) = \alpha > -2$, $\{G_i\}$ can not contain arbitrarily long paths S_n, because $\lim_{n\to\infty} \lambda^1(S_n) = -2$. So, for i large, G_i must contain at least one vertex v of large valence. By theorem 2.1, the vertices adjacent to v contain a large clique or large independent set; but the latter is impossible, because this would imply $K_{1,m} \subset G_i$ for m large, which contradicts $\lambda^1(G_1)$ of modest size (see section 3). Hence, we know that, for i large, G_i contains at least one large clique, say K_n. (This large clique eventually gets associated with the edge e of the tree which plays a special role.) The remainder of the proof uses theorems 2.2 and 2.1 to show that no vertex not in K_n can have large valence, in spirit similar to other applications already described.

We conjecture that $\lambda^1(\tilde{A}(L(T,e)))$ is always a limit point, but lack a proof at this time.

7. QUESTION (5)

To illustrate some of the fun in looking at symmetric matrices with entries ±1, (rather than (0,1)), we indicate the proof of

THEOREM 7.1. (cf. [1]). *There exists a function* f *such that, if* A *is any symmetric* ± 1 *matrix*, $|\lambda_i(A)| \leq f(\lambda^1(A))$, *for all* $i > 1$.

The argument consists of showing that there exists a matrix M of the form

$$M = \left(\begin{array}{c|c} + & - \\ \hline - & + \end{array}\right)$$

where the diagonal blocks are square (not necessarily of the same size; indeed, one may be empty), so that A-M = P consists of a matrix in which only $\frac{1}{2}f(\lambda^1)$ entries per row are non-zero (i.e., ± 2), after which matrix arguments produce the desired inequality. To construct M, it turns out to be sufficient to prove that given a lower bound on $\lambda^1(A)$, A cannot have two rows (say 1 or 2) such that $a_{1j}=a_{2j}$ for many j, and also $a_{1j}=-a_{2j}$ for many j. Assume otherwise. Then we have a right to assume A has a principal submatrix

$$\tilde{A} = \left(\begin{array}{cc|c|c} \pm 1 & \pm 1 & +1\ldots+1 & +1\ldots+1 \\ \pm 1 & \pm 1 & +1\ldots+1 & -1\ldots-1 \\ \hline +1 & +1 & & \\ \vdots & \vdots & A_{11} & A_{12} \\ +1 & +1 & & \\ \hline +1 & -1 & & \\ \vdots & \vdots & A_{21}=A_{12} & A_{22} \\ +1 & -1 & & \end{array}\right) \begin{array}{l} \\ \\ \\ \left.\vphantom{\begin{array}{c} 1\\1\\1 \end{array}}\right\} n \\ \left.\vphantom{\begin{array}{c} 1\\1\\1 \end{array}}\right\} n \end{array}$$

By theorem 2.2, $\tilde{A}$ (and hence A) has a principal submatrix (which we call $\tilde{\tilde{A}}$) of order n of the same form, where $A_{12} = J$ or $A_{12} = -J$. In either case, $\tilde{\tilde{A}}$ has a least eigenvalue which goes to $-\infty$ for large n.

REFERENCES

[1] HOFFMAN, A.J., *On eigenvalues of symmetric (+1,-1) matrices*, Israel J. Math., to appear.

[2] HOFFMAN, A.J., *On spectrally bounded graphs*, in: *A survey of combinatorial theory*, North-Holland Publ. Cy., Amsterdam, 1973, pp.277-284.

[3] HOFFMAN, A.J., *Eigenvalues and partitionings of the edges of a graph*, Linear Algebra and Appl., 5 (1972) 137-146.

[4] HOFFMAN, A.J., *The change in the least eigenvalue of the adjacency matrix of a graph under imbedding*, SIAM J. Appl. Math., 17 (1969) 664-671.

[5] HOFFMAN, A.J., *On limit points of spectral radii of non-negative symmetric integral matrices*, in: *Graph theory and its applications*, Springer-Verlag, Berlin etc., 1972, pp.165-172.

[6] HOWES, L., *On subdominantly bounded graphs; summary of results*, in: *Recent trends in graph theory*, Springer-Verlag, Berlin etc., 1971, pp.181-183.

SOME RECENT DEVELOPMENTS IN RAMSEY THEORY

R.L. GRAHAM

Bell Laboratories, Murray Hill, New Jersey 07974, USA

B.L. ROTHSCHILD

University of California, Los Angeles, Cal. 90024, USA and Bell Laboratories, Murray Hill, New Jersey 07974, USA

Recently a number of striking new results have been proved in an area becoming known as RAMSEY THEORY. It is our purpose here to describe some of these. Ramsey Theory is a part of combinatorial mathematics dealing with assertions of a certain type, which we will indicate below. Among the earliest theorems of this type are RAMSEY's theorem, of course, VAN DER WAERDEN's theorem on arithmetic progressions and SCHUR's theorem on solutions of $x+y = z$.

To make our task easier, we will introduce the "arrow notation" of ERDÖS and RADO. This was originally used for generalizations of Ramsey's Theorem to infinite cardinals, but can be easily adapted to other cases as well. The meaning of the arrow notation will become clear by its use in the examples throughout this paper.

As our first example, consider:

$$n \underset{k}{\rightarrow} (\ell_1,\ldots,\ell_r) .$$

This expression is just an abbreviation for the following assertion: if the k-element subsets of an n-element set are partitioned into r classes, then for some i there is an ℓ_i-element subset L_i of the n-element set such that all the k-element subsets of L_i are in the i-th class.

<u>THEOREM</u>. (RAMSEY). *For all positive integers* $k,r,\ell_1,\ldots,\ell_r$, *there exists an* $N = N(k,r,\ell_1,\ldots,\ell_r)$ *such that if* $n \geq N$, *then* $n \underset{k}{\rightarrow} (\ell_1,\ldots,\ell_r)$.

In fact, RAMSEY considered only the case where all the ℓ_i are equal.

M. Hall, Jr. and J. H. van Lint (eds.), Combinatorics, 261-276.

He also proved $\aleph_0 \underset{k}{\rightarrow} (\underbrace{\aleph_0, \ldots, \aleph_0}_{r})$ which actually is stronger than the finite theorem above. The consideration of such statements with large cardinals or ordinals is a subject in itself and will not be discussed here. For the large cardinals the subject is fairly complete and will be covered in a forthcoming book of ERDÖS, HAJNAL & RADO. For ordinals, the theory is developing rapidly, although there are still many open questions. To give the flavor of a result of this type, we mention one of the most interesting recent ones.

THEOREM. (CHANG, LARSEN, MILNER). $\omega^\omega \underset{2}{\longrightarrow} (\omega^\omega, k)$

This theorem asserts that if the pairs (i.e., 2-element subsets) of a set of order type ω^ω are partitioned into two classes, then either the first class contains all the pairs of a subset with induced order type ω^ω, or the second class contains all the pairs of some k-element subset.

This last example illustrates the arrow notation in a case where we deal with sets with structure (here the structure is that of order).

In general, in a Ramsey Theorem an assertion of the form $A \underset{B}{\rightarrow} (C_1, \ldots, C_r)$, where the symbols A, B and C_i denote objects with a certain structure. For example, as above, they could be sets or sets with order. Other examples include graphs, finite vector spaces, sets containing solutions to systems of linear equations, Boolean algebras and partitions of finite sets.

In the remainder of the paper, we will consider six examples of Ramsey theorems. The first two concern graphs and are due to W. DEUBER and to J. NEŠETŘIL & V. RÖDL. The next three concern systems of linear equations and their solution sets. These are results of N. HINDMAN, E. SZEMEREDI and W. DEUBER. Finally, we will discuss some results of K. LEEB on abstract categories which are *"Ramsey"*.

GRAPHS

Recalling the previous statement of Ramsey's Theorem, we see that the first non-trivial case is

$$6 \underset{2}{\rightarrow} (3,3) \ .$$

This can be restated as follows: if the edges of the complete graph K_6 on six vertices are 2-colored arbitrarily, then some monochromatic triangle K_3 must be formed. This graphical form leads to several general considerations. The most natural of these, an immediate consequence of Ramsey's Theorem (with k = 2), is simply:

For every finite graph H, *there is a finite graph* G *such that* $G \underset{2}{\rightarrow} (H,H)$.

Here, the arrow notation means that if the *edges* of G (represented by the 2 below the arrow) are 2-colored arbitrarily, then G will contain a monochromatic subgraph isomorphic to H.

It would be stronger to require that the monochromatic subgraph above be an *induced* subgraph of G. We could write $G \underset{2}{\rightarrow} (H,H)$ also in this case, provided we understand that we mean induced subgraphs here. Actually, to be rigorous, we should use a "different" kind of arrow for each different meaning. The proper setting for this is in terms of category theory as originally indicated by LEEB. We will elaborate on this when we discuss LEEB's recent results at the end of this paper.

We now turn our attention to the first result, which concerns *induced* subgraphs of graphs.

THEOREM. (DEUBER [2]). *For every finite graph* H, *there exists a finite graph* G *such that* $G \underset{2}{\rightarrow} (H,H)$.

SKETCH OF PROOF. What DEUBER actually proves is the equivalent but more convenient statement: for every choice of finite graphs G and H there exists a finite graph K such that $K \underset{2}{\rightarrow} (G,H)$. The proof is by induction on $|G|+|H|$ where $|G|$ denotes the number of vertices of G. The small cases are trivial. Let g be a vertex of G, $\bar{G} = G-\{g\}$, and let S be the subset of G to which g is connected. Also, let h in H, $\bar{H}$ and T be defined similarly.

By induction we can find G^* and H^* such that $G^* \underset{2}{\rightarrow} (\bar{G},H)$ and $H^* \underset{2}{\rightarrow} (G,\bar{H})$. We now form a large graph K as follows: Start with G^*. Let $\bar{G}_1,\ldots,\bar{G}_m$ be all the occurrences of $\bar{G}$ as an induced subgraph of G^* and let $S_1,\ldots,S_m$ be the corresponding subsets S (there may be more than one choice for an S_i; any one is allowed). Now replace each vertex of $\mathcal{S} = S_1 \cup \ldots \cup S_m = \{x_1,\ldots,x_\ell\}$ by a complete copy of H^*, with the copy of H^* replacing x_i denoted by H_i^*. Connect a vertex of H_i^* to a vertex of H_j^* iff x_i and x_j are connected in G^*. Also, if some vertex v is not in $\mathcal{S}$, connect v to all the vertices of H_i^* iff v and x_i are connected in G^*. Thus, we have essentially "exploded" some of the vertices of G^* into H^*'s.

Suppose, in the simplest case, that all the S_i are disjoint. Let $\bar{H}_1,\ldots,\bar{H}_n$ be the occurrences of $\bar{H}$ in H^* and let $T_1,\ldots,T_n$ denote the corresponding subsets T. For each fixed S_i, consider the associated H^*'s and choose one T_j from each H^*. For each i and such choice of T_j's, we introduce a new vertex connected exactly to these T_j's. Hence, if $|S_i| = k$, then for this i we have added n^k new vertices. Since we have m disjoint S_i, then there are altogether mn^k new vertices. This completes the definition of K.

Suppose now the edges of K are 2-colored, say, using the colors red and blue. By the construction of H^*, each H^* in K has either a red copy of G or a blue copy of $\overline{H}$. If the first alternative holds, then we are done. So assume each H^* in K contains a blue copy of $\overline{H}$. Let $y_1,\ldots,y_m$ be the new vertices corresponding to the subsets T_j for these copies of $\overline{H}$ (i.e., one y_i for each S_i). If any of the y_i are connected to any of the T_j by all blue edges, we are done since in this case we have a blue copy of H. Thus, we may assume that each y_i is connected by a red edge to some vertex $T \in T_j$ for each T_j to which it is connected. Let y_i be connected by red edges to $t_{i1},\ldots,t_{iw}$. Consider the graph $\tilde{G}$ obtained from K by deleting all the vertices of all the copies of H^* except for the t_{ij}, and deleting all the new vertices except $y_1,\ldots,y_m$. By construction, $\tilde{G}$ is isomorphic to G^* together with the y_i. Also, it is an induced subgraph of K. Since each y_i is connected to the corresponding S_i by only red edges, we are done. For either $G^* \subseteq \tilde{G}$ contains a blue copy of H or, it contains a red copy of $\overline{G}$, say $\overline{G}_i$, which together with y_i forms a red copy of G. This completes the argument for the case that the S_i are disjoint.

The only obstruction preventing this from being completely general is that it usually happens that for some a and b, $S_a \cap S_b \neq \emptyset$ in G. This in turn would prevent us from choosing the same t_{ij} for both S_a and S_b when necessary. To get around this, we add another step to the construction. Namely, after replacing the vertices of $\mathcal{S}$ by copies of H^*, we take those in the $S_a \cap S_b$ and replace each vertex of the H^* itself by a copy of H^*, connecting it up in the same way as before. We can then be certain of obtaining a vertex connected by only red edges to some copy of H^*, and we can proceed essentially in the same way as before. □

Of course, the graphs K resulting from this construction are usually much larger than are actually required. For example, the graph K constructed this way for the assertion $K \underset{2}{\rightarrow} (K_3,K_3)$ is $K = K_{81}$. Note also the high clique number K_{81} has relative to that of K_3.

F. GALVIN had asked if for each finite graph H with clique number $cl(H) = k$ (where $cl(H) = \max\{n \mid K_n \text{ is a subgraph of } H\}$), there is a graph G also having $cl(G) = k$ such that $G \underset{2}{\rightarrow} (H,H)$. As above, we consider induced subgraphs here. This question has been very recently answered in the affirmative by J. NEŠETŘIL & V. RÖDL. One sees easily that this implies DEUBER's result.

FOLKMAN, in response to a question of ERDÖS and HAJNAL, had earlier shown that there exists a graph G with cl(G) = k such that $G \underset{2}{\not\rightarrow} (K_k, K_k)$. FOLKMAN also proved that for any G and H, there exists a K with cl(K) = = max{cl(G), cl(H)} such that $K \underset{1}{\rightarrow} (G,H)$ (where the 1 below the arrow indicates that we are coloring the vertices of K instead of the edges). In fact, NEŠETŘIL & RÖDL also make use of this theorem.

The second result we discuss is the following:

THEOREM. (NEŠETŘIL & RÖDL). *For every finite graph* H *there exists a finite graph* G *such that* $G \underset{2}{\rightarrow} (H,H)$ *and* cl(G) = cl(H).

SKETCH OF PROOF. The proof uses the ingenious idea of letting the vertices of G be subsets of a large set. By appropriately defining when edges occur between them, and applying Ramsey's Theorem to certain subsets, a large subset is obtained with the vertices and edges determined by it being very well behaved.

We will make some definitions first, and then indicate somewhat how the proof goes, especially for the case of cl(H) = 2, which is considerably simpler and more direct than the general case. We begin with the definition of the graphs (n,T,p).

Let A,B be two p-subsets of [1,n] = {1,2,...,n}. The *type* (or *p-type*) t(A,B) of A and B is the pattern of their relative order, defined as follows: List the elements of $A \cup B$ in increasing order assuming $\min\{A-B\} < \min\{B-A\}$, say $x_1, x_2, \ldots, x_\ell$, $\ell \le 2p$. If $x_i \in A \cap B$ replace it by two copies of itself. The new list thus obtained, say $y_1, y_2, \ldots, y_{2p}$, is of length 2p. The type t(A,B) is then defined to be the sequence $(\overline{y}_1, \overline{y}_2, \ldots, \overline{y}_{2p})$, where $\overline{y}_i = 2$ if $y_i \in A \cap B$, $\overline{y}_i = 0$ if $y_i \in A-B$, and $\overline{y}_i = 1$ if $y_i \in B-A$. We let t(B,A) = t(A,B).

Let T be a set of p-types. The graph (n,T,p) is defined by having as vertices all $\binom{n}{p}$ p-subsets of [1,n], and as edges, all pairs A,B of p-subsets with $t(A,B) \in T$. We define the clique number of T by cl(T) = $= \sup_n cl((n,T,p))$. (Not all T have finite clique number, e.g., {(0,1)} = T, although some do.)

The beautiful construction of (n,T,p) has the property that for large n it is extremely rich in induced subgraphs (m,T,p), for m < n. This enables us to use Ramsey's Theorem ultimately to obtain very well behaved subgraphs.

The first result in the general case is to show that for each H there exist p and T, so that H is an induced subgraph of (n,T,p) for all large n, and with cl(T) = cl(H). T and p are defined inductively, in general, and are quite complicated. However, for cl(H) = 2, we can describe T much more simply and, in fact, we can assert even more. Namely, for each H, let its vertices be ordered arbitrarily, say, $x_1, x_2, \ldots, x_k$. Then there is a mapping $\Phi: H \to (n,T,p)$ for a suitable n,T,p such that cl(T) = 2, and Φ maps H isomorphically into an induced subgraph of (n,T,p) with $t(\Phi(x_i),\Phi(x_j))$ depending only on j, if i < j. In the general case cl(H) = k, a similar result holds, but the proof is much more complicated. For the remainder of the discussion, we restrict ourselves to cl(H) = 2. The mapping Φ is defined inductively. T is the set of all types starting with some 0's, two 2's, then 0's and 1's only, e.g., (0,0,0,2,2,1,1,0,1,0,1,1). It is easy to see that cl(T) = 2.

Now suppose for large N that the edges of (N,T,p) are 2-colored. For each (2p-1)-subset S of [1,N] there are $\frac{(2p-1)}{2}\binom{2p-2}{p-1}$ pairs of p-subsets A,B with $A \cup B = S$. Of those pairs, some number m have their type in T. If we list these in some canonical order, say lexicographically, then we get for each $A \cup B$ a list of m types, corresponding to m edges, and thus m colors. But this produces a 2^m-coloring of the (2p-1)-subsets of [1,N]. Thus, for any n, if N is large enough, Ramsey's Theorem implies that there is a subgraph (n,T,p) of (N,T,p) with all edges of a given type having the same color.

Let H be an arbitrary graph with cl(H) = 2, and let G^* be such that $G^* \underset{1}{\to} (H,H)$, which exists by FOLKMAN's result. Letting Φ be as above, we have $\Phi(G^*) \subseteq (n,T,p) \subseteq (N,T,p)$. Each vertex x_j of G^* is associated with a single type $t(\Phi(x_i),\Phi(x_j))$ for i < j, and thus with a single color. By choice of G^*, then, we obtain a subgraph H all of whose vertices have the same color. But by the definition of this coloring, all edges of H have the same color. This completes the case cl(H) = 2, since by letting G = (N,T,p) we have $G \underset{2}{\to} (H,H)$. As previously remarked, the proof for the general case cl(H) = k is similar in spirit but with somewhat more complicated details. □

LINEAR EQUATIONS

Let $L = L(x_1,\ldots,x_n)$ denote a finite system of homogeneous linear equations in the variables $x_1,\ldots,x_n$ with integer coefficients. For a set S of integers, we write $S \to (\underbrace{L,\ldots,L}_{r})$, if L always has a monochromatic solution for any r-coloring of S. A system L is said to be *regular*, if, for all r, $\mathbb{P} \rightarrow (\underbrace{L,\ldots,L}_{r})$, where $\mathbb{P}$ denotes the set of positive integers.

R. RADO has characterized all regular L by generalizing the properties of the two best known examples. These are, respectively, L_2: $x+y = z$ and $L(k)$: $x_1-x_2 = x_2-x_3 = \ldots = x_{k-1}-x_k$. That L_2 is regular is SCHUR's theorem. Of course, the regularity of $L(k)$ is trivial (by choosing all the x_i equal). However, if we rule out this possibility, then a solution of $L(k)$ determines an arithmetic progression of length k. This restricted regularity of $L(k)$ for all k is just VAN DER WAERDEN's well-known theorem.

Unfortunately, however, this surprising result still does not specify *which* color these progressions have. It was conjectured some 40 years ago by ERDÖS and TURÁN that a solution must always occur in the most frequently occurring color. More precisely, they conjectured that if R is an infinite sequence of integers with positive upper density, i.e.,

$$(*) \qquad \overline{\lim_{n\to\infty}} \frac{|R \cap [1,n]|}{n} > 0 ,$$

then R contains arbitrarily long arithmetic progressions. No progress was made on this problem until 1954 when K.F. ROTH showed that if R satisfies (*), then R at least contains a *three*-term arithmetic progression. In fact, he showed more, namely, that for some $c > 0$, if $|R \cap [1,n]| > \frac{cn}{\log\log n}$ then R must contain a three-term arithmetic progression. The next significant step was not made until 1967 when SZEMERÉDI proved that (*) implies that R contains a *four*-term progression. However, SZEMERÉDI's most recent result, which must be considered an achievement of the first magnitude, finally settles the original conjecture of ERDÖS and TURÁN in the affirmative.

THEOREM. (SZEMERÉDI). (*) *implies* R *contains arbitrarily long arithmetic progressions.*

SKETCH OF SKETCH OF PROOF. SZEMERÉDI's proof is completely combinatorial in nature and is based on a lemma on bipartite graphs which is of considerable importance in its own right. We shall give a very brief discussion of the flavor of the proof (which runs just under 100 pages in length), although we can only hint at the extreme ingenuity used in the proof itself.

Let G denote a bipartite graph with vertex sets A and B. We call G regular if all vertices in A have the same degree and all vertices in B have the same degree. We would like to assert that every sufficiently large bipartite graph can be decomposed into a relatively small number of regular bipartite subgraphs, but unfortunately this is not true. However, it is true if the subgraphs are only required to be "approximately" regular and if we are allowed to ignore a small fraction of the vertices in A and B. More precisely, for $X \subseteq A$, $Y \subseteq B$, let $k(X,Y)$ denote the number of edges in the graph induced by the vertex sets X and Y and let $\beta(X,Y)$ denote $\frac{k(X,Y)}{|X||Y|}$, the density of edges in this induced subgraph. Then SZEMERÉDI proves the following:

<u>LEMMA</u>. *For all* $\varepsilon_1,\varepsilon_2,\delta,\rho,\sigma$ *strictly between* 0 *and* 1, *there exist integers* m_0,n_0,M,N *such that for all bipartite graphs* G *with* $|A| = m > M$, $|B| = n > N$ *there exist disjoint* $C_i \subseteq A$, $0 \le i < m_0$, *and for each* $i < m_0$, *disjoint* $C_{i,j}$, $j < n_0$, *such that:*

(a) $|A - \bigcup_{i<m_0} C_i| < \rho m, |B - \bigcup_{j<n_0} C_{i,j}| < \sigma n$ *for any* $i < m_0$;

(b) *for all* $i < m_0$, $j < n_0$, $S \subseteq C_i$, $T \subseteq C_{i,j}$, *with* $|S| > \varepsilon_1|C_i|$, $|T| > \varepsilon_2|C_{i,j}|$, *we have* $\beta(S,T) \ge \beta(C_i,C_{i,j})-\delta$;

(c) *for all* $i < m_0$, $j < n_0$ *and* $x \in C_i$, $\beta(\{x\},C_{i,j}) \le \beta(C_i,C_{i,j})+\delta$.

Condition (a) says that we have not omitted too many vertices in the decomposition. Conditions (b) and (c) express the approximate regularity of the subgraphs induced by the vertex sets C_i and $C_{i,j}$.

The basic objects dealt with in the proof are not just arithmetic progressions, but more general structures known as *configurations*. A 1-configuration is just a finite arithmetic progression; an m-configuration is a finite arithmetic progression of (m-1)-configurations.

Let R be an arbitrary fixed set of integers having positive upper density. The basic idea is to show inductively that there exist very long m-configurations which have an extremely restricted manner in which they intersect R. This is done by recursively defining certain special classes of higher order configurations in terms of rather well-behaved progressions of lower order configurations. Essentially, by showing that there exist extremely long configurations of some order which are moderately "regular", one can deduce the existence of configurations of a higher order which are even more "regular". This in turn is done by forming bipartite graphs based on the intersection patterns of the configurations with R and applying the decomposition lemma. Needless to say, the subtlety of the ideas used can only be appreciated by reading the actual proof. □

Turning our attention back to SCHUR's system L_2, we can generalize this to the system L_k defined as follows: for the variables x_s and y_S, L_k consists of all equations of the form $\sum_{s\in S} x_s = y_S$ where S ranges over all non-empty subsets of [1,k]. RADO's results imply that for all k and r,

$$\mathbb{P} \rightarrow (\underbrace{L_k,\ldots,L_k}_{r})$$

It is natural to ask what happens for the system

$$L_\infty = \{\sum_{s\in S} x_s = y_S \mid S \subseteq \mathbb{P},\ 1 \le |S| < \infty\}.$$

N. HINDMAN's remarkable theorem answers this question.

THEOREM. (HINDMAN). *For all* r, $\mathbb{P} \rightarrow (\underbrace{L_\infty,\ldots,L_\infty}_{r})$.

SKETCH OF PROOF. In the case L_k it is even true that for each r there is an $N = N(k,r)$ such that $N \rightarrow (\underbrace{L_k,\ldots,L_k}_{r})$. In other words, no matter which r-coloring we have, values $x_1,\ldots,x_k$ can be chosen from [1,N] so that all the sums $\sum_{s\in S} x_s$ have the same color. For a fixed r-coloring of $\mathbb{P}$ restricted to [1,N(k,r)] it was not known whether upper bounds for the x_i existed independent of k. The existence of such bounds would allow HINDMAN's theorem to be obtained directly by a "compactness" argument.

What HINDMAN proves is that for each coloring π of $\mathbb{P}$ with a finite number of colors, there is a function $f_\pi : \mathbb{P} \rightarrow \mathbb{P}$ such that for each m, $0 < m < \infty$, there is a set $x_1,\ldots,x_m$ with all its finite sums the same color, and in addition, such that $x_i \le f_\pi(i)$ for $1 \le i \le m$. That is, we get monochromatic solutions to L_m for arbitrarily large m, where the sizes of the variables x_i are bounded above independently of k but depending on the coloring π (of all of $\mathbb{P}$).

We can illustrate several of the ideas of the proof, but we need some notation first. Let π be a finite coloring of $\mathbb{P}$, say $\mathbb{P} = A_1 \cup \ldots \cup A_r$. For $1 \le k \le n$, we define

$$F_\pi(k,n) = \{x \in \mathbb{P} \mid x \ge n \text{ and } \exists i \text{ such that } k,x,x+k \in A_i,\ x \in F_\pi(j,n),\ j < k\}.$$

The $F_\pi(k,n)$ are sets which can be translated by k without changing color. If $x_1,x_2,\ldots$ is a sequence of integers, let $S(x_i)$ be the set of finite sums of the x_i.

The core of HINDMAN's proof is an "exceedingly technical" and quite clever argument, which establishes that for each π there is an infinite sequence $x_1, x_2, \ldots$ and $n \in \mathbb{P}$ such that

$$S(x_i) \subseteq \bigcup_{k=1}^{n-1} F_\pi(k,n).$$

To manipulate sequences and sums conveniently, it would be nice to know that the numbers in the sequences were representable to base 2 in the following manner, e.g.,

$$\begin{aligned} x_1 &= 10110111 \\ x_2 &= 110000100000000 \\ x_3 &= 110000000000000000 \\ &\vdots \end{aligned}$$

That is, the support of x_j should be all beyond the support of x_{j-1} for each j. Formally, if $2^{s-1} \le x_{j-1}$, then $2^s | x_j$. Such a sequence will be called a *good sequence*. Now for every sequence $x_1, x_2, \ldots$ there is a good sequence (not necessarily a subsequence of the x_i's) $y_1, y_2, y_3, \ldots$ with $S(y_i) \subseteq S(x_i)$. This follows from a compactness argument again.

Hence we basically need to deal only with good sequences. The nice property of these is that if $X = \{x_1, x_2, \ldots\}$ is a good sequence, there is a bijection $\tau_X \colon S(x_i) \to \mathbb{P}$ which preserves sums, namely,

$$\tau_X\left(\sum_{s \in S} x_i\right) = \sum_{s \in S} 2^{s-1}.$$

That is, each block of support corresponds under τ to a single binary place.

We use this fact crucially in the following construction. Suppose π is a coloring, and $x_{\pi 1}, x_{\pi 2}, x_{\pi 3}, \ldots$ is a good sequence with

$$S(x_{\pi i}) = \bigcup_{k=1}^{n(\pi)-1} F_\pi(k, n(\pi)).$$

Then using the map τ_π determined by this sequence, we can get a new coloring π' of $\mathbb{P}$ by letting two numbers have the same π'-color iff their images under τ_π^{-1} are in the same $F_\pi(k, n(\pi))$. This is an $(n(\pi)-1)$-coloring.

Suppose for *all* π we have defined $f_\pi(i)$ for $i \le \ell$ so that arbitrarily long finite good sequences have monochromatic sums and the i-th term is at most $f_\pi(i)$, $i \le \ell$, where we take $f_\pi(1) = n(\pi)-1$ (which works for $\ell = 1$ by the definition of $n(\pi)$). Then consider such a sequence for the coloring π' associated as above with π. Taking τ_π^{-1} of this sequence, we get a similar sequence which is constrained by the definition of π' to have its first ℓ terms respectively less than $\tau_\pi^{-1}(f_{\pi'}(i))$, $i \le \ell$, and all greater than $n(\pi)-1$. Further, they must all have the same π-color, and for some common $k \le n(\pi)-1$, adding k does not change this color. Then adjoining k as a first term gives us a new sequence with the first term not exceeding $f_\pi(1)$. Also, if we let $f_\pi(j) = \tau_\pi^{-1}(f_\pi(j-1))$, we have the j-th term not exceeding $f_\pi(i)$ for $j \le \ell+1$.

We have thus constructed, simultaneously for all π, the bounds $f_\pi(i)$. What we have shown, then, is that for each $\pi = A_1 \cup A_2 \cup \ldots \cup A_r$, and each k, there is a sequence $x_1,\ldots,x_k$ with all its sums in some $A_{i(k)}$ and $x_i \le$ $\le f_\pi(i)$, $1 \le i \le k$. As we noted above, a compactness argument now completes the proof. □

We remark that because the supports of the x_i in a good sequence are disjoint, we can interpret the x_i as disjoint subsets of $\mathbb{P}$ and their sums as disjoint unions. Thus, we obtain: *for every* r*-coloring of the finite subsets of* $\mathbb{P}$, *there exists an infinite sequence of finite disjoint sets* $A_1, A_2, A_3, \ldots$ *such that all the finite unions have the same color.*

The last of the results on equations is that of DEUBER, who settles a conjecture RADO raised in his original work. We recall that a system L of homogeneous linear equations is called regular if for any r, $\mathbb{P} \to (\underbrace{L,\ldots,L}_{r})$. RADO defined a *set* $S \subseteq \mathbb{P}$ to be *regular* if for every regular system L and any r, $S \to (\underbrace{L,\ldots,L}_{r})$. What RADO conjectured and what DEUBER proves is the following:

<u>THEOREM</u>. (DEUBER). *If* $S \subseteq \mathbb{P}$ *is regular and* $S = A \cup B$, *then either* A *or* B *is regular.*

<u>SKETCH OF PROOF</u>. The main idea of DEUBER's proof is to define certain sets, called (m,p,c)-sets, and to characterize regular sets in terms of (m,p,c)-sets. He then proves a finite RAMSEY theorem for these sets. Finally, by considering the nice structure of (m,p,c)-sets, he uses a compactness argument to establish the desired result.

We define (m,p,c)-sets below. However, we can describe them informally as a kind of ℓ-dimensional array of numbers (actually, certain subsets of these).

DEFINITION. For m,p,c positive integers, $p \geq c$, an *(m,p,c)-set* A is a set for which there exist m positive integers $a_1, a_2, \ldots, a_m$, such that $A = \{ \sum_{i=1}^m \lambda_i a_i \mid |\lambda_i| \leq p$, and the first non-zero coefficient λ_i has the value $c\}$.

Now using RADO's characterization of regular systems of equations, we can show the following two facts:

(a) for every regular system L there exist m,p,c such that every (m,p,c)-set contains a solution to L;

(b) for all m,p,c there is a regular system L such that every solution set for L contains an (m,p,c)-set.

As an example, consider the single equation $x+y = z$. Then a solution is any set of the form a_1, a_2, a_1+a_2, which is certainly contained in the (2,1,1)-set generated by a_1 and a_2. On the other hand, the equations $x+y = z_1$, $x-y = z_2$, have solutions exactly of the form $x,y,z_1,z_2 = a_1, a_2, a_1+a_2, a_1-a_2$, a (2,1,1)-set. These examples avoid $c \neq 1$, which can arise when the coefficients are more complicated.

By (a) and (b) we see that a regular set is any set containing (m,p,c)-sets for all m,p,c.

Suppose now that we know the following: for each (m,p,c) there is an (n,q,d) such that $(n,q,d) \rightarrow ((m,p,c),(m,p,c))$. That is, if the elements of any (n,q,d)-set are 2-colored, then there must be a monochromatic (m,p,c)-set. Thus for S regular, and $S = A \cup B$, either A or B must contain "arbitrarily large" (m,p,c)-sets and hence, by what we have noted, either A or B is regular. The main part of DEUBER's proof is concerned then with establishing the Ramsey property $(n,q,d) \rightarrow ((m,p,c),(m,p,c))$.

This result is similar to one of GALLAI concerning "n-dimensional arrays". For our purposes, we may consider an n-dimensional array as a set of the form

$$X_{n,p} = \{a_0 + \sum_{i=1}^{n} \lambda_i a_i \mid |\lambda_i| \leq p\}.$$

For these we have that for n, p and r there is an N such that $X_{N,p} \rightarrow (\underbrace{X_{n,p}, \ldots, X_{n,p}}_{r})$.

However, this isn't quite good enough for our purposes, since an (m,p,c)-set will contain sums of the form $ca_i + \sum_{j>i} \lambda_j a_j$ along with certain differences as well (e.g., a_1, a_1+a_2 and a_2 in the example above) while

$X_{N,p}$ may not contain any of its differences. To handle this problem we proceed iteratively.

First, we find a monochromatic

$$X_{N,p} = \{b_1 + \sum_{i=1}^{N} \lambda_i a_i \mid |\lambda_i| \le p\} = b_1 + Z_{N,p}$$

where

$$Z_{N,p} = \{ \sum_{i=1}^{N} \lambda_i a_i \mid |\lambda_i| \le p\}.$$

Then in $Z_{N,p}$ we find a monochromatic $b_2+Z_{N',p}$ etc. Continuing in this manner we can find $b_1, b_2, \ldots, b_\ell$ such that the color of the sum

$$b_i + \sum_{j>i} \lambda_j b_j$$

depends only on i. For large enough ℓ, we may select m of these b_j to generate a monochromatic (m,p,1)-set.

This completes the case c = 1. For c > 1, a similar argument can be applied where, however, at each step p must be adjusted to compensate for the effect of c. □

CATEGORIES

The notion of a category having the Ramsey property was introduced by K. LEEB. It has been used to prove the Ramsey property for the category of finite vector spaces, among others. A category $\mathcal{C}$ is said to be *Ramsey* if for any objects A,B and number r, there is an object C such that for any r-coloring of the A-subobjects of C, all the A-subobjects of some B-subobject of C have the same color. Formally this says:

$$\forall A,B,r\ \exists C \ni \forall C\binom{C}{A} \xrightarrow{f} [1,r]\ ,$$

$\exists$ a monomorphism, $B \xrightarrow{\phi} C$ and i such that the following diagram commutes:

$$\begin{array}{ccc} \mathcal{C}\binom{C}{A} & \xrightarrow{f} & [1,r] \\ \bar{\phi}\uparrow & & \uparrow \text{incl.} \\ \mathcal{C}\binom{B}{A} & \rightarrow & \{i\} \end{array} .$$

Here $\mathcal{C}\binom{C}{A}$ denotes the set of subobjects of C of isomorphism type A, and $\bar{\phi}$ the function induced by ϕ.

We could also abbreviate this by using the arrow notation. A category $\mathcal{C}$ is Ramsey if for every r and objects A,B there is an object C such that

$$C \rightarrow (\underbrace{B,B,\ldots,B}_{r}) .$$

To prove this property for a certain class of categories, including the category of sets (Ramsey's Theorem) and that of finite vector spaces, an elaborate induction is used. The induction is fundamentally determined by a generalization of the classical Pascal identity, $\binom{n+1}{k+1} = \binom{n}{k+1} + \binom{n}{k}$.

In his lecture notes on "Pascaltheorie", LEEB has developed more formally and generalized this kind of relationship and used it to prove some new Ramsey theorems, among other things. What we describe here is LEEB's generalization of the ordinary notion of labeled trees to that of trees labeled with objects from a category. A Ramsey theorem for these structures is then true if it was true in the original category.

Consider a category $\mathcal{C}$. Then the category $\mathcal{O}rd(\mathcal{C})$ is defined to be the category of finite sequences of objects from $\mathcal{C}$. That is, the objects of $\mathcal{O}rd(\mathcal{C})$ are finite sequences of objects of $\mathcal{C}$, and morphisms $(C_1,C_2,\ldots,C_k) \rightarrow$ $\rightarrow (D_1,D_2,\ldots,D_\ell)$, $k \le \ell$, are sequences $(\phi_1,\phi_2,\ldots,\phi_k)$ of morphisms from $\mathcal{C}$ such that $\phi_i: C_i \rightarrow D_{j(i)}$ for some $j(i)$, and $1 \le j(1) < j(2) < \ldots < j(k) \le \ell$.

We can define the category $\mathcal{T}rees(\mathcal{C})$ similarly. We consider rooted, labeled trees with an orientation, or ordering, of the branches at each vertex. We take the labels from the objects of $\mathcal{C}$. Morphisms are defined as follows. Let $\mathcal{T}_1,\mathcal{T}_2$ be two such objects, and let T_1,T_2 be their underlying rooted trees. First we "immerse" T_1 into T_2. An immersion $\psi: T_1 \rightarrow T_2$ is a monomorphic mapping from the vertices of T_1 to those of T_2 such that:

(a) For any two vertices x,y in T_1, $\psi(x\wedge y) = \psi(x) \wedge \psi(y)$, where for two vertices u,v in a rooted tree T, $u\wedge v$ denotes the last common vertex in the paths from the root to u and from the root to v, respectively.

(b) The order of the branches is preserved by ψ. That is, let $B_1,B_2,\ldots,B_k$ be the vertex sets of the branches at a vertex x in T_1, given in order, and let $D_1, D_2, \ldots, D_\ell$ be the vertex sets of the branches at $\psi(x)$ in T_2, given in order. Then for each i, $1 \le i \le k$, $\psi(B_i) \subseteq D_{j(i)}$ for some j(i), and $1 \le j(1) \le \ldots \le j(k) \le \ell$.

For example, the circled vertices in T_2 below indicate an immersion of T_1 into T_2:

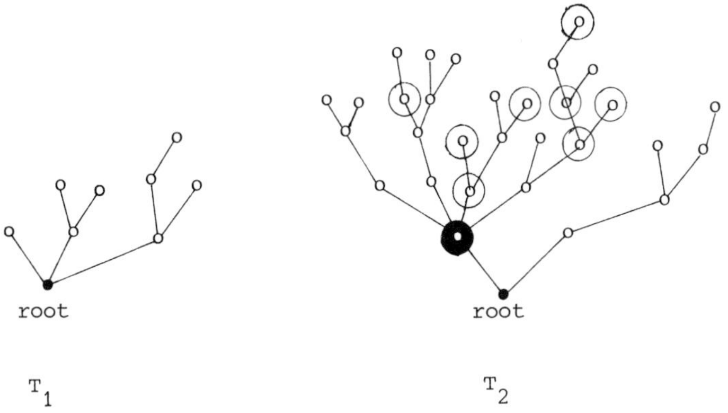

Once we have an immersion ψ of T_1 into T_2, we then find a set of morphisms from $\mathcal{C}$ taking the labels from T_1 into the corresponding labels (by the immersion) of T_2. Such sets of morphisms of $\mathcal{C}$ (with restrictions determined by (a) and (b)) are defined to be the morphisms of $\mathcal{Trees}(\mathcal{C})$. If we denote a $\mathcal{C}$-labeled tree by [a,B], where a is the root label and B the sequence of branches at the root (with labels), we get the Pascal identity:

$$\mathcal{Trees}(\mathcal{C})\binom{[a,B]}{[c,D]} = \coprod_{B_i \in B} \mathcal{Trees}(\mathcal{C})\binom{B_i}{[c,D]} + \mathcal{C}\binom{a}{c} \times \mathcal{Ord}(\mathcal{Trees}(\mathcal{C}))\binom{B}{D}.$$

What this says is that every subtree (labeled) of type [c,D] in a tree of type [a,B] either has its root at the root of [a,B], or lies entirely in one of the branches at the root, with labels mapped accordingly. If one considers the identity for trees with only one branch at each point, and $\mathcal{C}$ the category with only a single object, then this identity becomes the classical Pascal identity.

We say that a category $\mathcal{C}$ is *directed* if for any objects A and B, for some object C there exist monomorphisms A → C and B → C. What LEEB proves is the following:

THEOREM. (LEEB). *If C is Ramsey and directed, then Trees(C) is Ramsey and directed.*

The proof uses the Ramsey property for C, together with the standard "product" argument, also used to prove (among other things) the result of GALLAI mentioned in the previous section.

A related and less complicated result, using the same basic techniques, is that if C is Ramsey and directed, then so is Ord(C).

REFERENCES

[1] DEUBER, W., *Partitionen und lineare Gleichungssysteme*, Math. Z., to appear.

[2] DEUBER, W., *Generalizations of Ramsey's theorem*, Proc. Conference on Finite and Infinite Sets 1973, Keszthély (Hungary), 1974.

[3] HINDMAN, N., *Finite sums from sequences within cells of a partition of* ℕ, J. Combinatorial Theory A, to appear.

[4] LEEB, K., *Vorlesungen über Pascaltheorie*, Lecture Notes, Universität Erlangen, 1973.

[5] NEŠETŘIL, J. & V. RÖDL, *Ramsey properties of the categories of graphs with forbidden complete subgraphs*, to appear.

[6] SZEMERÉDI, E., *On sets of integers containing no* k *terms in arithmetic progression*, Acta Arith., to appear.

ON AN EXTREMAL PROPERTY OF ANTICHAINS IN PARTIAL ORDERS. THE LYM PROPERTY AND SOME OF ITS IMPLICATIONS AND APPLICATIONS *)

D.J. KLEITMAN

Massachusetts Institute of Technology, Cambridge, Mass. 02139, USA

I.

Let F be a collection of subsets of an n element set S, such that no member of F contains another. We call such a collection an *antichain*, in contrast to a collection that is totally ordered by inclusion which one usually calls a chain.

A well-known theorem of SPERNER states that an antichain F can have no more than $\binom{n}{[n/2]}$ members. YAMAMOTO and independently LUBELL and also MESHALKIN (and perhaps others) noticed that F satisfies a stronger property. If we denote the number of members of F of size k (having k elements) by f_k these numbers must satisfy

$$\sum_k f_k \Big/ \binom{n}{k} \leq 1 \ .$$

This result is stronger, in that the left-hand side is obviously greater than or equal to $\sum f_k \Big/ \binom{n}{[n/2]}$ so that we may immediately deduce that $\sum f_k \leq \binom{n}{[n/2]}$ which is SPERNER's theorem.

It is the purpose of this note first, to explore the class of partial orders (possessing a rank function) which satisfy this stronger property, called below the *LYM property*; that is in which an antichain under the partial order obeys a relation of the form $\sum f_k/N_k \leq 1$ where N_k represents the number of members of the order of rank k. Secondly, we show that this strong property is equivalent to a number of other properties of the partial order, which we shall shortly describe. Thirdly, we prove a number of general properties of such partial orders, and finally we give examples of application

---

*) Supported in part by ONR Contract N00014-67-A-0204-0063.

M. Hall, Jr. and J. H. van Lint (eds.), Combinatorics, 277-290.

of these properties.

II.

The LUBELL (and YAMAMOTO) proof of this property is remarkably simple. If we examine all the maximum sized chains, it is clear from symmetry among the n elements of S that each k element subset of S appears in exactly the same proportion of these chains as does any other, and hence in a proportion $1/\binom{n}{k}$ of them. Since no two members of F can lie in any maximal chain, the average number of members of F per chain, which is $\sum_{A\in F} 1/\binom{n}{|A|}$ (each A in F contributes $1/\binom{n}{|A|}$ to this average), cannot exceed one. This relation $\sum_{A\in F} 1/\binom{n}{|A|} \le 1$ is the LUBELL-YAMAMOTO inequality.

The form of the argument makes it immediately evident that an analogue holds for any partial order possessing a symmetry that is transitive on the members of each rank.

In any such order, one will again obtain that

$$\sum_{A\in F} 1/\binom{n}{|A|} \le 1$$

by the same argument. This implies that, for example, the lattice of subspaces of a vector space of finite dimension over a finite field, will possess this property. One can also deduce from these remarks, that they hold equally well if we weight the members of our partial order by a weighting function w that depends only upon rank. That is, we can deduce that

$$\sum_{A\in F} \frac{w(|A|)}{N_{|A|}} \le \max_k w(k),$$

or alternatively, for any weighting function w defined on the rank of the members of our order we must have

$$\sum_{A\in F} w(|A|) \le \max_k N_k w(k).$$

The form of the LUBELL-YAMAMOTO argument is in fact so simple that it leads one to expect that there is a necessary connection between the symmetry of the partial order and the workings of the argument.

However, the argument does not really require this; symmetry is unnecessary; the argument does not require that each set of a given rank occurs in the same proportion of all maximal chains.

What is required is that there exists some list of maximal chains that contains every set of each rank the same number of times. The argument will go through if we apply it not to the set of all possible maximal chains, but to this list of such chains.

A partial order will thus satisfy the LYM property whenever there exists a list of maximal chains such that each member of the order of rank k occurs in a proportion $1/N_k$ of the chains.

When then does such a list exist?

GRAHAM & HARPER introduced the concept of *normalized matching* in a partial order. We define it as follows. A partial order is said to have the normalized matching property, if, for every k, given any collection F of rank k members of the order there is a collection G_{k-1} of rank k-1 members of the order, such that every member of G_{k-1} is ordered with respect to at least one member of F, and such that

$$|G_{k-1}|/N_{k-1} \geq |F|/N_k.$$

We shall now show that this normalized matching property is equivalent to the LYM property under discussion. We prove the following theorem.

<u>THEOREM 1</u>. *Let* P *be a partial order with a rank function and* N_k *members of rank* k, *then the following four conditions are equivalent:*

1. *every antichain* F *satisfies*

$$\sum_{A\in F} 1/N_{k(A)} \leq 1, \qquad \textit{(LYM property)};$$

2. *every antichain* F *and real function* w *of* k, *the rank of* A, *satisfies*

$$\sum_{A\in F} w(|A|) \leq \max_k N_k w(k) \ ;$$

3. *there exists a list of maximal sized chains such that each member of* P *of rank* k *occurs in a proportion* $1/N_k$ *of the entries on the list;*
4. P *satisfies the normalized matching condition; that is, for every collection* F *of rank* k *members of* P *there is a collection* G_{k-1} *such that each member of* G_{k-1} *is ordered with respect to at least one member of* F

and

$$|G_{k-1}|/|F| \geq |N_{k-1}|/|N_k|.$$

PROOF. From the discussion previously, we have already noted that condition 3 implies conditions 1 and 2 and condition 2 implies condition 1 by choosing $w(k)$ to be $1/N_k$. We therefore need only prove that condition 4 implies condition 3, and that 1 implies 4.

We begin by proving the former. We assume that P satisfies the normalized matching condition. This condition can be seen to imply the ordinary PHILIP HALL matching condition if one takes a list of αN_{k-1} copies of the rank k members of P and a list of αN_k copies of the rank k-1 members for any integral α, and tries to match each member of the former list to one member of the latter ordered with respect to it in P. If we take any set of Q members of the former list, we will have at least $\lceil Q/\alpha N_{k-1} \rceil$ different members of rank k. By the normalized matching condition these will be ordered with respect to at least $\lceil Q/\alpha N_{k-1} \rceil N_{k-1}/N_k$ rank k-1 members of P and hence at least αN_k times as many or $\alpha N_{k-1} \lceil Q/\alpha N_{k-1} \rceil$ members (or at least Q members) of the second list. By the PHILIP HALL theorem there is a matching of the former to the latter.

We obtain a list of chains as required by condition 3 by starting at the highest rank r, starting with, for example, $(\Pi_{k=1}^{r-1} N_k)$ copies of the rank r members of P and performing the matching of the last paragraph for $k=r,r-1,\ldots,2$. Each entry on the list of rank r will be matched into an entry of rank r-1 ordered with respect to it, which will be matched into one of rank r-2, etc. The orbits under the matchings will be the desired maximal chains of condition 3.

To prove that condition 1 implies condition 4 we prove the contrapositive, that not 4 implies that 1 cannot hold.

Suppose therefore for some k there is a collection F of rank k members of P such that the collection G_{k-1} of rank k-1 members of P that are ordered with respect to one or more members of F satisfies

$$\frac{|G_{k-1}|}{|F|} < \frac{N_{k-1}}{N_k}.$$

Then consider the antichain connecting F and the rank k-1 members of P not in G. These satisfy, by trivial manipulation of the last inequality,

$$\frac{|F|}{N_k} + \frac{N_{k-1} - |G_{k-1}|}{N_{k-1}} = 1 + \left(\frac{|F|}{N_k} - \frac{|G_{k-1}|}{N_{k-1}}\right) > 1 .$$

Thus condition 1 implies condition 4 and the theorem is proved. □

III.

Granted all of this, what can we conclude about partial orders satisfying these conditions?

GRAHAM & HARPER, and also subsequently but independently HSIEH & KLEITMAN proved that direct products of partial orders each satisfying normalized matching and logarithmic convexity (for every k, $N_k^2 \geq N_{k-1}N_{k+1}$) also satisfy normalized matching and logarithmic convexity as well. Thus the lattice of divisors of an integer (with "divides" as order relation) satisfies this condition it being the direct product of chains. ANDERSON has also obtained these results for divisors of integers.

In consequence of our theorem, this lattice satisfies the LYM property: antichains obey the $\sum_{A \in F} 1/N_{k(A)} \leq 1$ inequality in the divisor-of-n lattice.

HSIEH & KLEITMAN proved further that if P_1 is a partial order satisfying normalized matching and P_2 is a chain, then the direct product $P_1 \otimes P_2$ satisfies normalized matching only if $P_1 \otimes P_2$ satisfies logarithmic convexity ($N_k^2 \geq N_{k-1}N_{k+1}$). If P_2 is an ordered pair, $P_1 \otimes P_2$ will satisfy normalized matching if and only if P_1 satisfies both normalized matching and logarithmic convexity.

On the other hand, the partial order consisting of partitions of an integer n ordered by refinement fails to satisfy normalized matching, at least for even n of magnitude at least 10. (For example, for n = 10, the partition 22222 is ordered with respect to only one partition into four integers (4222) while $N_4 = 9$ and $N_5 = 7$ where rank here is the number of blocks in the partition.) It is not known whether this partial order possesses the "Sperner property", that no antichain can have more than $\max_k N_k$ members.

GREENE & DILWORTH found an example of a geometric lattice which fails to satisfy either condition, Sperner or LYM. It is not known whether the lattice of partitions of a set satisfies either condition.

IV.

In a partial order satisfying any of the four conditions of theorem 1, a great many extremal properties of collections of order members can be deduced. We begin by enunciating a general theorem (which is itself a special case of the next one).

THEOREM 2. *Given a collection* F *of order members that obeys some restriction* R *and a weight function* w *of rank in* P, *with* P *a partial order satisfying the conditions of theorem 1, the sum of* w *over the collection is not greater than the maximum sum over chains satisfying* R *of the sum of* $w(|A|)N_{|A|}$

$$\sum_{A\in F} w(|A|) \le \max_{\substack{C \\ C \text{ chain, } C \text{ obeys } R}} \sum_{A\in C} w(|A|)N_{|A|}.$$

R *here and below is a restriction of the form "no set of members of* F *satisfies ...".*

This theorem has many powerful implications including a number of theorems of KATONA and of ERDÖS that we shall describe in the next section. Such theorems have a wide range of validity here - they hold in partial orders satisfying the LYM property. The proof is immediate; by the existence of our list of maximal chains we know that the average value $\sum_{A\in F} w(|A|)N_{|A|}$ over maximal chains is less than its maximal value for any maximal chain. Since A occurs in $1/N_{|A|}$ of the maximal chains, it contributes $w(|A|)$ to the average.

A still more general theorem is the analogue of the KLEITMAN-KATONA theorem. It is the following

THEOREM 3. *Given two partial orders* P_1 *and* P_2 *satisfying the LYM property; let* F *be a family of members of the direct product partial order* $P_1 \otimes P_2$, *subject to some restriction* R. *Let* w *be a function of rank in* P_1 *and* P_2. *Then the maximum value of the sum of* w(A) *for* A *in* F *cannot exceed the maximum over a subfamily of the direct product of two maximal chains, of* $w(r_1(A),r_2(A))\, N^{(1)}_{r_1(A)}\, N^{(2)}_{r_2(A)}$. *Here* $r_i(A)$ *is the rank of the* i*-th factor of* A. *That is*

$$\sum_{A\in F} w(r_1(A),r_2(A)) \le \max_{\substack{F'\subset C_1\otimes C_2 \\ F'\subset F,\ F' \text{ satisfying } R}} \sum_{A\in F'} w(A)N^{(1)}_{r_1(A)}\, N^{(2)}_{r_2(A)}.$$

This theorem has many important applications, as we shall discuss. It has an obvious generalization to direct products of k LYM-partial orders.

Its proof is also immediate, by exactly the argument previously applied to the lists of direct products of maximal chains in P_1 and P_2 whose existence follows from the LYM properties.

V.

We now examine some specific consequences of theorems 2 and 3. If in theorem 2 we let R be the restriction that if A,B are in F then $A \nmid B$, we have a portion of theorem 1. If we let R be the restriction that no k members of F form a chain, and set w = 1, we obtain a generalization of the theorem of ERDÖS that the maximum size of F is the sum of the largest (k-1) N_j's. If we let R be the restriction that no two members of F are ordered and differ by rank ≥k, we obtain the sum of the k largest consecutive N_j's. If we let R be the restriction that no two ordered members of F differ by less than (<) k in rank we obtain, if P is unimodular (N_j has only one local maximum), $\max_t \sum_{j=0}^{r/k} N_{t+jk}$ as a bound on the size of F (this is a generalization of a result of KATONA for divisors of an integer, and generalized somewhat differently by KATONA); likewise, if R restricts F to have no m+1 members within any rank interval k, for unimodular P one obtains the largest m values of $\sum_{j=0}^{r/k} N_{t+jk}$ for distinct t < j. This result again generalizes a result of KATONA for the lattice of divisors of an integer. Many other similar results follow as we have freedom to choose R and w as we please, always finding that the maximum sum over F can be evaluated by looking at the appropriately reweighted sum over a single chain.

Theorem 3 has a number of consequences that generalize known theorems as well as some new ones. Thus if R restricts F to contain no ordered pair we can deduce immediately that $P_1 \otimes P_2$ will satisfy the SPERNER property as long as P_1 and P_2 are both unimodular, while, by some detailed argument one can show using theorem 3, that $P_1 \otimes P_2$ will obey condition (1) of theorem 1 if both are logarithmically convex, thus providing a proof of the HARPER-GRAHAM theorem.

If R restricts F to contain no ordered pair that are identical in one factor, and if weight functions w_1, w_2 are chosen for P_1 and P_2 such that when arranged in decreasing order the $w_1(k)N_k^{(1)}$ are

$$w_1(k_1)N^{(1)}_{k_1},\ w_1(k_2)N^{(1)}_{k_2},\ldots,w_1(k_{r_1})N^{(1)}_{k_{r_1}}\ ,$$

and the $w_2(k)N^{(2)}_k$ arranged in decreasing order are

$$w_2(k'_1)N^{(2)}_{k'_1},\ w_2(k'_2)N^{(2)}_{k'_2},\ldots,w_2(k'_{r_2})N^{(2)}_{k'_{r_2}}\ ,$$

then the sum of the product of the weight functions over members of F can't exceed

$$\sum_j w_1(k_j)N^{(1)}_{k_j}\ w_2(k'_j)N^{(2)}_{k'_j}\,.$$

If $k_j+k'_j$ is a constant then this sum becomes a sum over a single rank in $P_1 \otimes P_2$. This will occur for example if after weighting both P_1 and P_2 are symmetric in rank and unimodular (a theorem of KLEITMAN and KATONA for subsets of a set). It will occur also if P_2 is P_1 turned upside down, or if P_1 has only two ranks and the largest rank values of $w_2(k)N^{(2)}_k$ are consecutive and appropriately ordered, and in a wide variety of other circumstances. Notice that even for subsets or divisors of an integer the result remains true here for non-trivial weight functions obeying certain rules.

Similar results hold if R restricts F to contain no ordered pair identical in one factor and differing by more than (>) k in the other. For unimodular symmetric P_1 and P_2 (after weighting), one obtains either the maximum of the sum of $N^{(1)}N^{(2)}w_1w_2$ over k consecutive ranks in $P_1 \otimes P_2$ or a related sum. Similar results hold for P_2 being P_1 upside down (i.e. with order relation reversed) or if (after weighting) the relative sizes of the wN's in P_1 and P_2 is as if P_2 was P_1 with order relations reversed.

One can obtain explicit best results when R is the restriction that no k members of F in $P_1 \otimes P_2$ that are identical in one factor form a chain. The results are not beautiful. For example if ranks in P_i run from 0 to r_i, and the population of rank k in P_i is $N^{(i)}_k$, we obtain for k = 3, and r_1 and r_2 even, that the size of F cannot exceed

$$\frac{2N^{(1+2)}_{\frac{r_1+r_2}{2}}}{2} - \left(\frac{N^{(1)}_{\frac{r_1}{2}}}{2} - \frac{N^{(1)}_{\frac{r_1+1}{2}}}{2}\right)\left(\frac{N^{(2)}_{\frac{r_2}{2}}}{2} - \frac{N^{(2)}_{\frac{r_2+1}{2}}}{2}\right) + 2\delta_{r_1r_2}N^{(1)}_0N^{(2)}_0$$

when P_1 and P_2 are symmetric and unimodular.

Such results can be used to obtain results for direct products of three or more partial orders; thus for P_1, P_2 symmetric and unimodular and r_1, r_2 even and $r_3 = 1$, the maximal size of a family F having no ordered pair of members differing in only one factor is, if $N_0^{(3)} < N_1^{(3)}$,

$$\frac{N_{r_1+r_2}^{(1+2)}}{2}(N_1^{(3)} + N_0^{(3)}) - N_0^{(3)}\left(\frac{N_{r_1}^{(1)}}{2} - \frac{N_{r_1+1}^{(1)}}{2}\right)\left(\frac{N_{r_2}^{(2)}}{2} - \frac{N_{r_2+1}^{(2)}}{2}\right) +$$

$$+ N_0^{(3)} \delta_{r_1 r_2}\left(N_0^{(1)} N_{r_2}^{(2)} + N_{r_1}^{(1)} N_0^{(2)}\right).$$

It may be possible to juggle the restrictions on F to obtain SPERNER like conclusions in triple products, particularly through use of the "no differences $\geq k$" result described above.

If R restricts F to contain no totally ordered members identical in one factor that differ by rank $<m$, then for symmetric unimodular orders P_1 and P_2 or for P_1 unimodular and P_2 being P_1 upside down, one obtains the maximum over j_0 of the sum of $N_i^{(1)} N_j^{(2)}$ over ranks whose sum is congruent to j_0 mod m. This is again a generalization of a divisor of n result of KATONA.

VI.

The basic method applied here to chains may be applied to other structures on orders defining appropriate classes of partial orders and developing properties of these. Similar ideas have been applied to partitions rather than chains. That is, given a list of partitions that contains members of identical rank the same number of times, one can draw similar conclusions about collections of subsets having disjointness (intersection) restrictions. Some examples of such results are described in [11].

If k divides n the ERDÖS-KO-RADO theorem follows immediately from this approach. It states that the number of non disjoint k element subsets of an n set cannot exceed a proportion k/n of them.

VII.

In this final section we give three examples of the application of some of the results in section V.

Consider sums of the form $\sum_{i=1}^{n} \varepsilon_i a_i$ for a_i vectors in two-dimensional Euclidean space of magnitude at least one, and $\varepsilon_i = \pm 1$. We could equally consider a wider coefficient set and remarks directly analogous to those that will follow can be made for such a set, and one in which the range of ε_i depends on i as well. For reasons of notational simplicity we shall confine ourselves to the present problem.

By elementary geometry, the sum of two or more vectors in any one quadrant each of magnitude at least one has magnitude at least $\sqrt{2}$. While the sum of three or more such vectors has magnitude at least $\sqrt{5}$, and the sum of k or more has magnitude at least $k\sqrt{2}$.

LITTLEWOOD & OFFORD raised the question, how many of the 2^n linear combinations considered above can lie inside a unit circle. KATONA raised (and solved) the same question for radius $\sqrt{2}$. If we arbitrarily divide the plane into quadrants, reverse the sign of enough a's (this doesn't effect the 2^n linear combinations) such that they all lie in two quadrants; we may use the facts of the last paragraph to bring these problems into the language of sets.

We can imagine the indices corresponding to vectors in the two quadrants forming sets S_1 and S_2. Each linear combination can be corresponded to a pair of subsets one of S_1, one of S_2 namely those for which $\varepsilon = +1$. If two linear combinations corresponding to the same subset in S_2 and to one subset containing the other in S_2 are to lie within a circle of radius r, they cannot differ by more than $[r\sqrt{2}]$ indices by those remarks, or for $r = \sqrt{5}$ by more than two indices.

We may conclude by one of the theorems of section V that the number of linear combinations lying in a circle of radius r cannot exceed the limits given in the following table. Details will be described in a subsequent paper.

$r = 1$	largest (KLEITMAN & KATONA) binomial coefficient,
$r = \sqrt{2}$	sum of largest two binomial coefficients (KATONA),
$r = \sqrt{5}$	sum of largest three binomial coefficients for $n \geq 5$,
$r = k$	sum of largest $2[k\sqrt{2}]$ binomial coefficients.

The results below $\sqrt{5}$ are best possible. The general k result, while not best possible, is interesting because it shows that for large n and reasonable sized k the number of linear combinations in a circle can grow *at most linearly* with the radius, *not* quadratically as does the area of the circle. A similar result in n dimensions would be quite interesting.

The results up to $\sqrt{2}$ have been obtained in arbitrary dimension by the present author by a different method; a best result is known also for $r \le \sqrt{3}$ in two dimensions, again proved by a different method.

A second application of the implications of theorem 3 is the proof of the HARPER-GRAHAM theorem itself. The theorem states that the direct product of two orders each satisfying LYM and each logarithmically convex will satisfy LYM and logarithmic convexity. Now LYM can be stated as the condition that for every chain $\sum_{A \in F} 1/N_a \le 1$. By theorem 3 this reduces to the inequality

$$\sum_{A \in F} \frac{N^{(1)}_{r_1(A)} \; N^{(2)}_{r_2(A)}}{N^{(1+2)}_{r_1(A)+r_2(A)}} \le 1$$

for F an antichain subfamily of the direct product of two chains. To prove the HARPER-GRAHAM theorem one must show that $P_1 \otimes P_2$ is logarithmically convex and that this inequality holds.

The former is a straightforward exercise (see HSIEH & KLEITMAN). We will outline an argument for the latter in the case that P_1 is itself a chain. In general, if P_1 has only two ranks, the inequality is easily seen to be equivalent to logarithmic convexity on P_2. It is relatively easy to show by induction that if C_j is a chain of length j, that

$$C_j \otimes P_1 \text{ satisfying LYM implies } C_{j+1} \otimes P_1 \text{ satisfies LYM}$$

(see HSIEH & KLEITMAN). Thus logarithmic convexity and LYM for P_2 implies that $C_j \otimes P_2$ satisfies LYM.

By a slighly more detailed argument one can verify LYM for all P_1 and P_2 satisfying the given conditions.

A final application makes use of the HARPER-GRAHAM theorem, extending a result of LEVINE & LUBELL. The LYM property is independent of weighting that is constant over rank; by the HARPER-GRAHAM theorem if the weighting maintains logarithmic convexity in each factor, the direct product partial order will possess both logarithmic convexity and LYM. Since logarithmic convexity is trivial for partial orders with only two ranks, we can conclude that the subset lattice which is a direct product of two rank chains will obey LYM and logarithmic convexity with a different weight function for each factor, the weight of a set being the product of the weight function over its elements. This is the LEVINE-LUBELL result. Now similar considerations apply to the divisors of an integer except that care must be taken to insure that logarithmic convexity holds in weighting each prime factor separately. This will hold if each factor p^k is weighted by $(X_p)^{\alpha_p(k)}$ for a convex function α_p. In the direct product the weighting is multiplicative over products of different prime factors.

We can conclude that for such weighting, the lattice of divisors of n still satisfies LYM. Thus, all the SPERNER like theorems of section IV hold here.

We content ourselves with three examples of such theorems.

1. Consider a collection of divisors of n no one dividing another. Then their sum cannot exceed the maximum of the same sum over collections having constant total degree.
2. Let n be a product of two relatively prime factors n_1 and n_2. Write any factor k of n as k_1k_2 with k_1/n_1, k_2/n_2. Then the sum of k_1/k_2 over a collection containing no chain of m+1 factors one dividing the next, cannot exceed the same sum over the m collections giving the largest value of this sum and having constant total degree.

   Many obvious generalizations may be made; the term k_1/k_2 can be replaced by k_1^a/k_2^b for any a,b; and one can subdivide n into more than two relatively prime factors with the same results.
3. In somewhat more generality, all of the results following from theorems 2 and 3 can be deduced for the divisor lattice with multiplicative weights of the form $(X_p)^{\alpha_p(k)}$ for each prime factor and convex α_p. With weight functions not constant over a rank the theorems take the following form.

THEOREM 2'. *In the lattice of divisors of an integer, if* $\alpha_p(k)$ *is a convex function of* k, *divisors containing only one prime factor* p *of degree* k *are weighted by a term* $(X_p)^{\alpha_p(k)}$ *and other divisors weighted with products of such terms over their prime factors, then*

$$\sum_{f\in F} w(f) \le \max_{C\subset F} \sum_{f\in C} \overline{w}(f)$$

with $\bar{w}(f) = \sum_{f'} w(f')$, $r(f') = r(f)$, r = *rank function of* f *in the lattice.*

THEOREM 3'. *If* n *is written as the product of two integers,* n_1 *and* n_2, *with similar weighting in the lattices of divisors of* n_1 *and* n_2 *the sum*

$$\sum_{f \in F} w(f)$$

cannot exceed the maximum over C_1, C_2, *chains of divisors of* n_1 *and* n_2, *of*

$$\sum_{f \in C_1 \otimes C_2} \bar{w}_1(f_1)\, \bar{w}_2(f_2)$$

with $\bar{w}_1(f) = \sum_{f'/n_i} w_i(f')$

$$r_i(f') = r_i(f).$$

Proofs of these theorems can be obtained by applying theorems 2 and 3 on partial orders obtained by replacing a given factor by a number of copies of it proportional to its weight, each ordered with respect to everything in it.

These theorems could be stated well for more general direct products.

The author would like to acknowledge many stimulating conversations with C. GREENE who suggested a number of improvements incorporated above.

REFERENCES

[1] ANDERSON, I., *On the divisors of a number*, J. London Math. Soc., 43 (1968) 410-418.

[2] DILWORTH, R.P. & CURTIS GREENE, *A counterexample to the generalization of Sperner's theorem*, J. Combinatorial Theory, 10 (1971) 18-21.

[3] ERDÖS, P., *On a lemma of Littlewood and Offord*, Bull. Amer. Math. Soc., 51 (1945) 898-902.

[4] GRAHAM, R. & L. HARPER, *Some results on matching in bipartite graphs*, preprint, 1968.

[5] HSIEH, W.N. & D. KLEITMAN, *Normalized matching in the direct product of partial orders*, preprint.

[6] KATONA, G., *On a conjecture of Erdös and a stronger form of Sperner's theorem*, Studies Su. Math. Hungar., 1 (1966) 59-63.

[7] KATONA, G., *A generalization of some generalizations of Sperner's theorem*, J. Combinatorial Theory B, 12 (1972) 72-81.

[8] KATONA, G., *Families of subsets having no subset containing another with small difference*, Nieuw Arch. Wisk., (3), 20 (1972) 54-67.

[9] KATONA, G., *Numbers of vectors in a circle of radius* $\sqrt{2}$, Akademiai Kiado, Budapest, Hungary, 1969.

[10] KATONA, G., *A simple proof of the Erdös-Ko-Rado theorem*, J. Combinatorial Theory B, 13 (1972) 183-184.

[11] KLEITMAN, D., M. EDELBERG & D. LUBELL, *Maximal sized antichains in partial orders*, J. Discrete Math., 1 (1971) 47-53.

[12] KLEITMAN, D., *Maximal number of subsets of a finite set no k of which are pairwise disjoint*, J. Combinatorial Theory, 5 (1968) 157-164.

[13] KLEITMAN, D., *On a lemma of Littlewood and Offord in the distribution of certain sums*, Math. Z., 90 (1968) 251-259.

[14] KLEITMAN, D., *How many sums of vectors can be in a circle of diameter* $\sqrt{2}$, Advances in Math., 9 (1972) 296.

[15] KLEITMAN, D., *On a lemma of Littlewood and Offord on the distribution of linear combinations of vectors*, Advances in Math., 5 (1970) 115-117.

[16] LEVINE, E. & D. LUBELL, *Sperner collections on sets of real variables*, Annals N.Y. Acad. Sciences, 75 (1970) 172-176.

[17] LUBELL, D., *A short proof of Sperner's lemmas*, J. Combinatorial Theory, 1 (1966) 299.

[18] MESHALKIN, L.D., *A generalization of Sperner's theorem on the number of subsets of a finite set*, Theory Probability Appl., 8 (1963) 203-204.

[19] SPERNER, E.S., *Ein Satz über Untermengen einer endlichen Menge*, Math. Z., 27 (1928) 544-548.

[20] YAMAMOTO, K., *Logarithmic order of free distributive lattices*, J. Math. Soc. Japan, 6 (1954) 343-353.

SPERNER FAMILIES AND PARTITIONS OF A PARTIALLY ORDERED SET *)

C. GREENE

Massachusetts Institute of Technology, Cambridge, Mass. 02139, USA

1. INTRODUCTION

This paper is a summary (without proofs) of the main results in a series of papers by the author and D.J. KLEITMAN [14] and the author [11, 12, 13] concerning subsets of a finite partially ordered set called *Sperner k-families*. If P is a finite partially ordered set, a subset $A \subseteq P$ is a *k-family* if A contains no chains of length k+1 (or, equivalently, if A can be expressed as the union of k 1-families in P). Maximum-sized k-families are called Sperner k-families of P.

The literature abounds with results about the maximum size of Sperner k-families for special classes of partially ordered sets. **) In this paper, however, we are not so much concerned with specific numbers as with structural properties. The results described here fall into one or more of the following categories:

(1) Bounds on the size of a k-family induced by partitions of P into chains.

(2) Relationships among numbers which arise for various values of k.

(3) Intersection theorems for collections of k-families.

(4) Complementary theorems obtained by interchanging the ideas of "chain" and "antichain".

(5) Properties of a lattice-ordering defined on k-families.

(6) Matching theorems and properties of certain submodular functions.

---

*) Supported in part by ONR N00014-67-A-0204-0063.

**) The name "Sperner k-family" comes from a generalization (due to ERDÖS [7]) of SPERNER's theorem on finite sets [17]. The generalization states that the maximum size of a k-family of subsets of an n-set is equal to the largest sum of k binomial coefficients $\binom{n}{j}$.

M. Hall, Jr. and J. H. van Lint (eds.), Combinatorics, 291-306. *All Rights Reserved.*

2. k-SATURATED PARTITIONS

If k = 1, k-families in P are also called *antichains*. Most of the results in this paper can be traced back to a deep theorem of DILWORTH [4], which states a basic relationship between chains and antichains of a partially ordered set:

THEOREM 2.1. *If every antichain of* P *has* d *or fewer elements, then* P *can be partitioned into* d *chains.*

The main object of [14] was to prove a similar statement about k-families. If $\mathcal{C} = \{C_1, C_2, \ldots, C_q\}$ is a partition of P into chains C_i, then, since chains meet k-families at most k times, it follows that no k-family can have more than

$$\sum_{i=1}^{q} \min\{k, |C_i|\}$$

members. Let $\beta_k(\mathcal{C})$ denote the bound induced by a partition $\mathcal{C}$ in this way, and let $d_k(P)$ denote the size of the largest k-family in P.

THEOREM 2.2. (cf. [14]). *For all* k, $d_k(P) = \min_{\mathcal{C}} \beta_k(\mathcal{C})$.

A partition $\mathcal{C}$ which satisfies $d_k(P) = \beta_k(\mathcal{C})$ is called a *k-saturated partition of* P. The fact that k-saturated partitions always exist is apparently much more difficult to prove than DILWORTH's theorem (which is a special case), and there are many interesting consequences.

Another way of stating theorem 2.2 is as follows:

THEOREM 2.3. (cf. [14]). *For all* k, $d_k(P) = \min_{S \subseteq P} \{|S| + k\, d_1(P-S)\}$.

Theorem 2.3 follows from the fact that a k-saturated partition remains k-saturated if all of the chains of length $\leq$ k are broken up into singletons.

Many important examples (Boolean algebras, integer divisors, subspaces of a vector space) have the property that P can be simultaneously k-saturated for all k. That is, one can find a partition $\mathcal{C}$ such that $\beta_k(\mathcal{C}) = d_k(P)$ for all k. However, this is not always possible if P is arbitrary.

<u>EXAMPLE 2.1</u>. Let P =

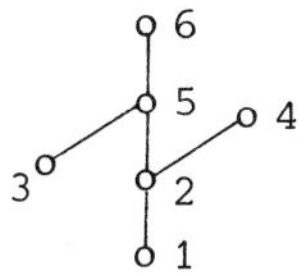

Then $d_1(P) = 2$, $d_2(P) = 4$, $d_3(P) = 5$ and $d_4(P) = 6$. $\mathcal{C} = \{\overline{1\ 2\ 4}, \overline{3\ 5\ 6}\}$ is 1-saturated and 2-saturated but not 3-saturated, while $\mathcal{C}' = \{\overline{3}, \overline{4}, \overline{1\ 2\ 5\ 6}\}$ is 2-saturated and 3-saturated but not 1-saturated. It is easy to see that no partition is simultaneously 1-, 2- and 3-saturated.

In view of this example, GREENE & KLEITMAN obtained the next best theorem on simultaneous k-saturation:

<u>THEOREM 2.4</u>. (cf. [14]). *For all* k, *there exists a partition which is simultaneously* k*-saturated and* (k+1)*-saturated.*

In fact, it was only by proving this stronger result that a proof of theorem 2.2 was obtained in [14]. The proof is by induction on $|P|$, and most of it is easy except for one critical step: for each i, define $\Delta_i(P) = d_i(P) - d_{i-1}(P)$, with $\Delta_1(P) = d_1(P)$ by convention. Then

<u>LEMMA 2.1</u>. (cf. [14]). *If* $\Delta_k(P) > \Delta_{k+1}(P)$, *there exists an element* $x \in P$ *which is contained in every Sperner* k*-family and every Sperner* (k+1)*-family of* P.

There is no reason to suppose that $\Delta_k(P) \geq \Delta_{k+1}(P)$ in general, but it turns out to be true. This result is more important (and less trivial) than it might seem at first glance:

<u>THEOREM 2.5</u>. (cf. [14]). $\Delta_k(P) \geq \Delta_{k+1}(P)$ *for all* k.

We know of no elementary proof of theorem 2.5, even when k = 2 (although it is trivial when k = 1). One difficulty is that there is apparently no combinatorial interpretation of $\Delta_k(P)$ if $k > 1$. It is not always true that a Sperner k-family can be obtained by adding $\Delta_k(P)$ elements to a Sperner (k-1)-family, as the following example shows:

<u>EXAMPLE 2.2</u>. Let P =

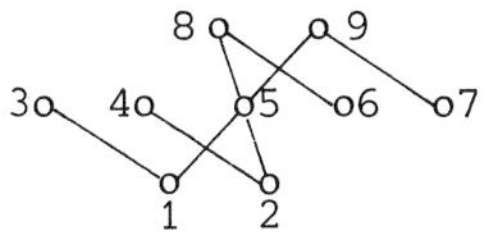

Then $d_1(P) = 5$ and $d_2(P) = 8$, but $\{3,4,5,6,7\}$ is the only 1-family of size 5 and $\{1,2,6,7\} \cup \{3,4,8,9\}$ is the only 2-family of size 8. Thus no 2-family of size 8 can be obtained by adding 3 elements to a 1-family of size 5.

If it is known that k-saturated partitions exist, theorem 2.5 is trivial, by the following easy lemma:

LEMMA 2.2. *Suppose that* $C = \{C_1,\dots,C_q\}$ *is a* k*-saturated partition of* P, *and exactly* h *of the chains have length* $\geq$ k. *Then* $\Delta_k(P) \geq h \geq \Delta_{k+1}(P)$.

Using theorem 2.4 and lemma 2.1, it is possible to completely characterize when the collection of all Sperner k-families of P has non-empty intersection.

THEOREM 2.6. (cf. [14]). *The following conditions are equivalent:*

(1) $\Delta_1(P) > \Delta_{k+1}(P)$.
(2) $d_{k+1}(P) > (k+1)d_1(P)$.
(3) *Every set of* k+1 *Sperner* k*-families has non-empty intersection.*
(4) *The collection of all Sperner* k*-families has non-empty intersection.*

The equivalence of the last two conditions suggests HELLY's theorem in k-dimensional euclidean space, which states that (3) and (4) are equivalent for any collection of convex sets. However, it is not true that Sperner k-families have the "Helly property" in a broad sense, since the property may not be inherited by subcollections.

3. COMPLEMENTARY PARTITIONS

DILWORTH's theorem (theorem 2.1) remains true if the words "chain" and "antichain" are interchanged. More surprising than the fact that this is true is its triviality by comparison with DILWORTH's theorem: define A_i to be the set of elements in P which have height i. (Define the *height* of an element x to be the length of the longest chain whose top is x.) If l is the length of the longest chain in P, then $A_1, A_2, \dots, A_l$ is a partition of P into antichains.

Thus it is natural to ask whether a similar transformation can be applied to lemma 2.1 (as well as the other results in section 2). It turns out that almost everything remains true, although at the present time the proofs do not seem to be trivial. These results were obtained by the author in [12].

We introduce the following terminology: if C is a subset of P which contains no antichains of size h+1, we call C an h-*cofamily* of P. By DILWORTH's theorem, $C = C_1 \cup \dots \cup C_h$ for some set of chains C_i. Let $\hat{d}_h(P)$ denote the size of the largest h-cofamily of P, and let $\hat{\Delta}_h(P) = \hat{d}_h(P) - \hat{d}_{h-1}(P)$. If $\mathcal{A} = \{A_1, A_2, \dots, A_k\}$ is a partition of P into antichains, let

$$\hat{\beta}_h(\mathcal{A}) = \sum_{i=1}^{k} \min\{h, |A_i|\}.$$

A partition $\mathcal{A}$ of P into antichains is h-*saturated* if $\hat{d}_h(P) = \hat{\beta}_h(\mathcal{A})$.

<u>THEOREM 3.1</u>. (cf. [12]). *For all* h *there exists a partition* $\mathcal{A}$ *of* P *into antichains which is both* h-*saturated and* (h+1)-*saturated.*

<u>THEOREM 3.2</u>. (cf. [12]). *For all* h, $\hat{\Delta}_h(P) \geq \hat{\Delta}_{h+1}(P)$.

By virtue of theorems 2.5 and 3.2, we can think of the numbers $\Delta_k(P)$ and $\hat{\Delta}_h(P)$ as forming the parts of a partition of the integer $|P|$, arranged in decreasing order. A remarkable relationship exists between these two sets of numbers: they are conjugate partitions.

<u>THEOREM 3.3</u>. (cf. [12]). *Define two partitions of* $|P|$ *as follows:* $\Delta(P) = \{\Delta_1(P) \geq \Delta_2(P) \geq \dots \geq \Delta_l(P)\}$, *where* l *is the length of the longest chain in* P, *and* $\hat{\Delta}(P) = \{\hat{\Delta}_1(P) \geq \hat{\Delta}_2(P) \geq \dots \geq \hat{\Delta}_d(P)\}$, *where* d *is the size of the largest antichain in* P. *Then* $\Delta(P)$ *and* $\hat{\Delta}(P)$ *are conjugate partitions. (That is,* $\hat{\Delta}_h(P)$ *equals the number of parts of* $\Delta(P)$ *of size* $\geq h$, *for all* h.*)*

As an illustration of theorem 3.3, consider the partially ordered set which appears in example 2.2. Since $d_1(P) = 5$, $d_2(P) = 8$ and $d_3(P) = 9$, the partition $\Delta(P)$ has shape

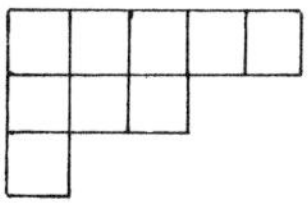

It is easy to check that $\hat{d}_1(P) = 3$, $\hat{d}_2(P) = 5$, $\hat{d}_3(P) = 7$, $\hat{d}_4(P) = 8$, and $\hat{d}_5(P) = 9$. (The partition $\mathcal{A} = \{1\ 2\ 6\ 7,\ 3\ 4\ 8\ 9,\ 5\}$ is 1-, 2- and 3-saturated, and the partition $\mathcal{A}' = \{1\ 2,\ 3\ 4\ 5\ 6\ 7,\ 8\ 9\}$ is 3-saturated, 4-saturated and 5-saturated.)

A more interesting example is obtained from the theory of permutations. Suppose that $\sigma = \langle a_1, a_2, \ldots, a_n \rangle$ is a sequence of distinct integers. Define P_σ to bet the set of pairs (a_i, i), with a partial order defined componentwise. It is easy to see that chains and antichains of P_σ correspond to increasing and decreasing subsequences of σ. Hence k-families (h-cofamilies) correspond to unions of k decreasing (h increasing) subsequences of σ. SCHENSTED [17] showed that the length of the longest increasing subsequence of σ could be computed by constructing a Young tableau (using what is now known as "Schensted's algorithm"), and counting the number of elements in the first row. Moreover, decreasing sequences can be considered by applying the same algorithm to σ in reverse order, in which case the tableau is transformed into its transpose.

In [11] the author extended SCHENSTED's theorem by giving a similar interpretation to the rest of the shape of the tableau associated with σ.

THEOREM 3.4. (cf. [11]). *Let σ be a sequence of distinct integers, and let P_σ be defined as above. If Schensted's algorithm maps σ onto a tableau of shape $\lambda = \{\lambda_1 \geq \lambda_2 \geq \ldots \geq \lambda_1\}$, then $\hat{d}_h(P_\sigma) = \lambda_1 + \lambda_2 + \ldots + \lambda_h$ for all h, and $d_k(P_\sigma) = \lambda_1^* + \lambda_2^* + \ldots + \lambda_k^*$ for all k. Hence $\hat{\Delta}_h(P_\sigma) = \lambda_h$ and $\Delta_k(P_\sigma) = \lambda_k^*$.*

Once it is proved that $\hat{\Delta}_h(P_\sigma) = \lambda_h$, it follows trivially that $\Delta_k(P_\sigma) = \lambda_k$, by reversing the order of σ. Hence theorem 3.3 is obvious in this case. It should be noted, however, that partially ordered sets of the form $P = P_\sigma$ are the *only* ones in which the relations of comparability and incomparability are interchangeable in this way.

Theorem 3.3 shows that we can associate a "shape" with every partially ordered set P, without actually constructing a tableau. When $P = P_\sigma$, this shape coincides with the one determined by Schensted's algorithm.

If $C = C_1 \cup \ldots \cup C_h$ is an h-cofamily of P, we can define a partition of P into chains by taking $C_1, C_2, \ldots, C_h$ and each of the elements in P-C as a singleton. Denote this partition by $\mathcal{C} = \{C_1, \ldots, C_h; P\text{-}C\}$.

THEOREM 3.5. (cf. [12]).

(i) *If $C = C_1 \cup \ldots \cup C_h$ is an h-cofamily of size $\hat{d}_h(P)$, then $\mathcal{C} = \{C_1, \ldots, C_h; P\text{-}C\}$ is a k-saturated partition for all k such that $\Delta_h^*(P) \geq k \geq \Delta_{h+1}^*(P)$.*

(ii) *If $\mathcal{C} = \{C_1, \ldots, C_h; T\}$ is a k-saturated partition, with each $|C_i| \geq k$,* *then* $\Delta_k(P) \geq h \geq \Delta_{k+1}(P)$, *and* $C = C_1 \cup \ldots \cup C_h$ *is an* h-*cofamily of size* $\hat{d}_h(P)$.

This result shows that, in a sense, h-cofamilies *are* k-saturated partitions provided that h and k are related properly. A similar statement holds for k-families and h-saturated partitions of P into antichains.

Next we mention a result which is the "complementary" analogue of theorem 2.6 (parts (3) and (4)):

THEOREM 3.6. (cf. [12]). *If every set of* h+1 h-*cofamilies of size* $\hat{d}_h(P)$ *has non-empty intersection, then there is an element* $x \in P$ *which is a member of every* h-*cofamily of size* $\hat{d}_h(P)$.

For example, if any two maximum-length chains have a common member, then they all have a common member.

The existence of "complementary" theorems makes one suspect that there might be a connection between these results and the theory of *perfect graphs*. A graph G is *perfect* if the analogue of DILWORTH's theorem holds for every subgraph of G. (We think of vertices connected by an edge as "comparable" and unconnected pairs as "incomparable". Hence chains correspond to complete subgraphs and antichains to independent sets.) If G is any graph, the complement G^* of G is obtained by interchanging the relations of "adjacent" and "non-adjacent". BERGE [1] conjectured and LOVÁSZ [16] proved (using ideas developed by FULKERSON [10]) that G^* is perfect whenever G is.

We have the following negative results:

(1) *Theorem 2.2 need not hold for perfect graphs. That is,* k-*saturated partitions (into complete subgraphs) do not always exist.*
(2) *If theorem 2.2 holds for all subgraphs of a graph* G, *it need not hold for* G^*.

An example which illustrates both observations is obtained by taking G and G^* to be the following graphs (both perfect):

$G =$ 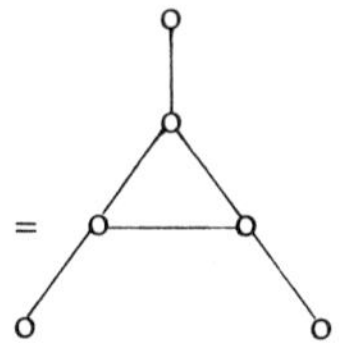 $G^* =$ 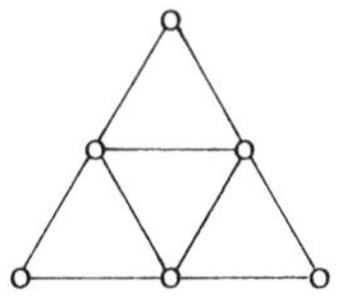

One easily computes

$$\begin{aligned} d_1(G) &= 3; & \hat{d}_1(G) &= d_1(G^*) = 3; \\ d_2(G) &= 5; & \hat{d}_2(G) &= d_2(G^*) = 4; \\ d_3(G) &= 6; & \hat{d}_3(G) &= d_3(G^*) = 6. \end{aligned}$$

It turns out that G^* has no 2-saturated partition, whereas G has 1-, 2- and 3-saturated partitions. Note that G^* also violates the condition $\Delta_1 \geq \Delta_2 \geq \Delta_3$.

On the other hand, G does not satisfy theorem 2.5. Although k-saturated partitions exist for k = 1,2,3, it is not possible to find a partition which is both 1-saturated and 2-saturated. The next theorem takes advantage of this loophole:

THEOREM 3.7. (cf. [12]). *Let* G *be a graph with the property that, for all* k, *there exists a partition of* G *into complete subgraphs which is both* k-*saturated and* (k+1)-*saturated. Then* G^* *also has this property.*

4. THE LATTICE OF k-FAMILIES

The technical details of [14] were based on a careful study of a natural ordering which can be defined on the set of all Sperner k-families of P. In particular, properties of this ordering were used to prove lemma 2.1, after which most of the other results in section 2 follow by relatively simple arguments.

Let $\mathcal{F}_k(P)$ denote the set of all k-families of P, and let $\mathcal{S}_k(P)$ denote the set of all Sperner k-families of P. If k = 1, we define an ordering on $\mathcal{F}_k(P)$ and $\mathcal{S}_k(P)$ as follows: if A and B are antichains, we say that $A \leq B$ if every element of A is $\leq$ some element of B. The following results are well-known:

THEOREM 4.1. (BIRKHOFF [2]). $\mathcal{F}_1(P)$ *is a distributive lattice.*

THEOREM 4.2. (DILWORTH [5]). $\mathcal{S}_1(P)$ *is a sublattice of* $\mathcal{F}_1(P)$ *(and hence is distributive).*

It is easy to describe the lattice operations in $\mathcal{F}_k(P)$. If U is any subset of P, define max [U] to be the set of maximal elements of U, and define $\overline{U}$ to be the order ideal generated by U (that is, $\overline{U}$ is the set of elements $\leq$ some member of U.)

LEMMA 4.1. *For any antichains* A *and* B $\in F_1(P)$,

(1) $A \leq B$ *if and only if* $\bar{A} \subseteq \bar{B}$;

(2) $A \vee B = \max[\bar{A} \cup \bar{B}] = \max[A \cup B]$;

(3) $A \wedge B = \max[\bar{A} \cap \bar{B}]$;

(4) $|A \vee B| + |A \wedge B| \geq |A| + |B|$.

Theorem 4.2 above is an immediate corollary of inequality (4). To prove (4), it is convenient to introduce an auxiliary operation $A \Delta B = ((A \cup B)-(A \vee B)) \cup (A \cap B)$. That is, $A \Delta B$ is the set of "non-maximal" elements of $A \cup B$ (plus all elements which occur twice). It is immediate that $A \Delta B \subseteq A \wedge B$ and $|A \vee B| + |A \Delta B| = |A| + |B|$, from which (4) follows. Using these properties alone, it is possible to obtain some interesting facts about antichains:

COROLLARY 4.1. (KLEITMAN, EDELBERG & LUBELL [15]). *There exists an antichain* A *of maximum size in* P *which is invariant under every automorphism of* P.

The proof is easy: take A to be the top element of $S_1(P)$. This argument (due to FREESE [8]) can be used to give a one-line proof of SPERNER's theorem for Boolean algebras.

Another application is the following: if A^+ and A^- denote the top and bottom elements of $S_1(P)$, then every member of $S_1(P)$ lies between them. Hence A^+ and A^- have non-empty intersection if and only if all of the members of $S_1(P)$ have non-empty intersection, and we have a special case of theorem 2.6:

COROLLARY 4.2. *The antichains of maximum size in* P *have non-empty intersection if and only if any two of them have non-empty intersection.*

Next we extend the ordering defined on antichains to $F_k(P)$ and $S_k(P)$. If A is a k-family, then A can always be partitioned into antichains $A_1, A_2, \ldots, A_k$ by taking $A_1 = \max[A]$, $A_2 = \max[A-A_1]$, $A_3 = \max[A-A_1-A_2]$, and so forth. That is, A_i is the set of elements of "depth" i in A. We call this partition the *canonical partition of* A.

If A and B are k-families, define $A \leq B$ if $A_i \leq B_i$ $(1 \leq i \leq k)$, where A_i and B_i denote antichains in the canonical partitions of A and B. It is clear that this definition makes both $F_k(P)$ and $S_k(P)$ into partially ordered sets. To show that both are lattices, we must define new operations:

$$A \vee B = \bigcup_{i=1}^{k} A_i \vee B_i,$$

$$A \,\Delta\, B = \bigcup_{i=1}^{k} A_i \,\Delta\, B_i,$$

$$A \overset{\wedge}{*} B = \bigcup_{i=1}^{k} A_i \wedge B_i.$$

THEOREM 4.3. (cf. [14]). *$F_k(P)$ is a lattice, in which the join of two* k-*families is given by* $A \vee B$.

However, it may not be true that either $A \,\Delta\, B$ or $A \overset{\wedge}{*} B$ coincides with $A \wedge B$ (the true g.l.b. of A and B in $F_k(P)$).

EXAMPLE 4.1. Let P =

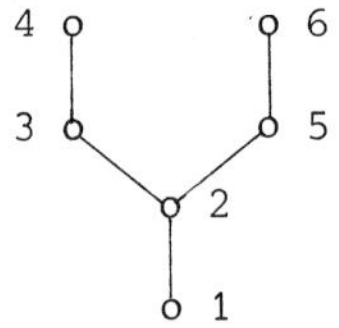

and $A = \{4,3\}$, $B = \{6,5\}$. Then $A \,\Delta\, B = \emptyset$ and $A \overset{\wedge}{*} B = \{2\}$. Yet $A \wedge B = \{2,1\}$.

A procedure for computing $A \wedge B$ was described in [14].

LEMMA 4.2. *For any* $A,B \in F_k(P)$

(1) $A \,\Delta\, B \subseteq A \overset{\wedge}{*} B \subseteq A \wedge B$;

(2) $|A \vee B| + |A \,\Delta\, B| = |A| + |B|$;

(3) $|A \vee B| + |A \wedge B| \geq |A| + |B|$.

It follows from inequality (3) that $S_k(P)$ is closed under $\vee$ and $\wedge$, and hence forms a sublattice of $F_k(P)$. If $k > 1$, it is no longer true in general that $F_k(P)$ is distributive (although it can be shown to be "locally distributive"). Hence one cannot conclude from (3) that $S_k(P)$ is a distributive lattice. Surprisingly, this turns out to be true anyway.

THEOREM 4.4. (cf. [14]). *$S_k(P)$ is a distributive sublattice of $F_k(P)$. Moreover, if $A,B \in S_k(P)$, then* $A \wedge B = A \,\Delta\, B = A \overset{\wedge}{*} B$.

The structure of $F_k(P)$ and $S_k(P)$ is discussed more carefully in [14]. We conclude this section by giving an application of theorem 4.4:

THEOREM 4.5. (cf. [14]). *If* P *is any partially ordered set, then for each* $k \geq 1$ *there exists a Sperner* k*-family* $A \in S_k(P)$ *which is invariant under every automorphism of* P.

This extends corollary 4.1 to k-families. The proof is exactly the same.

5. GRADED MULTIPARTITE GRAPHS

A different approach to the study of k-families was taken by the author in [13]. Essentially, the idea was to extend FULKERSON's method of obtaining DILWORTH's theorem from HALL's matching theorem [9].

If P is a partially ordered set, define $\Gamma_k(P)$ to be the *graded multipartite graph* obtained by taking k+1 copies of P (denoted by $P_1, P_2, \ldots, P_{k+1}$) and connecting $x \in P_i$ to $y \in P_{i+1}$ if $x < y$ in P. A *partial matching* in $\Gamma_k(P)$ is a collection of disjoint paths of length (k+1) which link some element of P_1 to some element of P_{k+1}. If $k = 1$, FULKERSON observed that the edges of a maximum partial matching can always be joined to form a minimum partition of P into chains. A proper interpretation of the HALL-ORE matching condition gives DILWORTH's theorem immediately. If $k > 1$, the situation is somewhat more complicated, since the first part of FULKERSON's argument is difficult to duplicate. Nevertheless, the second part carries over easily, and we obtain the following:

THEOREM 5.1. (cf. [13]). *The maximum number of paths in a partial matching in* $\Gamma_k(P)$ *is equal to* $|P| - d_k(P)$.

Theorem 5.1 is proved by showing that every minimal separating set in $\Gamma_k(P)$ is obtained by partitioning a set of the form P-A, where A is a Sperner k-family. Minimal separating sets can be found using a flow algorithm (or other methods in this case) and hence there is an effective procedure for constructing Sperner k-families.

If l is the length of the longest chain in P, and $k = l-1$, it is easy to see that disjoint paths in $\Gamma_k(P)$ correspond to disjoint maximum-length chains in P. Moreover A is a Sperner k-family if and only if P-A meets every chain of length l.

COROLLARY 5.1. *The maximum number of disjoint* 1-*chains in* P *is equal to the minimum number of elements in* P *which meet every* 1-*chain.*

This result is just another way of stating that (1-1)-saturated partitions exist. Hence we obtain an easier proof of theorem 2.2 in this case. It is interesting to note that corollary 5.1 remains true if chains are replaced by antichains (theorem 3.1), although there is apparently no analogous proof using flows.

If we attempt to extend FULKERSON's proof of DILWORTH's theorem, the difficulty for $k > 1$ is that partial matchings in $\Gamma_k(P)$ cannot be readily transformed into collections of chains in P. However, the converse problem is trivial: if $\mathcal{C} = \{C_1, C_2, \ldots, C_q\}$ is a partition of P into chains a partial matching in $\Gamma_k(P)$ is obtained by taking all consecutive segments of length (k+1) appearing in $\mathcal{C}$. Moreover, it is easy to see that the number of paths in the matching is exactly $|P| - \beta_k(\mathcal{C})$. Hence another (more constructive) proof of theorem 2.2 follows if we show that every maximum partial matching in $\Gamma_k(P)$ can be "straightened out", so that it corresponds to one obtained from a partition $\mathcal{C}$ by taking consecutive segments of length (k+1). This is easy for small values of k but becomes more difficult as k increases. A general algorithm was described by the author in [13].

6. SUBMODULAR FUNCTIONS

In this section we describe a class of combinatorial geometries which can be associated with a partially ordered set by means of submodular functions related to k-families. The basic tool is the identity

$$|A \vee B| + |A \Delta B| = |A| + |B| \qquad \text{(lemma 4.2)}.$$

LEMMA 6.1. (cf. [14]). *For any* k, *the function* d_k *is a super-modular function on order ideals of* P. *That is, if* $\overline{M}$ *and* $\overline{N}$ *are order ideals of* P, *then* $d_k(\overline{M} \cup \overline{N}) + d_k(\overline{M} \cap \overline{N}) \geq d_k(\overline{M}) + d_k(\overline{N})$.

If P is any partially ordered set, let P_0 be the set of maximal elements of P, and let $P^* = P - P_0$. If $X \subseteq P_0$ define

$$\delta_k(X) = d_k(P^* \cup X) - d_k(X), \quad \text{and} \quad r_k(X) = |X| - \delta_k(X).$$

THEOREM 6.1. (cf. [14]). *With the above notation,* r_k *is the rank function of a combinatorial geometry on* P_0. *That is*

(1) $r_k(\emptyset) = 0$;

(2) $r_k(X) \le r_k(X \cup p) \le r_k(X)+1, \qquad X \subseteq P_0,\ p \in P_0$;

(3) $r_k(X \cup Y) + r_k(X \cap Y) \le r_k(X) + r_k(Y), \qquad X,Y \subseteq P_0$.

If $k = 1$ and P has height two (so that P_0 and P^* are both antichains) r_k coincides with the usual rank function on bipartite graphs associated with the HALL-ORE matching theorem. In this case, a set X is independent if it forms the set of initial vertices of a matching. In general, one can give the following interpretation of what it means for a set to be independent:

THEOREM 6.2. (cf. [14]). *A subset* $X \subseteq P_0$ *satisfies* $r_k(X) = |X|$ *if and only if there exists a* k-*saturated partition of* P^* *and a matching of* X *into the set of tops of chains of length* $\ge k$.

We conjecture that this geometry is actually *induced* (in the sense of [3]) by another geometry on P^* defined by taking bases to be the sets of tops of chains of length $\ge k$ formed by k-saturated partitions of P^*. We have not been able to prove or disprove this.

If $k = 1$, however, it is true. Form the bipartite graph $\Gamma_1(P^*)$ (defined in section 5), and consider the standard transversal geometry which it determines on P^*. A subset $B \subseteq P^*$ is a basis if and only if for some 1-saturated partition of P^*, B is the set of elements which are not tops of chains. If we take the dual of this geometry, then the sets which are tops of chains become bases and we have proved the following:

THEOREM 6.3. *If* Q *is any partially ordered set, let* $\mathcal{B}(Q)$ *denote the collection of subsets of* Q *which are the tops of chains in some minimal partition of* Q *into chains. Then* $\mathcal{B}(Q)$ *is the set of bases of a combinatorial geometry.*

We conclude this paper with an application of theorems 6.1 and 6.2, giving a completely different proof of theorem 2.4 in the special case $k = 1$.

THEOREM 6.4. *For any partially ordered set* P, *there exists a partition of* P *into chains which is* 1-*saturated and* 2-*saturated.*

PROOF. Let A be an antichain of maximum size in P, and let P^+ and P^- denote the parts of P which lie above A and below A, respectively. Using the partially ordered set $P^+ \cup A$, define a geometry $G^+(A)$ on A, whose rank function

r^+ is obtained by taking k = 1 in the definition preceding theorem 6.1. Similarly, define a geometry $G^-(A)$ with rank function r^-, by using $P^- \cup A$ in the same way. Then $r^+(A) = |A| - (d_1(P^+ \cup A) - d(P^+)) = d_1(P^+)$, and $r^-(A) = |A| - (d_1(P^- \cup A) - d_1(P^-)) = d_1(P^-)$. Let X be a basis of $G^+(A)$ and let Y be a basis of $G^-(A)$. By theorem 6.2 we can partition $X \cup P^+$ into |X| chains, and $Y \cup P^-$ into |Y| chains. By linking these chains together and adding the remaining singletons of A, we obtain a partition of P into $d_1(P)$ chains which has exactly $|X \cup Y| = d(P^+) + d(P^-) - |X \cap Y|$ chains of length two or more. This partition (which is trivially 1-saturated) will be 2-saturated if the number of chains of length two or more is $d_2(P) - d_1(P)$. Hence we must show that X and Y can be chosen so that

$$|X \cap Y| = d_1(P) - d_2(P) + d_1(P^+) + d_1(P^-).$$

But this turns out to be a direct consequence of EDMONDS' matroid intersection theorem [6]. According to EDMONDS' theorem, there exists a set of size q which is independent in both $G^+(A)$ and $G^-(A)$ if and only if $q \leq r^+(U) + r^-(A-U)$ for all $U \subseteq A$. But

$$\begin{aligned}\min_{U\subseteq A} (r^+(U)+r^-(A-U)) &= \min_{U\subseteq A}\{(|U|-d_1(P^+ \cup U) + d_1(P^*)) + \\ &\quad + (|A-U|-d_1(P^- \cup (A-U)) + d_1(P^-))\} = \\ &= d_1(P) + d_1(P^+) + d_1(P^-) - \max_{U\subseteq A}\{d_1(P^+ \cup U) + \\ &\quad + d_1(P^- \cup (A-U))\} \geq \\ &\geq d_1(P) + d_1(P^+) + d_1(P^-) - d_2(P)\end{aligned}$$

as desired. (The last step follows if we observe that the sum in brackets is the size of some 2-family.) □

It seems likely that an extension of this argument could be used to prove theorem 2.4 for arbitrary k, but so far we have not been able to find such a proof.

ACKNOWLEDGEMENTS

The author would like to acknowledge helpful discussions with DANIEL J. KLEITMAN, JOEL SPENCER and RICHARD P. STANLEY.

REFERENCES

[1] BERGE, C., *Färbung von Graphen deren samtliche bzw. ungerade Kreise starr sind (Zusammenfassung)*, Wiss. Z. Martin Luther Univ. Halle Wittenberg, Math. Nat. Reihe, 1961, pp. 114.

[2] BIRKHOFF, G., *Lattice theory*, AMS Coll. Publ. 25, Amer. Math. Soc., Providence, 1967.

[3] BRUALDI, R., *Induced matroids*, Proc. Amer. Math. Soc., 29 (1971) 213-221.

[4] DILWORTH, R.P., *A decomposition theorem for partially ordered sets*, Ann. of Math., 51 (1950) 161-166.

[5] DILWORTH, R.P., *Some combinatorial problems on partially ordered sets*, in: *Combinatorial analysis*, R. BELLMAN & M. HALL Jr. (eds.), Proc. Symp. Appl. Math., Amer. Math. Soc., Providence, 1960, pp. 85-90.

[6] EDMONDS, J., *Submodular functions, matroids and certain polyhedra*, in: *Combinatorial structures and their applications*, Proc. Calgary Internat. Conference 1969, Gordon and Breach, New York, 1970, pp. 69.

[7] ERDÖS, P., *On a lemma of Littlewood and Offord*, Bull. Amer. Math. Soc., 51 (1945) 898-902.

[8] FREESE, R., *An application of Dilworth's lattice of maximal antichains*, Discrete Math., 7 (1974) 107-109.

[9] FULKERSON, D.R., *A note on Dilworth's theorem for partially ordered sets*, Proc. Amer. Math. Soc., 7 (1956) 701-702.

[10] FULKERSON, D.R., *Anti-blocking polyhedra*, J. Combinatorial Theory B, 12 (1972) 50-71.

[11] GREENE, C., *An extension of Schensted's theorem*, to appear.

[12] GREENE, C., *Partitions and conjugate partitions of a partially ordered set*, to appear.

[13] GREENE, C., *Construction of Sperner k-families and k-saturated partitions*, to appear.

[14] GREENE, C. & D.J. KLEITMAN, *On the structure of Sperner k-families*, to appear.

[15] KLEITMAN, D.J., M. EDELBERG & D. LUBELL, *Maximal sized antichains in partial orders*, Discrete Math., 1 (1971) 47-53.

[16] LOVÁSZ, L., *Normal hypergraphs and the perfect graph conjecture*, Discrete Math., 2 (1972) 253-267.

[17] SCHENSTED, C., *Longest increasing and decreasing subsequences*, Canad. J. Math., 13 (1961) 179-191.

[18] SPERNER, E., *Ein Satz über Untermengen einer endlichen Menge*, Math. Z., 27 (1928) 544-548.

COMBINATORIAL RECIPROCITY THEOREMS *)

R.P. STANLEY

Massachusetts Institute of Technology, Cambridge, Mass. 02139, USA

A *combinatorial reciprocity theorem* is a result which establishes a kind of duality between two related enumeration problems. This rather vague concept will become clearer as more and more examples of such theorems are given. We shall be content in this paper with explaining the meaning of various reciprocity theorems *via* mere statements of results, together with clarifying examples. A rigorous treatment with detailed proofs appears in [11].

1. POLYNOMIALS

A *polynomial reciprocity theorem* takes the following form. Two combinatorially defined sequences $S_1, S_2, \ldots$ and $\bar{S}_1, \bar{S}_2, \ldots$ of finite sets are given, such that the functions $f(n) = |S_n|$ and $\bar{f}(n) = |\bar{S}_n|$ are polynomials in n for all integers $n \geq 1$. One then concludes that $\bar{f}(n) = (-1)^d f(-n)$, where $d = \deg f$. Frequently the numbers $f(0)$ and $\bar{f}(0)$ will have a special significance.

EXAMPLE 1.1. Fix $p > 0$. Let $f(n)$ be the number of combinations *with repetitions* of n things taken p at a time. Let $\bar{f}(n)$ be the number of such combinations *without repetitions*. Thus $f(n) = \binom{n+p-1}{p}$ and $\bar{f}(n) = \binom{n}{p}$. Hence it can be verified by inspection that $f(n)$ and $\bar{f}(n)$ are polynomials in n of degree p, related by $\bar{f}(n) = (-1)^p f(-n)$.

EXAMPLE 1.2. (THE ORDER POLYNOMIAL). Let P be a finite partially ordered set of cardinality $p > 0$. Let $\omega: P \to [p]$ be a fixed bijection, where we use the "French notation" $[p] = \{1,2,\ldots,p\}$. Let $\Omega(n)$ denote the number of maps

*) Supported by NSF Grant #P36739 at M.I.T., Cambridge, Mass., USA.

M. Hall, Jr. and J. H. van Lint (eds.), Combinatorics, 307-318.

σ: P $\to$ [n] such that (i) $x \le y$ in P implies $\sigma(x) \le \sigma(y)$, and (ii) $x < y$ in P and $\omega(x) > \omega(y)$ implies $\sigma(x) < \sigma(y)$. Let $\bar{\Omega}(n)$ denote the number of maps τ: P $\to$ [n] such that (i) $x \le y$ in P implies $\tau(x) \le \tau(y)$, and (ii) $x < y$ in P and $\omega(x) < \omega(y)$ implies $\tau(x) < \tau(y)$. Then it can be shown [8, Proposition 13.2(i)] that Ω and $\bar{\Omega}$ are polynomial functions in n of degree p related by $\bar{\Omega}(n) = (-1)^p\Omega(-n)$. We call Ω the *order polynomial* of (P,ω). There are several ways to prove this reciprocity relationship between Ω and $\bar{\Omega}$, perhaps the simplest by a judicious use of the Principle of Inclusion-Exclusion which we leave to the reader. Note that if P is a p-element chain and ω is order-preserving, then $\Omega(n) = \binom{n+p-1}{p}$ and $\bar{\Omega}(n) = \binom{n}{p}$, so example 1.1 is a special case.

Several interesting consequences of the reciprocity between Ω and $\bar{\Omega}$ are derived in [8, §19]. For instance, if ω is order-preserving then for some integer ℓ we have $\Omega(n) = (-1)^p\Omega(-\ell-n)$ for all n if and only if every maximal chain of P has length ℓ.

EXAMPLE 1.3. (CHROMATIC POLYNOMIALS). Let G be a finite graph without loops or multiple edges, with vertex set V of cardinality p. Let $\chi(n)$ denote the number of pairs $(\mathcal{O},\sigma)$, where (i) $\mathcal{O}$ is an acyclic orientation of the edges of G, and (ii) σ: V $\to$ [n] is any map V $\to$ [n] such that if $u \to v$ in $\mathcal{O}$ (so $u,v \in V$ and uv is an edge of G) then $\sigma(u) > \sigma(v)$. Let $\bar{\chi}(n)$ be the number of such maps with the condition $\sigma(u) > \sigma(v)$ replaced with $\sigma(u) \ge \sigma(v)$. It is easily seen that $\chi(n)$ is the chromatic polynomial of G. In [9] two proofs are given of the reciprocity theorem $\bar{\chi}(n) = (-1)^p\chi(-n)$. In particular, $(-1)^p\chi(-1)$ is the number of acyclic orientations of G.

EXAMPLE 1.4. (ABSTRACT MANIFOLDS). Let Δ be a finite simplicial complex with vertex set V, with $|V| = p$. Thus Δ is a collection of subsets S of V such that $\{v\} \in \Delta$ for all $v \in V$, and if $S \in \Delta$ and $T \subset S$, then $T \in \Delta$. Let $f_i = f_i(\Delta)$ be the number of (i+1)-sets contained in Δ. Hence $f_{-1} = 1$ and $f_0 = p$. Define the polynomial $\Lambda(\Delta,n)$ by

$$\Lambda(\Delta,n) = \sum_{i\ge 0} f_i\binom{n-1}{i} .$$

Note that $\Lambda(\Delta,0) = f_0-f_1+f_2-\ldots = \chi(\Delta)$, the Euler characteristic of Δ.

Now suppose that the underlying topological space $|\Delta|$ of Δ is homeomorphic to a d-dimensional manifold with boundary. Hence deg $\Lambda(\Delta,n) = d$. Denote by $\partial\Delta$ those elements of Δ such that $|\partial\Delta| = \partial|\Delta|$, in the obvious sense. Hence $\partial\Delta$ is itself a simplicial complex, with vertex set contained in V. It follows from a result of MACDONALD [5, Proposition 1.1] that

$$(1.1)\qquad (-1)^d \Lambda(\Delta,-n) = \Lambda(\Delta,n) - \Lambda(\partial\Delta,n) .$$

For instance, let Δ consist of ABCD, BCDE, and all their subsets (ABCD is short for {A,B,C,D}, etc.) Then d = 3, $|\Delta|$ is a 3-ball, and $\partial\Delta$ consists of ABC,ABD,ACD,BCE,CDE,BDE, and all their subsets. Moreover,

$$\Lambda(\Delta,n) = 5 + 9\binom{n-1}{1} + 7\binom{n-1}{2} + 2\binom{n-1}{3}$$

and

$$\Lambda(\partial\Delta,n) = 5 + 9\binom{n-1}{1} + 6\binom{n-1}{2}.$$

It follows from (1.1) that

$$-\Lambda(\Delta,-n) = \binom{n-1}{2} + 2\binom{n-1}{3} .$$

A special case of particular interest occurs when $\partial\Delta = \emptyset$, i.e., when $|\Delta|$ is a manifold. We then have from (1.1) that

$$(1.2)\qquad (-1)^d \Lambda(\Delta,-n) = \Lambda(\Delta,n) .$$

Now (1.2) imposes certain constraints on the numbers f_i which define Λ. When $|\Delta|$ is a sphere, these constraints are simply the well-known DEHN-SOMMERVILLE equations [4, Chapter 9] [6, Chapter 2.4].

<u>EXAMPLE 1.5.</u> (CONCRETE MANIFOLDS). Let M be a subset of the s-dimensional euclidean space with the following properties: (i) M is a union of finitely many convex polytopes, any two of which intersect in a common face of both, (ii) the vertices of these convex polytopes have integer coordinates, and (iii) M is homeomorphic to a d-dimensional manifold with boundary. If $n > 0$, then let j(n) be the number of points $\alpha \in M$ such that $n\alpha$ has integer coordinates, and let i(n) be the number of such points not belonging to ∂M. Then a result due essentially to E. EHRHART [2] (for the generality considered here, one also needs [5, Proposition 1.1]) states that j(n) and i(n) are polynomial functions of n of degree d satisfying

(1.3) $j(0) = \chi(M),\ i(n) = (-1)^d j(-n)$.

We remark that condition (ii) can be replaced by the requirement (ii') the vertices have *rational* coordinates. In this case i and j need no longer be polynomials, but instead there is some $N > 0$ and polynomials $j_0, j_1, \ldots, j_{N-1}$ and $i_0, i_1, \ldots, i_{N-1}$ such that $j(n) = j_a(n)$ and $i(n) = i_a(n)$ whenever $n \equiv a (\mathrm{mod}\ N)$. We then have in place of (1.3) that $j_0(0) = \chi(M)$ and $i_a(n) = (-1)^d j_{-a}(-n)$, where the subscripts are taken modulo N.

An interesting application of (1.3) is to the problem of finding the volume $V(M)$ of a subset M satisfying conditions (i), (ii), (iii), and the additional condition that $s = d$. It is easy to see that then the leading coefficient of $j(n)$ is $V(M)$. Hence from (1.3) we see that if we know any $d+1$ of the numbers $\chi(M)$, $j(n)$, $i(n)$, $n \geq 1$, then we can compute $V(M)$. For a further discussion of this result (including references), see [11].

EXAMPLE 1.6. (MAGIC SQUARES). As a special case of example 1.5, take M to be the set of all doubly stochastic $N \times N$ matrices, so $s = N^2$ and $d = (N-1)^2$. It is well-known that M is a convex polytope whose vertices have integer coordinates, so $j(n)$ and $i(n)$ are polynomials in n of degree $(N-1)^2$. It is easy to see that $j(n)$ is the number of $N \times N$ matrices of non-negative integers with every row and column sum equal to n, while $i(n)$ is the number of such matrices with positive entries. Clearly $i(0) = i(1) = \ldots = i(N-1) = 0$ and $i(N+n) = j(n)$ for $n \geq 0$. There follows from (1.3),

$$j(-1) = j(-2) = \ldots = j(-N+1) = 0 ,$$

$$j(n) = (-1)^{N-1} j(-N-n) .$$

These results were first obtained in [10]. Another proof is given in [3].

2. HOMOGENEOUS LINEAR EQUATIONS

Consider the homogeneous linear equation $x = y$. Let $F(X,Y) = \sum X^\alpha Y^\beta$, where the sum is over all solutions $(x,y) = (\alpha,\beta)$ to $x = y$ in non-negative integers α,β. Let $\overline{F}(X,Y)$ be the corresponding sum over all solutions in positive integers. Clearly $F(X,Y) = 1/(1-XY)$ and $\overline{F}(X,Y) = XY/(1-XY)$. Hence as rational functions we have $\overline{F}(X,Y) = -F(1/X,1/Y)$. It is this result we wish to extend to more general systems of equations.

THEOREM 2.1. [10, Theorem 4.1]. *Let* E *be a system of finitely many linear homogeneous equations with integer coefficients, in the variables* $x_1, x_2, \ldots, x_s$. *Define*

$$(2.1)\qquad \begin{aligned} F(X_1,X_2,\ldots,X_s) &= \sum X_1^{\alpha_1} X_2^{\alpha_2} \cdots X_s^{\alpha_s}\,, \\ \overline{F}(X_1,X_2,\ldots,X_s) &= \sum X_1^{\beta_1} X_2^{\beta_2} \cdots X_s^{\beta_s}\,, \end{aligned}$$

where $(\alpha_1,\alpha_2,\ldots,\alpha_s)$ *ranges over all solutions* $x_i = \alpha_i$ *of* E *in non-negative integers* α_i, *while* $(\beta_1,\beta_2,\ldots,\beta_s)$ *ranges over all solutions in positive integers. Then* F *and* $\overline{F}$ *are rational functions of the* X_i*'s (in the algebra of formal power series, or for* $|X_i| < 1$*). A necessary and sufficient condition that*

$$\overline{F}(X_1,X_2,\ldots,X_s) = \pm F(1/X_1, 1/X_2, \ldots, 1/X_s)\,,$$

as rational functions, is for E *to possess a solution in positive integers. In this case the correct sign is* $(-1)^{\kappa}$, *where* κ *is the corank (=* s-rank E*) of* E.

Many of the results in section 1 can be deduced from the above theorem. We require a connection between evaluating polynomials at +n and -n, and substituting $1/X_i$ for X_i in a rational function. Such a connection is provided by the next result, which EHRHART [1] attributes to POPOVICIU [7].

PROPOSITION 2.1. *Let* H(n) *be a function from the integers* Z *to the complex numbers* C *of the form*

$$H(n) = \sum_{i=1}^{r} P_i(n)\alpha_i^n\,,$$

where the α_i*'s are fixed non-zero complex numbers and each* P_i *is a polynomial in* n. *Define*

$$F(X) = \sum_{n=0}^{\infty} H(n)X^n,\quad \overline{F}(X) = \sum_{n=1}^{\infty} H(-n)X^n\,.$$

Then F *and* $\bar{F}$ *are rational functions of* X, *related by* $\bar{F}(X) = -F(1/X)$.

Theorem 2.1 suggests that we try to find "rational function analogues" of examples 1.4 and 1.5.

PROPOSITION 2.2. *Let* Δ *be a finite simplicial complex with vertices* $v_1, v_2, \ldots, v_p$. *Suppose* $|\Delta|$ *is homeomorphic to a* d*-manifold with boundary. Define the generating functions*

$$F(V_1,V_2,\ldots,V_p) = \sum V_1^{\delta_1} V_2^{\delta_2} \ldots V_p^{\delta_p} + \chi(\Delta) - 1,$$

$$\bar{F}(V_1,V_2,\ldots,V_p) = \sum V_1^{\varepsilon_1} V_2^{\varepsilon_2} \ldots V_p^{\varepsilon_p},$$

where $(\delta_1,\delta_2,\ldots,\delta_p)$ *ranges over all* p*-tuples of non-negative integers such that* $\{v_i \mid \delta_i>0\} \in \Delta$, *while* $(\varepsilon_1,\varepsilon_2,\ldots,\varepsilon_p)$ *ranges over all* p*-tuples of non-negative integers such that* $\emptyset \neq \{v_i \mid \varepsilon_i>0\} \in \Delta-\partial\Delta$. *Then* F *and* $\bar{F}$ *are rational functions of the* v_i*'s related by*

$$\bar{F}(V_1,V_2,\ldots,V_p) = (-1)^{d+1} F(1/V_1,1/V_2,\ldots,1/V_p) .$$

Proposition 2.2 is a consequence of MACDONALD's result [5, Proposition 1.1] mentioned earlier. It is easily seen that

$$F(X,X,\ldots,X) = \sum_{n=0}^{\infty} \Lambda(\Delta,n)X^n ,$$

$$\bar{F}(X,X,\ldots,X) = \sum_{n=1}^{\infty} [\Lambda(\Delta,n)-\Lambda(\partial\Delta,n)]X^n ,$$

in the notation of example 1.4. Thus (1.1) follows from propositions 2.1 and 2.2.

PROPOSITION 2.3. *Let* M *satisfy properties* (i), (ii') *and* (iii) *of example 1.5. Define*

$$F(X_1,X_2,\ldots,X_s,Y) = \chi(M) + \sum X_1^{\alpha_1} X_2^{\alpha_2} \ldots X_s^{\alpha_s} Y^n ,$$

$$\bar{F}(X_1,X_2,\ldots,X_s,Y) = \sum X_1^{\beta_1} X_2^{\beta_2} \ldots X_s^{\beta_s} Y^n ,$$

where $(\alpha_1,\alpha_2,\ldots,\alpha_s,n)$ *ranges over all* (s+1)*-tuples of non-negative integers*

α_i *and positive integers* n *such that* $(\alpha_1/n,\alpha_2/n,\ldots,\alpha_s/n) \in M$, *while* $(\beta_1,\beta_2,\ldots,\beta_s,n)$ *ranges over all such* (s+1)*-tuples with* $(\beta_1/n,\beta_2/n,\ldots,\beta_s/n) \in M-\partial M$. *Then* F *and* $\overline{F}$ *are rational functions related by*

$$\overline{F}(X_1,X_2,\ldots,X_s,Y) = (-1)^{d+1}\, F(1/X_1,1/X_2,\ldots,1/X_s,1/Y) .$$

If we put each $X_i = 1$ and apply proposition 2.1, then we get (1.3).

3. RECIPROCAL DOMAINS

In theorem 2.1, we considered solutions $\alpha_i \geq 0$ $(i=1,2,\ldots,s)$ and $\beta_j > 0$ $(j=1,2,\ldots,s)$ to a system of homogeneous linear equations. It is natural to consider the following generalization. Let E be a system of finitely many linear homogeneous equations with integer coefficients, in the variables $x_1,x_2,\ldots,x_s$ (as in theorem 2.1). Let $S \subset [s]$. Define

$$F_S(X_1,X_2,\ldots,X_s) = \sum X_1^{\alpha_1}X_2^{\alpha_2}\cdots X_s^{\alpha_s} ,$$

(3.1)

$$\overline{F}_S(X_1,X_2,\ldots,X_s) = \sum X_1^{\beta_1}X_2^{\beta_2}\cdots X_s^{\beta_s} ,$$

where $(\alpha_1,\alpha_2,\ldots,\alpha_s)$ ranges over all solutions to E in non-negative integers such that $\alpha_i > 0$ if $i \in S$, while $(\beta_1,\beta_2,\ldots,\beta_s)$ ranges over all solutions to E in non-negative integers with $\beta_i > 0$ if $i \notin S$. Thus $\overline{F}_S = F_{[s]-S}$. Note that $F_\emptyset = F$ and $\overline{F}_\emptyset = \overline{F}$, where F and $\overline{F}$ are given by (2.1).

We now ask under what conditions do we have

$$(3.2) \qquad \overline{F}(X_1,X_2,\ldots,X_s) = (-1)^{\kappa}F(1/X_1,1/X_2,\ldots,1/X_s) ,$$

where κ is the corank of E. It seems plausible that (3.2) will hold whenever E has a solution in positive integers, as in theorem 2.1. In [11], however, we show that this is not the case; and we show why it is likely that there are no simple necessary and sufficient conditions for (3.2) to hold. There is, however, an elegant and surprising sufficient condition.

<u>THEOREM 3.1</u>. [11, Proposition 8.3]. *A sufficient condition for (3.2) to hold is that there exists a solution* $(\gamma_1,\gamma_2,\ldots,\gamma_s)$ *to* E *in integers* γ_i *such that*

$\gamma_i > 0$ *if* $i \in S$ *and* $\gamma_i < 0$ *if* $i \notin S$.

The proof of theorem 3.1 depends on a rather complicated geometric argument suggested by a result of EHRHART [1, p.22] on "reciprocal domains". It is much easier, on the other hand, to give a necessary condition for (3.2) to hold.

PROPOSITION 3.1. *If (3.2) holds, then either* $F = \overline{F} = 0$, *or else* E *has a solution in positive integers.*

PROOF. Assume (3.2) holds but not $F = \overline{F} = 0$. Then $F \neq 0$ and $\overline{F} \neq 0$, so E has solutions $\alpha = (\alpha_1, \alpha_2, \ldots, \alpha_s)$ and $\beta = (\beta_1, \beta_2, \ldots, \beta_s)$ as given in (3.1). Then $\alpha + \beta$ is a solution to E in positive integers. □

4. INHOMOGENEOUS EQUATIONS

Another way of extending theorem 2.1 besides theorem 3.1 is to consider *inhomogeneous* linear equations. Suppose we have a system

$$(4.1) \qquad \sum_{i=1}^{s} a_{ij} x_i = b_j \,, \qquad j \in [p] \,,$$

of p inhomogeneous linear equations with integer coefficients a_{ij} and integer constants b_j, in the variables $x_1, x_2, \ldots, x_s$. It turns out that the correct reciprocal notions to consider in this context are (i) solutions to (4.1) in non-negative integers, and (ii) solutions in positive integers to the "reciprocal system"

$$(4.2) \qquad \sum_{i=1}^{s} a_{ij} x_i = -b_j \,, \qquad j \in [p] \,.$$

Suppose, for example, that $S \subset [s]$ and that

$$b_j = -\sum_{i \in S} a_{ij} \,, \qquad j \in [p] \,.$$

Hence a solution $(\alpha_1, \ldots, \alpha_s)$ to (4.1) in non-negative integers corresponds to a solution $(\beta_1, \ldots, \beta_s)$ of the system $\sum a_{ij} x_i = 0$ in integers β_i satisfying $\beta_i \geq 0$ if $i \notin S$, $\beta_i > 0$ if $i \in S$ (set $\beta_i = \alpha_i$ if $i \notin S$, $\beta_i = \alpha_i + 1$ if $i \in S$). Moreover, a solution $(\alpha_1, \ldots, \alpha_s)$ to (4.2) in positive integers corresponds to a solution $(\beta_1, \ldots, \beta_s)$ of the system $\sum a_{ij} x_i = 0$ in integers β_i satisfying

$\beta_i > 0$ if $i \notin S$, $\beta_i \geq 0$ if $i \in S$ (set $\beta_i = \alpha_i$ if $i \notin S$; $\beta_i = \alpha_i - 1$ if $i \in S$). Hence our notion of reciprocity for inhomogeneous systems includes the reciprocity of section 3 as a special case.

We therefore define

$$(4.3)\qquad \begin{aligned} F(X_1,X_2,\ldots,X_s) &= \sum X_1^{\alpha_1}X_2^{\alpha_2}\ldots X_s^{\alpha_s}, \\ \overline{F}(X_1,X_2,\ldots,X_s) &= \sum X_1^{\beta_1}X_2^{\beta_2}\ldots X_s^{\beta_s}, \end{aligned}$$

where $(\alpha_1,\alpha_2,\ldots,\alpha_s)$ ranges over all solutions to (4.1) in non-negative integers, while $(\beta_1,\beta_2,\ldots,\beta_s)$ ranges over all solutions to (4.2) in positive integers. As usual, we seek conditions when $\overline{F}(X_1,X_2,\ldots,X_s) = (-1)^{\kappa}F(1/X_1,1/X_2,\ldots,1/X_s)$, where κ is the corank of (4.1) or (4.2). We shall say that (4.1) has the *R-property* if $\overline{F}(X_1,X_2,\ldots,X_s) = (-1)^{\kappa}F(1/X_1,1/X_2,\ldots,1/X_s)$. The possibility of obtaining reasonable necessary and sufficient conditions for E to have the R-property appears hopeless, and even reasonably general sufficient conditions are rather complex and not very edifying. We shall now discuss the nature of the sufficient conditions obtained in [11].

Let $\{i_1,i_2,\ldots,i_k\}$ be a set of $k < p$ elements from $[s]$ such that the determinant of coefficients taken from the first k rows and from columns $i_1,i_2,\ldots,i_k$ of (4.1) is non-zero. Hence we can solve the first k equations (i.e., $j \in [k]$) of (4.1) for $x_{i_1},x_{i_2},\ldots,x_{i_k}$ in terms of the remaining x_i's and substitute these values in the remaining p-k equations, obtaining p-k equations in s-k unknowns. Let $E(i_1,i_2,\ldots,i_k)$ denote the first of these p-k equations (i.e., the equation resulting from making the above substitution into the (k+1)-st equation of (4.1)). Thus in particular $E(\emptyset)$ is just the first equation $\sum a_{i1}x_i = b_1$ of (4.1). Note that the equations $E(i_1,i_2,\ldots,i_k)$ are really determined only up to a non-zero multiplicative constant. This need not concern us since we will be interested only in solutions to these equations.

<u>EXAMPLE 4.1</u>. Consider the system

$$\begin{aligned} x_1 - x_2 + 3x_3 \qquad &= b_1 \\ 2x_2 - x_3 - x_4 &= b_2. \end{aligned}$$

Then we obtain the equations

$$\begin{array}{llcl}
E(\emptyset): & x_1 - x_2 + 3x_3 & = & b_1 \\
E(1): & 2x_2 - x_3 - x_4 & = & b_2 \\
E(2): & 2x_1 + 5x_3 - x_4 & = & 2b_1 + b_2 \\
E(3): & x_1 + 5x_2 - 3x_4 & = & -b_1 + 3b_2 .
\end{array}$$

THEOREM 4.1. *A sufficient condition that the system (4.1) has the R-property is the following. For every set* $\{i_1,i_2,\ldots,i_k\} \subset [s]$ *for which* $E(i_1,i_2,\ldots,i_k)$ *is defined, the single equation* $E(i_1,i_2,\ldots,i_k)$ *should possess the R-property.*

It should be mentioned that in [11] theorem 4.1 is strengthened so that only a special subset of the equations $E(i_1,i_2,\ldots,i_k)$ need be considered. However, the definition of this subset is rather complicated and will be omitted here. Theorem 4.1 is proved in [11] using iterated contour integration. Contour integration may seem like an unwarranted artifice for a result like theorem 4.1. While it is undoubtedly possible to dispense with contour integration, the next results show that it is not too unnatural in the present context. We would like to complement theorem 4.1 by obtaining conditions for a single equation to possess the R-property.

THEOREM 4.2. *Let* $a_1x_1+a_2x_2+\ldots+a_sx_s = b$ *be a single linear equation* E *with integer coefficients* a_i *and integer constant term* b. *Then the following three conditions are equivalent.*

(i) *The rational functions*

$$\lambda^{b-1}/(1-\lambda^{-a_1})(1-\lambda^{-a_2})\ldots(1-\lambda^{-a_s}) \tag{4.4}$$

and

$$\lambda^{\bar{b}-1}/(1-\lambda^{-a_1})(1-\lambda^{-a_2})\ldots(1-\lambda^{-a_s}) \tag{4.5}$$

have zero residues at $\lambda = 0$. *Here* $\bar{b} = -b-a_1-a_2-\ldots-a_s$.

(ii) *The following two conditions are both satisfied.*

(a) *There does not exist a solution* $(\alpha_1,\alpha_2,\ldots,\alpha_s)$ *to* E *in integers such that*

$\alpha_t < 0$ *if* $a_t > 0$, *and*
$\alpha_t \geq 0$ *if* $a_t < 0$.

(b) *There does not exist a solution* $(\beta_1,\beta_2,\ldots,\beta_s)$ *to* E *in integers such that*

$$\beta_t \geq 0 \quad if \quad a_t > 0 \text{ , and}$$
$$\beta_t < 0 \quad if \quad a_t < 0 \text{ .}$$

(Note: *It is clear that at least one of (a) or (b) always holds.*)

(iii) E *has the R-property.*

THEOREM 4.3. *With the hypotheses of theorem 4.2, the following two conditions are equivalent.*

(i) *The rational functions of (4.4) and (4.5) have no poles at* $\lambda = 0$.

(ii) $\sum_{t-} a_t < -b < \sum_{t+} a_t$, *where* $\sum_{t-} a_t$ *(resp.* $\sum_{t+} a_t$*) denotes the sum of all* a_t *satisfying* $a_t < 0$ *(resp.* $a_t > 0$*).*

If, moreover, either of the two (equivalent) conditions (i) *or* (ii) *is satisfied, then* E *has the* R-*property.*

EXAMPLE 4.2. Consider the system E of example 4.1. By theorems 4.1 and 4.3, we see that E has the R-property if

$$\begin{aligned}
-1 &< -b_1 &&< 4\\
-2 &< \quad\quad - b_2 &&< 2\\
-1 &< -2b_1 - b_2 &&< 7\\
-3 &< b_1 - 3b_2 &&< 6 \text{ .}
\end{aligned}$$

These conditions hold if and only if $(b_1,b_2) = (0,-1),(0,0),(-1,-1),(-1,0),$ $(-2,-1)$ or $(-2,0)$.

Analogously to proposition 3.1, we have a simple necessary condition for a system (4.1) to have the R-property. The proof is essentially the same as the proof of proposition 3.1.

PROPOSITON 4.1. *Suppose the system (4.1) has the R-property. Then either* $F = \overline{F} = 0$, *or else the homogeneous system* $\sum_{i=1}^{s} a_{ij}x_i = 0$, $j \in [p]$, *has a solution in positive integers.*

We have given a sampling of what we believe to be the most interesting examples of combinatorial reciprocity theorems. Some additional types of reciprocity theorems are given in [11]. There are many other combinatorial relationships which can be viewed as reciprocity theorems and which we have not touched on. Examples include the inverse relationship between the Stirling numbers of the first and second kinds, and the MacWilliams identities of coding theory. We believe that many new interesting results and unifying principles are awaiting discovery in the field of combinatorial reciprocity.

REFERENCES

[1] EHRHART, E., *Sur un problème de géométrie diophantienne linéaire I, II*, J. Reine Angew. Math., 226 (1967) 25-49, and 227 (1967) 25-49.

[2] EHRHART, E., *Démonstration de la loi de réciprocité pour un polyèdre entier*, C. R. Acad. Sci. Paris A, 265 (1967) 5-7.

[3] EHRHART, E., *Sur les carrés magiques*, C. R. Acad. Sci. Paris A, 227 (1973) 651-654.

[4] GRUNBAUM, B., *Convex polytopes*, Wiley (Interscience), London, 1967.

[5] MACDONALD, I.G., *Polynomials associated with finite cell-complexes*, J. London Math. Soc. (2), 4 (1971) 181-192.

[6] MCMULLEN, P. & G.C. SHEPHARD, *Convex polytopes and the upper bound conjecture*, London Math. Soc., Lecture Note Series 3, Cambridge Univ. Press, 1971.

[7] POPOVICIU, T., *Studie si cercetari stiintifice*, Acad. R.P.R., Filiala Cluj, 4 (1953), 8.

[8] STANLEY, R., *Ordered structures and partitions*, Memoirs Amer. Math. Soc., 119 (1972).

[9] STANLEY, R., *Acyclic orientations of graphs*, Discrete Math., 5 (1973) 171-178.

[10] STANLEY, R., *Linear homogeneous diophantine equations and magic labelings of graphs*, Duke Math. J., 40 (1973) 607-632.

[11] STANLEY, R., *Combinatorial reciprocity theorems*, Advances in Math. (to appear).

P A R T 3

COMBINATORIAL GROUP THEORY

DIFFERENCE SETS *)

M. HALL, Jr.
California Institute of Technology, Pasadena, Cal. 91103, USA

1. INTRODUCTION

A symmetric block design D is a special kind of incidence structure [7,11,21] consisting of v points and v blocks, each block containing k distinct points, each point lying on k distinct blocks, and every pair of distinct points lying on λ different blocks. Counting the point pairs in two ways we have

(1.1) $$k(k-1) = \lambda(v-1).$$

A further consequence [11, p.104] is that any two distinct blocks have λ points in common, so that there is a duality between the points and blocks.

An automorphism α of D is a one-to-one mapping of points onto points and blocks onto blocks preserving incidence. It may happen that D has a group G of automorphisms which is transitive and regular on the points (and as can be shown also transitive and regular on the blocks). Identifying the points with elements of G and a block with the set of k points on it, a single block will determine all the rest. A set of k elements $B = \{d_1, \ldots, d_k\}$ from a group G of order v such that the translates $Bg = \{d_1 g, \ldots, d_k g\}$ form a symmetric block design D is called a *difference set*.

At first glance the concept of difference set might appear to be too restrictive to be of much interest. But difference sets are in fact quite numerous and have many interesting properties. For $k \le 100$ there are at least 85 difference sets with G a cyclic group [1], and there are a number of other cases in this range with G non-cyclic.

---

*) This research was supported in part by NSF grant GP 36230X.

M. Hall, Jr. and J. H. van Lint (eds.), Combinatorics, 321-346.

A remarkable fact, first discovered by the author [10] for finite projective planes is that if G is Abelian the design D often has a still larger group of automorphisms. It may happen that there is an integer t, necessarily prime to v, such that the mapping $x \to x^t$ is not only an automorphism of G but is also an automorphism of the design D. Such an integer is called a *multiplier*. There are a number of theorems proving the existence of multipliers. Every known difference set for which G is cyclic has non-trivial multipliers.

Studies of difference sets have led to very interesting connections with finite geometries, algebraic number theory, and group characters among other subjects.

Sections 2, 3, 4 of this paper give some properties of automorphisms of designs, a formal definition of a difference, and the simplest form of the multiplier theorem. Section 5 gives a list of the known types of difference sets and a few sporadic sets. Section 6 gives a brief sketch of the general theory and some of its results.

2. AUTOMORPHISMS OF DESIGNS

A general incidence structure S is a system ({p},{B},I) with two sets of objects, {p} a set of "points" and {B} a set of "blocks" together with an incidence relation I such that pIB for certain points p and certain blocks B. If T = ({q},{C},J) is another incidence structure, then an incidence preserving map of S into T is a mapping ϕ of {p} into {q} and {B} into {C} such that pIB implies $p\phi J B\phi$ for all $p \in \{p\}$ and $B \in \{B\}$.

In general the incidence preserving map is a *homomorphism* of S onto T. If ϕ is a one-to-one mapping then it is an *isomorphism* of S onto T. An isomorphism of S onto itself is called an *automorphism*. Clearly the automorphisms of an incidence structure form a group.

The incidence structures which will be considered here are the "partially balanced incomplete block designs" or more briefly "designs". For these if there are v points and b blocks, each block contains k distinct points, every point lies on r blocks, and every pair of distinct points lies on exactly λ blocks.

These parameters satisfy the two well-known relations

$$bk = vr, \quad r(k-1) = \lambda(v-1). \tag{2.1}$$

To avoid certain trivial designs we assume $2 < k < v-2$. With a design $D = D(v,b,r,k,\lambda)$ we associate an incidence matrix $A = [a_{ij}]$, $i=1,\dots,v$; $j=1,\dots,b$ where, if we number the points P_i, $i=1,\dots,v$ in an arbitrary way and blocks B_j, $j=1,\dots,b$ also arbitrarily, we put $a_{ij} = 1$ if $P_i \; I \; B_j$ and $a_{ij} = 0$ if $P_i \not I B_j$. Then it is well known that A satisfies

(2.2) $$AA^T = (r-\lambda)I_v + \lambda J_{vv}, \quad AJ_{bb} = rJ_{vb}, \quad J_{vv}A = kJ_{vb}.$$

Here A^T is the transpose of A, I_v is the $v \times v$ identity matrix, and J_{mn} the $m \times n$ matrix all of whose entries are 1's.

Writing $B = (r-\lambda)I_v + \lambda J_{vv}$ we can easily evaluate the determinant of B, obtaining

(2.3) $$\det B = (r-\lambda)^{v-1}(r+(v-1)\lambda).$$

The relations (2.1) together with the assumption $2 < k < v-2$ imply that $r > \lambda$ so that $\det B > 0$. From this it follows that the rank of the $v \times b$ matrix A is v, we obtain FISHER's inequality

(2.4) $$b \geq v.$$

If $b = v$ and so also $r = k$ the design is called a *symmetric design*.

If α is an automorphism of a block design let P_α be the permutation of the points $p \to (p)\alpha$ and Q_α the permutation of the blocks $B \to (B)\alpha$. The fact that α preserves incidences can be expressed in terms of the incidence matrix A

(2.5) $$P_\alpha^{-1} A Q_\alpha = A.$$

For if $A = [a_{ij}]$ the matrix on the left is $[a_{(i)\alpha(j)\alpha}]$ and as incidences are preserved this is identical with A. Conversely permutation matrices P_α and Q_α satisfying (2.5) determine an automorphism of the design with incidence matrix A.

THEOREM 2.1. (PARKER). *An automorphism of a symmetric block design* D *fixes the same number of blocks as points.*

PROOF. From (2.2) and (2.3) the incidence matrix of a symmetric block design is non-singular. Hence from (2.5) we may obtain

(2.6) $$Q_\alpha = AP_\alpha A^{-1}$$

and this gives for the traces

$$\text{(2.7)} \qquad tr(Q_\alpha) = tr(P_\alpha).$$

But $tr(P_\alpha)$ is the number of points fixed by α and $tr(Q_\alpha)$ is the number of blocks fixed by α and so the theorem is proved. □

COROLLARY. *If* G *is a group of automorphisms of a symmetric block design* D, *then the permutation representations of* $G = \langle\alpha,\beta,\ldots\rangle$ *on the points* $\langle P_\alpha, P_\beta,\ldots\rangle$ *and on the blocks* $\langle Q_\alpha, Q_\beta,\ldots\rangle$ *have the same character.*

A well-known property of symmetric designs is given by the following theorem.

THEOREM 2.2. *If* A *is the incidence matrix of a symmetric design* D *with parameters* v,k,λ, *then* A^T *is the incidence matrix of a dual symmetric design (interchanging points and blocks)* D^* *with the same parameters.*

Thus for A we have

$$\text{(2.8)} \qquad \begin{cases} AA^T = (k-\lambda)I + \lambda J, & AJ = kJ, \quad JA = kJ, \\ A^TA = (k-\lambda)I + \lambda J, & A^TJ = kJ, \quad JA^T = kJ. \end{cases}$$

Note that this implies that any two distinct blocks of D have exactly λ points in common.

3. DIFFERENCE SETS

Let D be a symmetric block design with parameters $b = v$, $r = k$, and λ. We shall suppose that D has a group G of automorphisms of order v which is transitive and regular on the points. Then if we take an arbitrary point P as a "base point", the points $(P)x = P_x$ as x ranges over G consist of all v points of D, each occurring exactly once. This gives a correspondence between the points of D and the elements of G making an arbitrary point P correspond to the identity of G. From theorem 2.1 and its corollary the same will be true of the blocks where taking an arbitrary block B the blocks $(B)x = B_x$ as x ranges over G will consist of all blocks of D each occurring exactly once.

The parameters v,k,λ necessarily satisfy

(3.1) $k(k-1) = \lambda(v-1)$.

We shall identify the points of D with the elements of G, following the correspondence above. Let B be an arbitrary block and consider the k points in D

(3.2) $B = \{d_1, d_2, \ldots, d_k\}$, $d_i \in G$.

Then for any block Bx

(3.3) $Bx = \{d_1x, d_2x, \ldots, d_kx\}$, $x \in G$.

Two points r and s will occur together in Bx if for some d_i and d_j we have $d_ix = r$, $d_jx = s$. Here $d_id_j^{-1} = rs^{-1}$. Conversely if $d_id_j^{-1} = rs^{-1}$ we may determine x uniquely by $d_ix = r$ and it will follow that $d_jx = s$. Hence for any $d \neq 1$ there are exactly λ choices $d_id_j^{-1} = d$ with $d_i, d_j \in \{d_1, \ldots, d_k\}$. Similarly with $d \neq 1$ the blocks B and Bd have exactly λ points in common. This means that for exactly λ choices of d_i there is a d_j with $d_i = d_jd$ or $d_j^{-1}d_i = d$.

THEOREM 3.1. *Let* $B = \{d_1, d_2, \ldots, d_k\}$ *be a set of* k *distinct elements in a group* G *of order* v, *and let* $k(k-1) = \lambda(v-1)$. *If either of the conditions* (1) *or* (2) *holds, both will hold.*

(1) *For every* $d \neq 1$ *there are exactly* λ *choices* $d_i, d_j \in B$ *such that* $d_id_j^{-1} = d$.

(2) *For every* $d \neq 1$ *there are exactly* λ *choices* $d_i, d_j \in B$ *such that* $d_j^{-1}d_i = d$.

Then the sets $Bx = \{d_1x, d_2x, \ldots, d_kx\}$ *will be the blocks of a symmetric block design* D *which has* G *as an automorphism group transitive and regular on the points of* D *and also on the blocks of* D.

PROOF. Clearly the v blocks Bx are all of size k. For a fixed d_i, the elements d_ix as x ranges over G give each element of G exactly once, so that each element of G is in exactly k blocks. For the incidence matrix A of the system D of these points and blocks condition (1) is equivalent to $AA^T = (k-\lambda)I + \lambda J$ while condition (2) is equivalent to $A^TA = (k-\lambda)I + \lambda J$ and as these are equivalent to each other the theorem is proved. □

DEFINITION. A set $B = \{d_1, \ldots, d_k\}$ of distinct elements in a group G of order v, where $k(k-1) = \lambda(v-1)$ is called *a difference set* if for any $d \neq 1$ in G there are exactly λ choices $d_i, d_j \in B$ such that $d_id_j^{-1} = d$.

If G is an Abelian group written additively then the condition is on differences $d_i - d_j$, and this is historically the origin of the term difference set.

4. THE MULTIPLIER THEOREM

The residues 0, 2, 3, 4, 8 (mod 11) are an (11,5,2) difference set in the additive group G of residues modulo 11. We check the difference property

$$(4.1)\qquad \begin{array}{ll} 1 \equiv 3-2 \equiv 4-3, & 6 \equiv 3-8 \equiv 8-2, \\ 2 \equiv 2-0 \equiv 4-2, & 7 \equiv 0-4 \equiv 4-8, \\ 3 \equiv 0-8 \equiv 3-0, & 8 \equiv 8-0 \equiv 0-3, \\ 4 \equiv 4-0 \equiv 8-4, & 9 \equiv 0-2 \equiv 2-4, \\ 5 \equiv 2-8 \equiv 8-3, & 10 \equiv 2-3 \equiv 3-4. \end{array} \qquad \text{(modulo 11)}$$

The corresponding design D is a symmetric design with parameters $(v,k,\lambda) = (11,5,2)$. We list the blocks and the points on them

$$(4.2)\qquad \begin{array}{ll} B_0: 0,2,3,4,8, & B_6: 6,8,9,10,3, \\ B_1: 1,3,4,5,9, & B_7: 7,9,10,0,4, \\ B_2: 2,4,5,6,10, & B_8: 8,10,0,1,5, \\ B_3: 3,5,6,7,0, & B_9: 9,0,1,2,6, \\ B_4: 4,6,7,8,1, & B_{10}: 10,1,2,3,7. \\ B_5: 5,7,8,9,2, & \end{array}$$

Here the design D has further automorphisms. Specifically the mapping $x \to tx$ of residues modulo 11 where t is one of 1,3,4,5,9 is an automorphism of D. Since a mapping $x \to tx$ fixes the point 0, from theorem 2.1 it must also fix a block, and in this case the block B_1 is fixed by all these automorphisms. In this case the full group of automorphisms includes still further elements one being given by the permutations

$$(0)(7)(1,2,8)(3,5,6)(4,10,9)(B_0,B_8,B_9)(B_1,B_2,B_6)(B_3)(B_4,B_{10},B_5)(B_7).$$

The full group of automorphisms of D is the simple group $L_2(11)$ of order 660.

It is a remarkable fact that many difference sets lead to designs D with further automorphisms beyond the given group G of order v. We will define the term *multiplier* in this context.

DEFINITION. An integer t is a *multiplier* of the difference set $\{d_1,d_2,\dots,d_k\}$ in the Abelian group G of order v if the mapping $x \to x^t$ is an automorphism of the design D determined by the difference set.

Note that multipliers are defined over Abelian groups. With G Abelian of order v, a multiplier t must necessarily be such that $x \to x^t$ is an automorphism of G since every element of G must be of the form x^t with x from G, and so $(v,t) = 1$. Clearly the multipliers form a multiplicative group of residues modulo v.

THEOREM 4.1. *The integer* t *is a multiplier of the Abelian difference set* $\{d_1,d_2,\dots,d_k\}$ *if and only if* $\{d_1^t,d_2^t,\dots,d_k^t\} = \{d_1w,d_2w,\dots,d_kw\}$ *for some* $w \in G$.

PROOF. If t is a multiplier of $B = \{d_1,\dots,d_k\}$ then $x \to x^t$ maps B into some block $Bw = \{d_1w,d_2w,\dots,d_kw\}$. Conversely if $\{d_1^t,d_2^t,\dots,d_k^t\} = \{d_1w,\dots,d_kw\}$ then $x \to x^t$ takes an arbitrary block $Bu = \{d_1u,\dots,d_ku\}$ into Bwu^t, and is an automorphism of the design. □

THEOREM 4.2. Multiplier theorem (HALL & RYSER [14]). *Let* $\{d_1,\dots,d_k\}$ *be a* (v,k,λ) *difference set over an Abelian group* G *of order* v. *Then if* p *is a prime such that* (i) $p\,|\,k-\lambda$, (ii) $(p,v) = 1$ *and* (iii) $p > \lambda$, *then* p *is a multiplier of the difference set.*

PROOF. We assume the group G to be written multiplicatively. We shall work with the group ring ZG over the rational integers Z. The elements of the group ring are formal sums $A = \sum_{g\in G} a(g)g$, $a(g) \in Z$. Addition and multiplication are defined by the rules

$$\sum_g a(g)g + \sum_g b(g)g = \sum_g (a(g) + b(g))g,$$

$$\left(\sum_g a(g)g\right)\left(\sum_g b(g)g\right) = \sum_k\left(\sum_{gh=k} a(g)b(h)\right) k.$$

With these rules ZG is an associative ring with identity.

With the difference set $\{d_1,d_2,\dots,d_k\}$ we associate the element $\theta(d)$ of the group ring

$$\theta(d) = d_1 + d_2 + \dots + d_k. \tag{4.3}$$

We also write, using a symbolic notation, for any integer t, defining $\theta(d^t)$

by

$$(4.4)\qquad \theta(d^t) = d_1^t + d_2^t + \dots + d_k^t.$$

Then from theorem 4.1 the proof of theorem 4.2 reduces to proving for some $w \in G$

$$(4.5)\qquad \theta(d^p) = w\theta(d).$$

Let us also define the element T of ZG by

$$(4.6)\qquad T = \sum_{x\in G} x.$$

With this notation the fact that $\{d_1,\dots,d_k\}$ is a (v,k,λ) difference set takes the form, writing $k-\lambda = n$,

$$(4.7)\qquad \theta(d)\theta(d^{-1}) = (k-\lambda) + \lambda T = n + \lambda T.$$

Here by $k-\lambda$ we mean $(k-\lambda)1$ where 1 is the identity of G. For in the left-hand side of (4.7) the identity appears k times as $d_i d_i^{-1}$, $i=1,\dots,k$, and every other $d \in G$ appears exactly λ times as $d_i d_j^{-1}$. Since the identity is one of the elements of T, its k occurrences are counted as $(k-\lambda) + \lambda$, and the λ occurrences of every $d \neq 1$ are counted in λT.

Since the binomial coefficients $\binom{p}{j}$, $j=1,\dots,p-1$ are multiples of the prime p we always have $(A+B)^p = A^p + B^p + pR$ in ZG with R some element of ZG. Hence

$$(4.8)\qquad \theta(d)^p = d_1^p + d_2^p + \dots + d_k^p + pW = \theta(d^p) + pW.$$

Multiplying (4.7) by $\theta(d)^{p-1}$ we have

$$(4.9)\qquad \theta(d)^p\theta(d^{-1}) = n\theta(d)^{p-1} + \lambda\theta(d)^{p-1}T.$$

Here since $xT = T$ for any $x \in G$ it follows that $\theta(d)T = kT$, and (4.9) becomes

$$(4.10)\qquad \theta(d)^p\theta(d^{-1}) = n\theta(d)^{p-1} + \lambda(k^{p-1}-1)T + \lambda T.$$

Now p divides $n = k-\lambda$. If p does not divide k then p divides $k^{p-1}-1$, while if p divides k, then also p divides λ, so that in all cases p divides $\lambda(k^{p-1}-1)$. Thus (4.10) takes the form

$$(4.11)\qquad \theta(d)^p\theta(d^{-1}) = pV + \lambda T$$

with some V in ZG. Combining (4.8) and (4.11) we have

$$\theta(d^p)\theta(d^{-1}) = pS + \lambda T. \tag{4.12}$$

If $x_1,\ldots,x_v$ are the elements of G then the left-hand side takes the form $\sum a_i x_i$ where the a_i are non-negative integers such that

$$\sum_i a_i = k^2. \tag{4.13}$$

Comparison with the right-hand side shows

$$a_i \equiv \lambda \pmod p), \qquad i=1,\ldots,v. \tag{4.14}$$

Also since we have assumed $p > \lambda$ it follows that $a_i \geq \lambda$ in every case. Thus if $S = \sum_i s_i x_i$ we have from (4.12)

$$a_i = ps_i + \lambda \tag{4.15}$$

so that $s_i \geq 0$ in every case and also

$$k^2 = \sum_i a_i = \sum_i ps_i + \lambda v. \tag{4.16}$$

But since $k(k-1) = \lambda(v-1)$ we have $k^2-\lambda v = k-\lambda = n$ so that

$$p \sum_i s_i = n, \qquad s_i \geq 0. \tag{4.17}$$

As a consequence

$$pST = p \sum_i s_i T = nT. \tag{4.18}$$

Applying the automorphism $x \to x^p$ of G to the relation (4.7) gives

$$\theta(d^p)\theta(d^{-p}) = n + \lambda T. \tag{4.19}$$

Applying the automorphism $x \to x^{-1}$ of G to (4.12) gives

$$\theta(d^{-p})\theta(d) = pS^* + \lambda T \tag{4.20}$$

where $S^* = \sum_i s_i x_i^{-1}$.

The product of the left-hand sides of (4.7) and (4.19) is the same as the product of the left-hand sides of (4.12) and (4.20). Equating the right-hand sides gives

(4.21) $$(pS + \lambda T)(pS^* + \lambda T) = (n + \lambda T)^2.$$

Since $pST = nT$ and $pS^*T = nT$ this simplifies to

(4.22) $$p^2SS^* = n^2$$

or

(4.23) $$p^2 \sum_i s_i x_i \sum_j s_j x_j^{-1} = n^2.$$

Since the coefficients s_i are all non-negative we cannot have $s_i > 0$ and $s_j > 0$ for $i \neq j$ since this would give $x_i x_j^{-1} \neq 1$ a positive coefficient on the left of (4.23). Hence only one s_i is different from 0 and as $p\sum s_i = n$ we conclude that for some $w \in G$

(4.24) $$pS = nw.$$

Now (4.12) takes the simpler form

(4.25) $$\theta(d^p)\theta(d^{-1}) = nw + \lambda T;$$

multiplying this by $\theta(d)$ we have

(4.26) $$\theta(d^p)\theta(d^{-1})\theta(d) = nw\theta(d) + \lambda\theta(d)T,$$

which becomes

(4.27) $$\theta(d^p)(n+\lambda T) = nw\theta(d) + \lambda kT$$

or

(4.28) $$n\theta(d^p) + \lambda kT = nw\theta(d) + \lambda kT,$$

whence

(4.29) $$\theta(d^p) = w\theta(d).$$

By theorem 4.1 this proves that p is a multiplier and completes the proof of theorem 4.2. □

This theorem has been very useful in proving that certain difference sets do not exist and in constructing others when they do exist. In this connection the following theorem is useful.

THEOREM 4.3. *If* t *is a multiplier of the design* $D(v,k,\lambda)$ *there is a block fixed by the multiplier. If* $(v,k) = 1$ *there is a block fixed by every multiplier.*

PROOF. The automorphism $x \to x^t$ given by the multiplier fixes the identity element of G. Since this is one of the points of the design, by theorem 2.1 there is a block fixed by this automorphism. Let $\{d_1,\ldots,d_k\}$ be the difference set and write $y = d_1 d_2 \ldots d_k$. Then in the block Bs the product of the k elements is ys^k. If $(v,k) = 1$ there is exactly one x with $yx^k = 1$ in G and clearly the block Bx will be fixed by every multiplier. □

Consider the difference set with $v = 111$, $k = 11$, $\lambda = 1$. Here $k-\lambda = n = 10$ and so 2 and 5 are multipliers. By theorem 4.3 there is a block B fixed by both of these multipliers. If c is a point of this block then c, c^2, c^4, c^5 are all in this block. As $c^2c^{-1} = c^5c^{-4} = c$ and $\lambda = 1$ this is possible only if $c^5 = c^2$ and $c^4 = c$ whence $c^3 = 1$. But in G there are only three elements satisfying $c^3 = 1$ and so we cannot find a block of 11 distinct elements of this kind. Hence no difference set exists for these parameters.

If $v = 73$, $k = 9$, $\lambda = 1$, as $n = k-\lambda = 8$, $p = 2$ is a multiplier. Let us write G in additive form as the group of residues modulo 73. Then in the block B fixed by the multiplier let c be an element not the 0 residue. Then the difference set will include

$$c,\ 2c,\ 4c,\ 8c,\ 16c,\ 32c,\ 64c,\ 55c,\ 37c \pmod{73}.$$

These will be all 9 elements of the difference set and without loss of generality we may take $c = 1$ to obtain the (73,9,1) difference set

(4.30) $$1,\ 2,\ 4,\ 8,\ 16,\ 32,\ 37,\ 55,\ 64 \pmod{73}.$$

In theorem 4.2 the condition (i) $p|k-\lambda$ is the source of the multiplier, while condition (ii) $(p,v) = 1$ is clearly necessary for p to be a multiplier. But in every known case the condition (iii) $p > \lambda$ appears to be unnecessary, though it is required in the proof using (4.15) to show $s_i \geq 0$. Indeed for every known difference set in which G is cyclic, there is a prime dividing $k-\lambda$ but not v and every such prime is a multiplier. There are however Abelian difference sets without multipliers. For example there is an Abelian difference set with parameters (16,6,2) the group G being the elementary Abelian group of order 16.

5. THE KNOWN DIFFERENCE SETS

There are several families of difference sets known. Most of these depend on arithmical properties of residues modulo primes or on finite fields.

Type S. (Singer difference sets [23]). These are hyperplanes in the n-dimensional projective geometry PG(n,q) over GF(q). The parameters are

$$v = \frac{q^{n+1}-1}{q-1}, \quad k = \frac{q^{n}-1}{q-1}, \quad \lambda = \frac{q^{n-1}-1}{q-1}.$$

Type Q. (Quadratic residues in GF(q), $q \equiv 3 \pmod 4$).

$$v = q = p^r = 4t-1, \quad k = 2t-1, \quad \lambda = t-1.$$

Type H_6. (p is a prime of the form $p = 4x^2+27$). There will exist a primitive root modulo p such that $\mathrm{Ind}_r(3) \equiv 1 \pmod 6$. The $(p-1)/2$ residues a_i such that $\mathrm{Ind}_r(a_i) \equiv 0$, 1 or 3 (mod 6) will form a difference set with parameters $v = p = 4t-1$, $k = 2t-1$, $\lambda = t-1$.

Type T. (Twin primes). Let p and $q = p+2$ be primes. Let r be a number such that r is a primitive root of p and also of q. Then $r^i \pmod{pq}$ $i=1,\dots,(p-1)(q-1)/2$ and $0,q,\dots,(p-1)q \pmod{pq}$ form a difference set with $v = pq = 4t-1$, $k = 2t-1$, $\lambda = t-1$.

Type B. (Biquadratic residues of primes $p = 4x^2+1$, x odd). Here $v = p = 4x^2+1$, $k = x^2$, $\lambda = (x^2-1)/4$.

Type B_0. (Biquadratic residues and zero modulo primes $p = 4x^2+9$, x odd). Here $v = 4x^2 + 9$, $k = x^2+3$, $\lambda = (x^2+3)/4$.

Type O. (Octic residues of primes $p = 8a^2+1 = 64b^2+9$ with a, b both odd). Here $v = p$, $k = a^2$, $\lambda = b^2$.

Type O_0. (Octic residues and zero for primes $p = 8a^2+49 = 64b^2+441$, a odd, b even). Here $v = p$, $k = a^2+6$, $\lambda = b^2+7$.

Type W_4. (A generalization of T developed by WHITEMAN [27]). Let p be a prime $p \equiv 1 \pmod 4$ and let $q = 3p+2$ also be a prime. Suppose also that $pq = v = 1+4x^2$ with x odd. Then take r to be a primitive root of both p and q. Writing $d = (p-1)(q-1)/4$ the residues $1,r,r^2,\dots,r^{d-1}$, $0,q,2q,\dots,(p-1)q$ (mod pq) are a difference set with $v = pq$, $k = (v-1)/4$, $\lambda = (v-5)/16$.

Type GMW. (GORDON, MILLS & WELCH [9]). The parameters are the same as those of the Singer type.

$$v = \frac{q^{n+1}-1}{q-1}, \quad k = \frac{q^n-1}{q-1}, \quad \lambda = \frac{q^{n-1}-1}{q-1}.$$

Here if we can write n+1 in the form n+1 = mM with $m \geq 3$ and if M is the product of r prime numbers, not necessarily distinct, then there are at least 2^r inequivalent difference sets with these parameters.

Type H(2). $v = 2^{2m}$, $k = 2^{2m-1} - 2^{m-1}$, $\lambda = 2^{2m-2} - 2^{m-1}$. Here G is the direct product of m groups of order 4 (some may be the cyclic group, other the four group). These difference sets and designs are most easily described by their relation to Hadamard matrices, as will be done in the next section.

L. BAUMERT [1] has listed the known 85 cyclic difference sets for which $3 \leq k \leq 100$. Most of these are special cases of the types listed above. There are 74 different possible parameters v, k, λ and for no other parameters with k in this range is a cyclic difference set possible. There is a cyclic difference set with v = 133, k = 33, λ = 8 but in all other cases the parameters are those of the listed types. For the projective planes with $v = n^2+n+1$, k = n+1, λ = 1 the writer has shown that the solution is unique when n = 2, 3, 4, 5, 7, 9, 11, 13, 16, 25, 27, 32, and there is certainly a Singer difference set whenever $n = q = p^r$ is a prime power, but for other prime powers in this range it is conceivable that other difference sets exist.

Cyclic difference sets

v = 133, k = 33, λ = 3

(5.1) 1, 4, 5, 14, 16, 19, 20, 21, 25, 38, 54, 56, 57, 64, 66, 70, 76, 80, 83, 84, 91, 93, 95, 98, 100, 101, 105, 106, 114, 123, 125, 126, 131 (mod 133).

For v = 121, k = 4, λ = 13 there is the Singer system of 3 spaces in

(5.2) 121A: 1, 3, 4, 7, 9, 11, 12, 13, 21, 25, 27, 33, 34, 36, 39, 44, 55, 63, 64, 67, 68, 70, 71, 75, 80, 81, 82, 83, 85, 89, 92, 99, 102, 103, 104, 108, 109, 115, 117, 119.

There are also three other difference sets with these parameters

121B: 1, 3, 4, 5, 9, 12, 13, 14, 15, 16, 17, 22, 23, 27, 32, 34, 36, 39, 42, 45, 46, 48, 51, 64, 66, 69, 71, 77, 81, 82, 85, 86, 88, 92, 96, 102, 108, 109, 110, 117.

121C: 1, 3, 4, 7, 8, 9, 12, 21, 24, 25, 26, 27, 34, 36, 40, 43, 49,
(5.3) 63, 64, 68, 70, 71, 72, 75, 78, 81, 82, 83, 89, 92, 94, 95, 97,
102, 104, 108, 112, 113, 118, 120.

121D: 1, 3, 4, 5, 7, 9, 12, 14, 15, 17, 21, 27, 32, 36, 38, 42, 45,
46, 51, 53, 58, 63, 67, 68, 76, 79, 80, 81, 82, 83, 96, 100, 103,
106, 107, 108, 114, 115, 116, 119.

For the parameters $v = 127$, $k = 63$, $\lambda = 31$ there are six non-isomorphic difference sets, three corresponding to the listed type, and three others.

127A: Type Q.
1, 2, 4, 8, 9, 11, 13, 15, 16, 17, 18, 19, 21, 22, 25, 26, 30,
31, 32, 34, 35, 36, 37, 38, 41, 42, 44, 47, 49, 50, 52, 60,
61, 62, 64, 68, 69, 70, 71, 72, 73, 74, 76, 79, 81, 82, 84,
87, 88, 94, 98, 99, 100, 103, 104, 107, 113, 115, 117, 120,
121, 122, 124.

127B: Type H_6.
1, 2, 3, 4, 5, 6, 7, 8, 10, 12, 14, 16, 19, 20, 23, 24, 25,
27, 28, 32, 33, 38, 40, 46, 47, 48, 50, 51, 54, 56, 57, 61,
63, 64, 65, 66, 67, 73, 75, 76, 77, 80, 87, 89, 92, 94, 95,
96, 97, 100, 101, 102, 107, 108, 111, 112, 114, 117, 119,
122, 123, 125, 126.

127C: Singer-hyperplanes in PG(6,2).
1, 2, 3, 4, 6, 7, 8, 9, 12, 14, 15, 16, 17, 18, 24, 27, 28,
29, 30, 31, 32, 34, 36, 39, 47, 48, 51, 54, 56, 58, 60, 61,
62, 64, 65, 67, 68, 71, 72, 77, 78, 79, 83, 87, 89, 94, 96,
(5.4) 97, 99, 102, 103, 105, 107, 108, 112, 113, 115, 116, 117,
120, 121, 122, 124.

127D: 1, 2, 3, 4, 6, 7, 8, 9, 12, 13, 14, 16, 17, 18, 19, 24, 25,
26, 27, 28, 31, 32, 34, 35, 36, 38, 47, 48, 50, 51, 52, 54,
56, 61, 62, 64, 65, 67, 68, 70, 72, 73, 76, 77, 79, 81, 87,
89, 94, 96, 97, 100, 102, 103, 104, 107, 108, 112, 115, 117,
121, 122, 124.

127E: 1, 2, 3, 4, 5, 6, 8, 9, 10, 12, 15, 16, 17, 18, 19, 20, 24,
25, 27, 29, 30, 32, 33, 34, 36, 38, 39, 40, 48, 50, 51, 54,
55, 58, 59, 60, 64, 65, 66, 68, 71, 72, 73, 76, 77, 78, 80,
83, 89, 91, 93, 96, 99, 100, 102, 105, 108, 109, 110, 113,
116, 118, 120.

127F: 1, 2, 3, 4, 5, 6, 8, 10, 11, 12, 16, 19, 20, 21, 22, 24, 25, 27, 29, 32, 33, 37, 38, 39, 40, 41, 42, 44, 48, 49, 50, 51, 54, 58, 63, 64, 65, 66, 69, 73, 74, 76, 77, 78, 80, 82, 83, 84, 88, 89, 95, 96, 98, 100, 102, 105, 108, 111, 116, 119, 123, 125, 126.

Two difference sets for $v = 36$, $k = 15$, $\lambda = 6$ have been given by P.K. MENON [18]. The non-Abelian group of order 6, isomorphic to the symmetric group S_3 can be given by $1, a, a^2, b, ab, a^2b$ where $a^3 = 1$, $b^2 = 1$, $ba^2 = ab$.

Taking $G = S_3 \times S_3$ the 15 elements

(5.5)
$$\begin{array}{lllll} (1,1), & (1,b), & (b,1), & (b,ab), & (ab,b), \\ (a,a^2), & (a,ab), & (ab,1), & (ab,a^2b), & (a^2b,ab), \\ (a^2,a), & (1,a^2b), & (a^2b,1), & (a^2b,b), & (b,a^2b), \end{array}$$

form the difference set. This is one of the few known difference sets for which G is non-Abelian.

Also for $G = Z_6 \times Z_6$ where Z_6 is the cyclic group of order 6, taken here as residues modulo 6

(5.6)
$$\begin{array}{lllll} (0,0), & (0,1), & (1,0), & (1,3), & (3,1), \\ (2,4), & (0,3), & (3,0), & (3,5), & (5,3), \\ (4,2), & (0,5), & (5,0), & (5,1), & (1,5). \end{array}$$

The writer has found a simpler form for such a difference set

(5.7)
$$\begin{array}{lllll} (1,1), & (2,2), & (3,3), & (4,4), & (5,5), \\ (0,1), & (0,2), & (0,3), & (0,4), & (0,5), \\ (1,0), & (2,0), & (3,0), & (4,0), & (5,0). \end{array}$$

These are examples of difference sets of Hadamard types.

6. GENERAL THEORY OF DIFFERENCE SETS

Let G be a finite Abelian group of order v. Then from the theory of representation of finite groups [8] it is well-known that over the complex field the irreducible representations are all of degree one. This is to say that if for each $x \in G$ there is a non-singular matrix $M(x)$, and if $M(xy) = M(x)M(y)$, then there is a matrix S such that $S^{-1}M(x)S = A(x)$ and $A(x)$ is a diagonal matrix for all $x \in G$. Thus $A(x) = \chi_1(x) + \dots + \chi_m(x)$ where for $x \in G$, $\chi_i(x)$ is a complex number and $\chi_i(xy) = \chi_i(x)\chi_i(y)$, $i=1,\dots,m$. We call these χ's characters. Since $\chi(1) = 1$, each character $\chi(x)$ is some r-th root of unity if $x^r = 1$ and so for all $x \in G$, $\chi(x)$ is a v-th root of unity. The characters themselves may be multiplied, defining $(\chi_i\chi_j)(x) = \chi_i(x)\chi_j(x)$, all $x \in G$. Under this rule the characters themselves form a group which is in fact isomorphic to G. In particular there are exactly v distinct characters. The character χ_0 with the property $\chi_0(x) = 1$ for all $x \in G$ is called the *principal character*.

The characters may readily be extended to the group ring ZG, where if

$$A = \sum_{g \in G} a(g) \cdot g \tag{6.1}$$

we put

$$\chi(A) = \sum_{g \in G} a(g)\chi(g). \tag{6.2}$$

Clearly for each character χ, $A \to \chi(A)$ is a ring homomorphism of ZG into the complex numbers. A simple but useful property involving all characters is

$$\sum_{\chi} \chi(g) = \begin{cases} v & \text{if } g = 1 \\ 0 & \text{if } g \neq 1. \end{cases} \tag{6.3}$$

A powerful application of this, if A is given by (6.1) is

$$\sum_{\chi} \chi(Ag^{-1}) = va(g). \tag{6.4}$$

Another simple property is

$$\sum_{g \in G} \chi(g) = \begin{cases} v & \text{if } \chi = \chi_0 \\ 0 & \text{if } \chi \neq \chi_0. \end{cases} \tag{6.5}$$

If $\{d_1, d_2, \dots, d_k\}$ is a difference set in G let us write

$$\begin{aligned} D &= d_1 + d_2 + \dots + d_k, \\ D^t &= d_1^t + d_2^t + \dots + d_k^t, \quad t \text{ any integer}, \\ T &= \sum_{g \in G} g. \end{aligned} \tag{6.6}$$

Then (4.7) takes the form

(6.7) $$DD^{-1} = n + \lambda T.$$

Hence if χ is any non-principal character of G, then

(6.8) $$\chi(D)\chi(D^{-1}) = n.$$

Here $\chi(D)$ is an algebraic integer in some subfield of the field of v-th roots of unity and $\chi(D^{-1})$ is its complex conjugate. Thus the existence of a difference set is related to the factorization of n in various cyclotomic fields.

An application of these methods is a proof of the following result due to MANN [17]. His original proof was more complicated.

THEOREM 6.1. *A difference set with* $1 < k < v-1$ *over an elementary Abelian 2-group necessarily has parameters*

$$v = 2^{2t+2}, \quad k = 2^{2t+1} - 2^t, \quad \lambda = 2^{2t} - 2^t$$

or the complementary parameters

$$v = 2^{2t+2}, \quad k = 2^{2t+1} + 2^t, \quad \lambda = 2^{2t} + 2^t.$$

PROOF. Let D be a difference set over the elementary Abelian group G of order 2^r. Then since $g^{-1} = g$ for every $g \in G$, in (6.7) we will have $D^{-1} = D$ and so

(6.9) $$D^2 = n + \lambda T.$$

If χ is any non-principal character of G then

(6.10) $$\chi(D)^2 = n.$$

Now $\chi(g) = \pm 1$ for every $g \in G$ and so $\chi(D)$ is a rational integer. Thus

(6.11) $$n = s^2, \quad \chi(D) = \pm s, \quad \chi \neq \chi_0, \quad \chi_0(D) = k.$$

With

(6.12) $$D = \sum_g a(g)g$$

k of the a(g) are +1 and v-k are 0. Using (6.4)

(6.13) $$2^r a(g) = \sum_\chi \chi(Dg^{-1}).$$

From (6.11) this gives for some integer c(g)

(6.14) $2^r a(g) = k + c(g)s.$

Taking some x for which $a(x) = 0$ we have

(6.15) $0 = k + c(x)s$

so that s divides k. Let us write

(6.16) $k = hs.$

Taking some y for which $a(y) = 1$ we have

(6.17) $2^r = k + c(y)s = hs + c(y)s$

whence s divides 2^r so that $s = 2^t$ for some exponent t. Here

(6.18) $\lambda = k-n = hs - s^2 = 2^t h - 2^{2t}.$

Also from $k(k-1) = \lambda(v-1)$ we have

(6.19) $2^t h(2^t h-1) = (2^t h-2^{2t})(2^r-1).$

This simplifies to

(6.20) $h^2 - 2^{r-t}h = - 2^r + 1,$

which gives

(6.21) $(h - 2^{r-t-1})^2 = 2^{2r-2t-2} - 2^r + 1.$

Here $2^{2r-2t-2} \geq 2^r$ so that $r \geq 2t+2$. If $r = 2t+2$ then $h = 2^{r-t-1}\pm 1 = 2^{t+1}\pm 1$ and $k = 2^t h = 2^{2t+1}\pm 2^t$, $\lambda = k-n = 2^{2t}\pm 2^t$ and $v = 2^r = 2^{2t+2}$, the parameters of the theorem. If $r > 2t+2$, then $2^{2r-2t-2}-2^r+1 \geq 1+2^r$, and also $2^{2r-2t-2}-2^r+1 \equiv 1 \pmod{2^r}$. But if $z^2 \equiv 1 \pmod{2^r}$ then $z \equiv \pm 1 \pmod{2^{r-1}}$ and if $z \neq \pm 1$, then $|z| \geq 2^{r-1}-1$. Thus (6.21) yields

(6.22) $|h - 2^{r-t-1}| \geq 2^{r-1} - 1.$

If $h-2^{r-t-1} \geq 2^{r-1}-1$ then $h \geq r^{r-1}$ and $k = 2^t h \geq 2^{r+t-1}$, but as $v = 2^r > k$ this is possible only if $t = 0$. On the other hand if $2^{r-t-1}-h \geq 2^{r-1}-1$, then $2^{r-t-1} \geq 2^{r-1}$ and again this is possible only if $t = 0$. In either case $t = 0$ and $k = h$. Then (6.21) becomes

(6.23) $(k-2^{r-1})^2 = (2^{r-1}-1)^2,$

so that $k-2^{r-1} = \pm(2^{r-1}-1)$ giving $k = 1$ or $k = 2^r-1 = v-1$ the trivial solutions excluded by assumption. □

The difference sets of theorem 6.1 all exist. They are special cases of Hadamard difference sets, so called because of their relation to Hadamard matrices. An Hadamard matrix $H = [h_{ij}]$ is a square matrix of order N with $h_{ij} = \pm 1$ which satisfies

$$HH^T = H^TH = NI. \tag{6.24}$$

The matrices

$$\begin{bmatrix} 1 \end{bmatrix}, \quad \begin{bmatrix} 1 & 1 \\ 1 & -1 \end{bmatrix}, \quad \begin{bmatrix} 1 & -1 & -1 & -1 \\ -1 & 1 & -1 & -1 \\ -1 & -1 & 1 & -1 \\ -1 & -1 & -1 & 1 \end{bmatrix} \tag{6.25}$$

are Hadamard matrices of orders 1, 2, 4 respectively. It is easy to show that the order of an Hadamard matrix is 1, 2, or 4m for m=1,2,... and it is conjectured that Hadamard matrices exist for all these orders. At present the first undecided order is 188 with m = 47.

For square orders $4m^2$, a symmetric block design with $v = 4m^2$, $k = 2m^2-m$, $\lambda = m^2-m$ or its complement with $v = 4m^2$, $k = 2m^2+m$, $\lambda = m^2+m$ can be used to determine a Hadamard matrix of order $N = 4m^2$ by putting $h_{ij} = +1$ if the j-th point is in the i-th block and putting $h_{ij} = -1$ otherwise. And it is not difficult to show that if a Hadamard matrix of order v has exactly k elements which are +1 in every row, the rows will determine a symmetric block design with the parameters above. The third matrix in (6.25) of order 4 is of this type with the trivial design $v = 4$, $k = 1$, $\lambda = 0$.

In particular a difference set in a group G of order $v = 4m^2$ with $k = 2m^2 \pm m$, $\lambda = m^2 \pm m$ determines a Hadamard matrix of this kind. The difference sets with $v = 36$, $k = 15$, $\lambda = 6$ in (5.5), (5.6) and (5.7) are of this kind.

If H and K are Hadamard matrices of orders N and M respectively then the Kronecker product $H \times K$ (sometimes called the direct product or tensor product) is also a Hadamard matrix of order MN. Here

$$(6.26) \qquad H \times K = \begin{bmatrix} h_{11}K & h_{12}K & \cdots & h_{1N}K \\ h_{21}K & h_{22}K & \cdots & h_{2N}K \\ \cdot & \cdot & \cdots & \cdot \\ \cdot & \cdot & \cdots & \cdot \\ \cdot & \cdot & \cdots & \cdot \\ h_{N1}K & h_{N2}K & \cdots & h_{NN}K \end{bmatrix}$$

The proof of the following theorem is straightforward and will be omitted. It is due to MENON [18] who, however, did not recognize the relation to Hadamard matrices.

THEOREM 6.2. *If* H *and* K *are Hadamard matrices given by difference sets over groups* G_1 *and* G_2 *respectively, then* $H \times K$ *is a Hadamard matrix given by a difference set over* $G_1 \times G_2$. *If* D_1 *is the difference set for* H *over* G_1 *and* D_2 *for* K *over* G_2 *then in the direct product* $G = G_1 \times G_2 = (G_1, G_2)$ *the difference set* D *is the union of* (D_1, D_2) *and* $(\bar{D}_1, \bar{D}_2)$ *where* $\bar{D}_i$, $i=1,2$, *is the complement of* D_i *in* G_i.

COROLLARY. *There are difference sets with* $v = 4m^2$, $k = 2m^2 \pm m$, $\lambda = m^2 \pm m$ *when* $m = 2^{a+b-1}3^b$.

PROOF. This follows from the theorem since such sets exist for $v = 4$ and $v = 36$. □

The designs with $v = 2^{2t+2}$ have been extensively investigated, in particular by BLOCK [2] and recently by KANTOR [15].

In a finite field GF(q), $q = p^r$, p a prime, the multiplicative group H of non-zero elements is cyclic of order p^r-1. If $p^r-1 = ef$ the e-th powers of elements form a subgroup of H of order f and index e in H. If D is a difference set over G, the additive group of GF(q), it may happen that the e-th powers are multipliers of D. Several of the types listed in section 5 are of this kind.

If g is a primitive root of GF(q) the e-th power cyclotomic numbers are the numbers (i,j) where (i,j) is defined to be the number of solutions g^s of

$$(6.27) \qquad g^2 + 1 = g^t, \qquad s \equiv i \pmod{e}, \qquad t \equiv j \pmod{e}.$$

If D has the e-th powers of elements of GF(q) as multipliers, then (considering D to be the block fixed by these multipliers) D will consist of one or more cosets of H^e and possibly the zero element. A knowledge of the cyclotomic numbers (i,j) will determine the number of differences in each coset of H^e, and so will determine which combinations will give a difference set. The author [13] has shown that the cyclotomic numbers (i,j) may be determined by the character table of the group G^* of transformations $x \to a^e x+b$, $a \neq 0$ where G^* is of order $p^r f$. For e = 2, 4, 6, 8 difference sets of types Q, B, B_0, H_6, O, O_0 exist when $q = p^r$ satisfies the appropriate conditions. Except for the case Q where e = 2, arithmetical considerations show that for $q = p^r$ the conditions are only satisfied when r = 1 and q = p is a prime. This was shown by the writer for e = 4, 6, and by STORER [24] also for e = 8.

The multiplier theorem can be generalized, though the generalizations would be trivial if the condition $p > \lambda$ could be dropped. One such generalization is

THEOREM 6.3. *Let* D *be a difference set of* k *elements in the Abelian group* G *of order* v. *Let* $n_1 = p_1 p_2 \cdots p_s$ *be a divisor of* $n = k-\lambda$ *where* $p_1,\ldots,p_s$ *are distinct primes. If* $(n_1,v) = 1$, $n_1 > \lambda$ *and if* t *is an integer such that* $t \equiv p_i^{e_i} \pmod v$ *for an appropriate power* $p_i^{e_i}$, $i=1,\ldots,s$, *then the automorphism* α *of* G *defined by* $x \to x^t$ *is a multiplier of the difference set.*

The proof is almost identical with the proof of the multiplier theorem. In other cases for some divisor w of v and for a group G^* of order w which is a homomorphic image of G, it may be possible to find a multiplier t with $\theta^*(d^t) = g^*\theta^*(d)$ holding in G^*, the asterisks denoting homomorphic images. Such a t is called a w-*multiplier*. For example with v = 177, k = 33, $\lambda = 6$ it can be shown that 3 is a multiplier for w = 59 and this can be used to show that there is no difference set. There are also non-numerical multipliers α, where α is an automorphism of G which is also an automorphism of the v-k-λ design. This concept was introduced by BRUCK [4] but no multiplier theorem has been found for these.

It is conjectured that conditions (i) $p|k-\lambda$, (ii) $(p,v) = 1$ are sufficient for p to be a multiplier, and the condition (iii) $p > \lambda$ is unnecessary If conditions (i) and (ii) are not sufficient for what values of v will p fail to be a multiplier? R. MCFARLAND [16] has proved a theorem which sheds some light on this question.

He defines a quantity M(m) for every positive integer m as follows: $M(1) = 1$, $M(2) = 7$, $M(3) = 3\cdot 11\cdot 13$, $M(4) = 2\cdot 3\cdot 7\cdot 31$. For $m \geq 5$ let

$u = (m^2-m)/2$ and let p be a prime divisor of m with p^e the highest power of p dividing m. Then if m is not a square let M(m) be the product of the distinct odd prime factors of

$$(6.28) \qquad m, M(m^2/p^{2e}), p-1, p^2-1, \ldots, p^u-1.$$

If m is a square let M(m) be the product of the distinct prime factors in (6.28), including 2.

THEOREM 6.4. (MCFARLAND). *Let* D *be a difference set with parameters* (v,k,λ,n) *in an Abelian group* G *of order* v *and exponent* v^*. *Let*

$$n_1 | n, \qquad (n_1,v) = 1, \quad n_1 = p_1^{e_1} \ldots p_s^{e_s}$$

for some integer n_1 *where the* p_i's *are distinct primes. Suppose there are integers* $t, f_1, \ldots, f_s$ *such that*

$$t \equiv p_1^{f_1} \equiv \ldots \equiv p_s^{f_s} \pmod{v^*}.$$

If either

$$n_1 > \lambda \quad \text{or} \quad (M(n/n_1), v) = 1$$

then t *is a multiplier of* D.

There are a number of results based on the factorization (6.8) of n in various cyclotomic subfields of the v-th roots of unity. These tend to be highly technical and depend on the theory of the prime ideal factorizations in these fields. Nevertheless, many of the consequences can be described in relatively simple terms. Most of these results are due to work of MANN [17], TURYN [25,26] and YAMAMOTO [28].

These results are best described by some special terminology. If a, b, c are integers ($c \geq 0$) and a^c divides b while a^{c+1} does not, then a^c is said to *strictly divide* b. Let p be a prime and let p^e strictly divide w, so that $w = p^e w_1$ with $(p, w_1) = 1$. If there exists an integer $f > 0$ such that $p^f \equiv -1 \pmod{w_1}$ then p is said to be *self-conjugate modulo* w. If all the prime divisors of an integer m are self-conjugate modulo w, then m is said to be *self-conjugate modulo* w.

THEOREM 6.5. (MANN [17]). *Let* $w > 1$ *be a divisor of* v *and assume a non-trivial* v,k,λ *difference set exists with* w*-multiplier* $t \geq 1$. *Let* p *be a prime divisor of* n *for which* $(p,w) = 1$. *If there exists an integer* $f \geq 0$ *such that* $tpf \equiv -1 \pmod{w}$, *then* n *is strictly divisible by an even power of* p. *If* v^* *is the exponent of* G *and* $v^* = w$, *then there is only the trivial difference set with* $k = v$.

The BRUCK-RYSER-CHOWLA theorem [5,6] asserts that for the existence of a symmetric design with parameters v, k, λ it is necessary that

(i) *if* v *is even,* $n = k-\lambda$ *is a square;*
(ii) *if* v *is odd there exists a solution in integers* x, y, z *not all zero of*

$$x = ny + (-1)^{(v-1)/2}\lambda z^2.$$

Condition (i) was first found by SCHÜTZENBERGER [22].

If there is a v, k, λ difference set then further equations of this type must be solvable.

THEOREM 6.6. (HALL & RYSER [14]). *If there is a cyclic* v, k, λ *difference set then the following equation has solutions in integers* x, y, z *not all zero*

$$x^2 = ny^2 + (-1)^{(w-1)/2}wz^2 \tag{6.29}$$

where w *is any odd divisor of* v.

THEOREM 6.7. (YAMAMOTO [28]). *If there is a* v, k, λ *Abelian difference set, if* q *is an odd divisor of* v, *and if* r *is a prime such that* r^e *strictly divides* n, *then the following equation is solvable in integers* x, y, z *not all zero*

$$x^2 = r^e y^2 + (-1)^{(q-1)/2}qz^2. \tag{6.30}$$

THEOREM 6.8. (TURYN [26]). *Assume a non-trivial Abelian* v, k, λ *difference set exists. Let* m^2 *divide* n *and suppose that* $m > 1$ *is self conjugate modulo* w *for some divisor* $w > 1$ *of* v. *If* $(m,w) > 1$ *then* $m \leq v/w$. *If* $(m,w) > 1$ *then* $m \leq 2^{r-1}v/w$, *where* r *is the number of distinct prime factors of* (m,w).

THEOREM 6.9. (MANN [17]). *If* $p|(v,n)$, *and if* $v = p^e v_1$, *and* $p^f \equiv -1 \pmod{v_1}$ *for some* $f \geq 0$ *there is no cyclic* v, k, λ *difference set.*

THEOREM 6.10. (TURYN [26]). *There is no cyclic difference set with* $v = 4m^2$, $k = 2m^2 \pm m$, $\lambda = m^2 \pm m$ *if* m *is a prime power.*

Most of these are non-existence theorems. The author [12] in 1956 studied cyclic difference sets with $3 \le k \le 50$ and was able to determine existence or non-existence in all but 12 cases whose parameters are given here:

(6.31)

v	k	λ	n	v	k	λ	n
45	12	3	9	120	35	10	25
36	15	6	9	288	42	6	36
96	20	4	16	100	45	20	25
64	28	12	16	208	46	10	36
175	30	5	25	189	48	12	36
171	35	7	28	176	50	14	36

These 12 cases in part inspired the efforts to find non-existence theorems. Theorems 6.9 and 6.10 exclude all of these except $(v,k,\lambda,n) = 171,35,7,28$ which is ruled out by theorem 6.5 with $p = 2$, $t = 1$ and the congruence $2^9 \equiv -1 \pmod{171}$ with $w = v^* = v$, and $(v,k,\lambda,n) = 120,35,10,25$ which is ruled out by theorem 6.8 with $m = 5$, $w = 30$ since $5 \equiv -1 \pmod 6$, $(m,w) = 5$ and so $r = 1$, but we do not have $5 \le 2^0 \cdot 120/30 = 4$.

REFERENCES

[1] BAUMERT, L.D., *Cyclic difference sets*, Lecture Notes in Mathematics 182, Springer-Verlag, Berlin etc., 1971.

[2] BLOCK, R.E., *Transitive groups of collineations of certain designs*, Pacific J. Math., 15 (1965) 13-19.

[3] BRAUER, A., *On a new class of Hadamard determinants*, Math. Z., 58 (1953) 219-225.

[4] BRUCK, R.H., *Difference sets in a finite group*, Trans. Amer. Math. Soc., 78 (1955) 464-481.

[5] BRUCK, R.H. & H.J. RYSER, *The non-existence of certain finite projective planes*, Canad. J. Math., 1 (1949) 88-93.

[6] CHOWLA, S. & H.J. RYSER, *Combinatorial problems*, Canad. J. Math., 2 (1950) 93-99.

[7] DEMBOWSKI, P., *Finite geometries*, Ergebnisse der Mathematik 44, Springer-Verlag, Berlin etc., 1968.

[8] FEIT, W., *Characters of finite groups*, Benjamin, New York, 1967.

[9] GORDON, B., W.H. MILLS & L.R. WELCH, *Some new difference sets*, Canad. J. Math., 14 (1962) 614-625.

[10] HALL Jr., M., *Cyclic projective planes*, Duke Math. J., 14 (1947) 1079-1090.

[11] HALL Jr., M., *Combinatorial theory*, Blaisdell, Waltham, Mass., 1967.

[12] HALL Jr., M., *A survey of difference sets*, Proc. Amer. Math. Soc., 7 (1956) 975-986.

[13] HALL Jr., M., *Characters and cyclotomy*, in: Proc. Symp. in Pure Math., vol. 8, Amer. Math. Soc., 1965.

[14] HALL Jr., M. & H.J. RYSER, *Cyclic incidence matrices*, Canad. J. Math., 3 (1951) 495-502.

[15] KANTOR W., *Symplectic groups, symmetric designs, and line ovals*, to appear.

[16] MCFARLAND, R.L., *On multipliers of Abelian difference sets*, thesis, The Ohio State University, 1970.

[17] MANN, H.B., *Balanced incomplete block designs and Abelian difference sets*, Illinois J. Math., 8 (1964) 252-261.

[18] KESAVA MENON, P., *Difference sets in Abelian groups*, Proc. Amer. Math. Soc., 11 (1960) 368-376.

[19] KESAVA MENON, P., *On difference sets whose parameters satisfy a certain relation*, Proc. Amer. Math. Soc., 13 (1962) 739-745.

[20] PARKER, E.T., *On collineations of symmetric designs*, Proc. Amer. Math. Soc., 8 (1957) 350-351.

[21] RYSER, H.J., *Combinatorial mathematics*, Carus Math. Monograph No. 14 Math. Assoc. Amer., Wiley, New York, 1963.

[22] SCHÜTZENBERGER, M.P., *A non-existence theorem for an infinite family of symmetrical block designs*, Ann. Eugenics, 14 (1949) 286-287.

[23] SINGER, J., *A theorem in finite projective geometry and some applications to number theory*, Trans. Amer. Math. Soc., 43 (1938) 377-385.

[24] STORER, T., *Cyclotomy and difference sets*, Markham, Chicago, 1967.

[25] TURYN, R.J., *The multiplier theorem for difference sets*, Canad. J. Math., 16 (1964) 386-388.

[26] TURYN, R.J., *Character sums and difference sets*, Pacific J. Math., 15 (1965) 319-346.

[27] WHITEMAN, A.L., *A family of difference sets*, Illinois J. Math., 6 (1962) 107-121.

[28] YAMAMOTO, K., *Decomposition fields of difference sets*, Pacific J. Math., 13 (1963) 337-352.

INVARIANT RELATIONS, COHERENT CONFIGURATIONS AND GENERALIZED POLYGONS *)

D.G. HIGMAN

University of Michigan, Ann Arbor, Mich. 48104, USA

A high point in the combinatorial approach to the theory of finite permutation groups is WIELANDT's theory of invariant relations, culminating in his theorem on groups of degree p^2 [16]. In section 1 we give a few rudiments of WIELANDT's theory in the context of the theory of G-spaces, illustrating the concepts by a proof, which seems first to have been made explicit by R. LIEBLER [12], of a theorem of ALPERIN [1].

In section 2 we axiomatize certain combinatorial aspects of the theory of G-spaces, defining the class of combinatorial structures which we call *coherent configurations* [7,8,9] and listing some results about these from [9]. Association schemes as defined by BOSE & SHIMAMOTO [2] are a special class of coherent configurations.

In section 3 we turn to generalized polygons, which were introduced by J. TITS in connection with the problem of classifying finite groups with (B,N)-pair (cf. [3,6,14]). Here we apply the results listed in section 2 to generalized polygons, obtaining a proof (essentially that of KILMOYER & SOLOMON [11]) of the FEIT-HIGMAN theorem, and a proof that $s \leq t^2$ for generalized quadrangles and octagons having s+1 points on each line and t+1 lines through each point, with $t > 1$.

The author is happy to express thanks to J.E. MCLAUGHLIN for suggestions which simplified section 3, especially for the use of Lagrange interpolation.

1. G-SPACES AND INVARIANT RELATIONS

X and Y will be finite non-empty sets. We regard a subset f of $X \times Y$ as a *relation* from X to Y, and put $f^{\cup} = \{(y,x) \mid (x,y) \in f\}$ (the *converse*

---

*) Research supported in part by the National Science Foundation.

M. Hall, Jr. and J. H. van Lint (eds.), Combinatorics, 347-363.

of f) and $f(x) = \{y \in Y \mid (x,y) \in f\}$ for $x \in X$.

An *action* of a group G on X is a map $X \times G \to X$, $(x,g) \mapsto xg$, such that $x(gh) = (xg)h$ and $x1 = x$ for all $x \in X$ and $g,h \in G$, where 1 is the identity element of G. Specifying an action of G on X is equivalent to specifying a homomorphism of G into the symmetric group Σ_X on X.

A G-space is partitioned into its G-*orbits*, which are the equivalence classes under the equivalence relation: $x \sim y$ if and only if $xg = y$ for some $g \in G$. A G-space is *transitive* if there is just one G-orbit.

A subset Y of a G-space X is *invariant* under G if $Yg \subseteq Y$ for all $g \in G$. Then G acts on Y and the G-orbits in Y are G-orbits in X. The G-orbits in a G-space X are transitive G-spaces. For $x \in X$, the subgroup $G_x = \{g \in G \mid xg = x\}$ of G is called the *stabilizer* of x in G.

If X and Y are G-spaces, then so is $X \times Y$ under the componentwise action $((x,y),g) \mapsto (xg,yg)$. A relation $F \subseteq X \times Y$ which is invariant under this action is called an *invariant relation* from X to Y. If $\mathcal{O}$ is the totality of G-orbits in $X \times Y$, then the invariant relations from X to Y are just the unions of members of $\mathcal{O}$.

Assume from now on in this section that X *and* Y *are transitive* G-*spaces*. Choose $x \in X$ and $y \in Y$. Then $\{f(x) \mid f \in \mathcal{O}\}$ is the partition of Y into G_x-orbits and $\{f^\cup(y) \mid f \in \mathcal{O}\}$ is the partition of X into G_y-orbits. Thus

$$f(x) \leftrightarrow f^\cup(y), \qquad (f \in \mathcal{O})$$

is a one-to-one correspondence between the G_x-orbits in Y and the G_y-orbits in X. The *lengths* $|f(x)|$ and $|f^\cup(y)|$ of corresponding orbits are proportional since $|X||f(x)| = |Y||f^\cup(y)| = |f|$.

We illustrate these concepts as follows. Suppose that $F \subseteq X \times Y$ is an invariant relation. Then F is partitioned into G-orbits $F = \bigcup_{f \in \mathcal{O}_F} f$ with $\mathcal{O}_F \subseteq \mathcal{O}$, and $F^\cup = \bigcup_{f \in \mathcal{O}_F} f^\cup$. Taking $(x,y) \in F$ we have that

$$f(x) \leftrightarrow f^\cup(y), \qquad (f \in \mathcal{O}_F)$$

is a one-to-one correspondence between the G_x-orbits in $F(x)$ and the G_y-orbits in $F^\cup(y)$.

To apply this, start with a transitive G-space X, choose $a \in X$ and a subgroup H of G_a, and construct a transitive G-space Y and an invariant relation F as follows:

Y is the totality of conjugates $H^g = g^{-1}Hg$ ($g \in G$) of H in G, with G acting on Y by conjugation,

$$F = \{(x,H^g) \mid x \in X,\ g \in G,\ H^g \subseteq G_x\}.$$

Then $(a,H) \in F$,

F(a) is the totality of conjugates of H which are contained in G_a,

G_H is the normalizer $N_G(H)$ of H in G ($N_G(H) = \{g \in G \mid H^g = H\}$),

and

$F^{\cup}(H)$ is the set of fixed points of H in G.

It follows therefore, that

THEOREM 1.1. (ALPERIN [1]). *If* X *is a transitive* G-*space*, $a \in X$, *and* H *is a subgroup of* G_a, *then there is a one-to-one correspondence between the conjugate classes in* G_a *of conjugates of* H *which are contained in* G_a *and the* $N_G(H)$-*orbits of fixed points of* H.

Finiteness plays no essential role in this proof of ALPERIN's theorem. In addition to the corollary in [1], theorems of JORDAN [15; 3.5-3.7], MANNING [15; 3.6'] and WITT [15; 9] are immediate corollaries of theorem 1.1. This proof of ALPERIN's theorem seems first to have been made explicit by R. LIEBLER [12].

2. COHERENT CONFIGURATIONS

As suggested by section 1, we consider *configurations* $(X,\mathcal{O})$ consisting of a (finite) non-empty set X and a set $\mathcal{O}$ of binary relations on X^2. Thus $\mathcal{O}$ is a subset of the power set $\mathcal{P}(X^2)$ of the cartesian square X^2 of X. There is no loss in generality in the restriction to relations on a single set, since, for example, a relation from X to Y can be regarded as a relation on the disjoint union of X and Y. We put $\mathcal{R}$ equal to the boolean subalgebra of $\mathcal{P}(X^2)$ generated by $\mathcal{O}$. If $f_1,f_2,\dots,f_s \in \mathcal{O}$ and $x,y \in X$, an $(f_1,\dots,f_s)$-*path* from x to y is an (s+1)-tuple $(x_1,\dots,x_{s+1})$ such that $x_1 = x$, $x_{s+1} = y$ and $(x_i,x_{i+1}) \in f_i$, $1 \le i \le s$. We call $(X,\mathcal{O})$ *coherent* if the following axioms (I) through (IV) are satisfied.

(I) $\mathcal{O}$ *is a partition of* X^2.

(II) $I = \{(x,x) \mid x \in X\} \in \mathcal{R}$.

(III) $f \in \mathcal{O}$ *implies* $f^{\cup} \in \mathcal{O}$.

(IV) *For all* $f,g,h \in \mathcal{O}$ *and* $(x,y) \in h$, *the number of* (f,g)*-paths from* x *to* y *is independent of the choice of* $(x,y) \in h$.

The number of (f,g)-paths from x to y is $|f(x) \cap g^{\cup}(y)|$ and can be denoted by a_{fgh} for $(x,y) \in h$ if $(X,\mathcal{O})$ is coherent, in which case the non-negative integers a_{fgh} are the *intersection numbers* of $(X,\mathcal{O})$. The number $r = |\mathcal{O}|$ is called the rank.

If X is a G-space for a group G and $\mathcal{O}$ is the set of G-orbits in X^2, then $(X,\mathcal{O})$ is coherent. We refer to this situation as the *group case*.

A coherent configuration is *homogeneous* if $I \in \mathcal{O}$ and *symmetric* if the *pairing* $f \mapsto f^{\cup}$ $(f \in \mathcal{O})$ is trivial, i.e. if every $f \in \mathcal{O}$ is symmetric. A symmetric configuration is necessarily homogeneous. In the group case, homogeneity is equivalent to transitivity of the G-space.

Symmetric coherent configurations are equivalent to *association schemes* as defined by BOSE & SHIMAMOTO [2].

The boolean algebra $\mathcal{R}$ of a coherent configuration is a semigroup under composition and the idempotents in $\mathcal{R}$ are of particular interest. We do not go into this here but turn at once to the *adjacency algebra*, which is the centralizer algebra in the group case.

We denote by $\mathrm{Mat}_{\mathbb{C}}\, X$ the totality of matrices $\phi\colon X^2 \to \mathbb{C}$, regarded as an algebra over $\mathbb{C}$ with respect to matrix (i.e. pointwise) addition and matrix multiplication. For $f \in \mathcal{O}$, $\Phi_f\colon X^2 \to \mathbb{C}$ will be the characteristic function of f, or, otherwise thought of, Φ_f is the adjacency matrix of the graph (X,f). If $(X,\mathcal{O})$ is coherent, then the set $\mathcal{B} = \{\Phi_f \mid f \in \mathcal{O}\}$ is a basis of a subalgebra $\mathcal{A}$ of $\mathrm{Mat}_{\mathbb{C}}\, X$, called the *adjacency algebra* of $(X,\mathcal{O})$.

$(X,\mathcal{O})$ will be called *commutative* if $\mathcal{A}$ is commutative. A symmetric configuration is necessarily commutative and a commutative configuration is necessarily homogeneous. As used by DELSARTE [4], the term association scheme is equivalent to commutative configuration. Section 3 will illustrate the importance of the non-commutative case.

For applications we often need the following translation of the axioms (I) through (IV).

THEOREM 2.1. *Let* $\mathcal{B}$ *be a set of non-zero (0,1)-matrices in* $\mathrm{Mat}_{\mathbb{C}}\, X$ *such that*

(1) $\Phi = \sum_{\phi\in\mathcal{B}} \phi$ *is the all 1 matrix*, $\Phi(x,y) = 1$ *for all* $x,y \in X$,

(2) *the identity matrix is a sum of members of* $\mathcal{B}$,

(3) $\phi \in \mathcal{B}$ *implies that* $\phi^t \in \mathcal{B}$, *and*

(4) $\mathcal{B}$ *spans a subalgebra* $\mathcal{A}$ *of* $\mathrm{Mat}_{\mathbb{C}}\, X$.

Then $(X,\mathcal{O})$ *with* $\mathcal{O} = \{\text{supp}\,\phi \mid \phi \in \mathcal{B}\}$, *is a coherent configuration with adjacency algebra* $\mathcal{A}$.

Now we list some basic facts about $\mathcal{A}$, referring to [9] for proofs.

THEOREM 2.2. $\mathcal{A}$ *is semisimple.*

THEOREM 2.3. *There is a unitary matrix* U *effecting a complete reduction of* $\mathcal{A}$, *i.e. such that for all* $\phi \in \mathcal{A}$,

$$U^*\phi U = \text{diag}(\Delta_1(\phi) \times I_{z_1}, \ldots, \Delta_m(\phi) \times I_{z_m}),$$

where $\Delta_1, \ldots, \Delta_m$ *are the inequivalent irreducible representations of* $\mathcal{A}$.

We put e_i = degree of Δ_i and call z_i the *multiplicity* of Δ_i. The *character* ζ_i *of* $\mathcal{A}$ *afforded by* Δ_i is defined by $\zeta_i(\phi) = \text{trace}\,\Delta_i(\phi)$. We have $\zeta_i(I) = e_i$ and $\sum_{i=1}^m e_i^2 = r$, $\sum_{i=1}^m e_i z_i = |X|$.

We write $\Delta_\alpha(\phi) = (a_{ij}^\alpha(\phi))$ and list the a_{ij}^α: $a_1, a_2, \ldots, a_r$. If $a_\lambda = a_{ij}^\alpha$, we write $a_{\bar\lambda} = a_{ji}^\alpha$ and $h_\lambda = z_\alpha$. For $f \in \mathcal{O}$ and $x \in$ domain f, $|f(x)|$ is independent of x and we put $v_f = |f(x)|$ and $\tilde\Phi_f = \frac{1}{|f|}\Phi_{f^\cup}$.

There is a distinguished irreducible character of $\mathcal{A}$ of degree 1 called the *principal character*. In the homogeneous case this means that if we choose a notation so that ζ_1 is the principal character, then $e_1 = z_1 = 1$ and $\zeta_1(\Phi_f) = v_f$ for all $f \in \mathcal{O}$.

Of fundamental importance are the Schur relations

$$\sum_{f\in\mathcal{O}} a_\lambda(\tilde\Phi_f)\, a_\mu(\Phi_f) = \delta_{\lambda\bar\mu}\frac{1}{h_\lambda}, \tag{2.1}$$

which imply the orthogonality relations

$$\sum_{f\in\mathcal{O}} \zeta_\alpha(\tilde\Phi_f)\, \zeta_\beta(\Phi_f) = \delta_{\alpha\beta}\frac{e_\alpha}{z_\alpha}. \tag{2.2}$$

Assume that $\Delta_\alpha(\phi^*) = \Delta_\alpha(\phi)^*$ for all α and ϕ, or equivalently, that

$$a_\sigma(\Phi_{f^\cup}) = a_{\bar\sigma}(\Phi_f) \qquad \text{for all } f \in \mathcal{O}, 1 \le \sigma \le r. \tag{2.3}$$

This will hold if the complete reduction of $\mathcal{A}$ is afforded by a unitary matrix. Then we have the *Krein condition*

$$(2.4)\begin{cases}\text{Choose } \lambda \text{ and } \mu,\ 1 \le \lambda,\ \mu \le r, \text{ such that } \lambda = \bar{\lambda} \text{ and } \mu = \bar{\mu}. \text{ If } a_\nu = a_{ij}^\alpha, \\ \text{put} \\ \qquad c_{ij}^\alpha = \sum_{f \in \mathcal{O}} \dfrac{a_\lambda(\Phi_f)\, a_\mu(\Phi_f)\, \overline{a_\nu(\Phi_f)}}{|f|^2}. \\ \text{Then } C_\alpha = (c_{ij}^\alpha) \text{ is hermitian positive semidefinite.}\end{cases}$$

(Actually (2.3) is needed only for the a_σ occurring in the definition of C_α to have this conclusion for a particular C_α.) (2.4) extends a result of L.L. SCOTT Jr. [13].

3. GENERALIZED POLYGONS

An *incidence structure* (P,L,F) consists of two disjoint non-empty sets P and L and a relation $F \subseteq P\times L$. The members of the sets $P \cup L$, P, L and F will be called *elements*, *points*, *lines* and *flags* respectively. Two elements x and y are *incident* if $(x,y) \in F \cup F^{\cup}$, so the flags are the incident point-line pairs. An incidence structure can be represented by a bipartite graph in which the vertices are the elements and the edges are the flags.

A *path of length* d from an element x to an element y is a (d+1)-tuple $(x_0,x_1,\ldots,x_d)$ of elements such that x_{i-1} and x_i are incident for all i, $1 \le i \le d$. The *distance* $\rho(x,y)$ between two elements is the length of the shortest path from x to y, or ∞ if no such path exists.

A *generalized* n-*gon*, where n is an integer > 0, is an incidence structure (P,L,F) satisfying the following two conditions for all elements x and y:

(A) for each x, the maximum distance $\rho(x,y)$ is n;

(B) if $\rho(x,y) < n$, then there is exactly one path of length < n from x to y.

A *generalized polygon* is a generalized n-gon for some n. The generalized polygons considered here will be assumed to satisfy the following additional condition:

(C) each line is incident with the same number s+1 of points and each point is incident with the same number t+1 of lines.

The generalized polygons with s = t = 1 are just the ordinary polygons. *We assume from now on that* st > 1. According to the FEIT-HIGMAN theorem (to be proved below as theorem 3.1), $n \in\{3,4,6,8,12\}$ and s = 1 or t = 1 if n = 12. A generalized triangle is the same thing as a projective plane. There is a fairly extensive literature about generalized quadrangles, but very little seems to be known about generalized n-gons with n > 4. Simple groups of Lie type of rank 2 act on generalized polygons. These groups are listed at the end of the section.

Note that the *dual* (L,P,F) of a generalized n-gon is a generalized n-gon.

We now construct a coherent configuration based on the set F of flags of a generalized n-gon (P,L,F), and systematically apply the results outlined in section 2. For this we need to determine the irreducible representations of the adjacency algebra.

If $x \in F$, then $x = (x_1, x_2)$ with $x_1 \in P$ and $x_2 \in L$. We start with the symmetric relations

$$S = \{(x,y) \in F^2 \mid x_1 \neq y_1 \text{ and } x_2 = y_2\}$$

and

$$T = \{(x,y) \in F^2 \mid x_1 = y_1 \text{ and } x_2 \neq y_2\} .$$

Composing relations in the usual way we see that $S^2 = I$ or $S \cup I$ according as $s = 1$ or $s > 1$, $T^2 = I$ or $T \cup I$ according as $t = 1$ or $t > 1$, and that the 2n relations

$$\begin{array}{lll} & S,\ ST,\ STS,\ \ldots,\ \underbrace{STS\ldots}_{n-1}, & \\ I & & \underbrace{STS\ldots}_{n} = \underbrace{TST\ldots}_{n} \\ & T,\ TS,\ TST,\ \ldots,\ \underbrace{TST\ldots}_{n-1}, & \end{array}$$

constitute a partition $\mathcal{O}$ of F^2. This uses only the conditions (A) and (B).

Put $A = \Phi_S$ and $B = \Phi_T$. We readily verify that for each $f = \ldots STS\ldots$ in $\mathcal{O}$, $\Phi_f = \ldots ABA\ldots$. In particular, therefore

$$\text{(3.1)} \qquad \underbrace{ABA\ldots}_{n} = \underbrace{BAB\ldots}_{n} .$$

At this point we invoke condition (C) to obtain the relations

$$\text{(3.2)} \qquad A^2 = (s-1)A + sI \quad \text{and} \quad B^2 = (t-1)B + tI.$$

It follows that the matrices Φ_f, $f \in \mathcal{O}$, constitute a basis of a subalgebra $\mathcal{A}$ of $\mathrm{Mat}_{\mathbb{C}}\,F$, and hence by theorem 2.1 that $(F,\mathcal{O})$ is a homogeneous coherent configuration with adjacency algebra $\mathcal{A}$. Moreover, $A \mapsto U$ and $B \mapsto V$ will determine a matrix representation of $\mathcal{A}$ if and only if U and V are matrices such that the conditions (3.1) and (3.2) are satisfied with U and V in place of A and B.

First we consider the 1-dimensional representations of A.

Case n = 2m. In this case A has four distinct linear characters $\zeta_1,\ldots,\zeta_4$:

(3.3)

	A	B
ζ_1	s	t
ζ_2	-1	-1
ζ_3	s	-1
ζ_4	-1	t

The principal character is ζ_1, so

$$|F| = \sum_{f\in\mathcal{O}} \zeta_1(\Phi_f) ,$$

that is:

(3.4) *if* $n = 2m$, *then* $|F| = (1+s)(1+t)\,\dfrac{(st)^m - 1}{st - 1}$.

Case n = 2m+1. Here there are two distinct linear characters ζ_1 and ζ_2 as in the first two rows of (3.3). From the relation (3.1),

$$s^{m+1}t^m = \zeta_1((AB)^m A) = \zeta_1(B(AB)^m) = s^m t^{m+1}, \text{ so}$$

(3.5) *if* n *is odd, then* $s = t$.

Next we determine the 2-dimensional irreducible representations of A. It turns out to be sufficient to find the real irreducible representations with composition factors affording ζ_1 and ζ_2. Thus we look for real matrices

$$U = \begin{pmatrix} -1 & 0 \\ 0 & s \end{pmatrix} \quad \text{and } V = \begin{pmatrix} b & u \\ u & c \end{pmatrix},$$

with $u \neq 0$, such that V has trace t-1 and determinant -t, i.e. such that

(3.6) $b+c = t-1$ and $u^2 = bc+t$.

Since (3.6) implies (3.2), $A \mapsto U$, $B \mapsto V$ will be a representation if and only if

(3.7) $(UV)^m = (VU)^m$ if $n = 2m$ and $(UV)^m U = V(UV)^m$ if $n = 2m+1$.

We need the products

$$UV = \begin{pmatrix} -b & -u \\ su & sc \end{pmatrix}, \qquad VU = \begin{pmatrix} -b & su \\ -u & sc \end{pmatrix},$$

$$UVU = \begin{pmatrix} b & -su \\ -su & s^2c \end{pmatrix}, \qquad VUV = \begin{pmatrix} -b^2+su^2 & (-b+sc)u \\ (-b+sc)u & -u^2+sc^2 \end{pmatrix}.$$

Assume that U and V satisfy (3.6) and (3.7), $u \neq 0$. The matrices UV and VU do not commute, otherwise we would have $b^2+u^2 = b^2+s^2u^2$ and $bsu - usc = bu + s^2uc$, whence $s = t = 1$, contrary to assumption.

By (3.7), $(UV)^r = (VU)^r$ with $r = n/2$ or n according as n is even or odd. But $(UV)^r = aI + b(UV)$, $a,b \in \mathbb{C}$, so $b \neq 0$ would imply that UV and VU commute. Hence $(UV)^r$ is a scalar matrix, so the similar matrices UV and VU have distinct eigenvalues λ and $\mu = \xi\lambda$ with $\xi^r = 1$, $\xi \neq 1$.

Now the determinant of UV is $su^2-bsc = s(u^2-bc) = st$, so $\lambda = \sqrt{st}\,\theta$ and $\mu = \sqrt{st}\,\theta^{-1}$ with $\theta^{2r} = 1$ and $\theta \neq \theta^{-1}$. Since trace $UV = -b+sc = \sqrt{st}(\theta+\theta^{-1})$,

$$\text{(3.8)} \qquad \begin{cases} (s+1)b = s(t-1) - \sqrt{st}(\theta+\theta^{-1}), \\ (s+1)c = t-1 + \sqrt{st}(\theta+\theta^{-1}) . \end{cases}$$

If X is a 2×2 matrix with distinct eigenvalues λ, μ, then for $k=1,2,\ldots,$

$$X^k = \frac{\lambda^k}{\lambda - \mu}(X-\mu I) + \frac{\mu^k}{\mu - \lambda}(X-\lambda I) =$$

$$= \frac{\lambda^k - \mu^k}{\lambda - \mu} X - \frac{\lambda\mu(\lambda^{k-1}-\mu^{k-1})}{\lambda - \mu} I.$$

In particular, therefore, for X = UV or VU and $k=1,2,\ldots$

$$\text{(3.9)} \qquad X^k = \frac{(\sqrt{st})^{k-1}}{\theta - \theta^{-1}}\{(\theta^k-\theta^{-k})X - \sqrt{st}(\theta^{k-1}-\theta^{-(k-1)})I\}.$$

We use (3.9) to determine the character of our representation. We have trace $U = s-1$, trace $V = t-1$, trace UV = trace $VU = -b+sc = \sqrt{st}(\theta+\theta^{-1})$, trace $UVU = b+s^2c = s(t-1) + (s-1)\sqrt{st}(\theta+\theta^{-1})$, and trace $VUV = -b^2 + (s-1)u^2 + sc^2 = -b^2 + (s-1)(bc+t) + sc^2 = (s-1)t + (b+c)(-b+sc) =$

$= (s-1)t + (t-1)\sqrt{st}(\theta+\theta^{-1})$. Hence by (3.9),

$$\text{trace }(UV)^k = \frac{(\sqrt{st})^{k-1}}{\theta-\theta^{-1}}\{(\theta^k-\theta^{-k})\sqrt{st}(\theta+\theta^{-1}) - 2\sqrt{st}(\theta^{k-1}-\theta^{-(k-1)})\} =$$

$$= \frac{(\sqrt{st})^{k}}{\theta-\theta^{-1}}\{(\theta^{k+1}-\theta^{-(k+1)}) - (\theta^{k-1}-\theta^{-(k-1)})\} =$$

$$= (\sqrt{st})^k(\theta^k+\theta^{-k})$$

and

$$\text{trace }(UV)^kU = \frac{(\sqrt{st})^{k-1}}{\theta-\theta^{-1}}\{(\theta^k-\theta^{-k})(s(t-1)+(s-1)\sqrt{st}(\theta+\theta^{-1})) + - \sqrt{st}(\theta^{k-1}-\theta^{-(k-1)})(s-1)\} =$$

$$= \frac{(\sqrt{st})^{k}}{\theta-\theta^{-1}}\{(s-1)[(\theta^k-\theta^{-k})(\theta+\theta^{-1}) - (\theta^{k-1}-\theta^{-(k-1)})] + + \frac{s}{\sqrt{st}}(t-1)(\theta^k-\theta^{-k})\} =$$

$$= \frac{(\sqrt{st})^{k}}{\theta-\theta^{-1}}\{(s-1)(\theta^{k+1}-\theta^{-(k+1)}) + \frac{s}{\sqrt{st}}(t-1)(\theta^k-\theta^{-k})\}.$$

The trace of $V(UV)^k$ is obtained from that of $(UV)^kU$ by interchanging s and t. Thus, if ζ is the character afforded by our representation, then

$$(3.10)\qquad \begin{cases} \zeta((AB)^k) = \zeta((BA)^k) = (\sqrt{st})^k(\theta^k+\theta^{-k}), \\ \zeta((AB)^kA) = \dfrac{(\sqrt{st})^{k}}{\theta-\theta^{-1}}\{(s-1)(\theta^{k+1}-\theta^{-(k+1)}) + \dfrac{s}{\sqrt{st}}(t-1)(\theta^k-\theta^{-k})\}, \\ \zeta(B(AB)^k) = \dfrac{(\sqrt{st})^{k}}{\theta-\theta^{-1}}\{(t-1)(\theta^{k+1}-\theta^{-(k+1)}) + \dfrac{t}{\sqrt{st}}(s-1)(\theta^k-\theta^{-k})\}. \end{cases}$$

Now choose b and c according to (3.8), with $\theta^{2r} = 1$, $r = n/2$ or n according as n is even or odd, $\theta = \cos\alpha+i\sin\alpha \neq \pm 1$. Then $-b+sc = \sqrt{st}(\theta+\theta^{-1}) = 2\sqrt{st}\cos\alpha$, so $bc+t = a \leq 0$ would imply that $-b^2 + s(-t+a) = 2b\sqrt{st}\cos\alpha$, or $(b+\sqrt{st}\cos\alpha)^2 - st\sin^2\alpha = sa \leq 0$, and hence $\sin\alpha = 0$, or

$\theta = \pm 1$, contrary to assumption. Hence the solutions of $u^2 = bc+t$ are real. For such a u the matrices U and V satisfy $(UV)^r = (VU)^r$. We consider the cases of even and odd n separately.

Case $n = 2m$. In this case we have $(UV)^m = (VU)^m = (\sqrt{st})^m \theta^m I$ and hence $A \mapsto U$, $B \mapsto V$ is a representation. We obtain m-1 inequivalent irreducible representations of degree 2 on taking ε a primitive n-th root of unity and $\theta = \varepsilon^i$, $i=1,2,\ldots,m-1$. The sum of the squares of the degrees of the irreducible representations of A obtained so far is $4+4(m-1) = 2n$, the dimension of A, so all irreducible representations are accounted for in this case.

We now apply the orthogonality relations (2.2) to determine the multiplicity z of each irreducible character ρ. In our present case we have

$$\frac{\rho(1)}{z}|F| = \sum_{k=0}^{m-1} \{2\frac{\rho((AB)^k)^2}{(st)^k} + \frac{\rho((AB)^k A)^2}{s(st)^k} + \frac{\rho(B(AB)^k)^2}{(st)^k t}\} + \frac{\rho((AB)^m)^2}{(st)^m} - \rho(1)^2 .$$

For the linear characters $\zeta_1,\ldots,\zeta_4$ the respective multiplicities are

$$(3.11)\quad \begin{cases} z_1 = 1,\ z_2 = (st)^m , & \\ z_3 = z_4 = \dfrac{s}{m} \cdot \dfrac{s^{2m}-1}{s^2-1} & \text{if } s = t , \\ z_3 = t^m \dfrac{s-t}{s^m-t^m} \dfrac{(st)^m - 1}{st-1} & \\ z_4 = s^m \dfrac{s-t}{s^m-t^m} \dfrac{(st)^m-1}{st-1} & \text{if } s \neq t . \end{cases}$$

Now take $\rho = \zeta$ as in (3.10). Then

$$\begin{aligned} \frac{2|F|}{z} = \sum_{k=0}^{m-1} \{ & 2(\theta^k+\theta^{-k})^2 + \\ & + \frac{1}{s} \frac{1}{(\theta-\theta^{-1})^2} [(s-1)(\theta^{k+1}-\theta^{-(k+1)}) + \frac{s}{\sqrt{st}} (t-1)(\theta^k-\theta^{-k})]^2 + \\ & + \frac{1}{t} \frac{1}{(\theta-\theta^{-1})^2} [(t-1)(\theta^{k+1}-\theta^{-(k+1)}) + \frac{t}{\sqrt{st}} (s-1)(\theta^k-\theta^{-k})]^2 \} + \\ & + (\theta^m+\theta^{-m})^2 - 4 = \end{aligned}$$

$$= 4m + \frac{1}{(\theta-\theta^{-1})^2}\sum_{k=0}^{m-1}\{[\frac{(s-1)^2}{s}+\frac{(t-1)^2}{t}][(\theta^{k+1}-\theta^{-(k+1)})^2+(\theta^k-\theta^{-k})^2] +$$

$$+ \frac{4(s-1)(t-1)}{\sqrt{st}}(\theta^{k+1}-\theta^{-(k+1)})(\theta^k-\theta^{-k})\}.$$

Hence

$$\frac{|F|}{z} = n\{1 - \frac{1}{(\theta-\theta^{-1})^2}[\frac{(s-1)^2}{s}+\frac{(t-1)^2}{t}] - \frac{\theta+\theta^{-1}}{(\theta-\theta^{-1})^2}\frac{(s-1)(t-1)}{\sqrt{st}}\}. \tag{3.12}$$

Let ε be a primitive n-th root of unity, take $\theta = \varepsilon$ and $\theta = -\varepsilon$ in (3.12) and add to obtain that

$$\frac{1}{(\varepsilon-\varepsilon^{-1})^2}[\frac{(s-1)^2}{s}+\frac{(t-1)^2}{t}] \in \mathbb{Q}$$

and hence that $(\varepsilon-\varepsilon^{-1})^2 \in \mathbb{Q}$. Since ε is a primitive n-th root of unity with $n = 2m \geq 4$, it follows that $n = 4,6,8$ or 12.

Now

$$\frac{\theta+\theta^{-1}}{(\theta-\theta^{-1})^2}\frac{(s-1)(t-1)}{\sqrt{st}} \in \mathbb{Q}.$$

Assume that $s > 1$ and $t > 1$, then $\frac{\theta+\theta^{-1}}{\sqrt{st}} \in \mathbb{Q}$ and we obtain the indicated solution on taking the indicated choice of θ:

	θ	conclusion
$n = 6$	primitive 6-th root	st a square
$n = 8$	primitive 8-th root	2st a square
$n = 12$	primitive 12-th root	3st a square
	primitive 6-th root	st a square

In particular, if $n = 12$, then $s = 1$ or $t = 1$.

The multiplicities of the irreducible characters of degree 2 in the cases n = 4, 6 and 8 are as follows:

n = 4.

$$z = \frac{st|F|}{s^2t + t^2s + s + t} .$$

n = 6.

$$z_{\pm} = \frac{st|F|}{2\{s^2t + t^2s - st + s + t \pm (s-1)(t-1)\sqrt{st}\}} .$$

n = 8.

$$z_{\pm} = \frac{st|F|}{4\{s^2t + st^2 - 2st + s + t \pm (s-1)(t-1)\sqrt{2st}\}}$$

$$z = \frac{st|F|}{2\{s^2t + st^2 + s + t\}} .$$

Now we turn to the case of odd n.

Case n = 2m+1. Here we can verify directly using (3.9) that $(UV)^mU = V(UV)^m$ if and only if $\theta^n = 1$. Taking ε to be a primitive n-th root of unity and $\theta = \varepsilon^i$, i=1,2,...,m, we obtain m inequivalent irreducible representations of degree 2. With the two irreducible representations of degree 1, this accounts for all irreducible representations of A.

Since s = t, the formulas (3.10) become

$$\zeta((AB)^k) = \zeta((BA)^k) = s^k(\theta^k+\theta^{-k}),$$

$$\zeta((AB)^kA) = \zeta(B(AB)^k) = \frac{s^k(s-1)}{\theta - \theta^{-1}} (\theta^{k+1}-\theta^{-(k+1)} + \theta^k-\theta^{-k}) .$$

By (2.2)

$$\frac{2|F|}{z} = 2\sum_{k=0}^{m} \frac{\zeta((AB)^k)^2}{s^{2k}} - 4 + 2\sum_{k=0}^{m} \frac{\zeta((AB)^kA)^2}{s^{2k+1}} - \frac{\zeta((AB)^mA)^2}{s^{2m+1}} =$$

$$= 2\sum_{k=0}^{m} (\theta^{k}+\theta^{-k})^2 - 4 +$$

$$+ \frac{2}{(\theta-\theta^{-1})^2}\frac{(s-1)^2}{s}\sum_{k=0}^{m} (\theta^{k+1}-\theta^{-(k+1)} + \theta^{k}-\theta^{-k})^2 +$$

$$- \frac{1}{(\theta-\theta^{-1})^2}\frac{(s-1)^2}{s} (\theta^{m+1}-\theta^{-(m+1)} + \theta^{m}-\theta^{-m})^2 =$$

$$= 2(n+2) - 4 - \frac{2n}{(\theta-\theta^{-1})^2} (\theta+\theta^{-1}+2)\frac{(s-1)^2}{s} =$$

$$= 2n\{1 - \frac{1}{\theta+\theta^{-1}-2} \frac{(s-1)^2}{s}\} .$$

Hence $\theta+\theta^{-1} \in \mathbb{Q}$, and taking ε to be a primitive n-th root of unity, this implies that $n = 3$.

Of course the main conclusion from our discussion so far is the celebrated theorem of WALTER FEIT and GRAHAM HIGMAN [5,11].

THEOREM 3.1. *If a generalized* n*-gon has* $s+1$ *points on each line and* $t+1$ *lines through each point, with* $st > 1$, *then* $n = 3,4,6,8$ *or* 12. *If* $s > 1$ *and* $t > 1$, *then*

(1) st *is a square in case* $n = 6$,

(2) $2st$ *is a square in case* $n = 8$, *and*

(3) $n \neq 12$.

The methods under discussion here do not give any results for projective planes, so we assume from now on that n is even.

To apply the Krein condition to the linear character ζ_3 we need the values of ζ_1 and ζ_3:

	n=4								n=6				n=8			
ζ_1	1	s	t	st	st	s^2t	st^2	s^2t^2	s^2t^2	s^3t^2	s^2t^3	s^3t^3	s^3t^3	s^4t^3	s^3t^4	s^4t^4
ζ_3	1	s	-1	$-s$	$-s$	$-s^2$	s	s^2	s^2	s^3	$-s^2$	$-s^3$	$-s^3$	$-s^4$	s^3	s^4

The condition is

$$
\begin{aligned}
0 \le \sum_{f\in\mathcal{O}} \frac{\zeta_3(\Phi_f)^3}{\zeta_1(\Phi_f)^2} = 1 + \frac{s^3}{s^2} - \frac{1}{t^2} - 2\,\frac{s^3}{s^2t^2} - \frac{s^6}{s^4t^2} + \frac{s^3}{s^2t^4} + \frac{s^6}{s^4t^4} & \quad \Big| \; n = 4 \\
+ \frac{s^6}{s^4t^4} + \frac{s^9}{s^6t^4} - \frac{s^6}{s^4t^6} - \frac{s^9}{s^6t^6} & \quad \Big| \; n = 6 \\
- \frac{s^9}{s^6t^6} - \frac{s^{12}}{s^8t^6} + \frac{s^9}{s^6t^8} + \frac{s^{12}}{s^8t^8} & \quad \Big| \; n = 8
\end{aligned}
$$

where the sum stops as indicated in the respective cases.

In case n = 8 this becomes

$$1 + s - \frac{1}{t^2}(1+s)^2 + \frac{1}{t^4}s(1+s)^2 - \frac{1}{t^6}s^2(1+s)^2 + \frac{1}{t^8}s^3(1+s) \ge 0 ,$$

i.e.

$$t^8 - t^6(1+s) + t^4 s(1+s) - t^2s^2(1+s) + s^3 \ge 0 ,$$

i.e.

$$(t^8-t^6) - (t^6-t^4)s + (t^4-t^2)s^2 - (t^2-1)s^3 \ge 0.$$

Assuming $t > 1$, therefore, $t^6-t^4s+t^2s^2-s^3 \ge 0$, i.e. $(t^2-s)(t^4+s^2) \ge 0$. Hence $s \le t^2$. In case n = 4 we obtain the same inequality, but for n = 6 there is no conclusion. We have proved

THEOREM 3.2. *If a generalized quadrangle or octagon has* $s+1$ *points on each line and* $t+1$ *lines through each point, with* $t > 1$*, then* $s \le t^2$.

The simple groups of Lie type of rank 2 act on generalized polygons. We list these groups, their Weyl groups W and the parameters n,s,t for the corresponding generalized polygons which we refer to as generalized polygons of *Lie type*.

type	identification	W	n	s	t
$A_2(q)$	$PSL_3(q)$	Σ_3	3	q	q
$B_2(q)$	$PSP_4(q)$	D_8	4	q	q
$A_3'(q)$	$PSU_4(q)$	D_8	4	q^2	q
$A_4'(q)$	$PSU_5(q)$	D_8	4	q^2	q^3
$G_2(q)$	Dickson's group	D_{12}	6	q	q
${}^3D_4(q)$	triality group	D_{12}	6	q^3	q
$F_4'(q)$	Ree group	D_{16}	8	q^2	q

$(q = 3^{2a+1})$

From the table we see that theorem 2 gives the right inequality to quadrangles and octagons. The irreducible representations of A have been obtained in a form satisfying (2.3), and, using (3.9) we can easily write out the full matrix $A = (a_\lambda(\Phi_f))$ and apply the full force of the Krein condition in case $n = 6$. Unfortunately there is no conclusion for this case, and worse yet, we have no way of determining failure short of carrying through the entire procedure. We originally proved theorem 3.2 for quadrangles by a quite different method [10] which can be extended to give the result for octagons, but also gives no result for hexagons. Maybe there are hexagons with $s > t^3$. Although there are many known quadrangles which are not of Lie type, the only known generalized hexagons and octagons (satisfying (C)) seem to be those of Lie type.

REFERENCES

[1] ALPERIN, J.L., *On a theorem of Manning*, Math. Z., 88 (1965) 434-435.

[2] BOSE, R.C. & T. SHIMAMOTO, *Classification and analysis of partially balanced incomplete block designs with two associate classes*, J. Amer. Statist. Assoc., 47 (1952) 151-184.

[3] CARTER, R.W., *Simple groups of Lie type*, J. Wiley & Sons, New York, 1972.

[4] DELSARTE, P., *An algebraic approach to the association schemes of coding theory*, thesis, Univ. Catholique de Louvain, Fac. des Sciences Appliquées, 1973.

[5] FEIT, W. & G. HIGMAN, *The non-existence of certain generalized polygons*, J. Algebra, 1 (1964) 114-138.

[6] FONG, P. & G.M. SEITZ, *Groups with (B,N)-pair of rank 2, I*, Invent. Math., 21 (1973) 1-57.

[7] HIGMAN, D.G., *Coherent configurations*, Rend. Sem. Mat. Univ. Padova, 44 (1970) 22-42.

[8] HIGMAN D.G., *Combinatorial considerations about permutation groups*, Lecture Notes, Oxford, 1972.

[9] HIGMAN, D.G., *Coherent configurations, part I: Ordinary representation theory*, to appear in Geometriae Dedicata.

[10] HIGMAN, D.G., *Partial geometries, generalized quadrangles and strongly regular graphs*, in: Atti Convegno di Geometria e sue Applicazioni, Perugia, 1971.

[11] KILMOYER, R. & L. SOLOMON, *On the theorem of Feit-Higman*, J. Comb. Theory 15 (1973) 310-322.

[12] LIEBLER, R., *On finite planes and collineation groups of low rank*, thesis, Univ. of Michigan, 1970.

[13] SCOTT Jr., L.L., *A condition on Higman's parameters*, Notices Amer. Math. Soc., 20 (1973) A-97.

[14] TITS, J., *Buildings of spherical type*, Lecture Notes in Mathematics, Springer-Verlag, Berlin, 1974.

[15] WIELANDT, H., *Finite permutation groups*, Acad. Press, New York, 1964.

[16] WIELANDT, H., *Permutation groups through invariant relations and invariant functions*, Lecture Notes, Ohio State Univ., 1969.

2-TRANSITIVE DESIGNS *)

W.M. KANTOR

University of Oregon, Eugene, Oregon 97403, USA

INTRODUCTION

A great deal of work was done on 2-transitive groups during the last century and the beginning of this one. There has been a recent resurgence of interest in them for several reasons. First of all, many finite simple groups either have 2-transitive permutation representations or are closely related to groups that do. Also, recent work on finite simple groups has made the study of permutation groups more accessible. Finally, the close relationship between these groups and finite geometries has been recognized and has benefitted both group theory and geometry.

This survey will be concerned with designs having 2-transitive automorphism groups. A complete account of the relationship between designs and groups, as it was known in 1968, is contained in the beautiful book of DEMBOWSKI [40]. However, quite a lot has been done since then.

Since this is a combinatorics conference, I will try to minimize the group theory. However, the interplay between the groups and the designs they act on is fundamental to the subject: the fact that the automorphism group G of a design $\mathcal{D}$ permutes both the points and blocks of $\mathcal{D}$ suggests that these two actions should be played off against one another. Moreover, the manner in which designs occur in group-theoretic situations is a basic source for geometric problems and geometric theorems.

The difference between the study of 2-transitive designs and 2-transitive groups seems to be as follows. In the former case, one makes an assumption concerning the set stabilizer (or point-wise stabilizer) of a block: its transitivity properties, index in G, etc. In the latter case, one assumes structural properties of the stabilizer of one or more points. Just how fine a distinction this is can be seen from papers of O'NAN [128, 135], HARADA [63], ASCHBACHER [2,5], SHULT [149], KANTOR, O'NAN & SEITZ [107], and HERING, KANTOR & SEITZ [66], where designs are explicitly or

*) The preparation of this paper was supported in part by NSF Grant GP 37982X.

M. Hall, Jr. and J. H. van Lint (eds.), Combinatorics, 365-418. *All Rights Reserved.*

implicitly obtained in the course of "purely" group-theoretic investigations.

So as not to give a false impression, it should be noted that the relationship between permutation groups, geometry and combinatorics has been known for a long time - see the books of BURNSIDE [18] and CARMICHAEL [32].

There are also important relationships between projective planes and groups. However, I will not discuss collineation groups of projective, affine, or inversive planes at all - that would require a survey paper of its own. Incidentally, most of the problems and methods considered here become meaningless or trivial in the case of such planes. I hope to demonstrate the richness of the geometric nature of a subject spawned in part by, but quite different from, projective planes.

The organization of this paper is as follows. Section 1 consists of little more than geometric and group-theoretic notation. Section 2 discusses the elementary, well-known construction of designs from 2-transitive groups.

In the remaining sections, G will be an automorphism group 2-transitive on the points of a design $\mathcal{D}$. One natural approach is to first try to find $\mathcal{D}$, and then find G. Unfortunately, even if $\mathcal{D}$ is known to be a projective or an affine space, it is still very difficult to determine G (see section 3). This fact is, in turn, undoubtedly partly to blame for the difficulties encountered in the situations described in sections 6-10.

Section 4 contains a brief discussion of the geometry of the Mathieu groups. These designs and groups will arise in later sections.

The subject matter of this survey properly begins in section 5. There, and in the remaining sections, a variety of possible restrictions on 2-transitive designs are discussed. In each case, classical projective or affine spaces satisfy the additional hypothesis and partly motivate its study. With the exception of section 5, the goal will be the determination of $\mathcal{D}$, not of G.

Section 5 is devoted almost exclusively to results of O'NAN. The main geometric application of his striking classification theorems is to the subgroup of G fixing all the blocks through a point of $\mathcal{D}$.

HALL [58] considered the case where the 2-transitive design $\mathcal{D}$ is a Steiner triple system. In section 6, a more general situation is studied: $\lambda = 1$, and the stabilizer of two points fixes all points on the line through them.

In section 7, it is assumed that the pointwise stabilizer of a block of $\mathcal{D}$ is transitive on the complement of the block. An added combinatorial bonus here is the relevance of geometric lattices.

However, the richest combinatorial structure occurs in section 8, where $\mathcal{D}$ is assumed to be a symmetric design. As the length of the section indicates, more work has been done in this case than in any other. There are also several applications, which are discussed in sections 8 and 9; these include difference sets (section 8), Hadamard matrices (section 9), symmetric 3-designs (section 9), the suborbit structure of permutation groups (section 8), and the reducibility of certain complex polynomials (section 8).

Section 9 briefly discusses symmetric 3-designs. Finally, section 10 contains a variety of miscellaneous topics. An appendix lists the known 2-transitive groups.

Throughout the paper -and especially in section 10- I have occasionally digressed slightly from the main topic. In most cases, geometric questions related to 2-transitive groups are raised, even if designs are not involved. In fact, it would be absurd to claim that the only relationship between combinatorics and 2-transitive groups is through designs. The best examples of this, which will not be described here, are the graph extension theorem of SHULT [147] and the growing theory of 2-graphs (SEIDEL [143]; HIGMAN [71]; TAYLOR [155,156]). Also, if G is 2- but not 3-transitive, G_x determines graphs on $S - \{x\}$ which have yet to be studied. Probably the most basic problem in the combinatorial approach to 2-transitive groups is to find ways to use groups, designs, and graphs simultaneously. Thus far, this problem has been considered briefly in only two papers: SIMS [150] and O'NAN [133].

I am indebted to the following people for comments which helped in the preparation of this survey: P. CAMERON, M. FRIED, N. ITO, W. KNAPP, H. NAGAO, P. NEUMANN, and R. NODA.

1. BACKGROUND

A. Designs

A *design* $\mathcal{D}$ consists of a set S of points ("varieties" of wheat in the original statistical context), together with certain subsets called *blocks*, such that the following conditions hold for some integers v,b,k,r,λ: there are v points, b blocks, k points per block, r blocks per point, and λ blocks through any two distinct points. The following non-degeneracy conditions will also be assumed: $v \geq k+2 > 4$, and some k-subset of S is not a block. The *parameters* v, b, k, r, λ satisfy $vr = bk$ and $\lambda(v - 1) = r(k - 1)$. Also, $b \geq v$ (FISHER's inequality).

$\mathcal{D}$ is a *symmetric design* if $b = v$, or equivalently, if $r = k$. The parameters v, k, λ, and $n = k - \lambda$ then satisfy further restrictions (see DEMBOWSKI [40, § 2.1]), but these will not be needed. A *Hadamard design* is a symmetric design with $v = 2k+1$.

If x is a point of a design $\mathcal{D}$, then $\mathcal{D}_x$ denotes the set $S - \{x\}$ of points together with the sets $B - \{x\}$, where B is a block on x. $\mathcal{D}$ is called an *extension* of $\mathcal{D}_x$.

If B is a block of $\mathcal{D}$, then $\mathcal{D}_B$ denotes the set of points of B and the sets $B \cap C$, where C is any block other than B. This is again a design if $\mathcal{D}$ is symmetric.

The *complementary design* $\mathcal{D}'$ of $\mathcal{D}$ is the design having the same point set as $\mathcal{D}$, and whose blocks are the complements of those of $\mathcal{D}$.

A t-*design* is a design $\mathcal{D}$ such that each set of t points is in the same number $\lambda_t > 0$ of blocks. If $\lambda_t = 1$, $\mathcal{D}$ is also called a *Steiner system* $S(t,k,v)$.

A *line* of $\mathcal{D}$ consists of the intersection of all the blocks containing two given points. Two points are contained in a unique line. While lines of a design can usually have different sizes, they will automatically have the same size in this paper. Note that, when $\lambda = 1$, blocks are lines; in this case, I will use the more suggestive term line. Also, if $\lambda = 1$, a *subspace* of $\mathcal{D}$ is a set Δ of points such that, whenever x and y are distinct points of Δ, their line is contained in Δ.

An *automorphism* of $\mathcal{D}$ is a permutation of the points which also permutes the blocks. The automorphisms of $\mathcal{D}$ form a group Aut $\mathcal{D}$, the *automorphism group* of $\mathcal{D}$. The fact that Aut $\mathcal{D}$ permutes both the points and blocks is crucial.

If $\mathcal{D}$ is a symmetric design, the *dual* design $\tilde{\mathcal{D}}$ has the roles of points and blocks interchanged. $\tilde{\mathcal{D}}$ is symmetric, with the same parameters as $\mathcal{D}$. An *antiautomorphism* (or *correlation*) of $\mathcal{D}$ is an isomorphism $\theta: \mathcal{D} \to \tilde{\mathcal{D}}$. Then θ induces an isomorphism $\tilde{\mathcal{D}} \to \mathcal{D}$, also called θ, by acting on the points and blocks of $\tilde{\mathcal{D}}$ as θ does on the blocks of $\mathcal{D}$. θ is a *polarity* if $\theta^2 = 1$ (i.e. if $x \in y^\theta$ implies $y \in x^\theta$). If g is in Aut $\mathcal{D}$, so is $\theta^{-1}g\theta$. The group

(Aut $\mathcal{D}$) <θ> contains Aut $\mathcal{D}$ as a subgroup of index 2, and contains all anti-automorphisms of $\mathcal{D}$.

The following notation will be used for the classical geometries:

$PG_e(d,q)$, $1 \le e \le d-1$, denotes the design of points and e-spaces of PG(d,q); and

$AG_e(d,q)$, $1 \le e \le d-1$, denotes the design of points and e-spaces of AG(d,q).

As usual, $PG(2,q) = PG_1(2,q)$ and $AG(2,q) = AG_1(2,q)$. The automorphism group of PG(d,q) is $P\Gamma L(d+1,q)$.

In section 7, *geometric lattices* will arise. These are (finite) lattices L such that each element is a join of points (i.e., atoms), and which satisfy the exchange condition: if x and y are points, and $X \in L$, then $x \not< X$ and $y < x \vee X$ imply $x < y \vee X$. Each $X \in L$ then has a dimension dim(X), where dim(0) = -1, dim(X) = dim(Y) - 1 if X < Y is maximal in Y, and $\dim(X \vee Y) + \dim(X \wedge Y) \le \dim(X) + \dim(Y)$ (for all $X,Y \in L$). Moreover, *bases* of X can be introduced as sets of dim(X) + 1 points of L, none of which is in the join of the rest. The usual replacement conditions then hold for bases.

B. Permutation groups

Let H be a group *inducing* a group of permutations on a finite set S of points. It is essential to allow the possibility that non-trivial elements of H induce the identity on S. H(S) denotes the (normal) subgroup of H consisting of those $h \in H$ fixing every point of S, that is, the pointwise stabilizer of S. H^S denotes the group of permutations of S induced by H. Thus, $H^S \cong H/H(S)$.

x^h denotes the image of $x \in S$ under $h \in H$. X^h denotes the image of $X \subseteq S$ under $h \in H$: $X^h = \{x^h \mid x \in X\}$.

$H_X = \{h \in H \mid X^h = X\}$ is the (set) *stabilizer* of X in H. Clearly, H_X contains the *pointwise stabilizer* H(X) of X, and H_X induces the permutation group $H_X^X \cong H_X/H(X)$ on X. It is convenient to abbreviate $H_{\{x\}} = H_x$. If, say, $X,Y \subseteq S$ then $H_{XY} = H_X \cap H_Y$.

x^H denotes the *orbit* of x under H: $x^H = \{x^h \mid h \in H\}$. The orbits of H partition S.

H is *transitive* if $x^H = S$ for some (and hence each) x. Clearly, H is transitive on each of its orbits. H is *regular* if it is transitive and

$H_x = 1$ for some (and hence each) x. H is *primitive* on S if H is transitive on S and H_x is a maximal subgroup of H. H is t-*transitive* on S if it acts transitively on the ordered t-subsets of S. In this case, H_x is (t-1)-transitive on S - {x}. H is *sharply* t-*transitive* if it is regular on the set of ordered t-subsets of S; for t ≥ 2, all such H have been determined (ZASSENHAUS [171,172]; JORDAN [89, pp.345-361]; HALL [57, pp.72-73]).

The *rank* of a transitive group H is the number of orbits of H_x. Thus, having rank 2 is the same as being 2-transitive. An *involution* in H is an element of order 2.

C. Preliminary lemmas

(1) ORBIT THEOREM. *If* $G \le \mathrm{Aut}\,\mathcal{D}$, *then* G *has at least as many block-orbits as point-orbits. If* $\mathcal{D}$ *is symmetric, these numbers are the same* (see DEMBOWSKI [40, p.78]).

(2) *If* $\mathcal{D}$ *is a symmetric design, then each* $g \in \mathrm{Aut}\,\mathcal{D}$ *fixes the same number of points and blocks* (see DEMBOWSKI [40, p.81]).

(3) *If* $\mathcal{D}$ *is symmetric and* $1 \ne g \in \mathrm{Aut}\,\mathcal{D}$, *then* g *fixes at most* $\frac{1}{2}v$ *points* (FEIT [44]). As noted in KANTOR [102], FEIT's proof shows that, if g fixes exactly $\frac{1}{2}v$ points, then g is an involution and v = 4n.

(4) *Let* H *act as a permutation group on* S. *Let* $K \le H$. *Then the normalizer* $N_H(K)$ *of* K *is contained in the set-stabilizer* $H_{\Omega(K)}$ *of the set* $\Omega(K)$ *of fixed points of* K. *Moreover, if* $g \in G$ *then* $\Omega(K^g) = \Omega(K)^g$.

(5) ORBIT LENGTH. *If* $X = x^H$ *is an orbit of* H *on* S, *then* $|X| = |H:H_x|$ *is the index in* H *of the stabilizer of* x.

(6) *Suppose* H *is as in* 1B *and let* X *and* Y *be orbits of* H. *If* d *is the g.c.d of* $|X|$ *and* $|Y|$, *and* $x \in X$, *then each orbit of* H_x *on* Y *has size divisible by* $|Y|/d$. *In particular, if* d = 1 *then* G_x *is transitive on* Y.

2. CONSTRUCTIONS

A. Basic construction

(1) *Let* G *be* 2-*transitive on the finite set* S. *Let* B *be any* k-*subset of* S,

and assume that G *is not transitive on (unordered)* k-*subsets of* S.. *Then the distinct sets* B^g, g ∈ G, *are the blocks of a design* $\mathcal{D} = \mathcal{D}(G,S,B)$.

PROOF. Each B^g has $|B| = k$ points. If $x^g = y$, then g sets up a 1-1-correspondence between the blocks on x and those on y; this provides us with r. The same proof yields λ. □

(2) $G \le \mathrm{Aut}\,\mathcal{D}(G,S,B)$, *and* G *is transitive on blocks. Hence,* $\mathcal{D}(G,S,B)$ *has* $b = |G:G_B|$ *blocks* (see 1C(5)). *In particular,* $\mathcal{D}(G,S,B)$ *is symmetric if and only if* $|G:G_B| = v$.

Of course, the trouble with this construction is that B, and hence $\mathcal{D}$, may be totally unrelated to the action or structure of G. It is necessary to choose B carefully if $\mathcal{D}$ is to provide information about G. This is what will be done in later sections. One can, for example, assume that B is the set of fixed points of G_{xy}, $x \ne y$, or that G(B) is transitive on S - B.

In almost every case of interest, B is an orbit of some subgroup of G, so that G_B is transitive on B. Note that, *if* λ = 1, *then necessarily* G_B *is 2-transitive on* B.

PROOF. If $x,y,x',y' \in B$, $x \ne y$ and $x' \ne y'$, then any $g \in G$ such that $x^g = x'$ and $y^g = y'$ must fix B. □

If G is t-transitive, then $\mathcal{D}(G,S,B)$ is clearly a t-design.

B. When is λ = 1?

Suppose $\mathcal{D} = \mathcal{D}(G,S,B)$, where B is the set of fixed points of some subset W of G_{xy} (where $x \ne y$). In this situation (as in the general one) it is natural to ask when λ = 1. The simplest answer is due to WITT [169]:

(1) λ = 1 *if* $W^g \subseteq G_{xy}$ *and* $g \in G$ *imply* $W^g = W$.

PROOF. If $x,y \in B^g$, then W^g fixes x and y by 1C(4). That is, $W^g \subseteq G_{xy}$, so $W^g = W$. Thus, $B^g = B$. □

This result has been used in a variety of circumstances. For example, if G_{xy} is cyclic, it applies to every subgroup $W \ne 1$ of G_{xy} fixing more than two points; this was very useful in the determination of all such groups (KANTOR, O'NAN & SEITZ [107]). The designs and groups that arise here are very interesting. Assume that G does not have a regular normal subgroup and that G_{xy} has such a subgroup W. Then $v = q^3 + 1$, $r = q^2$, and $k = q + 1$ for some prime power q. (In the terminology of DEMBOWSKI [40, p.104], these are the parameters of a *unital*.) There are just two possibilities. One is that G is PSU(3,q) or PGU(3,q), and the design consists of the absolute points and non-absolute lines of a unitary polarity of $PG(2,q^2)$. (See O'NAN [128] for a detailed study of this design.) In the other case, $q = 3^{2e+1}$ for some $e \geq 0$, and G is a group of Ree type (see WARD [163] and KANTOR, O'NAN & SEITZ [107] for some properties of G and the design); the case q = 3 will arise again in section 6, where the design is called $\mathcal{D}(4)$.

There is, of course, an obvious t-design analogue of WITT's result.

There are some other interesting conditions which imply $\lambda = 1$. The most striking one is due to O'NAN [130]:

(2) *Suppose* B *is the set of fixed points of* $W \leq G_{xy}$. *Assume that no element of* $G_{xy} - W$ *is conjugate in* G_x *to an element of* W. *Then* $\lambda = 1$.

It is worthwhile to compare this with 2B(1). The main hypothesis there concerns conjugates of W, while in 2B(2) it concerns conjugates of elements of W. On the other hand, 2B(1) considers *all* conjugates, while 2B(2) only considers conjugates in G_x.

The proof of 2B(2) is elementary, but not straightforward. The main application is as follows:

(3) *Suppose* N *is a normal subgroup of* G_x, $y \neq x$, $N_y \neq 1$, *and* N_y *fixes more than two points. Then* 2B(2) *applies to* $W = N_y$ (O'NAN [130]).

<u>PROOF</u>. Suppose $g \in G_x$. Then $W^g \cap G_{xy} \leq N^g \cap G_{xy} = N_y = W$. □

Note that N fixes every block through x. Both 2B(2) and 2B(3) are crucial in the proofs of the theorems in section 5.

(4) *Suppose* $1 \neq G_{xy} < K < G_x$ *and* $B = \{x\} \cup y^K$. *Then*

(i) $\mathcal{D}(G,S,B)$ *has* $\lambda = 1$ *if for any three points* x,y,z, G_x *has an element interchanging* x *and* y (O'NAN, unpublished; ATKINSON [6]; a t-design version has been found by NEUMANN [122]); *and*

(ii) *if* $|y^K| \leq 3$ *then* G *acts on a design with* $\lambda = 1$ *as a 2-transitive automorphism group* (O'NAN, unpublished; ATKINSON [6]).

A few comments are needed concerning 2B(4). If $\lambda = 1$ for a given $\mathcal{D}(G,S,B)$, let $x,y \in B$, $x \neq y$, and set $K = G_{xB}$. Then $G_{xy} < K < G_x$. This makes the hypothesis of 2B(4) seem more reasonable.

One example of 2B(4i) is provided by the following unpublished result of SHULT (applied to $H = G_x$ acting on $X = S - \{x\}$). Suppose H is transitive on X, and some involution $t \in H$ fixes exactly one point. Then, if $y,z \in X$, $y \neq z$, there is a conjugate of t interchanging y and z.

3. COLLINEATION GROUPS

A. Projective spaces

Let G be a collineation group of PG(d,q) which is 2-transitive on points. The only known examples are: $G \geq PSL(d+1,q)$, the group of projective collineations of determinant 1; and the peculiar but fascinating example $G \cong A_7$ acting on PG(3,2).

It seems unlikely that other examples exist, but this has been verified in only a few cases. WAGNER [162] proved this for $d \leq 4$, D.G. HIGMAN (unpublished) for d = 5,6, and KANTOR [102] for d = 7,8, or when $d = s^{\alpha}$ for a prime divisor s of q-1. The same conclusion holds if some non-trivial element of G fixes a (d-2)-space pointwise (WAGNER [162], HIGMAN [68], KANTOR [102]).

Here are two interesting properties of G.

(1) *If* E *is a plane, then* G_E^E *contains* PSL(3,q) (WAGNER [162]).

(2) *If* H *is a hyperplane, then* G_H *is 2-transitive on* S - H (KANTOR [93]).

Additional (but technical) properties of G are found in KANTOR [102].

(3) Since the example $G \cong A_7$ will arise in sections 6-8, it is perhaps worthwhile to discuss it in some detail. By one of the flukes of nature, $A_8 \cong PSL(4,2)$ (see 4A(2) for a proof). Thus, PG(3,2) does indeed have a collineation group $G \cong A_7$. Thus, A_8 can be regarded as acting on the 8 cosets of G, or on the $15 \cdot 14/2$ 2-sets of points of PG(3,2). By 1C(6), G is transitive on these 2-sets. It follows that G is indeed 2-transitive.

By 1C(5), if $x \neq y$ then $|G_{xy}| = 12$. Take a point z not on the line L through x and y. It is easy to see that G cannot contain any non-trivial elation (= transvection), so $G_{xyz} = 1$. Again by 1C(5), G_{xy} must be transitive (and hence even regular) on the 12 points not in L.

It is now not difficult to prove $G_x \cong PSL(3,2)$ and $G_{xy} \cong A_4$.

B. PERIN's results

What happens if some kind of additional transitivity is assumed in 3A? This question was posed and almost completely answered by PERIN [139]:

Suppose G *is transitive on the triangles of points of* PG(d,q). *Then* G ≥ PSL(d+1,q), *except perhaps if* q = 2 *and* d *is odd.* (In 3A(3), the collineation group A_7 of PG(3,2) was shown to be transitive on triangles.) *If* G *is transitive on tetrahedra, then* G ≥ PSL(d+1,q).

The proof is ingenious and surprisingly easy. It depends solely on elementary number theory and elementary group theory.

PERIN's results are certainly the strongest and most useful ones concerning 2-transitive collineation groups of finite projective spaces. They arise several times in later sections of this survey. They have also been useful elsewhere: they were involved in one of the first proofs used for the determination of the 2-transitive permutation representations of the groups PSL(n,q). This in turn led CURTIS, KANTOR & SEITZ [36] to the determination of the 2-transitive representations of all the finite Chevalley groups.

C. Affine spaces

(1) Now let G be a collineation group of AG(d,q) 2-transitive on points. Here, the question is whether G must contain the translation group V of the space. The only known counterexample is G ≅ PSL(3,2) ≅ PSL(2,7) acting on AG(3,2).

Suppose, for a given d and q, G must contain V. Then by 3A(2), each 2-transitive collineation group of PG(d,q) must contain PSL(d+1,q). The only d and q for which it is known that G > V must hold is $d = s^{\alpha}$ for a prime divisor s of q-1 (KANTOR [102]).

It is important to note that there are many 2-transitive groups G > V. The classification of these groups is equivalent to the classification of finite groups of semilinear transformations transitive on non-zero vectors.

(2) If G is also transitive on triangles, it can be shown that G > V, except perhaps if q = 2 and d is odd. If G is transitive on tetrahedra, then G > V.

D. Generalizations

PERIN [139] studied the more general situation in which G is a collineation group of PG(d,q) transitive on e-spaces. When combined with KANTOR [100], the result is that G is 2-transitive on points if $2 \le e \le d-2$, except for groups of order $31\cdot 5$ line-transitive on PG(4,2); if $3 \le e \le d-3$, then $G \ge PSL(d+1,q)$ except perhaps if $q = 2$ and d is odd; if $4 \le e \le d-4$, then $G \ge PSL(d+1,q)$.

Suppose next that G is a collineation group of AG(d,q) transitive on e-spaces, where $1 \le e \le d-1$. It is then easy to see (by 1C(6)) that G_x is transitive on the e-spaces through x. If $2 \le e \le d-2$, this essentially reduces the problem to the one of the preceding paragraph.

4. THE MATHIEU GROUPS

The Mathieu groups will appear several times in the remainder of this survey. The following brief description of these groups and some of their properties is based primarily on WITT [169,170] and LÜNEBURG [111].

A. M_{22}, M_{23}, and M_{24}

(1) There are unique Steiner systems $\mathcal{W}_{22} = S(3,6,22)$, $\mathcal{W}_{23} = S(4,7,23)$, and $\mathcal{W}_{24} = S(5,8,24)$, discovered by WITT [169]. If x is a point of $\mathcal{W}_v$, then $(\mathcal{W}_v)_x = \mathcal{W}_{v-1}$ for $v = 24, 23$, and $(\mathcal{W}_{22})_x = PG(2,4)$.

Aut $\mathcal{W}_v$ is $(v-19)$-transitive on points and transitive on blocks. Write $M_{24} = \text{Aut } \mathcal{W}_{24}$ and $M_{23} = \text{Aut } \mathcal{W}_{23}$. If x and y are in $\mathcal{W}_{24}$, $x \ne y$, then $(M_{24})_{\{x,y\}} = \text{Aut } \mathcal{W}_{22}$ contains $M_{22} = (M_{24})_{xy}$ as a subgroup of index 2. The three groups M_{24}, M_{23} and M_{22} are simple groups, the "large" Mathieu groups.

If x,y and z are distinct points of $\mathcal{W}_{24}$, then $(M_{24})_x = M_{23}$, $(M_{24})_{xy} = M_{22}$, $(M_{24})_{\{x,y\}} = \text{Aut } \mathcal{W}_{22} = \text{Aut } M_{22}$, $(M_{24})_{xyz} = PSL(3,4)$, and $(M_{24})_{\{x,y,z\}} = P\Gamma L(3,4)$. Suppose B is a block of $\mathcal{W}_{24}$. Then $(M_{24})_B^B \cong A_8$ and $(M_{24})(B)$ is regular on S-B; here, $(M_{24})(B)$ induces an elation group of $(\mathcal{W}_{24})_{xyz} = PG(2,4)$ if $x,y,z \in B$.

(2) M_{24} provides an easy proof that $A_8 \cong PSL(4,2)$. Namely, consider $G = (M_{24})_B$. By 4A(1), $G^B \cong A_8$. But G(B) is elementary abelian of order

16, and is normal in G. It follows readily that A_8 is isomorphic to a subgroup of the automorphism group PSL(4,2) of G(B). Since $|A_8| = |PSL(4,2)|$, this proves $A_8 \cong PSL(4,2)$.

(3) Any two blocks B and C of $\mathcal{W}_{24}$ meet in 0, 2 or 4 points. M_{24} is transitive on the ordered pairs of blocks whose intersections have a fixed size. Also, if $|B \cap C| = 4$, then the symmetric difference B + C is a block.

Any two blocks of $\mathcal{W}_{23}$ meet in 1 or 3 points. M_{23} is again transitive on the ordered pairs of blocks meeting in 1 or in 3 blocks.

Any two blocks of $\mathcal{W}_{22}$ meet in 0 or 2 points. M_{22} is transitive as above.

(4) If B_0 is a block of $\mathcal{W}_{22}$, there are 16 points outside B_0 and 16 blocks missing B_0. These form a symmetric design with parameters v = 16, k = 6, $\lambda = 2$ and full automorphism group $(\text{Aut } M_{22})_{B_0} \cong S_6 \cdot V \cong Sp(4,2) \cdot V$, where $V = M_{22}(B_0)$ is an elementary abelian group of order 16. This design is $\mathcal{S}^{-1}(4)$ in the notation of 8B(4).

(5) The remarks in 4A(4) can be interpreted in $\mathcal{W}_{24}$ as follows (CAMERON [24]). Fix a block B^* of $\mathcal{W}_{24}$ and set $S^* = S - B^*$. If $x,y \in B^*$, $x \neq y$, let S_{xy} be the set of all blocks B such that $B \cap B^* = \{x,y\}$. By 4A(4), $|S^*| = |S_{xy}| = 16$, and S^* and S_{xy} yield a symmetric design.

Let $z \in B^* - \{x,y\}$. Then S_{xy} and S_{xz} also determine a symmetric (16,6,2)-design: call $B \in S_{xy}$ and $C \in S_{xz}$ *incident* if $|B \cap C \cap S^*| = 1$. All the resulting symmetric designs are isomorphic (they are $\mathcal{S}^{-1}(4)$ in the notation of 8B(4)).

B. M_{11} and M_{12}

(1) There are unique Steiner systems $\mathcal{W}_{11} = S(4,5,11)$ and $\mathcal{W}_{12} = S(5,6,11)$. If x, y and z are three points of $\mathcal{W}_{12}$, then $(\mathcal{W}_{12})_x = \mathcal{W}_{11}$, $(\mathcal{W}_{11})_{xy}$ is the miquelian inversive plane of order 3, and $(\mathcal{W}_{11})_{xyz} = AG(2,3)$. Write $M_v = \text{Aut } \mathcal{W}_v$, v = 11,12. Then M_v is sharply (v-7)-transitive on the points of $\mathcal{W}_v$, and is transitive on blocks. M_{11} and M_{12} are both simple; $(M_{12})_{\{x,y\}} = P\Gamma L(2,9)$.

$G_B \cong G_B^B \cong S_6$ if B is a block of $\mathcal{W}_{12}$. Also, S-B is another block, and $G_{\{B,S-B\}} \cong \text{Aut } S_6$.

(2) $\mathcal{W}_{12}$ is obtained from $\mathcal{W}_{24}$ as follows. Let B and C be blocks of the latter design such that $|B \cap C| = 2$. Then their symmetric difference B+C has size 12, and $(M_{24})_{B+C}$ is just M_{12}. Note that $|B \cap (B+C)| = 6$; the blocks of $\mathcal{W}_{12}$ are precisely the intersections of size 6 of B+C with blocks of $\mathcal{W}_{24}$.

If $x,y,z \in B+C$, then $(\mathcal{W}_{12})_{xyz} = AG(2,3)$ is embedded in $(\mathcal{W}_{24})_{xyz} = PG(2,4)$ as the unital preserved by $P\Gamma U(3,2) = (\mathcal{W}_{12})_{\{x,y,z\}}$. The latter group is precisely the full collineation group of AG(2,3).

Moreover, the complement of B+C again has the form B_1+C_1 (where $|B_1 \cap C_1| = 2$), and $(M_{24})_{\{B+C,B_1+C_1\}} = \text{Aut } M_{12}$ contains M_{12} as a subgroup of index 2.

(3) In the notation of 4B(2), fix a point $p \notin B+C$. Then $M_{11} = (M_{24})_{B+C,p}$ is 3-transitive on B+C (as well as on $(B_1+C_1) - \{p\})$. The $\mathcal{W}_{12}$ determined by B+C has exactly 22 blocks through p. Together with the points of $\mathcal{W}_{12}$ (i.e., B+C), these form a 3-design $\mathcal{D}$ with $v = 12$ and $k = 6$. If $x \in B+C$, then $\mathcal{D}_x$ is a symmetric (11,5,2)-design. Aut $\mathcal{D} = M_{11}$, and Aut $\mathcal{D}_x = (M_{11})_x \cong PSL(2,11)$. The designs $\mathcal{D}_x$ and $\mathcal{D}$ will reappear in sections 8 and 9.

C. Applications and characterizations

(1) The Mathieu groups are intimately linked to the sporadic simple groups of CONWAY [34,35], HIGMAN & SIMS [72], MCLAUGHLIN [119] and FISCHER [48]. For descriptions of these groups, see the above papers, LÜNEBURG [111], and SEIDEL [143].

Several characterizations of Mathieu groups will appear in sections 7 and 9. The following characterizations do not, however, fit into the framework of those sections.

(2) Let $\mathcal{D}$ be a Steiner system S(t,k,v). Suppose $G \leq \text{Aut } \mathcal{D}$ is transitive on the ordered (t+1)-tuples of points not contained in a block, and also on the ordered (t+2)-tuples of points no t+1 of which are contained in a block. Then $\mathcal{D}$ is PG(2,q), $AG_2(d,2)$, $\mathcal{W}_{22}$, $\mathcal{W}_{23}$, or $\mathcal{W}_{24}$ (TITS [158]).

(3) Let $\mathcal{D}$ be a Steiner system S(t,2t-2,v). Assume that, whenever B and C are distinct blocks and $|B \cap C| = t-1$, necessarily B+C is a block. Then $\mathcal{D}$ is $AG_2(d,2)$ or $\mathcal{W}_{24}$. This striking result is due to CAMERON [29]. Actually, CAMERON proves a stronger theorem characterizing $PG_1(d,2)$ and $\mathcal{W}_{23}$.

(4) I know of no satisfactory characterizations of the designs $\mathcal{W}_{11}$ or $\mathcal{W}_{12}$ in terms of the action of M_{11} or M_{12} on them. These 2-transitive designs do not seem to fit into any known general design setting as do the other three WITT designs. (There is, however, a lattice-theoretic setting; see KANTOR [103].)

5. NORMAL SUBGROUPS OF G_x

A. Situation

G is 2-transitive, and N is a non-trivial normal subgroup of G_x. Of course, I have in mind placing some restrictions on G_x of a geometric nature. Nevertheless, many purely group-theoretic results have been proved recently which are very useful to geometry. Here are two of these: if N is regular on $S - \{x\}$, then G is of known type (HERING, KANTOR & SEITZ [66], SHULT [146]); if $|S|$ is odd, $|N|$ is even, and all involutions in N fix only x, then G is either known or has a regular normal elementary abelian subgroup (ASCHBACHER [3]).

B. O'NAN's results

The best work presently being done on 2-transitive groups is due to O'NAN. Some of his general results are described in 5B and then applied in 5C.

(1) *Suppose* N *is abelian and not semiregular (i.e.,* $N_y \neq 1$ *for some* $y \in S-\{x\}$). *Then* $P\Gamma L(n,q) \geq G \geq PSL(n,q)$ *for some* $n \geq 3$ *and* q (O'NAN [130]).

(2) *Suppose* $N \cap N^g = N$ *or* 1 *for all* $g \in G$, *and* N *is not semiregular. Then* $P\Gamma L(n,q) \geq G \geq PSL(n,q)$ *for some* $n \geq 3$ *and* q (O'NAN [132]).

(3) *Suppose* N *is cyclic. Then either* G *has a regular normal subgroup, or* $G \geq PSL(2,q)$ *or* $PSU(3,q)$ *for a prime* q, *or* G *is* $P\Gamma L(2,8)$ (O'NAN, unpublished; ASCHBACHER [4]).

(4) *If* N *is abelian, and* $|N|$ *and* $|\Omega|$ *are odd, then* G *has a regular normal subgroup* (O'NAN [135]).

Further beautiful results are found in O'NAN [133]. While these are not strictly geometric, he finds very ingenious ways to use designs and graphs in his arguments.

O'NAN [134] considered the 3-transitive analogue of the above situation. He classified those 3-transitive groups G such that G_{xy} has a non-trivial abelian normal subgroup.

C. Applications

O'NAN's applications of his results are also basic for his proofs. Let $\mathcal{D}$ be a design and suppose $G \leq \text{Aut}\ \mathcal{D}$ is 2-transitive on points. Let N be the group of $g \in G_x$ fixing all blocks on x. Then N is normal in G_x.

Clearly, N is a very natural geometric subject. It corresponds to groups of central collineations of projective spaces, and dilatation groups of affine ones.

By 5B(2), $\mathcal{D}$ is a projective space if $N_y \neq 1$ for some $y \neq x$. By 5B(3,4), $\mathcal{D}$ is severely restricted if N is cyclic or if N is abelian and $|N|$ and v are both odd. The same is true if $|N|$ is even but each involution in N fixes only x (ASCHBACHER [3]). However, the case N abelian, $|N|$ odd, and v even has not yet been settled.

A slightly different application of 5B(2) is found in 8E(1).

There are, of course, analogous applications to t-designs with $t > 2$.

6. G_{xy} FIXES k POINTS

A. Situation

G is 2-transitive on S. If $x \neq y$, G_{xy} fixes precisely k points, where $2 < k < v$.

Let L be the set of fixed points of G_{xy}. By WITT's result 2B(1) (with $W = G_{xy}$), $\{L^g \mid g \in G\}$ yields a design $\mathcal{D}$ with $\lambda = 1$. Moreover, G_L^L is sharply 2-transitive on L, from which it follows that k is a prime power.

Possibly the main property of $\mathcal{D}$ and G is that the set of fixed points of any subgroup of G is a subspace of $\mathcal{D}$ (defined in 1A). In spite of all the subspaces of $\mathcal{D}$ this usually guarantees, it is very hard to get solid information about $\mathcal{D}$.

B. Known examples of $\mathcal{D}$

(1) $AG_1(d,k)$.

(2) $PG_1(d,2)$.

(3) A unique design $\mathcal{D}(4)$ with $v = 28$, $k = 4$. In this case, necessarily $G \cong P\Gamma L(2,8)$.

Note that, even if $\mathcal{D}$ is $AG_1(d,k)$ or $PG_1(d,2)$, in view of section 3 it is still very difficult to determine G. This fact is undoubtedly one of the major obstacles to the study of $\mathcal{D}$ itself. Note also that $G \cong A_7$ occurs here for $PG_1(3,2)$, in which case $G_{xy} \cong A_4$ is regular on S-L (cf. 3A(3)).

C. Classification theorems

The study of the present situation was initiated by HALL [58] in the case $k = 3$. His and all subsequent results have depended on 2-subgroups of G.

(1) *Suppose* $k = 3$ *and some line is the set of fixed points of an involution. Then* $\mathcal{D}$ *is* $AG_1(d,3)$, $PG(2,2)$, *or* $PG_1(3,2)$. (M. HALL [58] combined with J. HALL [55] or TEIRLINCK [157].)

(2) *Suppose some involution fixes just one point. Then* G *has a regular normal elementary abelian* p*-subgroup, where* p *is an odd prime and* $p|k$. (This is an easy consequence of GLAUBERMAN [53] and FEIT & THOMPSON [46]. The case $k = 3$ is in HALL [58], and is very elementary.)

The best result known is due to HARADA [63]:

(3) *Assume that all involutions fix at most* k *points. Then one of the following holds:*

(i) $\mathcal{D}$ *is* $AG(2,k)$, $PG(2,2)$ *or* $PG_1(3,2)$;

(ii) $\mathcal{D}$ *is* $AG_1(3,k)$ *with* k *odd; or*

(iii) $\mathcal{D}$ *is an affine translation plane of odd order* k.

(Actually, this is slightly different from HARADA's original formulation; see the Appendix of KANTOR [105].)

The only known non-desarguesian examples of (iii) have order $k = 9$. Results of HUPPERT [77] imply that there is a unique such example with G solvable. If G is non-solvable, results contained in CZERWINSKI [37] and HERING [65] show that the "exceptional" nearfield plane of order 9 is the only example possible; unfortunately, as of the writing of the present survey, these results had not quite been completely proved.

(4) *Assume that* G *is transitive on non-incident point-line pairs. Then* $\mathcal{D}$ *is* $AG_1(d,k)$ *or* $PG_1(d,2)$. (HALL and BRUCK for k = 3; see HALL [60] or DEMBOWSKI [40, pp.100-101]; KANTOR [99] in general. Other special cases are due to ITO [84] and OSBORN [136]. A variation on this theme is found in BUEKENHOUT [14].)

(5) *If some non-trivial element of* G_x *fixes all lines through* x, *then either* $\mathcal{D}$ *is* $PG_1(d,2)$ *or* $\mathcal{D}(4)$, *or* G *has a regular normal elementary abelian subgroup.*

PROOF. 5B(2) or 5B(3) applies to a non-trivial normal subgroup of G_x minimal with respect to fixing all lines through x. □

It is easy to see that 6C(5) contains 6C(2); however, 6C(2) is the far more useful result.

D. Subplanes

(1) In [58], HALL showed that, when k = 3, $\mathcal{D}$ has a subspace PG(2,2) or AG(2,3). Because of the 2-transitivity of G, $\mathcal{D}$ has many such concrete subplanes. What is lacking is a way to tie these subplanes together into a projective or affine space.

KANTOR [105] proved the following awkward result, which both generalizes HALL's result and implies 6C(4). $\mathcal{D}$ *must have a subspace such that either*

(i) $|\Delta| = k^i$, $i \geq 2$, *and* G_Δ^Δ *is 2-transitive, has a regular normal subgroup, and has no involution fixing more than one point;*

(ii) k = 3 *and* Δ *is* PG(2,2);

(iii) Δ *is an affine translation plane, and* G_Δ^Δ *contains the translation group and is flag-transitive or has exactly two flag-orbits; or*

(iv) k *is a power of* 2, *and* Δ *is the design* $\mathcal{D}(k)$ *obtained from the dual of the complement of a completed conic in* PG(2,2k).

Note that, if k is prime, Δ is $AG_1(i,k)$ in (i) and AG(2,k) in (iii), so $\mathcal{D}$ must have AG(2,k) as a subplane if k > 3 is prime. As in 6C(2), it is very likely that CZERWINSKI [37] and

HERING [65] will imply that the only non-desarguesian planes which might arise in (iii) are the exceptional nearfield plane of order 9 or a semifield plane of odd order.

Once again, a method is needed for tying all these subplanes together.

(2) In this context, it is natural to recall the standard methods of gluing planes together to form projective or affine spaces: the axioms of VEBLEN & YOUNG [161], and the theorem of BUEKENHOUT [11]. Groups are not needed for these (nor even finiteness).

Let $\mathcal{D}$ be a design with $\lambda = 1$. If each triangle is contained in a subspace which is a projective plane, then $\mathcal{D}$ consists of the points and lines of a projective space (VEBLEN & YOUNG [161]).

If each triangle of $\mathcal{D}$ is contained in an affine plane of order > 3, then $\mathcal{D}$ consists of the points and lines of an affine space (BUEKENHOUT [11]). This is false if $k = 3$ (see HALL [58]). But here, if Aut $\mathcal{D}$ is primitive on points (e.g., if Aut $\mathcal{D}$ is 2-transitive), then $\mathcal{D}$ is an affine space. (This is contained in FISCHER [47]; it is also an easy consequence of HALL [58] and GLAUBERMAN [53]).

J. HALL [55] and TEIRLINCK [157] have also handled the case where each triangle of $\mathcal{D}$ is in a projective or affine plane (a situation which arises in proving 6C(1)).

There are further interesting geometric questions of this sort that can be asked, with or without a group present; see BUEKENHOUT & DEHERDER [17].

E. Higher transitivity

It is natural to modify the situation under consideration as follows: G is t-transitive on S, and the stabilizer of t points fixes exactly k points, where $2 < t < k < v$. This time, the design $\mathcal{D}$ is a Steiner system $S(t,k,v)$. If B is a block, G_B^B is sharply t-transitive.

(1) Suppose that $t = 3$. The only known examples of $\mathcal{D}$ are:
(i) $AG_2(d,2)$, and (ii) if $PGL(2,q^i)$, $i \geq 2$, is regarded as acting on $GF(q^i) \cup \{\infty\}$, the blocks of $\mathcal{D}$ are the sets $(GF(q) \cup \{\infty\})^g$, $g \in PGL(2,q^i)$. Note that miquelian inversive planes are special cases of (ii).

It is not difficult to prove that the designs in (ii) have $P\Gamma L(2,q^i)$ as their full automorphism groups. For this reason, it seems as if the present situation should be much easier than that of 6A: if $k > 4$, G should be small.

Unfortunately, nothing is known here other than variations on the 2-transitive results of 6C and 6D. Thus, G_x acts on $S-\{x\}$ as a group satisfying the condition 6A. There is a natural definition for subspaces: sets Δ of points such that the block of $\mathcal{D}$ through any three points of Δ is again contained in Δ. There is always a subspace which is $AG_2(3,2)$ or is as in (ii) (where $k = q+1$); see KANTOR [105]. BUEKENHOUT [12,13] has proved other design versions of results related to 6C and 6D.

(2) According to a remarkable result of NAGAO [120], the case $t \geq 4$ does not occur. I will outline a proof, using an approach somewhat simpler than NAGAO's.

Suppose G exists; without loss of generality, $t = 4$. This time, G_B^B is sharply 4-transitive. There are thus just three cases (JORDAN [89, pp.245-361]; HALL [57, pp.72-73]):

(I) $k = 5$, $G_B^B \cong S_5$;

(II) $k = 6$, $G_B^B \cong A_6$; and

(III) $k = 11$, $G_B^B \cong M_{11}$.

(I) Here it is straightforward to use arguments of HALL [58] to find a subspace which is an extension of $AG_2(3,2)$ or the (miquelian) inversive plane of order 3 having an involution fixing a block pointwise. However, no such extensions exist. (This elementary, highly combinatorial approach was not used by NAGAO. In fact, case (I) was the hardest for him, requiring a complicated argument and involving the FEIT-THOMPSON theorem.)

(II) Let $t \in G$ be an involution and let f be its number of fixed points. Fix a 2-cycle (x,x^t) of t. If (y,y^t) is any other 2-cycle, then $\{x,x^t,y,y^t\}$ belongs to a unique block B, and t fixes B. Since t^B is in A_6, it fixes exactly two points z_1,z_2 of B. Conversely, any two fixed points z_1,z_2 of t uniquely determine a 2-cycle (y,y^t). Hence, t has exactly $\frac{1}{2}(v-f) - 1 = \frac{1}{2}f(f-1)$ 2-cycles other than (x,x^t). Thus, $v = f^2+2$. In particular, $f > 2$.

On the other hand, there are exactly $(v-3)/(k-3)$ blocks containing three fixed points of t, of which $(f-3)/(k-3)$ consist entirely of fixed points. Thus, $f^2+2 = v \equiv 0 \equiv f \pmod 3$, which is impossible.

(III) The same type of argument as in (II) shows that each involution t has exactly $f = \sqrt{v-2}$ fixed points. If (x, x^t) is a 2-cycle, then t commutes with some involution $u \in G_{xy}$. Here, t and u fix exactly $g < f$ common points.

Let Δ be the set of fixed points of t. Then Δ is a subspace of the design, and again as in (II), u fixes exactly $g = \sqrt{f-2}$ points of Δ. Here $g \geq 2$. There are $(v-2)(v-3)/9 \cdot 8$ blocks containing two points fixed by t and u, of which $(f-2)(f-3)/9 \cdot 8$ are fixed pointwise by t and $(g-2)(g-3)/9 \cdot 8$ are fixed pointwise by both involutions. However, the conditions $v = f^2+2$, $f = g^2+2$, and $(v-2)(v-3) \equiv (f-2)(f-3) \equiv (g-2)(g-3) \equiv 0 \pmod 9$ cannot be met.

This contradiction proves NAGAO's theorem. Note that, in (II) and (III), the arguments were purely combinatorial, almost not requiring G.

7. JORDAN GROUPS

A. Situation

$\mathcal{D}$ is a design, $G \leq \text{Aut}\,\mathcal{D}$ is 2-transitive on points and transitive on blocks, and G(B) is transitive on S-B. Intuitively, this means that $\mathcal{D}$ has many "axial automorphisms".

JORDAN [88] (= [89, pp.313-338]) initiated the study of essentially this situation from the point of view of permutation groups. Almost 100 years later, HALL [58] noticed the geometric content of JORDAN's assumptions.

B. Examples

(1) $PG_e(d,q)$, $1 \leq e \leq d-1$.

(2) $AG_e(d,q)$, $1 \leq e \leq d-1$ if $q \neq 2$, and $2 \leq e \leq d-1$ if $q=2$.
(This restriction is needed to eliminate the degenerate case $q = 2$, $e = 1$, where lines have only two points.)

(3) The Witt designs $\mathcal{W}_{22}$, $\mathcal{W}_{23}$ and $\mathcal{W}_{24}$ (see section 4).

For the latter designs, G must be M_{22}, Aut M_{22}, M_{23}, or M_{24}. By PERIN's results (see 3B), if $\mathcal{D}$ is $PG_e(d,q)$ then $G \geq PSL(d+1,q)$, except perhaps if $e = 1$, $q = 2$, and d is odd. (The collineation group $G \cong A_7$ of PG(3,2) is, in fact, an example of this exceptional situation; see 3A(3).) Similarly, 3C applies when $\mathcal{D}$ is $AG_e(d,q)$.

C. Basic properties

(1) First of all, $v \geq 2k$.

W. KNAPP has been kind enough to look into the history of this result. That $v \leq 2k$ implies the 3-transitivity of G was first proved by JORDAN [88, Théorème 1] (and not by MARGGRAFF [114], as stated on p.34 of WIELANDT [166]). KNAPP found that, in his two inaccessible papers, MARGGRAFF [114,115] proved the impossibility of $v < 2k$ (see WIELANDT [166, pp.34-38] for a proof), and also showed that $v \geq \frac{5}{2}k$ if v-k is not a power of 2 (but obtained no characterizations of this exceptional case). Finally, KNAPP noted inaccuracies in the reference to MARGGRAFF in WIELANDT's bibliography.

For the case $v \leq 6k$, see 7D(2).

(2) *Now let L consist of the set of intersections of families of blocks. Certainly, L is a lattice* (this has nothing to do with $\mathcal{D}$). *In fact, L is a geometric lattice* (see 1A for the terminology). *Moreover,* G *is transitive on bases of L, and, if* $X \in L$, *then* G(X) *is transitive on* S-X (KANTOR [105], using different terminology).

PROOF. Let $\emptyset \neq X \in L$ and $X \subset B,C$ with B and C different blocks. Then $|S-(B \cup C)| = v-2k+|B \cap C| > 0$ by (1). Since G(B) is transitive on S-B and G(C) is transitive on S-C, $G(B \cap C)$ is transitive on $S-(B \cap C)$. It follows that G(X) is transitive on S-X. Consequently, G(X) is transitive on those $Y \in L$ in which X is maximal, so that X is maximal in $X \vee y$ for all $y \in S-X$. This proves that L is a geometric lattice, and the remaining assertions follow easily. □

(3) There is a great deal of information contained in (2). For example, G is 3-transitive if and only if lines have just two points, and is 4-transitive if and only if planes have just three points.

(4) If $X \in L$, $G(X)$ induces an automorphism group $\overline{G(X)}$ on the interval $[X,S] = \{Y \in L \mid X \leq Y\}$. $\overline{G(X)}$ is 2-transitive on those elements of $[X,S]$ of dimension $1 + \dim(X)$. If $\dim(X) \leq \dim(G) - 3$, then $\overline{G(X)}$ and the blocks in $[X,S]$ provide a group and a design satisfying the same conditions as G and $\mathcal{D}$.

Similarly, suppose for simplicity that G is not 3-transitive. Let $X \in L$ be neither $\emptyset$, a point, nor a line. Then G_X^X also acts on the interval $[\emptyset,X]$ as in 7A.

(5) By (4) and some classical geometry (or KANTOR [105], or DOYEN & HUBAUT [43]), if suitable intervals $[X,S]$ or $[\emptyset,X]$ are of known type, $\mathcal{D}$ is essentially known. (See KANTOR [105, 6.5] for a precise statement.) This fact provides very nice inductive possibilities.

D. Characterizations

(1) *If* $k = 3$, *then* $\mathcal{D}$ *is* $PG_1(d,2)$ *or* $AG_1(d,3)$. (This is the HALL-BRUCK theorem; see 6C(4).)

(2) *If* $v \leq 6k$, *then* $\mathcal{D}$ *is a projective or affine space*, W_{22}, W_{23}, *or* W_{24} (KANTOR [105]). *Moreover, in this case*, G *is even known.*

(3) *If* G_B *is 2-transitive on* $S-B$, *then* $\mathcal{D}$ *is* $PG_{d-1}(d,q)$, $AG_{d-1}(d,2)$, W_{22}, W_{23}, *or* W_{24}, *and* G *is known* (KANTOR [105]).

(4) *If* $G(B)$ *has an abelian subgroup transitive on* $S-B$, *the conclusions of* (3) *hold.*

PROOF. BY 7C(5), without loss of generality G is not 3-transitive, so lines have $h > 2$ points. Fix $x \in B$. Then the given abelian group $A \leq G(B)$ is transitive on the $(v-k)/(h-1) < |A|$ lines on x not in B. It follows that some $a \neq 1$ in A fixes all lines through x. Now a result of O'NAN [130] (see 5B(2)) completes the proof. □

Special cases of (4) are found in KANTOR [105,106], and MCDONOUGH [117,118].

(5) *If* $v-k$ *is a prime power, the conclusions of* (3) *hold.* (KANTOR [104]; special cases are in KANTOR [105,106], and MCDONOUGH [117,118]. Stronger results are proved in KANTOR [104].)

PROOF. By 7C(5), without loss of generality $\lambda = 1$. Let p be the prime dividing v-k. Let $B \cap C = x$. A Sylow p-subgroup P of G(B) is transitive on S-B. Since $|P:P_C| = r-1 < v-k$, P_C fixes no point of S-B. Since P_C normalizes a Sylow p-subgroup Q of G(C), it centralizes some $q \neq 1$ in the center of Q. Then q fixes the set B of fixed points of P_C. Now the transitivity of Q on S-C implies that q fixes all lines through x. Once again, O'NAN's theorem 5B(2) completes the proof. □

(6) *If* G(B) *has a subgroup normal in* G_B *and regular on* S-B, *then the conclusions of* (3) *hold or* $\mathcal{D}$ *is* $PG_1(3,2)$ *or* $AG_2(4,2)$. (KANTOR [97]; special cases have already been mentioned in 6C(4). This result, and its proof, were motivated by the HERING-KANTOR-SEITZ-SHULT theorem, already mentioned in 5A.)

E. Applications

(1) KANTOR & MCDONOUGH [106] showed that, *if* G *is a permutation group of degree* $v = (q^n-1)/(q-1)$ *containing the 2-transitive group* PSL(n,q), $n \geq 3$, *then either* G *contains the alternating group or* $PSL(n,q) \leq G \leq P\Gamma L(n,q)$.

PROOF. If G is as much as $k = (q^{n-1}-1)/(q-1)$ transitive, results of WIELANDT [164] imply that G is alternating or symmetric. If G is not k-transitive, let $\mathcal{D}$ have as blocks $\{H^g \mid g \in G\}$, where H is a hyperplane. Now use any one of D(3, 4, or 5). □

Unfortunately, the preceding proof does not apply when n = 2. That case is far more interesting than the case $n \geq 3$, since $PSL(2,11) < M_{12} < A_{12}$ and $PSL(2,23) < M_{24} < A_{24}$. In fact, the study of groups G satisfying $PSL(2,p) < G < A_{p+1}$, with p prime, is precisely what led MATHIEU to the discovery of M_{12} and M_{24}. NEUMANN [124] has recently proved that G is necessarily 4-transitive here. For an application of this problem to coding theory, see SHAUGHNESSY [144].

(2) Several of the classification theorems concerning Jordan groups can be interpreted as stating that certain natural attempts at generalizing M_{22}, M_{23} and M_{24} lead to nothing new.

(3) PRAEGER [141] has recently used D(2) in the course of proving some general results concerning 2-transitive groups. Another recent appli-

cation of JORDAN's original situation is made in the beautiful paper of SCOTT [142].

F. Problem

Besides the obvious problem of determining all designs admitting Jordan automorphism groups, there is a natural, interesting type of problem these designs lead to.

First, can G be acting on the set S of points of PG(d,q) or AG(d,q) without $\mathcal{D}$ being $PG_e(d,q)$ or $AG_e(d,q)$ for some e? The answer is no for PG(d,q), q > 2, by results of PERIN [139] (see 3B).

Now let's forget the group, and just consider the remaining geometric situation. Can a design with $\lambda = 1$ be constructed using all the points, and some but not all e-spaces, of a projective or affine space? Such designs are probably rare. There is an obvious generalization of this question in which a generalization of t-designs is involved.

Next, can a design with $\lambda > 1$ be constructed using some but not all e-spaces, in which the lines of the design consist of all the lines of the underlying geometry? I conjecture that this is impossible.

8. 2-TRANSITIVE SYMMETRIC DESIGNS

A. Situation

$\mathcal{D}$ is a symmetric design, and $G \leq \mathrm{Aut}\,\mathcal{D}$ is 2-transitive on points.

2A(2) indicates the group-theoretic interpretation of this situation. Note that the complementary design $\mathcal{D}'$ satisfies the same conditions as $\mathcal{D}$.

B. Examples

There are several very interesting examples of 2-transitive symmetric designs. It is only necessary to describe one of $\mathcal{D},\mathcal{D}'$. In each case, $\mathcal{D}$ has polarities.

(1) Projective spaces: $PG_{d-1}(d,q)$. Of course, $\mathrm{Aut}\,\mathcal{D} = P\Gamma L(d+1,q)$. In view of section 3, from this example it should already be clear that there will be serious obstacles to the study of G.

(2) The unique 11-point Hadamard design $\mathcal{W}_{11}$. Here $v = 11$, $k = 5$, $\lambda = 2$ (compare 4B(3)). The only possible G is $G = \text{Aut } \mathcal{W}_{11} \cong PSL(2,11)$. Here, $G_B \cong A_5$ acts as A_5 on B and as PSL(2,5) on S-B. $\mathcal{W}_{11}$ has polarities θ, and $G\langle\theta\rangle \cong PGL(2,11)$.

(3) G. HIGMAN's design $\mathcal{W}_{176}$ (see G.HIGMAN [73]; SIMS [150]; SMITH [152, 153]; CONWAY [35]). Here, $v = 176$, $k = 50$, and $\lambda = 14$. The only possible G is G = HS, the sporadic simple group of D.G. HIGMAN & C.C. SIMS [72]. $G_B \cong PSU(3,5)$ has rank 3 on B (and $G_{xB} \cong A_7$ if $x \in B$), while G_B acts on S-B in its usual 2-transitive representation of degree 5^3+1. Also, $\mathcal{W}_{176}$ has polarities ϕ, and $G\langle\phi\rangle \cong$ Aut HS.

$\mathcal{W}_{176}$ has a fascinating property: there is a 1-1-correspondence θ from 2-sets of points to 2-sets of blocks which is preserved by G. Here, θ is not induced by a polarity of $\mathcal{W}_{176}$. Moreover, $G_{\{x,y\}} = G_{\{x,y\}^\theta} \cong Z_2 \times \text{Aut } A_6$.

(4) The symplectic symmetric designs $\mathcal{S}^\varepsilon(2m)$, one for each $m \geq 2$ and $\varepsilon = \pm 1$. Here $v = 2^{2m}$, $k = 2^{m-1}(2^m+\varepsilon)$, $\lambda = 2^{m-1}(2^{m-1}+\varepsilon)$. $\mathcal{S}^1(2m)$ and $\mathcal{S}^{-1}(2m)$ are complementary designs.

Set $G = \text{Aut } \mathcal{S}^\varepsilon(2m)$. Then G has a regular normal elementary abelian 2-subgroup V of order $v = 2^{2m}$, and $G = VG_x$, $V \cap G_x = 1$, where $G_x \cong Sp(2m,2)$ is a symplectic group acting on V in the usual way. $G_B \cong Sp(2m,2)$ is 2-transitive on B and S-B. If $x \in B$, then G_{xB} is the orthogonal group $GO^\varepsilon(2m,2)$.

Moreover, by 2A, the preceding properties of G completely determine $\mathcal{S}^\varepsilon(2m)$. It is remarkable that these properties were implicitly contained in work of JORDAN 100 years ago (see JORDAN [89, pp.XXI-XXIII] and [90, pp.229-249]).

Any subgroup of G of the form VT, with $T \leq G_x$ transitive on $V - \{1\}$, is 2-transitive on $\mathcal{S}^\varepsilon(2m)$; for example, T can be $Sp(2e,2^f)$ whenever $ef = m$. The question of whether every 2-transitive automorphism group necessarily contains V leads to the same difficulties as in 3C.

In view of the action of G_x on V, there is an involution $t \in G_x$ fixing exactly $\frac{1}{2}v$ points (t is a transvection). If x_1 and x_2 are distinct points, there is a unique conjugate of t interchanging x_1 and x_2.

$\mathcal{S}^\varepsilon(2m)$ has interesting combinatorial properties. Let + denote the symmetric difference of sets of points. If B, C and D are any blocks, then B+C+D is either a block or the complement of a block. (This property alone does not quite characterize these designs.) If $B \neq C$, then V_{B+C} is transitive on B+C. (This property does characterize $\mathcal{S}^{\varepsilon}(2m)$, assuming only that V is an automorphism group of a symmetric design regular on points; see KANTOR [101].)

Here's another description of $\mathcal{S}^{1}(2m)$. Consider the dual of a completed conic in $PG(2,2^m)$. Use the dual of the knot as the line at infinity of $AG(2,2^m)$. Let B be the union (in $AG(2,2^m)$) of the remaining 2^m+1 lines. Then the translates of B are the blocks of $\mathcal{S}^{1}(2m)$.

A similar description of $\mathcal{S}^{1}(4(2e+1))$ can be given in terms of the LÜNEBURG-TITS affine planes of order $2^{2(2e+1)}$ (defined in LÜNEBURG [110,111]): once again, the dual of a suitable oval can be used, in which the dual of the line at infinity is the knot. I know of no other planes which yield any designs $\mathcal{S}^{1}(2m)$ in this manner, but such planes undoubtedly exist (and merit study).

A (-1,1)-incidence matrix of $\mathcal{S}^{\varepsilon}(2m)$ is a Hadamard matrix known since the last century: the tensor product of m Hadamard matrices of size 4. BLOCK [9] first noticed (using this incidence matrix) that Aut $\mathcal{S}^{-1}(2m)$ is 2-transitive on points for each m. He pointed this out to me in 1968. All the properties of $\mathcal{S}^{\varepsilon}(2m)$ just described were proved at that time, and eventually appeared in KANTOR [101]. The designs were later rediscovered by RUDVALIS (1969, unpublished), HILL [74], and CAMERON & SEIDEL [30]. The latter paper provides an interesting relationship between these designs and coding theory.

C. Basic properties

The most famous result concerning 2-transitive symmetric designs is the beautiful theorem of OSTROM & WAGNER [137]: *if* $\lambda = 1$, *then* $\mathcal{D}$ *is a desarguesian projective plane*. Consequently, I will assume $\lambda > 1$ throughout this section.

(1) G *is* 2-*transitive on blocks. If* B *is a block, then* G_B *is transitive on* B *and* S-B, *and dually. Moreover, if* $(v,k) = 1$, *then* G_{xB} *is transitive on* S-B (by 1C(6)), *and dually.*

(2) *If* G_B *is* 2-*transitive on both* B *and* S-B, *then the dual statements hold and* G_x *has rank* 3 *on* S - {x}. (More generally, in KANTOR [93] it is proved that, *if* G *is an automorphism group of a design* 2-*transitive on*

points and transitive on blocks, and if G_B *is 2-transitive on both* B *and* S-B, *then the rank* ρ *of* $G_x^{S-\{x\}}$ *satisfies* $\rho \le 5$, *and even* $\rho \le 3$ *if* $v \ne 2k$.)

(3) *If* $\mathcal{D}$ *is a Hadamard design,* G_B *is necessarily 2-transitive on* S-B. This will be proved in 8C(5) below. Further special transitivity properties are found in KANTOR [93], especially Lemma 4.2.

(4) In KANTOR [93], a great deal of attention is paid to the case $k|v-1$ (which is equivalent to $(k,\lambda) = 1$, and which holds in $PG_{d-1}(d,q)$ and H_{11}). Assume this condition. *Then* G_B *must be primitive on* S-B. (In view of KANTOR [91, 4.7 and 4.8], the same conclusion holds under much weaker numerical restrictions.) Also, G has a simple normal subgroup 2-transitive on points.

Of course, the example $S^{\varepsilon}(2m)$ shows that the last assertion does not hold in general. KANTOR [93] showed that $\mathcal{D}$ has the parameters of $S^{\varepsilon}(2m)$ for some m,ε if G has a regular normal subgroup.

(5) As an example of the proofs of transitivity properties, I will prove: *if* $k-1|v-1$ *(or equivalently, if* $\lambda|k$*), then* G_B *is 2-transitive on* B. (Note that this implies 8C(2) when $\mathcal{D}$ is the complementary design of a Hadamard design.)

PROOF. G_x is transitive on the $v-1$ points $\ne x$, and on the k blocks B on x. By 1C(6), each orbit of G_{xB} on $S - \{x\}$ has size divisible by $(v-1)/(v-1,k)$. But $k = \lambda\cdot(v-1)/(k-1)$ implies that $(v-1,k) = (v-1)/(k-1)$. Thus, G_{xB} has an orbit on $B - \{x\}$ of size divisible by $k-1$. □

In the next section it will be seen how desirable it is to have sufficiently strong transitivity results.

D. The DEMBOWSKI-WAGNER theorem

This theorem provides the basic characterization of projective spaces needed for the study of symmetric designs. Namely:

$\mathcal{D}$ is a projective space if any one of the following holds:

(i) *every line meets every block;*

(ii) *every line has at least* $1 + (v-1)/k$ *points; or*

(iii) G *is transitive on ordered triples of non-collinear points.*

Slightly stronger combinatorial characterizations are found in DEMBOWSKI [40, pp.65-67], and KANTOR [92,93]; in particular, the latter reference describes the relationship with geometric lattices.

PROOF. If L is a line of $\mathcal{D}$ (the intersection of the λ blocks containing two points), there are $v-\lambda-|L|(k-\lambda)$ blocks missing L. Since $(v-\lambda)/(k-\lambda) = 1+(v-1)/k$, this implies that (i) and (ii) are equivalent; assume both of them. If $x \notin L$, and if ρ blocks contain x and L, then there are $k-\rho = |L|(\lambda-\rho)$ blocks on x not containing L. Thus, ρ is a constant, so planes can be defined, and each is determined by any triangle in it. Suppose L and M are distinct lines of a plane E. Then some block $B \supset L$ does not contain E. Since B meets M, $L \cap M = E \cap (B \cap M) \neq \emptyset$. Thus, E is a projective plane, so $\mathcal{D}$ is a projective space (VEBLEN & YOUNG [161]).

Now assume (iii). Then G_L is transitive on L and S-L. By the Orbit Theorem 1C(1), G_L has just two block-orbits. Since these must be the blocks containing L and the blocks meeting L once, (i) holds. □

E. Classification theorems

Many theorems have been proved classifying 2-transitive symmetric designs under suitable additional conditions. A catalogue of these follows.

(1) *If* $G(B) \neq 1$, *then* $\mathcal{D}$ *is a projective space* (ITO [81]). Thus, in the remainder of this section it may be assumed that $G(B) = 1$.

PROOF. G is 2-transitive on blocks. G(B) is a non-trivial normal subgroup of G_B. Each non-trivial element of G(B) fixes more than one point, and hence more than one block (1C(2)). A theorem of O'NAN [132] (see 5B(2)) now applies. (Of course, this wasn't ITO's original proof.) □

(2) *If* $\mathcal{D}$ *has the same parameters as* $PG_{d-1}(d,q)$, *then* $\mathcal{D}$ *is* $PG_{d-1}(d,q)$ (KANTOR [98]).

(3) *If* k *is prime, then* $\mathcal{D}$ *is* $\mathcal{W}_{11}$ *or a projective space* (KANTOR [93]; the case where v and k are prime is due to ITO [3]). From this it follows easily that $\mathcal{D}$ is $\mathcal{W}_{11}$ or a projective space if (v-1)/2 is prime.

(4) *If* $n = k-\lambda$ *is prime,* $\mathcal{D}$ *is* $\mathcal{W}_{11}$, $(\mathcal{W}_{11})'$, *or* PG(2,n)' (KANTOR [93]).

(5) *If* k/2 *is prime, then* $\mathcal{D}$ *is a projective space,* PG(2,2)', $(\mathcal{W}_{11})'$, $S^1(4)$, *or* $S^{-1}(4)$ (ITO & KANTOR [87]).

(6) *If* n/2 *is prime, then* $\mathcal{D}$ *is* $S^1(4)$ *or* $S^{-1}(4)$.

(7) *If* k-1 *is prime and* $\lambda > 2$, *then* $\mathcal{D}$ *is* $(\mathcal{W}_{11})'$ *or* $PG_{d-1}(d,2)'$.

PROOF. Write k-1 = p. Then $\lambda(v-1) = p(p+1)$ and $k > \lambda+1$ imply $p \mid v-1$, so $p \mid |G|$. A Sylow p-subgroup of G fixes a block B and a point x, and is transitive on B - {x}. Thus, G_B is 2-transitive on B. By 8E(10) (see below), it may be assumed that G_B is not 3-transitive on B. Also, by 8E(1), G(B) = 1. BURNSIDE [18, p.341] and classification theorems now yield the precise structure of G_B, from which $\mathcal{D} = (\mathcal{W}_{11})'$ is readily deduced. □

(8) *If* k-1 *and* v *are prime, then* $\mathcal{D}$ *is* $(\mathcal{W}_{11})'$ (ITO [83]).

Note that theorems 8E(2)-(8) all assume nothing more than *numerical* restrictions. In theorems 8E(9)-(14), further *transitivity* conditions will be imposed.

(9) *If* $\mathcal{D}$ *is a Hadamard design, and* G_B *is 2-transitive on* B, *then* $\mathcal{D}$ *is* $\mathcal{W}_{11}$ *or a projective space* (KANTOR [93]).

(10) *If* G_B *is 3-transitive on* B *and* $\lambda > 2$, *then* $\mathcal{D}$ *is* $PG_{d-1}(d,2)'$.

This is an unpublished result of CAMERON and KANTOR. The idea of the proof is as follows. As usual, $\mathcal{D}_B$ consists of the points of B and blocks ≠ B. Here, $\mathcal{D}_B$ is a 3-design. If $x \in B$, then $\mathcal{D}_B$ has the same number k-1 of points ≠ x as blocks on x. Thus, $\mathcal{D}_B$ is a symmetric 3-design, so a theorem of CAMERON [22] (see 10A,B) yields $k = 4\mu+4$, $\lambda = 2\mu+2$ or $k = (\mu+2)(\mu^2+4\mu+2) + 1$, $\lambda = \mu^2+3\mu+2$ (compare CAMERON [25]).

In the first case, $\lambda(v-1) = k(k-1)$ implies $v = 2k-1$, and 8E(9) applies to $\mathcal{D}'$. In the second case, if $x \notin B$ then G_{xB} has rank 3 on the blocks through x, and the parameter restrictions of HIGMAN [69] yield a contradiction.

(11) *If* G_B *is 2-transitive on both* B *and* S-B (compare 8C(2)), $\lambda > 2$, *and 3 points exist lying on no block, then* $\mathcal{D}$ *is* PG(d,2)'.

The proof is very similar to that of 8E(10). Note that the desired 3 points are easily shown to exist if $k \geq \lambda^2-\lambda+1$, except when $\mathcal{D}$ is PG(3,λ-1).

(12) *If* $\lambda = 2$ *and* G_B *is* 2-*transitive on* S-B, *then* $\mathcal{D}$ *is* PG(2,2)', $\mathcal{W}_{11}$ *or* $\mathcal{S}^{-1}(4)$ (CAMERON [29] and KANTOR [93]).

(13) *If* G_B *is* 4-*transitive on* B, *then* $\mathcal{D}$ *is* PG(2,2)', $\mathcal{W}_{11}$ *or* $\mathcal{S}^{-1}(4)$. (This easy consequence of 8E(10) and 8E(12) is due to CAMERON [29].)

Further results of these types are found in KANTOR [3]. The following is quite a different sort of result, which (in spite of its technical nature) will be used in 8G.

(14) *Suppose* $k \mid v-1$, $x \notin B$, *and* G_{xB} *has a cyclic subgroup* A *regular on the points on* B *and the blocks on* x. *Then* $\mathcal{D}$ *is* $\mathcal{W}_{11}$ *or a projective space if either* (i) k *has no proper divisor* $\equiv 1 \pmod{\lambda}$, *or* (ii) $k < (\lambda+1)^2$ (KANTOR [93]). (In the projective space case, the given cyclic group is a Singer cycle of B.)

Some characterizations are also known for the designs $\mathcal{S}^{\varepsilon}(2m)$ and $\mathcal{W}_{176}$.

(15) *If some* $g \neq 1$ *in* G *fixes at least* $\frac{1}{2}v$ *points, then* $\mathcal{D}$ *is* $\mathcal{S}^{\varepsilon}(2m)$ (KANTOR [101]).

(16) *If some* $g \neq 1$ *fixes* S-(B+C) *pointwise for some* $B \neq C$, *then* $\mathcal{D}$ *is* $\mathcal{S}^{\varepsilon}(2m)$ *or* $PG_{d-1}(d,2)$ (KANTOR [101]).

Both 8E(15) and 8E(16) rely heavily on FEIT's result 1C(3) and the DEMBOWSKI-WAGNER theorem 8C. The only possible automorphisms g which actually occur in 8E(15) and 8E(16) are elations of the underlying classical geometry.

(17) *If* G *has a regular normal subgroup, and if* G_B *is* 2-*transitive on both* B *and* S-B, *then* $\mathcal{D}$ *is* $\mathcal{S}^{\varepsilon}(2m)$ (KANTOR [101]).

(18) *Suppose* G *preserves a* 1-1-*correspondence from* 2-*sets of points to* 2-*sets of blocks. If* $n = (\lambda-2)^2/4$, *then* $\mathcal{D}$ *is* $\mathcal{W}_{176}$ *or* $\mathcal{S}^{1}(4)$. (KANTOR, unpublished; this was proved under additional transitivity assumptions by SMITH [152]).

F. Prime v and linked systems

(1) One of the main sources of interest in 2-transitive symmetric designs is permutation groups G of prime degree v. These are necessarily

solvable or 2-transitive (BURNSIDE [18, p.341]. Very few 2-transitive examples are known: $P\Gamma L(d+1,q) \geq G \geq PSL(d+1,q)$ acting on PG(d,q), for rare pairs d,q; PSL(2,11) with v = 11; and A_v, S_v, M_{11} and M_{23}. Here, the first two types yield symmetric designs (see 10A for the sense in which M_{23} produces a generalization of a symmetric design). This naturally leads to the study of symmetric designs with prime v. The reader is referred to NEUMANN [123] for an excellent survey of the general question of 2-transitive groups of prime degree.

(2) If $\mathcal{D}$ is a symmetric design, v is prime, and Aut $\mathcal{D}$ is transitive, then $\mathcal{D}$ is obviously a difference set design. See HALL [56,61] (and his talk at this conference)*), MANN [113], and DEMBOWSKI [40] for the definitions and basic properties of difference set designs.

Of importance in the present context is the well-known fact that, if A is an abelian automorphism group regular on the points of a symmetric design $\mathcal{D}$, and if v is odd, then the map $a \to a^{-1}$, $a \in A$, does not induce an automorphism of $\mathcal{D}$. More generally: *an involutory automorphism of a design cannot fix just one block* (NEUMANN [121]).

Also, if $\mathcal{D}$ and A are as above, then $\mathcal{D}$ admits polarities.

In the case of 2-transitive symmetric designs with v prime, the only other known way of using the primality of v is through modular character theory (as in ITO [82,83]).

(3) In 1955, WIELANDT posed the following problem: can a 2-transitive group of prime degree v have more than two conjugacy classes of subgroups of index v? Certainly, two are possible, as has been noted in 8F(1).

Thus, suppose G is 2-transitive on each of the sets $S_1,\ldots,S_\mu$, $\mu > 2$, $|S_i| = v$ for each i, and the stabilizer of a point x_i in S_i fixes no point in any S_j, $j \neq i$. By 2A(2), each pair (S_i,S_j), $i \neq j$, determines a 2-transitive symmetric design. By 8C(1), G_{x_i} has two orbits on S_j. Thus:

(*) if $x_i \in S_i$ and $x_j \in S_j$, $i \neq j$, then the number of $x_h \in S_h$, $h \neq i,j$, incident with both x_i and x_j, depends only on i,j,h and whether x_i and x_j are incident or not.

CAMERON [24] considered this situation from a purely combinatorial point of view. *A system of linked symmetric designs* consists of sets $S_1,\ldots,S_\mu$, $\mu > 2$, and an incidence relation between each pair of sets turning each pair into a symmetric design, such that (*) holds.

*) This volume, pp. 321-346.

Needless to say, there is a lot of arithmetic information in this situation. CAMERON rediscovered some such unpublished information due to WIELANDT and to ITO, but in the more general combinatorial setting. The conditions proved there are, however, too technical to reproduce. Additional numerical information has been obtained by ITO. For example, very recently, ITO [68] has shown that if v is prime, then for some design (S_i, S_j) neither k nor v-k can divide v-1.

Furthermore, NEUMANN [123] used a computer to show that WIELANDT's original situation cannot occur if $p < 2{,}000{,}000$. The proof of this provided a test for the available numerical data.

(4) WIELANDT has proved that, in the original situation in 8F(3), G can be the full automorphism group of at most one of the designs. (A proof is found in CAMERON [24].)

(5) The combinatorial setting is as interesting as WIELANDT's group-theoretic one: examples exist.

(a) Let V be a 2m-dimensional vector space over GF(q), $q = 2^e$. Let Sp(2m,q) act on V as usual. Then $G = V\cdot Sp(2m,q)$ has exactly q classes of complements to V (POLLATSEK [140]). Clearly, the scalar transformations act on this family of q sets, and it is not hard to see that Aut G is 2-transitive on these q sets. Since each pair of sets determines an $S^{\varepsilon}(2me)$, this is a linked system of designs having $v = 2^{2me}$ and $\mu = q$.

(b) A much larger system is possible for a given $v = 2^{2m}$. Namely, a system of linked symmetric designs with $\mu = 2^{2m-1}$ has been constructed by GOETHALS from the KERDOCK [108] codes (see CAMERON [24] and CAMERON & SEIDEL [30]).

(c) CAMERON [24] notes the following construction for examples (a) and (b) when v = 16. In the notation of 4A(5), S^*, S_{xy}, S_{xz}, S_{yz} (with x,y,z three points of B^*) form example (a) with m = 1, e = 2. S^*, together with the seven sets S_{xy}, $y \in B^* - \{x\}$, for a fixed $x \in B^*$, form example (b) with m = 2.

In each of examples (a)-(c), each symmetric design is isomorphic to $S^{\varepsilon}(2\ell)$ for some ℓ. No other examples are known of symmetric designs arising in linked systems.

(6) If $S_1,\dots,S_\mu$ is a linked system, its *automorphism group* H consists of those permutations of $S_1 \cup \dots \cup S_\mu$ which preserve both the partition and incidence. In example (a), H is 2-transitive on the q systems; the subgroup of H fixing each set S_i is 2-transitive on each S_i.

In example (b), H is known only for m = 2. Namely, from (c) it is clear that H contains $(M_{24})_{xB^*} \cong A_7 \cdot V$, where $V = M_{24}(B^*)$ is elementary abelian of order 16. In fact (CAMERON & SEIDEL [30]), $H \cong A_8 \cdot V \cong SL(4,2) \cdot V$, where V fixes S^* and each S_{xy}, while A_8 acts as usual on these 8 sets. The subgroup of H fixing 2 of the 8 sets is $A_6 \cdot V$, and induces an automorphism group of the resulting design $\mathcal{S}^{-1}(4)$.

Some properties of H for certain types of linked systems (e.g., when v is prime) are found in WIELANDT [167] and CAMERON [24].

G. Some difference set designs

In this section, a special class of difference set designs will be considered. These are of interest for both combinatorial and number-theoretic reasons (see HALL [61] and MANN [113]).

(1) Let v be an odd prime power, and set F = GF(v). Let $1 < k < v-1$ and $k \mid v-1$, and let $B = B(v,k)$ be the subgroup of F^* of order k. Let $\mathcal{D}(v,k)$ have the elements of F as points and the translates B+a, a ∈ F, as blocks. B is a difference set in F^+ if and only if $\mathcal{D}(v,k)$ is a symmetric design.

The designs $\mathcal{D}(v,\frac{1}{2}(v-1))$ are the Hadamard designs of PALEY [138], where $v \equiv 3 \pmod 4$ can be any prime power.

By DEMBOWSKI [40, p.35] (or an easy Singer cycle argument), $\mathcal{D}(v,k)$ cannot be a projective space if $\lambda > 1$. If $\lambda = 1$, the only desarguesian exceptions are PG(2,2) and PG(2,8).

(2) PROBLEM: what is Aut $\mathcal{D}(v,k)$?

Clearly, Aut $\mathcal{D}(v,k)$ contains the group S(v,k) of all mappings $x \to bx^\sigma + a$, b ∈ B, a ∈ F, σ ∈ Aut F. In only three cases is Aut $\mathcal{D}(v,k) > S(v,k)$ known, namely, $\mathcal{D}(11,5) = \mathcal{W}_{11}$, $\mathcal{D}(7,3)$ = PG(2,2) and $\mathcal{D}(73,9)$ = PG(2,8). These are almost certainly the only possibilities.

This problem can be reformulated in terms of permutation polynomials. Let f(x),g(x) ∈ F[x], and assume that both polynomials act as permutations of F. If

$$f(x+b) - g(x) \in B \qquad \forall x \in F,\ \forall b \in B,$$

then the pair (f,g) determines an automorphism of $\mathcal{D}(v,k)$. Conversely, each automorphism determines such a pair (f,g), where f is the permutation induced on blocks and g the one on points.

(3) Write G = Aut $\mathcal{D}(v,k)$, and assume $G > S(v,k)$. If v is prime, then G is 2-transitive on points by BURNSIDE's theorem on groups of prime degree (see BURNSIDE [18, p.341]). If $k = \frac{1}{2}(v-1)$, G must also be 2-transitive (KANTOR [93]; compare CARLITZ [31]; MCCONNEL [116]; BRUEN & LEVINGER [10]).

However, it is not known in any other cases that G must be 2-transitive if $G > S(v,k)$.

(4) *If* $G > S(v,k)$ *and* $k = \frac{1}{2}(v-1)$, *then* $\mathcal{D}$ = PG(2,2) *or* $\mathcal{W}_{11}$ (KANTOR [93]; for some small values of v, this was proved by TODD [159] and F. HERING [67]).

More generally, *if* G *is* 2-*transitive then* $\mathcal{D}$ = PG(2,2) *or* $\mathcal{W}_{11}$ *provided that either* $1 + \sqrt{k} > (v-1)/k$ *or* k *has no proper divisor* $\equiv 1 \pmod{\lambda}$.

PROOF. 8E(14) applies with $A = \{x \to bx \mid b \in B\}$. □

Further information when G is 2-transitive (but when the above numerical conditions do not hold) is found in KANTOR [93]. The fact that, even for these specific designs, it is not known whether Aut $\mathcal{D}$ can be 2-transitive, indicates the sad state of affairs concerning 2-transitive symmetric designs!

H. An application to the irreducibility of polynomials

A very unexpected sort of occurrence of 2-transitive symmetric designs has recently been found by M. FRIED. Let K be a subfield of the complex field C. If $f(x) \in K[x]$ and $g(x) \in C[x]$, it is natural to study the irreducibility of $f(x) - g(y)$ in $C[x,y]$. This question leads to difference set designs having 2-transitive automorphism groups!

The following discussion is based primarily on FRIED [49,50] (see also CASSELS [33]). f(x) is called *indecomposable* over K if it is not possible to write $f(x) = f_1(f_2(x))$ with $f_i \in K[x]$ and deg $f_i > 1$, i=1,2; assume that this is the case. Assume further that g(x) cannot be written $g(x) = f(ax+b)$

for some $a,b \in C$, $a \neq 0$. Finally, assume that $f(x) - g(y) = \prod_{i=1}^{t} h_i(x,y)$ with $h_i(x,y) \in C[x,y]$ irreducible and $t > 1$.

FRIED shows that it may be assumed that deg f = deg g = v, say. Then g(x) is indecomposable over C. Moreover, t = 2. Write $k = \deg h_1(x,y)$. Then there is a difference set mod v with k elements. The corresponding symmetric design $\mathcal{D}$ admits a 2-transitive automorphism group G. (Here, G can be interpreted as the Galois group of a suitable extension field of C(x).)

Furthermore, G is generated by permutations $s_1,\ldots,s_\mu$, with $\mu \leq 3$, such that (i) $s_1\ldots s_\mu$ is a v-cycle on points, and (ii) $\sum_i \ell(s_i) = v-1 = \ell(s_1\ldots s_\mu)$. (Here, $\ell(s_i)$ is the smallest integer ℓ such that s_i is the product of ℓ transpositions.)

Of course, $PG_{d-1}(d,q)$ and $\mathcal{W}_{11}$ are the only known cyclic difference set designs $\mathcal{D}$ for which Aut $\mathcal{D}$ is 2-transitive. (Examples 8B(3) and 8B(4) do not admit transitive cyclic automorphism groups.) FEIT [50] enumerated all cases in which these designs can arise in FRIED's situation; each case produces a pair of polynomials f(x), g(x).

Needless to say, conditions (i) and (ii) are weird from a geometric or group-theoretic point of view. Nevertheless, it should be clear that they merit further study.

Note that the study of the polynomial f(x) - g(y) is remarkably reminiscent of the situation in 8G(2).

In more recent work of FRIED [51], 2-transitive designs have arisen in which b = 2v and some element of order v has one v-cycle on points and two on blocks.

I. 2-transitive suborbits

One recent occurrence of 2-transitive symmetric designs has been in work of CAMERON [19,20,21,26], on multiply-transitive suborbits (i.e., orbits of G_x) of primitive permutation groups. Since these will be discussed in CAMERON's talk at this conference, the reader is referred to that talk *) and the above papers.

*) This volume, pp. 419-450.

J. Problems

(1) The case $\lambda = 2$ should be feasible. The combinatorial structure here is extremely rich (see HUSSAIN [78,79], HALL [62], and CAMERON [23,29]). So, for that matter, is the permutation structure: G_B must be 2-transitive on B; if $x,y \in B$, $x \neq y$, then either G_B is 3-transitive on B, or G_{xyB} has two orbits of length $(k-2)/2$ on $B - \{x,y\}$ (KANTOR [3], CAMERON [23]). CAMERON [23,29] has indicated a possible approach to this problem.

Note that only three examples are known: PG(2,2)', $\mathcal{W}_{11}$ and $\mathcal{S}^{-1}(4)$.

(2) In the situation of 8C(2), there is a natural strongly regular graph structure on $S - \{x\}$. Unfortunately, the parameter restrictions on this graph and the tactical decomposition relations of DEMBOWSKI [38; 40, pp.60-61] involve too many unknowns. The latter relations were studied by KANTOR [93,101]; the former, in a purely combinatorial setting, by CAMERON [25] (using a method of GOETHALS & SEIDEL [54]). All the results thus far are very inconclusive.

(3) Prove that $\mathcal{D}$ is $\mathcal{S}^{\varepsilon}(2m)$ if G has a regular normal subgroup. As already mentioned in 8C(4), in this case $\mathcal{D}$ has the same parameters as some $\mathcal{S}^{\varepsilon}(2m)$.

(4) No satisfactory characterization of $\mathcal{W}_{176}$ is known. $\mathcal{W}_{176}$ and $(\mathcal{W}_{176})'$ are probably the only 2-transitive symmetric designs with $\lambda > 2$ and $v-2k+\lambda > 2$ in which G preserves a correspondence θ as in 8E(18); no numerical restrictions should be needed. (The main reason for the restriction in 8E(18) is to prevent k from being too large relative to λ.) If such a θ exists, $\mathcal{D}$ can be replaced (if necessary) by $\mathcal{D}'$ in order to obtain $\{x,y\} \subset X \cap Y$ if $\{x,y\}^{\theta} = \{X,Y\}$. Then $2(v-1)/k$ is an integer τ (so this situation is similar to the one considered in KANTOR [93], where $k|v-1$). If τ is odd, $G_{\{x,y\}}$ is transitive on $\{x,y\}^{\theta}$, and if $x \in B$, G_{xB} is transitive on the τ points $y \in B - \{x\}$ for which $B \in \{x,y\}^{\theta}$.

SMITH [152] has proposed a reasonable axiom one can assume in addition to the existence of θ in order to try to characterize $\mathcal{W}_{176}$, but this is too technical to state.

(5) Each of the known 2-transitive symmetric designs has polarities. Study these, and find some way to use them in the characterization of self-dual designs.

When v is prime, $\mathcal{D}$ automatically has "natural" polarities. However, no effective use has been found for them.

(6) The proof of 8E(2) in KANTOR [93] indicates that, when n is a power of a prime not dividing λ, $\mathcal{D}$ should be $\mathcal{W}_{11}$ or a projective space.

(7) Remove the numerical restrictions (i) and (ii) of 8E(14) and 8G(4).

(8) Answer WIELANDT's question (see 8F(3)). More generally, decide exactly what parameters can occur for linked systems (compare 8F(5)).

9. SYMMETRIC 3-DESIGNS

A. CAMERON's theorem

A *symmetric 3-design* is a 3-design $\mathcal{D}$ such that $\mathcal{D}_x$ is a symmetric design for each x. CAMERON [22] proved that the parameters of $\mathcal{D}$ must satisfy one of the following conditions (where μ is the number of blocks on any three points):

(i) $v = 4\mu + 4$, $k = 2\mu + 2$ (Hadamard 3-design);

(ii) $v = (\mu+2)(\mu^2+4\mu+2) + 1 = (\mu+1)(\mu^2+5\mu+5)$, $k = \mu^2+3\mu+2$;

(iii) $v = 112$, $k = 12$, $\mu = 1$ (extension of a projective plane $\mathcal{D}_x$ of order 10); or

(iv) $v = 496$, $k = 40$, $\mu = 3$.

Note that the λ for $\mathcal{D}$ is given by $\lambda = k-1$. Case (i) occurs if and only if there is a v×v Hadamard matrix. The only other case known to occur is $\mu = 1$ in (ii), when $\mathcal{D}$ is $\mathcal{W}_{22}$.

For a generalization of CAMERON's theorem, see CAMERON [27].

B. 3-transitive automorphism groups

(1) Now suppose $G \leq \mathrm{Aut}\,\mathcal{D}$ is 3-transitive on points. Then G_x is a 2-transitive automorphism group of the symmetric design $\mathcal{D}_x$ (cf. section 8).

It is not hard to show that cases (iii) and (iv) cannot occur. Cases (i) and (ii) remain open. Some special values of μ have, however, been ruled out by CAMERON [19], such as when $2 \le \mu < 103$ or $\mu+1$ is a prime power.

For a remarkable occurrence of case (ii) -which originally led CAMERON to his theorem- see CAMERON [20,26].

(2) *If* G_B *is* 3-*transitive on* B, *then* $\mathcal{D}$ *is* $AG_{d-1}(d,2)$, *the unique Hadamard* 3-*design with* 12 *points, or* $\mathcal{W}_{22}$. (This follows readily from 8E(9) and 8E(11).)

(3) Suppose next that $\mathcal{D}$ is a Hadamard 3-design. NORMAN [127] proved that $v = 12$ if μ is even. A slight modification of his argument shows that the same conclusion holds if G is 3-transitive on parallel classes of blocks. Note that, by 5B(4), the unique Hadamard 3-design having 12 points satisfies these conditions. The case n even -where $\mathcal{D}$ should be $AG_{d-1}(d,2)$- remains open.

C. Hadamard matrices

An automorphism of a Hadamard matrix H of size n is a pair (P,Q) of monomial $n \times n$ matrices such that PHQ = H. The automorphisms form a group G = Aut H containing 1 = (I,I) and -1 = (-I,-I) in its center. $\bar{G} = G/\langle -1\rangle$ acts faithfully as a permutation group on the union of the sets of rows and columns of H.

It may be assumed that the first row r and column c of H consist of 1's. Deleting columns 1 and n+1 of (H,-H) produces the (-1,1) incidence matrix of a Hadamard 3-design $\mathcal{D}$. Then $\bar{G}_c$ is the automorphism group of $\mathcal{D}$. In view of this, the results in B(2) and B(3) apply to $\mathcal{D}$. These in turn yield results about H. For example, if G is 4-transitive on rows, then n = 4 or 12. Another characterization of the case n = 12 follows from 6G(4) (KANTOR [94]).

Suppose n = 12. Then B(2) and the discussion of $\bar{G}_c$ imply that $\bar{G}_c \cong M_{11}$, from which $\bar{G} \cong M_{12}$ follows easily. However, $G \not\cong M_{12} \times \langle -1\rangle$. At the end of 4B(2) it was noted that $|\mathrm{Aut}\, M_{12}| = 2|M_{12}|$. The resulting outer automorphism can be visualized in the present context as follows. $(P,Q) \in G$ implies that PHQ = H, and hence (since H is symmetric) that $Q^tHP^t = H$, so $(Q^t,P^t) \in G$. Thus, $(P,Q) \to (Q^t,P^t)$ is an automorphism of G, and induces one of $\bar{G}$; these are both outer automorphisms (see HALL [59]).

10. FURTHER TOPICS AND PROBLEMS

A. Block intersections

Let $\mathcal{D}$ be a t-design, $t \geq 2$. According to a generalization of FISHER's inequality $b \geq v$, if $v \geq k+\frac{1}{2}t$ then $b \geq \binom{v}{[\frac{1}{2}t]}$ (WILSON & RAY-CHAUDHURI [168]). Equality holds only if $t = 2s$ for an integer s, and then $\mathcal{D}$ is called a *tight* t-*design*. (This is evidently a generalization of symmetric designs.) WILSON & RAY-CHAUDHURI also proved that, if $\mathcal{D}$ is a 2s-design, then $\mathcal{D}$ is tight if and only if there are at most s different intersection sizes $|B \cap C|$, where B and C run through all pairs of distinct blocks (cf. CAMERON [25]).

It is natural to consider 2s-transitive automorphism groups of tight 2s-designs. Partly motivated by the group-theoretic context, ITO [85] has just completed a proof that the only tight 4-designs are degenerate ($v = k-2$), $\mathcal{W}_{23}$, or its complementary design $(\mathcal{W}_{23})'$. The case $s > 2$ remains completely open in both the combinatorial and group-theoretic contexts.

One way to guarantee that a t-design $\mathcal{D}$ has few intersection sizes $|B \cap C|$ is to assume that $G = \mathrm{Aut}\,\mathcal{D}$ is block-transitive and has small block-rank ρ; thus, G_B has exactly ρ block orbits (so there are at most $\rho-1$ different sizes $|B \cap C|$ with $B \neq C$). This was considered by NODA [126] when $\mathcal{D}$ is a Steiner system $S(t,k,v)$. He assumed $t = 3$ or 4 and $\rho = 3$ or 4, and showed $\mathcal{D}$ must be $\mathcal{W}_{22}$, $\mathcal{W}_{23}$, $\mathcal{W}_{24}$ or $AG_2(3,2)$. The proofs are very similar to tight design arguments. (In fact, the case $t = 4$, $\rho = 3$ follows from the aforementioned results of WILSON & RAY-CHAUDHURI.)

It should also be possible to handle the case $t = 2$, $\lambda = 1$ and $\rho = 3$. Here, G_B is transitive on the lines disjoint from B, and G_x is 2-transitive on the lines through x. Presumably, $\mathcal{D}$ must be $AG(2,k)$ or $PG_1(d,k-1)$. NODA has observed that $\mathcal{D}$ is $AG(2,k)$ if G is not line-primitive; moreover, in unpublished work, he has used an argument of HIGMAN [70] to show that $\mathcal{D}$ is $PG_1(d,k-1)$ if $v > k^2(k-1)^2(k-2)^2 + k^2 - k + 1$.

B. Parallel relations

Let $\mathcal{D}$ be a design. A parallel relation on $\mathcal{D}$ is an equivalence relation $\parallel$ partitioning the blocks into classes, each of which partitions the points of $\mathcal{D}$. Each parallel class has v/k blocks, and there are exactly r parallel classes.

Relatively little is known about subgroups G of Aut $\mathcal{D}$ which preserve $\parallel$. If the classical affine space (or plane) case is excluded, little is known beyond NORMAN's theorem (see 9B(3)) and the following result of CAMERON [28].

(1) *Let* $\mathcal{D}$ *be the degenerate design with* $k = 2$ *and* $\lambda = 1$, *whose blocks are just the 2-sets of points. Assume that* $v > 3$, G *is 3-transitive, and* G *preserves* $\parallel$. *Then either* $v = 6$ *and* $G \cong PGL(2,5)$, *or* $v = 2^d$ *for some* d *and* $\mathcal{D}$ *can be regarded as the design* $AG_1(d,2)$ *with the obvious parallel relation.*

PROOF. Let x,y,z be any three points. Then G_{xyz} fixes the block through z parallel to $\{x,y\}$. Hence, G_{xyz} fixes $k \geq 4$ points. If $k = v$ then ZASSENHAUS [172] can be used to show that $v = 6$ and G is PGL(2,5). If $k < v$, 6D(1) can be applied to yield $k = 4$. If B and C are two blocks of this $S(3,4,v)$, and if $|B \cap C| = 2$, then $B - B \cap C$, $B \cap C$, and $C - B \cap C$ are parallel. Hence, $B+C$ is a block of the $S(3,4,v)$. It follows easily that the $S(3,4,v)$ is $AG_2(d,2)$ (compare 4C(3)). □

Actually, CAMERON's proof does not use 6D(1). In fact, it was while I was eliminating one case of CAMERON's situation that 6D(1) and 6E(1) were born.

More recently, CAMERON has obtained a generalization of 10B(1) to groups preserving a parallelism of the trivial design of all k-sets of a v-set $(1 < k < v)$.

The natural extension of 10B(1) to the case of triangle-transitive automorphism groups of more general designs $\mathcal{D}$ (with $\parallel$) remains open.

(2) *If* $\mathcal{D}$ *and* $\parallel$ *are as before, then* $b \geq v+r-1$; *moreover,* $b = v+r-1$ *if and only if any two blocks meet in* 0 *or* k^2/v *points* (see DEMBOWSKI [40, pp.72-73]). When $b = v+r-1$, $\mathcal{D}$ is called an *affine design*. Clearly, affine designs provide a common generalization of Hadamard 3-designs and affine spaces. A theorem of DEMBOWSKI [40, p.74] characterizes affine spaces $AG_{d-1}(d,q)$, $q > 2$, among affine designs; this result is similar to the DEMBOWSKI-WAGNER theorem (see 8C). But relatively little attention has been paid to automorphism groups, so perhaps a few additional remarks are worthwhile.

Consider $\mathcal{D}$, $\parallel$, and $G \leq$ Aut $\mathcal{D}$ preserving $\parallel$. Let G have t_p point-orbits, t_b block-orbits, and $t_\parallel$ parallel-class orbits. If $\mathcal{D}$ is an affine design, then $t_b + 1 = t_p + t_{\|}$ (NORMAN [127]). In general, it turns out that one can at least say $t_b + 1 \geq t_p + t_{\|}$. Also, if $\mathcal{D}$ is affine and $g \in G$, then $f_p + 1 = f_b + f_{\|}$, where f_p, f_b and $f_{\|}$ are the numbers of points, blocks and parallel-classes fixed by g. From these facts, further results can be deduced as in KANTOR [91].

Incidentally, it should be noted that the arguments on pp.113-114 of DEMBOWSKI [40] show that the number of non-isomorphic affine designs having the same parameters as $AG_{d-1}(d,q)$, $d \geq 3$, is enormous (and in fact $\to \infty$, as $d \to \infty$ or $q \to \infty$). However, I conjecture that affine spaces are the only affine designs which are not Hadamard 3-designs and whose automorphism groups are transitive on ordered pairs of non-parallel blocks.

C. Transitive extensions

Let H be a given group, possibly given together with a specific transitive permutation representation on a set S'. A *transitive extension* of H is a 2-transitive group G on a set S such that, for some $x \in S$, $G_x \cong H$; if, moreover, H is given as acting on S', then it is also required that $|S| = |S'|+1$ and that G_x acts on $S-\{x\}$ as H does on S'.

A basic open problem concerning 2-transitive groups is: if H is known as an abstract group, find all transitive extensions of H. Needless to say, very few groups H have transitive extensions.

Transitive extensions have been studied geometrically by DEMBOWSKI [39], HUGHES [75,76], and TITS [158]. Their approach was to extend designs associated with groups such as the collineation group of AG(d,q) or PG(d,q), given as acting 2-transitively on the points of the corresponding affine or projective space.

Much more generally, TITS (unpublished) has shown that a Chevalley group over GF(q), acting on a class of parabolic subgroups, has no transitive extensions if q is not very small. Still more generally, SEITZ (unpublished) has obtained the same conclusion if H is isomorphic to a Chevalley group over GF(q) and $(q,|S|-1) = 1$.

D. Some maximal subgroups of alternating or symmetric groups

Let H be a transitive permutation group on S, about which a lot is known. PROBLEM: determine all permutation groups G on S containing H.

Here, I have in mind some "geometric" group H and set S. The case $H = PSL(n,q)$, $n \geq 2$, with S the set of points of $PG(n-1,q)$, has been discussed in 7E(1). In general, if H is chosen "large" enough, and $G > H$, then G will presumably have to be 2-transitive. PROBLEM: handle the case $H = PSL(n,q)$, $n \geq 4$, and S the set of e-spaces of $PG(n-1,q)$, where $1 \leq e \leq n-2$.

I have settled the case $H = Sp(2m,2)$, in its 2-transitive representations of degree $2^{m-1}(2^m \pm 1)$: if $G > H$ then G is alternating and symmetric. The elementary proof uses transvections and the geometry of $GO^{\pm}(2m,2)$.

The reader should have no difficulty in listing many other, similar questions. Perhaps the most intriguing general question of this type concerns a Chevalley group H acting on a set S of parabolic subgroups.

E. Sp(2m,2) and .3

SHULT [148] has obtained some graph-theoretic characterizations of $Sp(2m,2)$ in its 2-transitive representations of degree $2^{m-1}(2^m \pm 1)$. However, no characterization is known in terms of designs. The difficulty is that no really interesting designs seem to have $Sp(2m,2)$ as a 2-transitive automorphism group.

Precisely the same difficulty occurs in the case of CONWAY's smallest group .3, in its 2-transitive representation of degree 276 (see CONWAY [35]). In both cases, the 2-graph approach seems more relevant than the design one (cf. SEIDEL [143]).

APPENDIX

The known 2-transitive groups

The following is a list of all the known 2-transitive groups G having no regular normal subgroup.

(1) $G = A_n$ or S_n, $|S| = n$.

(2) $PSL(d+1,q) \leq G \leq P\Gamma L(d+1,q)$; S is the set of points or hyperplanes of $PG(d,q)$.

(3) $PSU(3,q) \leq G \leq P\Gamma U(3,q)$; S is the set of q^3+1 points of the corresponding unital.

(4) G has a normal Ree subgroup; S is the set of q^3+1 points of the corresponding unital ($q = 3^{2e+1}$). When $e = 0$, $G \cong P\Gamma L(2,8)$, acting on the points of $\mathcal{D}(4)$ (see 6B(3)).

(5) $Sz(2^{2e+1}) \le G \le \mathrm{Aut}\, Sz(2^{2e+1})$; S is the set of $(2^{2e+1})^2 + 1$ points of the corresponding inversive plane or ovoid (see LÜNEBURG [111]).

(6) $G = Sp(2m,2)$, $|S| = 2^{m-1}(2^m \pm 1)$, $G_x = GO^{\pm}(2m,2)$.

(7) $G = PSL(2,11)$ acting on the 11 points or blocks of $\mathcal{W}_{11}$ (see 8B(2)).

(8) $G = A_7$ acting on the 15 points or planes of PG(3,2) (see 4A).

(9) The Mathieu groups M_{11}, M_{12}, M_{22}, Aut M_{22}, M_{23} and M_{24} in their usual representations on the points of the corresponding Steiner systems.

(10) $G = M_{11}$ acting 3-transitively on the 12 points of a Hadamard 3-design (see 4B(3), 9B and 9C).

(11) $G = HS$ acting on the 176 points or blocks of $\mathcal{W}_{176}$ (see 8B(4)).

(12) $G = .3$, $|S| = 276$.

REFERENCES

[1] ASCHBACHER, M., *On doubly transitive permutation groups of degree $n \equiv 2 \bmod 4$*, Illinois J. Math., 16 (1972) 276-279.

[2] ASCHBACHER, M., *Doubly transitive groups in which the stabilizer of two points is abelian*, J. Algebra, 18 (1971) 114-136.

[3] ASCHBACHER, M., *A condition for the existence of a strongly embedded subgroup*, Proc. Amer. Math. Soc., 38 (1973) 509-511.

[4] ASCHBACHER, M., *F-Sets and permutation groups*, to appear.

[5] ASCHBACHER, M., *2-Transitive groups whose 2-point stabilizer has 2-rank 1*, to appear.

[6] ATKINSON, M.D., *Doubly transitive but not doubly primitive permutation groups I*, J. London Math. Soc. (2), 7 (1974) 632-634; *II*, ibid., to appear.

[7] BENDER, H., *Endliche zweifache transitive Permutationsgruppen, deren Involutionen keine Fixpunkte haben*, Math. Z., 104 (1968) 175-204.

[8] BENDER, H., *Transitive Gruppen gerader Ordnung, in denen jede Involution genau einen Punkt festlasst*, J. Algebra, 17 (1971) 527-554.

[9] BLOCK, R.E., *Transitive groups of collineations of certain designs*, Pacific J. Math., 15 (1965) 13-19.

[10] BRUEN, A. & B. LEVINGER, *A theorem on permutations of a finite field*, Canad. J. Math., 25 (1973) 1060-1065.

[11] BUEKENHOUT, F., *Une caractérisation des espaces affins basée sur la notion de droite*, Math. Z., 111 (1969) 367-371.

[12] BUEKENHOUT, F., *A characterization of affine spaces of order two as 3-designs*, Math. Z., 118 (1970) 83-85.

[13] BUEKENHOUT, F., *An axiomatic of inversive spaces*, J. Combinatorial Theory A, 11 (1971) 208-212.

[14] BUEKENHOUT, F., *On 2-designs whose group of automorphisms is transitive on the ordered pairs of intersecting lines*, J. London Math. Soc., 5 (1972) 663-672.

[15] BUEKENHOUT, F., *Transitive groups whose involutions fix one or three points*, J. Algebra, 23 (1972) 438-451.

[16] BUEKENHOUT, F., *Doubly transitive groups in which the maximum number of fixed points of involutions is four*, Arch. Math. (Basel), 23 (1972) 362-369.

[17] BUEKENHOUT, F. & R. DEHERDER, *Espaces linéaires finis à plans isomorphes*, Bull. Soc. Math. Belg., 23 (1971) 348-359.

[18] BURNSIDE, W., *Theory of groups of finite order*, Dover, New York, 1955.

[19] CAMERON, P.J., *Structure of suborbits in some primitive permutation groups*, thesis, Oxford Univ., 1971.

[20] CAMERON, P.J., *Permutation groups with multiply transitive suborbits I*, Proc. London Math. Soc. (3), 25 (1972) 427-440; *II*, Bull. London Math. Soc., to appear.

[21] CAMERON, P.J., *Primitive groups with most suborbits doubly transitive*, Geometriae Dedicata, 1 (1973) 434-446.

[22] CAMERON, P.J., *Extending symmetric designs*, J. Combinatorial Theory A, 14 (1973) 215-220.

[23] CAMERON, P.J., *Biplanes*, Math. Z., 131 (1973) 85-101.

[24] CAMERON, P.J., *On groups with several doubly transitive permutation representations*, Math. Z., 128 (1972) 1-14.

[25] CAMERON, P.J., *Near-regularity conditions for designs*, Geometriae Dedicata, 2 (1973) 213-224.

[26] CAMERON, P.J., *Permutation groups with multiply-transitive suborbits II*, to appear.

[27] CAMERON, P.J., *Locally symmetric designs*, to appear.

[28] CAMERON, P.J., *On groups of degree n and n-1, and highly symmetric edge colourings*, to appear.

[29] CAMERON, P.J., *Characterizations of some Steiner systems, parallelisms, and biplanes*, to appear.

[30] CAMERON, P.J. & J.J. SEIDEL, *Quadratic forms over GF(2)*, Indag. Math., 35 (1973) 1-8.

[31] CARLITZ, L., *A theorem on permutations of a finite field*, Proc. Amer. Math. Soc., 11 (1960) 456-459.

[32] CARMICHAEL, R.D., *Introduction to the theory of groups of finite order*, Dover, New York, 1956.

[33] CASSELS, J.W.S., *Factorization of polynomials in several variables*, in: Proc. 15th Scandinavian Congress 1968, Lecture Notes in Mathematics 118, Springer-Verlag, Berlin, 1970.

[34] CONWAY, J.H., *A group of order 8,315,553,613,086,720,000*, Bull. London Math. Soc., 1 (1969) 79-88.

[35] CONWAY, J.H., *Three lectures on exceptional groups*, in: *Finite simple groups*, Academic Press, New York, 1971, pp.215-247.

[36] CURTIS, C.W., W.M. KANTOR & G.M. SEITZ, *The 2-transitive permutation representations of the finite Chevalley groups*, to appear.

[37] CZERWINSKI, T., *Collineation groups containing no Baer involutions*, Proc. Internat. Conference on Projective Planes, Washington State Univ. Press, Pullman, 1973.

[38] DEMBOWSKI, P., *Verallgemeinerungen von Transitivitätsklassen endlichen projektiver Ebenen*, Math. Z., 69 (1958) 59-89.

[39] DEMBOWSKI, P., *Die Nichtexistenz von transitiven Erweiterungen der endlichen affinen Gruppen*, J. Reine Angew. Math., 220 (1965) 37-44.

[40] DEMBOWSKI, P., *Finite geometries*, Ergebnisse der Mathematik 44, Springer-Verlag, Berlin etc., 1968.

[41] DEMBOWSKI, P. & A. WAGNER, *Some characterizations of finite projective spaces*, Arch. Math. (Basel), 11 (1960) 465-469.

[42] DICKSON, L.E., *Linear groups*, Dover, New York, 1955.

[43] DOYEN, J. & X. HUBAUT, *Finite regular locally projective spaces*, Math. Z., 119 (1971) 83-88.

[44] FEIT, W., *Automorphisms of symmetric balanced incomplete block designs*, Math. Z., 118 (1970) 40-49.

[45] FEIT, W., *On symmetric balanced incomplete block designs with doubly transitive automorphism groups*, J. Combinatorial Theory A, 14 (1973) 221-247.

[46] FEIT, W. & J.G. THOMPSON, *Solvability of groups of odd order*, Pacific J. Math., 13 (1963) 771-1029.

[47] FISCHER, B., *Eine Kennzeichnung der symmetrischen Gruppen von Grade 6 und 7*, Math. Z., 95 (1967) 288-298.

[48] FISCHER, B., *Finite groups generated by 3-transpositions, I*, Invent. Math., 13 (1971) 232-246.

[49] FRIED, M., *On the diophantine equation $f(y)-x = 0$*, Acta Arith., 19 (1971) 79-87.

[50] FRIED, M., *The field of definition of function fields and a problem in the reducibility of polynomials in two variables*, Illinois J. Math., 17 (1973) 128-146.

[51] FRIED, M., *On Hilbert's irreducibility theorem*, J. Number Theory, to appear.

[52] FRIED, M. & D.J. LEWIS, *Solution spaces to diophantine problems*, Bull. Amer. Math. Soc., to appear.

[53] GLAUBERMAN, G., *Central elements in core-free groups*, J. Algebra, 4 (1966) 403-420.

[54] GOETHALS, J.M. & J.J. SEIDEL, *Strongly regular graphs derived from combinatorial designs*, Canad. J. Math., 22 (1970) 597-614.

[55] HALL, J.I., *Steiner triple systems with geometric minimally generated subsystems*, to appear.

[56] HALL, Jr., M., *A survey of difference sets*, Proc. Amer. Math. Soc., 7 (1956) 975-986.

[57] HALL, Jr., M., *The theory of groups*, MacMillan, New York, 1959.

[58] HALL, Jr., M., *Automorphisms of Steiner triple systems*, IBM J. Res. Develop., 4 (1960) 460-472. (Also in Proc. Symp. Pure Math. 6 (1962) 47-66).

[59] HALL, Jr., M., *Note on the Mathieu group* M_{12}, Arch. Math. (Basel), 13 (1962) 334-340.

[60] HALL, Jr., M., *Group theory and block designs*, in: Proc. Internat. Conf. Theory of Groups, Gordon & Breach, New York, 1967, pp.115-144.

[61] HALL, Jr., M., *Combinatorial theory*, Blaisdell, Waltham, Mass., 1967.

[62] HALL, Jr., M., *Symmetric block designs with* $\lambda = 2$, in: *Combinatorial mathematics and its applications*, Univ. of North Carolina Press, 1969, pp.175-186.

[63] HARADA, K., *On some doubly transitive groups*, J. Algebra, 17 (1971) 437-450.

[64] HERING, C., *Zweifach transitive Permutationsgruppen, in denen zwei die maximale Anzahl von Fixpunkten von Involutionen ist*, Math. Z., 104 (1968) 150-174.

[65] HERING, C., *On linear groups which contain an irreducible subgroup of prime order*, in: Proc. Internat. Conf. Projective Planes, Washington State Univ. Press, Pullman, 1973.

[66] HERING, C., W.M. KANTOR & G.M. SEITZ, *Finite groups having a split BN-pair of rank 1*, J. Algebra, 20 (1972) 435-475.

[67] HERING, F., *Über die Kollineationsgruppen einiger Hadamard-Matrizen*, unpublished manuscript.

[68] HIGMAN, D.G., *Flag transitive collineation groups of finite projective spaces*, Illinois J. Math., 6 (1962) 434-446.

[69] HIGMAN, D.G., *Finite permutation groups of rank 3*, Math. Z., 86 (1964) 145-156.

[70] HIGMAN, D.G., *Characterization of families of rank 3 permutation groups by the subdegrees II*, Arch. Math. (Basel), 21 (1970) 353-361.

[71] HIGMAN, D.G., *Remark on Shult's graph extension theorem*, in: *Finite groups '72*, T. GAGEN, M.P. HALE & E.E. SHULT (eds.), North-Holland Publ. Coy., Amsterdam, 1973, pp. 80-83.

[72] HIGMAN, D.G. & C.C. SIMS, *A simple group of order 44,352,000*, Math. Z., 105 (1968) 110-113.

[73] HIGMAN, G., *On the simple group of D.G. Higman and C.C. Sims*, Illinois J. Math., 13 (1969) 74-84.

[74] HILL, R., *Rank 3 permutation groups with a regular normal subgroup*, thesis, Univ. of Warwick, 1971.

[75] HUGHES, D.R., *On t-designs and groups*, Amer. J. Math., 87 (1965) 761-778.

[76] HUGHES, D.R., *Extensions of designs and groups: Projective, symplectic, and certain affine groups*, Math. Z., 89 (1965) 199-205.

[77] HUPPERT, B., *Zweifach transitive, auflösbare Permutationsgruppen*, Math. Z., 68 (1957) 126-150.

[78] HUSSAIN, Q.M., *Impossibility of the symmetrical incomplete block design with* $\lambda = 2$, $k = 7$, Sankhya, 7 (1946) 317-322.

[79] HUSSAIN, Q.M., *Symmetrical incomplete block designs with* $\lambda = 2$, $k = 8$ *or* 9, Bull. Calcutta Math. Soc., 37 (1945) 115-123.

[80] ITO, N., *Über die Gruppen* $PSL_n(q)$, *die eine Untergruppe von Primzahl-index enthalten*, Acta Sci. Math. (Szeged), 21 (1960) 206-217.

[81] ITO, N., *On a class of doubly, but not triply transitive permutation groups*, Arch. Math. (Basel), 18 (1967) 564-570.

[82] ITO, N., *On permutation groups of prime degree p which contain (at least) two classes of conjugate subgroups of index p*, Rend. Sem. Mat. Padova, 38 (1967) 287-292.

[83] ITO, N., *On permutation groups of prime degree p which contain at least two classes of conjugate subgroups of index p, II*, Nagoya Math. J., 37 (1970) 201-208.

[84] ITO, N., *A theorem on Jordan groups*, in: *Theory of finite groups*, Benjamin, New York, 1969, pp.47-48.

[85] ITO, N., *Tight 4-designs*, to appear.

[86] ITO, N., *On the Wielandt number of transitive permutation groups of prime degree*, to appear.

[87] ITO, N. & W.M. KANTOR, *2-Transitive symmetric designs with $k = 2p$*, Notices Amer. Math. Soc., 16 (1969) 774.

[88] JORDAN, C., *Théorèmes sur les groupes primitifs*, J. Math. Pures Appl., 16 (1871) 383-408.

[89] JORDAN, C., *Oeuvres*, J. DIEUDONNÉ (ed.), Gauthier-Villars, Paris, 1961.

[90] JORDAN, C., *Traité des substitutions*, (new edition), Gauthier-Villars, Paris, 1957.

[91] KANTOR, W.M., *Automorphism groups of designs*, Math. Z., 109 (1969) 246-252.

[92] KANTOR, W.M., *Characterizations of finite projective and affine spaces*, Canad. J. Math., 21 (1969) 64-75.

[93] KANTOR, W.M., *2-Transitive symmetric designs*, Trans. Amer. Math. Soc., 146 (1969) 1-28.

[94] KANTOR, W.M., *Automorphisms of Hadamard matrices*, J. Combinatorial Theory, 6 (1969) 279-281.

[95] KANTOR, W.M., *Jordan groups*, J. Algebra, 12 (1969) 471-493.

[96] KANTOR, W.M., *Elations of designs*, Canad. J. Math., 22 (1970) 897-904.

[97] KANTOR, W.M., *On a class of Jordan groups*, Math. Z., 118 (1970) 58-68.

[98] KANTOR, W.M., *Note on symmetric designs and projective spaces*, Math. Z., 122 (1971) 61-62.

[99] KANTOR, W.M., *On 2-transitive groups in which the stabilizer of two points fixes additional points*, J. London Math. Soc., 5 (1972) 114-122.

[100] KANTOR, W.M., *Line-transitive collineation groups of finite projective spaces*, Israel J. Math., 14 (1973) 229-235.

[101] KANTOR, W.M., *Symplectic groups, symmetric designs, and line ovals*, to appear.

[102] KANTOR, W.M., *On 2-transitive collineation groups of finite projective spaces*, Pacific J. Math., 48 (1973) 119-131.

[103] KANTOR, W.M., *Some highly geometric lattices*, to appear.

[104] KANTOR, W.M., *Primitive groups having transitive subgroups of smaller, prime power degree*, to appear.

[105] KANTOR, W.M., *Plane geometries associated with certain 2-transitive groups*, to appear.

[106] KANTOR, W.M. & T.P. MCDONOUGH, *On the maximality of PSL(d+1,q), d ≥ 2*, to appear.

[107] KANTOR, W.M., M.E. O'NAN & G.M. SEITZ, *2-Transitive groups in which the stabilizer of two points is cyclic*, J. Algebra, 21 (1972) 17-50.

[108] KERDOCK, A.M., *A class of low-rate nonlinear binary codes*, Information and Control, 20 (1972) 182-187.

[109] KING, J., *Doubly transitive groups in which involutions fix one or three points*, Math. Z., 111 (1969) 311-321.

[110] LÜNEBURG, H., *Über projektive Ebenen, in denen jede Fahne von einer nichttrivialen Elation invariant gelassen wird*, Abh. Math. Sem. Univ. Hamburg, 29 (1965) 37-76.

[111] LÜNEBURG, H., *Die Suzukigruppen und ihre Geometrien*, Lecture Notes in Mathematics 10, Springer-Verlag, Berlin etc., 1965.

[112] LÜNEBURG, H., *Transitive Erweiterungen endlicher Permutationsgruppen*, Lecture Notes in Mathematics 84, Springer-Verlag, Berlin etc., 1969.

[113] MANN, H.B., *Addition theorems*, Tracts in Pure and Applied Math. 18, Interscience, New York, 1965.

[114] MARGGRAFF, B., *Über primitive Gruppen, welche eine transitive Gruppe geringeren Grades enthalten*, thesis, Giessen, 1889.

[115] MARGGRAFF, B., *Primitive Gruppen, welche eine transitive Gruppe geringeren Grades enthalten*, Wiss. Beilage zu den Jahresberichten des Sophiengymnasiums zu Berlin, Berlin, 1895.

[116] MCCONNEL. R., *Pseudo-ordered polynomials over a finite field*, Acta Arith., 8 (1963) 127-151.

[117] MCDONOUGH, T.P., *On Jordan groups*, J. London Math. Soc., 6 (1972) 73-80.

[118] MCDONOUGH, T.P., *On Jordan groups - addendum*, to appear.

[119] MCLAUGHLIN, J.E., *A simple group of order 898,128,000*, in: *Theory of finite groups*, Benjamin, New York, 1969, pp.109-111.

[120] NAGAO, H., *On multiply transitive groups IV*, Osaka J. Math., 2 (1965) 327-341.

[121] NEUMANN, P.M., *Transitive permutation groups of prime degree*, J. London Math. Soc., 5 (1972) 202-208.

[122] NEUMANN, P.M., *Generosity and characters of multiply transitive permutation groups*, to appear.

[123] NEUMANN, P.M., *Transitive permutation groups of prime degree*, in: Proc. Conf. Theory of Groups, Canberra 1973, Lecture Notes in Mathematics, Springer-Verlag, Berlin etc., to appear.

[124] NEUMANN, P.M., *Transitive permutation groups of prime degree IV*, to appear.

[125] NODA, R., *Doubly transitive groups in which the maximal number of fixed points of involutions is four*, Osaka Math. J., 8 (1971) 77-90.

[126] NODA, R., *Steiner systems which admit block-transitive automorphism groups of small rank*, Math. Z., 125 (1972) 113-121.

[127] NORMAN, C.W., *A characterization of the Mathieu group M_{11}*, Math. Z., 106 (1968) 162-166.

[128] O'NAN, M.E., *Automorphisms of unitary block designs*, J. Algebra, 20 (1972) 495-511.

[129] O'NAN, M.E., *A characterization of $U_3(q)$*, J. Algebra, 22 (1972) 254-296.

[130] O'NAN, M.E., *A characterization of $L_n(q)$ as a permutation group*, Math. Z., 127 (1972) 301-314.

[131] O'NAN, M.E., *The normal structure of the one-point stabilizer of a doubly-transitive group*, in: *Finite groups '72*, T. GAGEN, M.P. HALE & E.E. SHULT, (eds.), North-Holland Publ. Coy., Amsterdam, 1973, pp.119-121.

[132] O'NAN, M.E., *Normal structure of the one-point stabilizer of a doubly-transitive permutation group, I*, to appear.

[133] O'NAN, M.E., *Normal structure of the one-point stabilizer of a doubly-transitive permutation group, II*, to appear.

[134] O'NAN, M.E., *Triply-transitive permutation groups whose two-point stabilizer is local*, to appear.

[135] O'NAN, M.E., *Doubly-transitive groups of odd degree whose one point stabilizer is local*, to appear.

[136] OSBORN, J.H., *Finite doubly-transitive permutation groups without subgroups fixing exactly two points*, thesis, Univ. of Wisconsin, 1972.

[137] OSTROM, T.G. & A. WAGNER, *On projective and affine planes with transitive collineation groups*, Math. Z., 71 (1959) 186-199.

[138] PALEY, R.E.A.C., *On orthogonal matrices*, J. Math. Phys., 12 (1933) 311-320.

[139] PERIN, D., *On collineation groups of finite projective spaces*, Math. Z., 126 (1972) 135-142.

[140] POLLATSEK, H., *First cohomology groups of some linear groups over fields of characteristic two*, Illinois J. Math., 15 (1971) 393-417.

[141] PRAEGER, C.E., *Finite permutation groups*, thesis, Oxford Univ., 1973.

[142] SCOTT, L.L., *A double transitivity criterion*, Math. Z., 115 (1970) 7-8.

[143] SEIDEL, J.J., *A survey of two-graphs*, to appear.

[144] SHAUGHNESSY, E.P., *Codes with simple automorphism groups*, Arch. Math. (Basel), 22 (1971) 459-466.

[145] SHULT, E.E., *On the fusion of an involution in its centralizer*, to appear.

[146] SHULT, E.E., *On a class of doubly transitive groups*, Illinois J. Math., 16 (1972) 434-445.

[147] SHULT, E.E., *The graph extension theorem*, Proc. Amer. Math. Soc., 33 (1972) 278-284.

[148] SHULT, E.E., *Characterizations of certain classes of graphs*, J. Combinatorial Theory B, 13 (1972) 142-167.

[149] SHULT, E.E., *On doubly transitive groups of even degree*, to appear.

[150] SIMS, C.C., *On the isomorphism of two groups of order 44,352,000*, in: *Theory of finite groups*, Benjamin, New York, 1969, pp.101-108.

[151] SIMS, C.C., *Computational methods in the study of permutation groups*, in: *Computational problems in abstract algebra*, J. LEECH (ed.), Pergamon Press, London, 1970, pp. 169-183.

[152] SMITH, M.S., *A combinatorial configuration associated with the Higman-Sims simple group*, to appear.

[153] SMITH, M.S., *On the isomorphism of two simple groups of order 44,352,000*, to appear.

[154] SYLVESTER, J.J., *Collected mathematical papers, vol. II*, Cambridge Univ. Press, 1908, pp.615-628.

[155] TAYLOR, D.E., *Some topics in the theory of finite groups*, thesis, Oxford Univ., 1971.

[156] TAYLOR, D.E., *Monomial representations and strong graphs*, in: Proc. First Australian Conf. Combinatorial Math., Univ. of Newcastle, 1972, pp. 197-201.

[157] TEIRLINCK, L., *On linear spaces in which every plane is either projective or affine*, to appear.

[158] TITS, J., *Sur les systèmes de Steiner associés aux trois "grands" groupes de Mathieu*, Rend. Mat. e Appl., 23 (1964) 166-184.

[159] TODD, J.A., *A combinatorial problem*, J. Math. Phys., 12 (1933) 321-333.

[160] TSUZUKU, T., *On doubly transitive permutation groups of degree* $1+p+p^2$ *where* p *is a prime number*, J. Algebra, 8 (1968) 143-147.

[161] VEBLEN, O. & J.W. YOUNG, *Projective geometry I*, Ginn, Boston, 1916.

[162] WAGNER, A., *On collineation groups of finite projective spaces, I*, Math. Z., 76 (1961) 411-426.

[163] WARD, H.N., *On Ree's series of simple groups*, Trans. Amer. Math. Soc., 121 (1966) 62-69.

[164] WIELANDT, H., *Abschätzungen für den Grad einer Permutationsgruppe von vorgeschriebenem Transitivitätsgrad*, Schr. Math. Sem. Univ. Berlin, 2 (1934) 151-174.

[165] WIELANDT, H., *Über den Transitivitätsgrad von Permutationsgruppen*, Math. Z., 74 (1960) 297-298.

[166] WIELANDT, H., *Finite permutation groups*, Acad. Press, New York, 1964.

[167] WIELANDT, H., *On automorphism groups of doubly-transitive permutation groups*, in: Proc. Internat. Conf. Theory of Groups, Gordon & Breach, New York, 1967, pp.389-393.

[168] WILSON, R.M. & D.K. RAY-CHAUDHURI, *Generalization of Fisher's inequality to t-designs*, Notices Amer. Math. Soc., 18 (1971) 805.

[169] WITT, E., *Die 5-fach transitiven Gruppen von Mathieu*, Abh. Math. Sem. Univ. Hamburg, 12 (1938) 256-264.

[170] WITT, E., *Über Steinersche Systeme*, Abh. Math. Sem. Univ. Hamburg, 12 (1938) 265-275.

[171] ZASSENHAUS, H., *Über endliche Fastkörper*, Abh. Math. Sem. Univ. Hamburg, 11 (1935) 187-220.

[172] ZASSENHAUS, H., *Kennzeichnung endlicher linearer Gruppen als Permutationsgruppen*, Abh. Math. Sem. Univ. Hamburg, 11 (1935) 17-40.

SUBORBITS IN TRANSITIVE PERMUTATION GROUPS

P.J. CAMERON

Merton College, Oxford OX1 4JD, England

With any graph we can associate a group, namely its automorphism group; this acts naturally as a permutation group on the vertices of the graph. The converse idea, that of reconstructing a graph (or a family of graphs) from a transitive permutation group, has been developed by C.C. SIMS, D.G. HIGMAN, and many other people, and is the subject of the present survey. In his lecture notes [23], HIGMAN has axiomatised the combinatorial objects that arise from permutation groups in this way, under the name *coherent configurations*; but I shall discuss only the case where a group is present. My own introduction to the theory was via the unpublished paper of P.M. NEUMANN [30].

In section 1, the construction and some of its basic properties are described. Also in this section I introduce the basis matrices and the centraliser algebra they generate. The theory of this algebra has played an important role in the study of permutation groups and of various combinatorial objects (in such papers as [20],[22],[30],[2],[12],[16]); but here I shall be more concerned with other aspects of the theory, those which may be described as more "graph-theoretic".

Section 2 is about paired suborbits. In graph-theoretic terms, we have a directed graph, and ask about the relations between the actions of the stabiliser G_α of a vertex α on the vertices joined "to" and "from" α.

In section 3 the subject is more group-theoretic. Suppose the action of G_α on the vertices adjacent to α (or perhaps just the number of such vertices) is given. What can be said about the structure of G_α? After pioneering work by TUTTE [41],[42] and SIMS [34], the most powerful results here are due to WIELANDT [47].

M. Hall, Jr. and J. H. van Lint (eds.), Combinatorics, 419-450.

Sections 4 and 5 are closely related. In section 4 we take once more the graph-theoretic viewpoint and investigate further transitivity properties of undirected graphs; in the following section these investigations are put back in group-theoretic context, and generalisations of them are studied. One of the themes of this section is the deduction of bounds for the rank of a primitive permutation group from hypotheses about the stabiliser of a point.

Section 6 discusses some aspects of algebraic relations (such as commutativity) between basis matrices. There are no general results here, since none seem to exist, and it appears to be an untilled field full of thorny problems; a special case is considered, to illustrate the ideas.

For the general theory of finite permutation groups, the reader is referred to the book by WIELANDT [45]. It should be mentioned that, unless specifically stated otherwise, all permutation groups are finite.

1. INTRODUCTION AND NOTATION

With the exception of section 4, the point of view throughout these notes is a group-theoretic one. We take a transitive permutation group, associate with it a class of directed or undirected graphs, and use the graphs to get information about the group. All the graphs involved have the property that their automorphism groups act transitively on vertices and on directed edges, and indeed any graph Γ with this property is covered by our remarks (simply by taking $G = \mathrm{Aut}\,\Gamma$ as a permutation group on the vertices of Γ). So the process is essentially a two-way one, and both points of view should be kept in mind. Of course, if we start with a graph, the general machinery produces a whole family of graphs, whose interrelations can be studied.

Suppose G is a transitive permutation group on a set Ω. G has a natural action on $\Omega \times \Omega$, defined by

$$(\alpha,\beta)g = (\alpha g,\beta g)$$

for all $\alpha,\beta \in \Omega$, $g \in G$. So $\Omega \times \Omega$ is partitioned into orbits $\Gamma_0,\Gamma_1,\ldots,\Gamma_{r-1}$ under the action of G. These are called *suborbits* of G, and their number r is the *rank*. (If $|\Omega| > 1$ then $r \geq 2$, since the diagonal $\Gamma_0 =$ $= \{(\alpha,\alpha) \mid \alpha \in \Omega\}$ is always an orbit. Note that $r = 2$ if and only if G is doubly transitive on Ω.) We can regard each suborbit Γ_i as a G-invariant relation on Ω, or alternatively (if $i > 0$) as the edge set of a directed graph which admits G as a group of automorphisms transitive on vertices and directed edges. The suborbit *paired* with a given suborbit Γ (or the *converse* relation to Γ) is

$$\Gamma^* = \{(\beta,\alpha) \mid (\alpha,\beta) \in \Gamma\} .$$

Other notation in common use is Γ' (WIELANDT [45]) and $\Gamma^{\cup}$ (HIGMAN [23]). Γ is *self-paired* or *symmetric* if $\Gamma = \Gamma^*$. If Γ is self-paired, we can regard it as an undirected graph.

The suborbits are the minimal G-invariant binary relations on Ω, and any G-invariant relation is the union of a subcollection of them. (Sometimes I shall call such a relation a *generalized suborbit*.) Of particular importance are the G-invariant equivalence relations. There are always at least two of these, the diagonal Γ_0 and the whole of $\Omega \times \Omega$. Given any non-diagonal suborbit Γ, there is a unique G-invariant equivalence E which is minimal subject to containing Γ. This is "generated" by Γ in a certain sense which we shall make precise. Consider the undirected graph corresponding to Γ (that is, the relation $\Gamma \cup \Gamma^*$). E contains $\Gamma \cup \Gamma^*$ (by symmetry) and so it contains the relation of being connected by a path in $\Gamma \cup \Gamma^*$ (by transitivity). However, this latter relation is an equivalence, and so is equal to E. SIMS showed that we can give a simpler description of E.

<u>THEOREM 1.1</u>. *The smallest* G-*invariant equivalence relation containing* Γ *is the relation of being connected by a directed path in* Γ.

<u>PROOF</u>. For $\alpha \in \Omega$, let $E'(\alpha)$ be the set of points reachable by directed Γ-paths from α. Clearly, if $\beta \in E'(\alpha)$, then $E'(\beta) \subseteq E'(\alpha)$. But if g is an element of G with $\alpha g = \beta$, then $E'(\alpha)g = E'(\beta)$, and so $|E'(\alpha)| = |E'(\beta)|$. It follows that $E'(\alpha) = E'(\beta)$, and in particular $\alpha \in E'(\beta)$. Thus E' is symmetric. It is obviously reflexive and transitive, so it is the minimal equivalence relation E containing Γ. □

G. GLAUBERMAN, who lives in Chicago, formulated this result as follows. The graph Γ (interpreted as a one-way system) has the property that, whenever it is possible to walk from α to β, it is possible to drive. Clearly not all directed graphs have this property! It is a consequence of our transitivity assumption. (This formulation was communicated to me by L. SCOTT.)

We can phrase the result in yet another way. The set of G-invariant relations on Ω, with the operation of composition, forms a semigroup. The union of all members of the subsemigroup generated by Γ is the smallest equivalence relation containing Γ. We may say this is the equivalence relation *generated* by Γ.

It is not difficult to show that if $(\alpha,\beta) \in \Gamma$ and $g \in G$ satisfy $\alpha g = \beta$, then the subgroup of G fixing the equivalence class of E containing α is the subgroup generated by G_α and g. (G_α is the subgroup of G fixing α.) In particular, Γ is connected if and only if $\langle G_\alpha, g\rangle = G$.

The group G is said to be *primitive* if the only G-invariant equivalence relations are the trivial ones Γ_0 and $\Omega \times \Omega$. By theorem 1.1, G is primitive if and only if the graph of every non-diagonal suborbit is connected; this occurs if and only if $\langle G_\alpha, g\rangle = G$ for all $g \notin G_\alpha$; that is, if and only if G_α is a maximal subgroup of G.

As defined in WIELANDT's book [45], a suborbit is an orbit of G_α in Ω. There is a natural one-to-one correspondence between the two concepts: if $\Gamma_0,\ldots,\Gamma_{r-1}$ are the G-orbits in $\Omega \times \Omega$, then $\Gamma_0(\alpha),\ldots,\Gamma_{r-1}(\alpha)$ are the G_α-orbits in Ω, where

$$\Gamma_i(\alpha) = \{\beta \mid (\alpha,\beta) \in \Gamma_i\} = \text{set of points joined "from" } \alpha \text{ in } \Gamma_i.$$

Conversely, Γ_i is the G-orbit containing (α,β), for some $\beta \in \Gamma_i(\alpha)$. The *subdegree* associated with Γ_i is $|\Gamma_i(\alpha)| = |\Gamma_i| \,/\, |\Omega|$, the valency of the graph Γ_i. It follows that paired suborbits have the same subdegree.

The graph-theoretic interpretation of suborbits often provides simple proofs of old theorems on suborbits and subdegrees. As an example I give SIMS' proof of Theorem 17.4 in WIELANDT's book.

THEOREM 1.2. *If* G *is primitive, with subdegrees* $1=n_0,n_1,\ldots,n_{r-1}$ *(in increasing order), then* $n_1 n_{i-1} \geq n_i$ *for* $i=1,\ldots,r-1$.

PROOF. Let Δ be the generalised suborbit $\Gamma_0 \cup \ldots \cup \Gamma_{i-1}$. Since G is primitive, there is a Γ_1-edge (γ,δ) from a point $\gamma \in \Delta(\alpha)$ to a point $\delta \notin \Delta(\alpha)$; say $\gamma \in \Gamma_j(\alpha)$, $\delta \in \Gamma_k(\alpha)$, with $j < i \leq k$. The number of Γ_1-edges with initial point in $\Gamma_j(\alpha)$ and terminal point in $\Gamma_k(\alpha)$ is at most $n_1 n_j \leq n_1 n_{i-1}$, and also at least $n_k \geq n_i$; so $n_1 n_{i-1} \geq n_i$. □

We do not really need primitivity; connectedness of Γ_1 will suffice. If Γ_1 is self-paired, a simple change in the argument gives the result $(n_1-1)n_{i-1} \geq n_i$ for $i=2,\ldots,r-1$.

If Γ_i and Γ_j are suborbits, their composition $\Gamma_i \circ \Gamma_j$ is the generalised suborbit consisting of the pairs (α,β) for which there exists a point γ with $(\alpha,\gamma) \in \Gamma_i$, $(\gamma,\beta) \in \Gamma_j$. For convenience later on, we make the additional arbitrary assumption that $\alpha \neq \beta$. (This is relevant only if $\Gamma_j = \Gamma_i^*$.)

Since G is transitive on directed edges of each Γ_k, the number of points γ for which $(\alpha,\gamma) \in \Gamma_i$, $(\gamma,\beta) \in \Gamma_j$, depends only on which suborbit contains the pair (α,β); we shall let a_{ijk} denote this number, if $(\alpha,\beta) \in \Gamma_k$. Note that $a_{ijk} = 0$ if $k \neq 0$ and $\Gamma_k \not\subseteq \Gamma_i \circ \Gamma_j$. These intersection numbers satisfy various identities, which can be verified by counting arguments:

$$
(1.1) \qquad \begin{cases}
a_{ij0} = n_i\, \delta_{ij^*}\,; & a_{i0j} = a_{0ij} = \delta_{ij}\,; \\
a_{ijk} = a_{j^*i^*k^*}\,; & n_k\, a_{ijk} = n_i\, a_{kj^*i}\,; \\
\displaystyle\sum_{j=0}^{r-1} a_{ijk} = n_i\,; & \\
\displaystyle\sum_{t=0}^{r-1} a_{lit}\, a_{tjm} = \sum_{k=0}^{r-1} a_{lkm}\, a_{ijk}\,, &
\end{cases}
$$

where n_i is the subdegree of Γ_i, δ_{ij} the Kronecker delta, and $\Gamma_{i^*} = \Gamma_i^*$.

The last of these relations is proved by counting quadrilaterals in two ways, as shown in fig. 1.1.

Fig. 1.1

The *basis matrix* C_i corresponding to Γ_i is the matrix with rows and columns indexed by Ω, with (α,β) entry 1 if $(\alpha,\beta) \in \Gamma_i$, 0 otherwise. The basis matrices (adjacency matrices of the graphs) satisfy

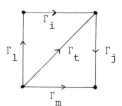

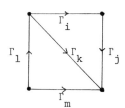

$$(1.2)\quad \begin{cases} C_0 = I, \qquad \sum_{i=0}^{r-1} C_i = J \text{ (the all 1 matrix)}, \\ C_i^T = C_{i^*}, \\ C_i\, C_j = \sum_{k=0}^{r-1} a_{ijk}\, C_k\,. \end{cases}$$

The last equation is proved by observing that the (α,β) entry of C_iC_j is a_{ijk} if $(\alpha,\beta) \in \Gamma_k$. It follows that the span of $C_0,\ldots,C_{r-1}$ (over $\mathbb{C}$) is an algebra of dimension r, which is semi-simple (by the second equation). Furthermore, G can be represented as a linear group on the vector space $\mathbb{C}\Omega$ with basis indexed by Ω (elements of G permute the basis vectors in the natural way); it is easy to verify that a given matrix commutes with all the permutation matrices if and only if its (α,β) entry depends only on the suborbit containing (α,β), that is, if and only if it lies in the span of $C_0,\ldots,C_{r-1}$. Thus these matrices span the *centraliser algebra* of the permutation matrices. The fifth equation of (1.1) can be interpreted as the associativity of basis matrices:

$$(C_lC_i)C_j = \sum_{m=0}^{r-1}\Big(\sum_{t=0}^{r-1} a_{lit}a_{tjm}\Big)\, C_m = \sum_{m=0}^{r-1}\Big(\sum_{k=0}^{r-1} a_{lkm}a_{ijk}\Big)\, C_m = C_l(C_iC_j).$$

This is an example of the interaction between the algebra and the graph theory. Another is the fact that C_i and C_j commute if and only if $a_{ijk} = a_{jik}$ for all k (see also section 6).

Since the centraliser algebra is semi-simple, it is a direct sum of matrix algebras of degrees $e_0,\ldots,e_s$ over $\mathbb{C}$. Since its dimension is r, we have $\sum e_i^2 = r$; and we can assume that $e_0 = 1$, since (n_i) is a 1-dimensional direct summand of C_i. $\mathbb{C}\Omega$ is a natural module for this algebra, and is a direct sum of irreducible submodules; suppose the natural module for the i-th summand of the algebra has multiplicity g_i. Then $|\Omega| = \sum e_ig_i$, and $g_0 = 1$ (by the PERRON-FROBENIUS theorem). Double centraliser theory shows that the numbers e_i and g_i are respectively multiplicities and degrees of the irreducible representations of G which occur in the permutation representation on $\mathbb{C}\Omega$.

If we know enough about the algebra (for example, all intersection numbers), we can in principle compute the numbers e_i and g_i. First we compute the irreducible representations of the algebra. This is made easier by the fact that there is an isomorphism from the centraliser algebra to the *intersection algebra* whose elements are $r \times r$ matrices (and so in general much smaller). The intersection matrix M_i of Γ_i is defined to have (l,m) entry a_{lim}. Then

$$(M_iM_j)_{lm} = \sum_{t=0}^{r-1} a_{lit}\, a_{tjm} = $$
$$= \sum_{k=0}^{r-1} a_{lkm}\, a_{ijk} =$$
$$= \left(\sum_{k=0}^{r-1} a_{ijk}\, M_k\right)_{lm} ,$$

so $M_iM_j = \sum_{k=0}^{r-1} a_{ijk} M_k$, and the map $C_i \mapsto M_i$ is an algebra isomorphism.

In the case where all e_i are equal to unity (so the centraliser algebra is commutative), an irreducible representation associates with each basis matrix an eigenvalue, and the g_i are the multiplicities of these eigenvalues. The equations $\mathrm{Trace}(C_i) = |\Omega|\delta_{i0}$ are now a set of linear equations for the g_i.

The main applications of the theory are in finding non-trivial "integrality conditions" on the intersection numbers, by computing the degrees and multiplicities and observing that they must be integers. ([20],[22], [4], for example.) Other papers (such as [43],[30]) start with the degrees and multiplicities and compute the intersection numbers. Neither of these will be our chief concern.

There is a little more information that can be gleaned from the permutation character π of G. π is the character of the permutation representation of G on $\mathbb{C}\Omega$, that is, $\pi(g)$ is the number of fixed points of g, for $g \in G$. If the irreducible constituents of π appear with multiplicities $e_1,\ldots,e_s$, we know that $r = \sum e_i^2$ and the principal character 1_G has multiplicity $e_0 = 1$. (Thus, G is doubly transitive if and only if $\pi = 1_G + \chi$, where χ is irreducible.) We can also obtain a formula for the number of self-paired suborbits. The FROBENIUS-SCHUR number n_χ of an irreducible character χ of G is defined to be +1 if χ is of the first kind (the character of a real representation), -1 if χ is of the second kind (a real character not afforded by any real representation), and 0 if χ is of the third kind (complex-valued). This can be extended to an arbitrary character by linearity; thus n_π is the number of constituents of π of the first kind minus the number of the second kind (counted with multiplicity). Now $n_\pi = (|G|)^{-1} \sum_{g\in G} \pi(g^2)$ (see [15; 3.5]), from which it follows that n_π is the number of self-paired suborbits of G (see [10]). (Note that this number is at least 1.) An irreducible of the second kind must occur with even multiplicity in any real representation. So if all the multiplicities e_i are 1, then # self-paired suborbits = # real irreducible constituents of π, # non-self-paired suborbits = # non-real irreducible constituents of π.

2. PAIRED SUBORBITS

One of our main concerns will be with relations among the different *subconstituents* (transitive constituents of the stabiliser of a point) in a transitive permutation group. In complete generality there can be no relationships at all, as the following example shows. Let H be a permutation group on a set Δ, not necessarily transitive. Let $V = F\Delta$ be a vector space over $F = GF(p)$ (p prime) with basis indexed by Δ; then H acts as a group of linear transformations on V. Let $G = \{x \mapsto xh+v \mid h \in H, v \in V\}$. G is transitive on V, and $G_0 \cong H$ has a union of orbits $\Delta(0)$ on which it acts as H does on Δ. Of course, if H is intransitive on Δ, then none of the graphs corresponding to suborbits in $\Delta(0)$ is connected, and a connected component of one of them contains none of the others. We might expect that two suborbits which generate the same equivalence relation might be better behaved. As shown in theorem 1.1, paired suborbits always fulfil this condition. And indeed paired subsconstituents always share at least one property: they have the same degree.

There is a general machine for producing bad behaviour in paired suborbits. Suppose H is a group, and K and L are isomorphic subgroups of H which are embedded "differently" in H. Embed H in G, the symmetric group of degree $|H|$, by means of its regular representation. K and L are each represented by $|H:K|$ times the regular representation, and so are isomorphic as permutation groups. Thus there is an element $g \in G$ such that $K^g = L$. Indeed there are many such g, and we can usually choose one such that $H \cap H^g = L$ (and then $H \cap H^{g^{-1}} = K$). Represent G as a permutation group on the right cosets of H (acting by right multiplication), and let α, γ, γ' be the cosets H, Hg, Hg^{-1}. Then $G_\alpha = H$, $G_{\alpha\gamma} = L$, $G_{\alpha\gamma'} = K$, and $(\gamma',\alpha)g = (\alpha,\gamma)$. So γ and γ' belong to paired suborbits affording the representations of H on the right cosets of L and K respectively.

EXAMPLES.

1. $H = S_4$, $K = <(12),(34)>$, $L = <(12)(34),(13)(24)>$. One suborbit affords the (rank 3) representation of S_4 on unordered pairs, while the other is not faithful (since $L \lhd H$) but affords the regular representation of $H/L \cong S_3$.
2. $H = M_{12}$, K and L are non-conjugate subgroups isomorphic to PSL(2,11); one is maximal and the other is not. (Thus one constituent is primitive and the other is imprimitive.)
3. $H = S_n \times S_k$ $(n > 2k)$, $K \cong L \cong S_{n-k} \times S_k$. One constituent is the representation of S_n on ordered k-tuples of distinct elements, while the other is faithful.

These examples, which could be multiplied, indicate that many elementary properties (such as order, rank, primitivity, and regularity) are not necessarily shared by paired subconstituents (or "preserved under pairing"). The first positive result was proved by SIMS (see [31]):

THEOREM 2.1. *If* G *is transitive on* Ω *and* Γ *is a suborbit with subdegree greater than* 1, *then* $G_\alpha^{\Gamma(\alpha)}$ *and* $G_\alpha^{\Gamma^*(\alpha)}$ *have a common non-trivial epimorphic image.*

I shall outline SIMS' proof of this result, which goes by contradiction. First, note that if a group G acts on Γ and Δ, and if K_Γ and K_Δ are the kernels of these actions, then $G^\Gamma \cong G/K_\Gamma$ and $G^\Delta \cong G/K_\Delta$; $G/K_\Gamma K_\Delta$ is a common epimorphic image of G^Γ and G^Δ. If the only such epimorphic image is trivial, then $G = K_\Gamma K_\Delta$, whence $G^\Delta = K_\Gamma^\Delta$ and *a fortiori* $G^\Delta = G_\gamma^\Delta$ for $\gamma \in \Gamma$.

Suppose G, acting on Ω with suborbit Γ, is a counterexample. We prove by induction the statements about the Γ-graph:

A_k : G permutes paths of length k transitively;

B_k : if $(\alpha_1,\ldots,\alpha_k)$ is a path of length k-1, then

$$G_{\alpha_1\ldots\alpha_k}^{\Gamma(\alpha_k)} = G_{\alpha_k}^{\Gamma(\alpha_k)} \text{ and } G_{\alpha_1\ldots\alpha_k}^{\Gamma^*(\alpha_1)} = G_{\alpha_1}^{\Gamma^*(\alpha_1)} .$$

A_1 and B_1 are trivially true. If A_k and B_k hold, and $(\alpha_1,\ldots,\alpha_k)$ is as in B_k, then $G_{\alpha_1\ldots\alpha_k}^{\Gamma(\alpha_k)}$ and $G_{\alpha_1\ldots\alpha_k}^{\Gamma^*(\alpha_1)}$ have no common non-trivial epimorphic image (by B_k and hypothesis), so $G_{\alpha_1\ldots\alpha_k}^{\Gamma(\alpha_k)} = G_{\alpha_0\ldots\alpha_k}^{\Gamma(\alpha_k)}$ and $G_{\alpha_1\ldots\alpha_k}^{\Gamma^*(\alpha_1)} = G_{\alpha_1\ldots\alpha_{k+1}}^{\Gamma^*(\alpha_1)}$ for $\alpha_0 \in \Gamma^*(\alpha_1)$, $\alpha_{k+1} \in \Gamma(\alpha_k)$. Thus A_{k+1} and B_{k+1} hold.

So G acts transitively on paths of length k for all k, a contradiction since the number of such paths is $|\Omega|\ |\Gamma(\alpha)|^k$ and goes to infinity with k.

SIMS' idea was employed by CAMERON [8,I] to show that a variety of properties of transitive groups are "preserved under pairing". The most important of these are double transitivity and the property of containing the alternating group. The first of these can be expressed thus. Let Γ be a directed graph whose automorphism group is transitive on vertices and directed edges. Then the properties of transitivity on the two kinds of figure shown in figure 2.1 are equivalent.

Fig. 2.1

The most general result in [8;I] asserts that a property P of transitive groups of degree d > 1 is preserved under pairing if it satisfies the two conditions

(i) $G^\Omega \in P$, $G \le H$ implies $H^\Omega \in P$;

(ii) if G acts transitively on Γ and Δ, with $|\Gamma| = |\Delta| = d$, $G^\Delta \in P$, $G^\Gamma \notin P$, and $\gamma \in \Gamma$, then $G_\gamma^\Delta \in P$.

More recently I have found a more general necessary condition; it is awkward to state, but implies (for example) that if n is large enough (for given k), the property of acting as S_n or A_n on k-element subsets is preserved under pairing. Compare this with examples 1 and 3 above.

A significant advance was made by KNAPP [26], who found an alternative method of proof. Again I shall illustrate by proving SIMS' result (theorem 2.1). Suppose a group G is a counterexample to the theorem; let Ω be a set affording a permutation representation of G of maximal degree for which the theorem is false, and Γ the suborbit involved. If $(\gamma,\alpha),(\alpha,\beta) \in \Gamma$, then as before $G_{\alpha\gamma}^{\Gamma(\alpha)} = G_\alpha^{\Gamma(\alpha)}$ and $G_{\alpha\beta}^{\Gamma^*(\alpha)} = G_\alpha^{\Gamma^*(\alpha)}$. Thus G is transitive on Γ, and $\Phi = \{((\gamma,\alpha),(\alpha,\beta)) \in \Gamma \times \Gamma\}$ is a suborbit, with $G_{(\gamma,\alpha)}^{\Phi(\gamma,\alpha)} \cong G_{\alpha\gamma}^{\Gamma(\alpha)} = G_\alpha^{\Gamma(\alpha)}$ and $G_{(\alpha,\beta)}^{\Phi^*(\alpha,\beta)} \cong G_{\alpha\beta}^{\Gamma^*(\alpha)} = G_\alpha^{\Gamma^*(\alpha)}$. Thus the theorem is false for the action of G on Γ, and $|\Gamma| = |\Omega||\Gamma(\alpha)| > |\Omega|$, contradicting the maximality of $|\Omega|$.

KNAPP was able to extend the argument to show

THEOREM 2.2. *Suppose P is a property of transitive groups of degree* $d > 1$ *such that*

(i) *P is preserved under pairing;*

(ii) *groups in P are primitive;*

(iii) *if* Y *is a regular normal subgroup of a group* $X \in P$, *then* X/Y *has no normal subgroup isomorphic to* Y.

Let G *be a transitive permutation group on* Ω *with suborbit* Γ *such that* $|\Gamma(\alpha)| = d$ *and* $G_\alpha^{\Gamma(\alpha)} \in P$. *Then* $G_\alpha^{\Gamma(\alpha)} \cong G_\alpha^{\Gamma^*(\alpha)}$ *(as abstract group).*

Note that (ii) is satisfied by most known properties satisfying (i), while (iii) is usually a very weak requirement. Note also that we cannot conclude that the representations of G_α on $\Gamma(\alpha)$ and $\Gamma^*(\alpha)$ are equivalent, even under much stronger hypotheses.

It might be expected that stronger results might hold about paired suborbits in primitive groups. However, nothing seems to be known except for consequences of the more general relations between subconstituents in primitive groups, to be discussed in the next section. Paired subconstituents need not be equivalent, even when they are multiply transitive; but the worst examples I know are very much milder than the general examples discussed previously.

Even when paired constituents are isomorphic as permutation groups, they may look quite different geometrically. For example, let G be a group doubly transitive on Ω, and Δ a fixed set of $G_{\{\alpha,\beta\}}$. Let $S_{(2)}$ denote for the moment the collection of 2-element subsets of a set S. G is transitive on $\Omega_{(2)}$, and has a generalized suborbit Γ defined by $\Gamma(\{\alpha,\beta\}) = \Delta_{(2)}$. We might expect that there would be a fixed set Δ' for $G_{\{\alpha,\beta\}}$ in Ω such that $\Gamma^*(\{\alpha,\beta\}) = \Delta'_{(2)}$, but this need not be the case; it can occur that the members of $\Gamma^*(\{\alpha,\beta\})$ are pairwise disjoint, even when this set is a single suborbit! See ATKINSON [1;II].

In multiply transitive groups there are possibilities of more general kinds of pairing. Suppose G is k-transitive on Ω, and $|\Omega| > k$. As in section 1 there is a natural correspondence between G-orbits on ordered (k+1)-tuples of distinct elements of Ω and $G_{\alpha_1 \ldots \alpha_k}$-orbits on $\Omega - \{\alpha_1, \ldots, \alpha_k\}$. Now there are potentially (k+1)! orbits "paired" with a given one, corresponding to the (k+1)! possible rearrangements of a (k+1)-tuple. (The number of these which are distinct is the index in S_{k+1} of the group of permutations induced by G on a (k+1)-tuple in the given orbit.) The results of this section apply to any two orbits "paired" by a transposition, and so to any two "paired" orbits, since transpositions generate the symmetric group. However, we might expect that for $k > 1$ stronger results would hold. This has not been investigated.

It is clear that SIMS' and KNAPP's proofs depend on the finiteness of Ω and G. If Ω is infinite, then any kind of bad behaviour is possible. Thus any two transitive permutation groups (not necessarily of the same degree) can occur as paired subconstituents in a transitive permutation group. The counterexample machine works even better. Instead of embeddings in symmetric groups, we can simply apply the HIGMAN-NEUMANN-NEUMANN construction [24]; BRITTON's lemma [6] guarantees that the element g exists. In bad cases, SIMS' argument often shows that G acts transitively on paths of length k for every k; yet G may have uncountably many orbits on one-way infinite paths. The very simplest example is the directed tree in which every vertex has one edge entering it and two leaving it.

3. MORE GENERAL RELATIONS BETWEEN SUBCONSTITUENTS

Suppose we are concerned with permutation groups with a given subdegree v (or with graphs of a given valency v having automorphism groups transitive on vertices and directed edges). We know then that the stabiliser of a point has a homomorphism into the symmetric group on v letters. It is very important to get hold of the kernel of this homomorphism; but this is a difficult problem. SIMS conjectured that at least the size of the kernel is bounded, provided the group is primitive.

SIMS' CONJECTURE. *There is a function* f *on the natural numbers with the property that, if* G *is a primitive permutation group with a suborbit* Γ *with* $|\Gamma(\alpha)| = v$, *then* $|G_\alpha| \leq f(v)$.

We might expect a similar function to exist under the weaker hypothesis that the Γ-graph is connected; but this is false, as can be seen by considering the directed graph with vertices (i,j), where $0 \leq i \leq n-1$, $0 \leq j \leq m-1$, and edges $((i,j),(i+1,k))$ for all i,j,k, where the first coor dinate is taken mod n. Here $|\Gamma(\alpha)| = m$ but $|G_\alpha| = (m!)^{n-1}(m-1)!$ The underlying undirected graph has a similar property.

We observe two simple cases: $f(1) = 1$, $f(2) = 2$. For if $|\Gamma(\alpha)| = 1$ then the Γ-graph is a directed polygon. If $|\Gamma(\alpha)| = 2$ and $\Delta = \Gamma^* \circ \Gamma$, then

$|\Delta(\alpha)| = 2$ and Δ is self-paired, so the Δ-graph is an undirected polygon. SIMS [34;I] showed that $f(3) = 48$, and QUIRIN [31] reports an unpublished result of SIMS and THOMPSON that $f(4) = 2^4 3^6$. No other values are known.

However, we can find structural restrictions on G_α when we are given $G_\alpha^{\Gamma(\alpha)}$. The first, discovered by JORDAN, is Theorem 18.4 of WIELANDT:

<u>THEOREM 3.1</u>. *If the Γ-graph is connected and p is a prime dividing $|G_\alpha|$, then p divides $|G_\alpha^{\Gamma(\alpha)}|$.*

<u>PROOF</u>. Suppose $p \nmid |G_\alpha^{\Gamma(\alpha)}|$, and suppose g is an element of order p fixing α. By assumption, g fixes every point of $\Gamma(\alpha)$; then g fixes every point at distance 2 from α; and so on. It follows that $g = 1$, a contradiction. So $p \nmid |G_\alpha|$. □

WIELANDT generalized this result to composition factors, assuming G is primitive; a composition factor of G_α is a composition factor of some subgroup of $G_\alpha^{\Gamma(\alpha)}$. It need not, however, be a composition factor of $G_\alpha^{\Gamma(\alpha)}$ itself. For example, let $G = S_{2n-1}$ act on the set of $(n-1)$-element subsets. $G_\alpha = S_{n-1} \times S_n$, but if $\Gamma(\alpha)$ is the set of $(n-1)$-element subsets disjoint from α, then $G_\alpha^{\Gamma(\alpha)} = S_n$, and (if $n \geq 6$) A_{n-1} is a composition factor of G_α but not of $G_\alpha^{\Gamma(\alpha)}$. Note that A_{n-1} is a composition factor of the stabiliser of a point in $G_\alpha^{\Gamma(\alpha)}$. This is quite general, as a result of WIELANDT shows (see [21, Theorem 1.1]).

Suppose K is a stable functor on finite groups, that is, K associates with any group a characteristic subgroup and has the property that $K(X) \leq Y \triangleleft X$ implies $K(Y) = K(X)$. KNAPP [26] shows

<u>THEOREM 3.2</u>. *If G is primitive with a suborbit Γ, and K is a stable functor such that $K(G_{\alpha\beta})$ acts trivially on both $\Gamma(\alpha)$ and $\Gamma^*(\beta)$ (where $(\alpha,\beta) \in \Gamma$), then $K(G_{\alpha\beta}) = 1$.*

The proof is very similar to that of theorem 3.1: $K(G_{\alpha\beta})$ fixes every point reachable from α by an "alternating" path of type $\Gamma, \Gamma^*, \Gamma, \Gamma^*, \ldots$, and this set includes a connected component of $\Gamma \circ \Gamma^*$. Note that connectedness of Γ is not enough here, as the previous example shows.

Examples of stable functors include

(i) O^Λ, where Λ is a set of finite simple groups ($O^\Lambda(X)$ is the smallest normal subgroup Y of X such that all composition factors of X/Y lie in Λ). This includes O^π, for π a set of primes (smallest normal subgroup whose index is a π-number);

(ii) $\mathcal{O}_\Lambda$, largest normal subgroup whose composition factors belong to Λ. This includes $\mathcal{O}_\pi$, largest normal π-subgroup;

(iii) F, the Fitting subgroup (largest normal nilpotent subgroup).

Apply theorem 3.2 to $\mathcal{O}^\Lambda$, where Λ is the set of composition factors of $G_{\alpha\beta}^{\Gamma(\alpha)}$ and $G_{\alpha\beta}^{\Gamma^*(\beta)}$, $(\alpha,\beta) \in \Gamma$. If $K(\alpha)$, $K^*(\alpha)$ are the kernels of the actions of G_α on $\Gamma(\alpha), \Gamma^*(\alpha)$ respectively, then $\mathcal{O}^\Lambda(G_{\alpha\beta}) \leq K(\alpha) \cap K^*(\beta)$, so $\mathcal{O}^\Lambda(G_{\alpha\beta}) = 1$. In particular, the composition factors of $K(\alpha)$ belong to Λ. This is the cited result of WIELANDT.

Deeper results on the structure of G_α in a primitive group G can be obtained using normaliser and centraliser theorems of WIELANDT for subnormal subgroups. In this way WIELANDT [47] was able to show that, if $M(\alpha)$ is the subgroup of G fixing $\{\alpha\} \cup \Gamma(\alpha) \cup (\Gamma \circ \Gamma^*)(\alpha)$ pointwise, and $M^*(\alpha)$ is similarly defined, then one at least of $M(\alpha)$ and $M^*(\alpha)$ is a p-group, for some prime p. This has the following consequence, closely related to SIMS' conjecture:

THEOREM 3.3. *If* G *is primitive and* Γ *is a suborbit with* $|\Gamma(\alpha)| = v$, *then, for some prime* p, G_α *has a normal* p-*subgroup of index at most* $v!((v-1)!)^v$.

The fact that G has a normal p-subgroup of bounded index (given v) was first proved by THOMPSON [39]. KNAPP [26] has shown that, if $G_\alpha^{\Gamma(\alpha)}$ is 2-homogeneous (transitive on unordered pairs) or if v is prime, then the bound can be improved to $v!(v-1)!$, and we have seen that this is best possible. In several cases KNAPP is able to determine the precise structure of G_α when $G_\alpha^{\Gamma(\alpha)}$ is given.

Sometimes it is possible to prove that the normal p-subgroup referred to must be trivial. Applying theorem 3.2 with $K = F$ (or $\mathcal{O}_p$), we see that, if $G_{\alpha\beta}^{\Gamma(\alpha)}$ and $G_{\alpha\beta}^{\Gamma^*(\beta)}$ have no normal nilpotent subgroup then neither does $G_{\alpha\beta}$; so in this case $M(\alpha)$ or $M^*(\alpha) = 1$, and we deduce that
$|G_\alpha| \leq v!((v-1)!)^v$, or (if $G_\alpha^{\Gamma(\alpha)}$ is 2-homogeneous or v is prime)
$|G_\alpha| \leq v!(v-1)!$

This is particularly useful when $G_\alpha^{\Gamma(\alpha)}$ and $G_\alpha^{\Gamma^*(\alpha)}$ are 2-primitive, in particular when $v = p+1$ for some prime p dividing $|G_\alpha|$. In this case, if $M(\alpha)$ and $M^*(\alpha)$ are non-trivial, then the stabiliser of a point in $G_\alpha^{\Gamma(\alpha)}$ or $G_\alpha^{\Gamma^*(\alpha)}$ has a regular normal subgroup. Doubly transitive groups with this property have been determined by HERING, KANTOR & SEITZ [19], so we can pin down the structure of $G_\alpha^{\Gamma(\alpha)}$ very precisely. This has been done by GARDINER [17;I] in the case where Γ is self-paired and $v = p+1$;

he was able to determine the structure of G_α. The result will be discussed further in the next section.

Application of these techniques, together with theorems about abstract finite groups, to the problem of primitive permutation groups with small subdegrees has resulted in a complete determination of such groups with a subdegree 3 (SIMS [34;I], WONG [48]), and of such groups with a subdegree 4 under the extra hypothesis that $G_\alpha^{\Gamma(\alpha)}$ is a 2-group (SIMS [34;II]) or the alternating group (QUIRIN [31]); partial results on subdegree 5 are obtained by QUIRIN [31] and KNAPP [26].

There are a number of results which assert that, under certain conditions, G_α acts faithfully on a generalized suborbit $\Delta(\alpha)$. This is true if G is primitive on Ω and one of the following holds:

(i) G_α is primitive on every suborbit not contained in $\Delta(\alpha)$ (MANNING [27]);

(ii) Δ contains Γ or Γ^* for every suborbit Γ (WIELANDT [44]);

(iii) G_α is 2-primitive on $\Delta(\alpha)$, $\Delta = \Delta^*$, and either G_α is primitive on $(\Delta \circ \Delta)(\alpha)$, or $|(\Delta \circ \Delta)(\alpha)| \neq |\Delta(\alpha)|(|\Delta(\alpha)|-1)$ (MANNING [27]).

A curious open conjecture asserts that the same result holds if G is primitive and G_α acts regularly on $\Delta(\alpha)$. This is known to be true if the stabiliser of a point in $\Delta^*(\alpha)$ fixes additional points there ([45, Theorem 18.6]), and so in particular if Δ is self-paired. It is false if we assume only that the Δ-graph is connected, even with the extra hypothesis; see example 1 of section 2.

Finally, I mention an extension of some of these results to doubly transitive groups due to SIMS [35]. Theorems like those of this section hold for the stabiliser of two points in a doubly transitive group G under a weaker assumption than 2-primitivity of G, namely the assumption that G is not an automorphism group of a block design with $\lambda = 1$. For example, suppose G is 2-transitive, $\Gamma(\alpha,\beta)$ is an orbit of $G_{\alpha\beta}$, and p is a prime dividing $|G_{\alpha\beta}|$ but not $|G_{\alpha\beta}^{\Gamma(\alpha,\beta)}|$. Then $O^{p'}(G_{\alpha\beta})$ is weakly closed in $G_{\alpha\beta}$ with respect to G, since it is generated by all the elements of p-power order in $G_{\alpha\beta}$, and it fixes $\Gamma(\alpha,\beta)$ pointwise; then WITT's lemma applies. (See result 2.B.1 of KANTOR's talk at this meeting [25] for discussion, or Theorem 9.4 of WIELANDT [45].)

4. DIGRESSION ON TRANSITIVITY IN GRAPHS

Throughout this section, Γ is a connected undirected graph whose automorphism group G is transitive on vertices and directed edges. (Thus we fix attention on a particular suborbit Γ, replace primitivity of G by connectedness of Γ, but assume that Γ is self-paired.)

Suppose that G is transitive on (ordered) paths of length 2. If Γ contains a triangle, then any two points joined by a path of length 2 are adjacent, and Γ is a complete graph K_{v+1}. We shall ignore this possibility, and assume that Γ contains no triangles. Let Δ be the graph with the same vertex set as Γ, in which two vertices are adjacent if and only if they are joined by a path of length 2 in Γ. (That is, $\Delta = \Gamma \circ \Gamma$.). Then G acts transitively on vertices and directed edges of Δ, and so Δ is regular with valency $v(v-1)/k$, where v is the valency of Γ and $k = |\Gamma(\alpha) \cap \Gamma(\beta)|$ for $(\alpha,\beta) \in \Delta$; k is a measure of "how many quadrilaterals Γ contains". Δ has at most two connected components, with exactly two if and only if Γ is bipartite. The first result shows how conditions on the structure of Δ influence Γ.

THEOREM 4.1. *If a connected component of Δ is a complete graph, then Γ is the incidence graph of a (possibly degenerate) self-dual symmetric design $\mathcal{D}$ satisfying*

(i) Aut $\mathcal{D}$ *is doubly transitive on the points of* $\mathcal{D}$;

(ii) *if* β *is a block, then* $\mathrm{Aut}(\mathcal{D})_\beta$ *is doubly transitive on the points incident with* β.

PROOF. Δ has two components; call the vertices of one component *points*, and those of the other *blocks*, and call a point and block *incident* if they are adjacent in Γ. The conditions are easily verified. □

At this meeting, KANTOR will discuss such designs ([25; section 8]). Of the known symmetric designs with doubly transitive automorphism groups, all are self-dual, and all except one (the HIGMAN design H_{176}) satisfy conclusion (ii) of theorem 4.1. These designs give us examples of such graphs. The degenerate designs have k = v or k = v-1; the corresponding graphs are $K_{v,v}$ and the graph obtained from $K_{v+1,v+1}$ by deleting the edges of a matching. (Note: our v and k are the design parameters k and λ.)

Strong conclusions about Δ can be drawn if k is large enough.

THEOREM 4.2. *Either* $k \le v/2$ *or a connected component of* Δ *is a complete graph.*

PROOF. Suppose $k > v/2$; take $\delta_1, \delta_2 \in \Delta(\alpha)$. Then

$$|\Gamma(\alpha) \cap \Gamma(\delta_1)| = |\Gamma(\alpha) \cap \Gamma(\delta_2)| > \tfrac{1}{2}|\Gamma(\alpha)|,$$

and so there is a point in $\Gamma(\alpha) \cap \Gamma(\delta_1) \cap \Gamma(\delta_2)$. Then $(\delta_1, \delta_2) \in \Delta$. □

To refine this result, we need to consider more systematically the sets $\Gamma(\alpha) \cap \Gamma(\delta)$. Call elements of $\Gamma(\alpha)$ and $\Delta(\alpha)$ *points* and *blocks* respectively, and call a point and block *incident* if they are adjacent in Γ. Provided $k > 1$, this gives a design $\mathcal{D}$ whose blocks can be identified with the sets $\Gamma(\alpha) \cap \Gamma(\delta)$; v and k agree with the design parameters denoted by the same letters, and $\lambda = k-1$.

THEOREM 4.3. *Either* $v(v-1)k(k^2-3k+v)^2 \le 2(v-k)^4(v-k-1)^2$ *(which implies* $k < (2v)^{4/5}$*), or a connected component of* Δ *is a complete* m*-partite graph* $K_{n,n,\dots,n}$ *for some* m,n *(that is,* K_{mn} *with the edges of* m *pairwise vertex-disjoint* K_n*'s deleted.)*

PROOF. Let G_1 and G_2 be the graphs with vertex set $\Delta(\alpha)$ defined thus. Two vertices are adjacent in G_1 if (as blocks) they are not disjoint and no block is disjoint from both; two vertices are adjacent in G_2 if and only if (as blocks) they are disjoint. Both graphs admit the vertex-transitive group $G_\alpha^{\Delta(\alpha)}$, and so are regular, with valencies d_1 and d_2 respectively. Two points which are not adjacent in G_1 are joined by a path of length at most 2 in G_2; so $1+d_2+d_2(d_2-1) \ge v(v-1)/k-d_1$. From design theory we find that $d_2 \le (v-k)^2(v-k-1)/k(k^2-3k+v)$. (See [8;II].) An easy counting argument shows that, if δ_1 and δ_2 are adjacent in G_1, then $\Delta(\delta_1)-\Delta(\alpha) = \Delta(\delta_2)-\Delta(\alpha)$. If G_1 is connected, then this holds for all δ_1, δ_2, whence the second alternative of theorem 4.3 holds. Otherwise $d_1 \le v(v-1)/2k - 1$. Putting the three inequalities together gives the result. □

Curiously, only five graphs with $k > 2$ are known which fail to satisfy the second conclusion of theorem 4.3. These are the HIGMAN-SIMS graph on 100 vertices with $v = 22$, $k = 6$, two of its subgraphs (one on 100 vertices with $v = 15$, $k = 5$, obtained by deleting the edges of two vertex-disjoint HOFFMAN-SINGLETON subgraphs; the other on 77 vertices with $v = 16$, $k = 4$, on the vertices not adjacent to a given vertex), and graphs obtained from the first and third by a "doubling" construction described below. It seems likely that the bound of theorem 4.3 can be improved substantially, and that graphs satisfying the second alternative can be described more precisely. (Note that such graphs must be bipartite.)

In passing, I note that a similar argument gives bounds for certain strongly regular graphs, such as those with no triangles and those associated with generalized quadrangles.

Our hypotheses imply that G_α acts doubly transitively on $\Gamma(\alpha)$. Stronger results can be obtained by increasing the degree of transitivity assumed. To state them, we need some definitions.

If H is a Hadamard matrix, define a graph Γ_H whose vertices are symbols (r_i,ε) and (c_j,ε), where i and j index rows and columns of H respectively and $\varepsilon = \pm 1$; (r_i,ε) and (c_j,ε') are adjacent if and only if the (i,j) entry of H is $\varepsilon\varepsilon'$. Γ_H and $\Gamma_{H'}$ are isomorphic if and only if H and H' are related by permuting and changing of rows and columns and (if necessary) transposing; thus Γ_H is a convenient "equivalence-invariant" of H.

If Γ is a non-bipartite graph satisfying our hypotheses, we can construct a bipartite graph satisfying them, with the same v and k, by the following "doubling" construction: vertices are symbols (α,ε), where α is a vertex of Γ and $\varepsilon = \pm 1$; (α,ε) and (β,ε') are adjacent if and only if α and β are adjacent in Γ and $\varepsilon\varepsilon' = -1$.

THEOREM 4.4. *Suppose that, with the hypotheses of this section,* G_α *is triply transitive on* $\Gamma(\alpha)$. *Then one of the following occurs*:

(i) $k = v$, $\Gamma = K_{v,v}$;

(ii) $k = v-1$, $\Gamma = K_{v+1,v+1}$ *with the edges of a matching removed*;

(iii) $v = 2^d$, $k = v/2$, Γ *is the incidence graph of the complementary design of* $PG_{d-1}(d,2)$;

(iv) $k = v/2$, $\Gamma = \Gamma_H$ *for some Hadamard matrix* H;

(v) $v = (\mu+1)(\mu^2+5\mu+5)$, $k = (\mu+1)(\mu+2)$ *for some positive integer* μ, *and* Γ *is strongly regular on* $(\mu+1)^2(\mu+4)^2$ *vertices, or is obtained from such a graph by "doubling"*;

(vi) $k \le 2$.

The main idea in the proof is this. We may suppose $2 < k < v-1$. The design $\mathcal{D}$ is now a 3-(v,k,μ) design, for some positive integer μ. Also, the number of blocks incident with a point $\gamma (= |\Gamma(\gamma) - \{\alpha\}|)$ is equal to the number of points different from $\gamma (= |\Gamma(\alpha) - \{\gamma\}|)$. So $\mathcal{D}$ is a symmetric

3-design (an extension of a symmetric design), so the main result of CAMERON [9] applies. (See [25; section 9], for this result and discussion of symmetric 3-designs.) If a connected component of Δ is a complete graph, then (iii) holds (see [25; 8.E.10]). Otherwise it can be shown that (iv) or (v) occurs.

THEOREM 4.5. *Suppose, with the hypotheses of this section, that* G_α *acts as the symmetric or alternating group on* $\Gamma(\alpha)$, *and that* Γ *contains a quadrilateral (that is,* $k > 1$*). Then one of the following holds*:

(i) $k = v$, $\Gamma = K_{v,v}$;

(ii) $k = v-1$, $\Gamma = K_{v+1,v+1}$ *with the edges of a matching removed;*

(iii) $k = 2$, $\Gamma = Q_v$ *(the* v*-dimensional cube)*;

(iv) $k = 2$, $v \geq 5$, Γ *is obtained from* Q_v *by identifying opposite vertices;*

(v) $k = 2$, $v = 4$, Γ *is a unique graph on* 14 *vertices;*

(vi) $k = 2$, $v = 5$, Γ *is a unique graph on* 22 *vertices.*

The graphs under (v) and (vi) are the incidence graphs of the unique (7,4,2) and (11,5,2) designs. Note that G_α acts on $\Gamma(\alpha)$ as the symmetric group in all cases except (vi). Theorems 4.3-4.5 are proved under the stronger assumption of primitivity in CAMERON [8].

Another direction in which the hypotheses can be strengthened is that of requiring transitivity on longer paths. We say $G = \mathrm{Aut}\,\Gamma$ is s-*path transitive* if it is transitive on paths of length s. If Γ is a circuit, then G is s-path transitive for every s; but for any other finite graph, there is an upper bound on the degree of path-transitivity. It has been conjectured that $s \leq 7$ for any graph which is not a circuit. If $s \geq 3$ and Γ contains a quadrilateral, then it is a complete bipartite graph $K_{v,v}$; so we shall assume that (in the previous notation) $k = 1$. The graph O_v whose vertices are the (v-1)-element subsets of a (2v-1)-element set, adjacent if disjoint, admits the 3-path transitive automorphism group S_{2v-1}; here $G_\alpha^{\Gamma(\alpha)}$ is the symmetric group S_v. However, it appears that s-path transitivity with $s \geq 4$ places severe restrictions on the structure of $G_\alpha^{\Gamma(\alpha)}$.

THEOREM 4.6. *If* $v = 3$ *and* G *is* s-*path transitive then* $s \leq 5$.

This was proved by TUTTE [41], [42]. TUTTE's work provided the inspiration for later research by SIMS on primitive permutation groups with a subdegree 3. SIMS [34] and DJOKOVIC [14] were able to extend it to the case v = p+1 (p prime) under the hypothesis that G contains a subgroup H which is s-*path regular* (that is, H is s-path transitive and only the identity fixes an s-path). In this case $|G_\alpha| = (p+1)p^{s-1}$, and $G_{\alpha\beta}$ is a p-group (for $\beta \in \Gamma(\alpha)$); calculations in this p-group show that $s \le 5$ or $s = 7$.

The general problem was attacked by GARDINER [17;I], by combining these methods with those of WIELANDT discussed in section 3. WIELANDT's result (preceding theorem 3.3) is used to produce a p-group, normal in $G_{\alpha\beta}$, in which SIMS' calculations can be carried out. GARDINER's result is

THEOREM 4.7. *If* v = p+1 (p *prime*) *and* G *is* s (*but not* s+1)-*path transitive, then* $s \le 5$ *or* $s = 7$.

In subsequent papers [18], [17;II], GARDINER has weakened the assumption on v. However, the general conjecture remains open.

EXAMPLES of graphs with s-path transitive groups.

For graphs of valency 3, all those with primitive groups have been determined by WONG [48]. Graphs of valency $v > 3$ with $s \ge 4$ seem much less common. The only known examples are the incidence graphs of certain self-dual generalized (s-1)-gons; in all cases v-1 is a prime power, and if s = 5 or s = 7 then v-1 is an odd power of 2 or 3 respectively. See TITS [40]. The resemblance of theorem 4.7 to the conclusions of FEIT & HIGMAN [16] is striking, since the methods are completely different.

Another unsolved problem is the exact relation between the degree of path-transitivity of G and the degree of transitivity of G_α on $\Gamma(\alpha)$. GARDINER has conjectured that, if G is 4-path transitive, then G_α is 2-primitive on $\Gamma(\alpha)$. The significance of this conjecture is clear from remarks in section 3.

A further kind of transitivity has been studied by BIGGS [4] and others. A graph Γ is called *distance-transitive* if it is connected and its automorphism group G acts transitively on (ordered) pairs of vertices at distance i for every i with $0 \le i \le d$, where d is the diameter of Γ. In particular, G is transitive on vertices and directed edges. However, G may not be 2-path transitive, since Γ may contain triangles. The main technique in the study of distance-transitive graphs is the computation of eigenvalues and multiplicities for the basis matrices. This is simplified by the facts that the basis matrix of Γ generates the centraliser algebra (and so it has d+1 distinct real eigenvalues), and the intersection matrix is tridiagonal. (Similar remarks apply if the transitivity of G is replaced by appropriate "coherence" or "metric regularity" conditions; graphs such as Moore graphs satisfy these conditions, and this method was used by BANNAI & ITO [2] and DAMARELL [12] to show the non-existence of such graphs with diameter and valency greater than 2. Graphs satisfying these conditions will be discussed at this meeting by DELSARTE [13] under the name of *metric* or P-*polynomial association schemes*. See also [4],[22],[23].)

Two other ideas are relevant to the study of distance-transitive graphs. The first is the observation of D.H. SMITH [36] that if Γ is distance-transitive and $G = \text{Aut}\,\Gamma$ is imprimitive, then Γ is either bipartite or antipodal. (Γ is *antipodal* if the relation of being equal or at distance d is an equivalence relation on the vertex set, where d is the diameter of Γ.) If Γ is antipodal but not bipartite, then by identifying vertices at distance d we obtain another distance-transitive graph with primitive automorphism group. Secondly, if Γ is distance-transitive with given valency, it may be possible to obtain a bound for $|G_\alpha|$ by the methods of section 3. Often such a bound can be converted into a bound for the diameter of Γ. (This is true if Γ has valency 3 [5] or is bipartite [37].) If both steps can be done, the complete determination of such graphs is reduced to a finite amount of calculation.

5. COMBINATORIAL RELATIONS AMONG SUBORBITS

Through this section we shall assume that G is primitive. We are concerned with the consequences of assumptions about the action of G_α on some or all of its orbits. The prototype is a theorem of MANNING [28], which asserts that, *if* G_α *is doubly transitive on* $\Gamma(\alpha)$, *where* $|\Gamma(\alpha)| > 2$, *then* G_α *has an orbit* $\Delta(\alpha)$ *with* $|\Delta(\alpha)| > |\Gamma(\alpha)|$, *or* G *is triply transitive*. The hypothesis implies that G acts transitively on figures of the first kind in figure 2.1 (or on paths of length 2 if Γ is self-paired), and hence that $\Delta = \Gamma^* \circ \Gamma$ is a single suborbit; $|\Delta(\alpha)| = v(v-1)/k$, where $v = |\Gamma(\alpha)|$ and $k = |\Gamma^*(\alpha) \cap \Gamma^*(\delta)|$, $(\alpha,\delta) \in \Delta$. Now our situation is very similar to that of the last section, and by similar arguments we can prove (as in [8]):

<u>THEOREM 5.1</u>. *Either* $v(v-1)k(k^2-3k+v)^2 \leq 2(v-k)^4(v-k-1)^2$ *or* G *is doubly transitive*.

This considerably strengthens MANNING's theorem, but is again probably not best possible; in all known cases except two, $k \le 2$ or G is doubly transitive. The two exceptions both have rank 3.

THEOREM 5.2. *If* G_α *is triply transitive on* $\Gamma(\alpha)$, *or if* G_α *has rank at most* 3 *on* $\Delta(\alpha)$, *then one of the following holds:*

(i) $k \le 2$;

(ii) $|\Omega| = (\mu+1)^2(\mu+4)^2$, $v = (\mu+1)(\mu^2+5\mu+5)$, $k = (\mu+1)(\mu+2)$, *for some positive integer* μ, *and* G *has rank* 3;

(iii) G *is doubly transitive.*

THEOREM 5.3. *If* $G_\alpha^{\Gamma(\alpha)}$ *is the symmetric or alternating group, then one of the following holds*:

(i) $k = 1$;

(ii) $|\Omega| = 2^{v-1}$, v *odd*, $G = V_{2^{v-1}} \cdot S_v$ *or* $V_{2^{v-1}} \cdot A_v$;

(iii) $G = S_{v+1}$ *or* A_{v+1}.

These results can be regarded as *rank-bounding theorems*; from a certain hypothesis we deduce either a (stronger) conclusion or a bound for the rank of the primitive group G. (This is valuable because of the existing techniques for studying multiply transitive groups and groups of small rank.) Such theorems have occurred from time to time in the literature; I shall digress to discuss some of them, classified according to the type of hypothesis.

1. *Hypotheses about the degree* $n = |\Omega|$

Let p denote a prime. A classic theorem of BURNSIDE [7] asserts that, if $n = p$, then G is soluble or doubly transitive. (In the former case, G is the group $\{x \mapsto a^m x+b \mid a,b \in GF(p),\ a \ne 0\}$, where m is a fixed divisor of p-1.) Related results are due to WIELANDT [43],[46]: If $n = 2p$ then G has rank at most 3, and G is doubly transitive unless $p = 2a^2+2a+1$ for some integer a; in the latter case, the subdegrees are $a(2a+1)$ and $(a+1)(2a+1)$, and the intersection numbers are also polynomials in a. (S_5 and A_5, acting on 2-element subsets, provide examples with $a = 1$.) Similar results for $n = 3p$ have been found by NEUMANN [30] and SCOTT [33], and for $n = 4p$ by COOPER, incomplete as yet. If $n = p^2$ then one of the following holds: G has a regular normal subgroup (so $G \le AGL(2,p)$); $G \le S_p \text{ wr } S_2$ (acting on the vertices of the square lattice graph); G is doubly transitive.

2. *Group-theoretic assumptions*

In this class may be put BENDER's theorem [3] on strongly embedded subgroups; a crucial part of it states that, if G_α is strongly embedded in G, then either G has a regular normal subgroup of odd order, or G is doubly transitive. (In the latter case BENDER has determined the possible groups.) Another such result is the FEIT-HIGMAN theorem [16] which asserts that, if G has a BN-pair of rank 2 and G_α is a maximal parabolic subgroup, then $|G_\alpha| = 2$ or G has rank 2, 3, 4, 5 or 7.

3. *Numerical conditions on subdegrees*

It follows from the theorem of BANNAI & ITO [2] and DAMERELL [12] on Moore graphs that, if G has subdegrees 1, a, a(a-1),..., $a(a-1)^{r-2}$, then either a = 2 or $r \leq 3$. Similar results are given by HIGMAN [22].

4. *Hypotheses about the action of* G *on some or all of its orbits*

Theorems 5.1-5.3 are of this type, and others follow.

MANNING's theorem implies that, in a primitive permutation group G, if G_α is doubly transitive on every suborbit different from $\{\alpha\}$, then $|G_\alpha| = 2$ or G is doubly transitive. This suggests defining the *subrank* of a transitive permutation group to be the maximum rank of the stabiliser of a point on its orbits, and making the conjecture that in a primitive group of subrank m, either the rank is bounded by a function of m, or $|G_\alpha| = m$. (As originally formulated, the second alternative was that G_α has a non-trivial regular orbit. This is implied by the condition $|G_\alpha| = m$, and is equivalent to it if the conjecture at the end of section 3 is correct. Frobenius groups, and many other examples, show that this possibility must be allowed.) If true, the conjecture implies the existence of functions f and g defined by

$$f(m) = \min\{r \mid G \text{ primitive \& subrank}(G) = m \text{ implies } \text{rank}(G) \leq r \text{ or } |G_\alpha| = m\},$$

$$g(m) = \min\{r \mid \exists \text{ finite set } S \text{ of permutation groups such that G primitive \& subrank}(G) = m \text{ implies rank}(G) \leq r \text{ or } |G_\alpha| = m \text{ or G is isomorphic to a group in } S\}.$$

It would be interesting to have exact values for f(m) and g(m) where they are defined. Clearly f(2) = g(2) = 2. Lower bounds for suitable values of m can be obtained from specific groups:

1. Let $F = GF(q^2)$, where $q = 2^n$, and
   $G = \{x \mapsto a^{q-1}x^{\sigma} + b \mid a,b \in F,\ a \neq 0,\ \sigma \in \mathrm{Aut}\ F,\ \sigma^2 = 1\}$.
   G_0 is a dihedral group of order $2(q+1)$ and has $q-1$ orbits of length $q+1$, one containing each non-zero element of $GF(q)$. So $f(2^{n-1}+1) \geq 2^n$.

2. Let $G = S_n \text{ wr } S_k$, in its representation of degree n^k. G has rank $k+1$ and subrank $[(k+2)(k+3)/6]$, independent of n. So $g([(k+2)(k+3)/6]) \geq k+1$.

For $m = 3$, these bounds are exact [10]:

THEOREM 5.4. *If* G *is a primitive permutation group with subrank* 3, *then one of the following holds*:
(i) G *has rank at most* 3;
(ii) $|G_{\alpha}| = 3$, G *is a Frobenius group*;
(iii) G *is the group of example 1, with rank* 4 *and degree* 16.

PROOF. If all subdegrees are at most 5, or if $|G_{\alpha}|$ is odd, then the result can be proved by *ad hoc* arguments. If $|\Gamma(\alpha)| \geq 6$ and $|G_{\alpha}^{\Gamma(\alpha)}|$ is even for some suborbit Γ, then the complete graph on $\Gamma(\alpha)$ is partitioned into subgraphs corresponding to at most two suborbits; by RAMSEY's theorem, one of these contains a triangle. So $\Delta \subseteq \Delta \circ \Delta$ for some suborbit Δ. Then it is shown that the Δ-graph has diameter at most 2. Regarding the application of RAMSEY's theorem, it is worth noting that the three graphs in case (iii) are non-degenerate with respect to the next case of the theorem (that is, none of them contains a triangle). □

Primitive groups with rank and subrank 3 have been investigated by character-theoretic methods by M.S. SMITH [38], who has found strong restrictions on their parameters. A number of examples exist, including $S_n \text{ wr } S_2$, the split extension $V_{2^{2n}} \cdot O^{\pm}(2n,2)$, $P\Gamma U(4,q^2)$ (for prime power q), and the HIGMAN-SIMS and MCLAUGHLIN groups. The graph-theoretic analogue of this situation is a strongly regular graph Γ with the property that the restrictions of Γ to the points adjacent and non-adjacent to any point are both strongly regular. The arguments of [38] extend to this situation, using the algebras associated with "coherent configurations" in place of the centraliser algebras of permutation groups. The result is that either such a graph is of pseudo-Latin square or negative Latin square type [29], or the parameters of it or its complement are given by

$$n = 2(2r-s)^2(r(r-1)-(2r+1)s)/(s+r^2-r)(s-r^2-3r),$$
$$k = (r-s)((2r+1)s-r(3r+1))/(s-r^2-3r),$$
$$l = (r-1-s)((2r+1)s-r(3r+1))/(s+r^2-r),$$
$$\lambda = r(r-1-s)(s+r^2+r)/(s-r^2-3r),$$
$$\mu = (r+1)(r-s)(s+r^2-r)/(s-r^2-3r),$$

where s and r are integers satisfying

$$-s > r(r+1),$$
$$(s-r^2-3r) \mid 2r^2(r+1)^2(r+2),$$
$$(s+r^2-r) \mid 2r^2(r-1)(r+1)^2.$$

(Here, as in [20], n is the degree, k and l the subdegrees, and λ and μ the intersection numbers a_{111} and a_{112}.) SMITH remarks that these conditions have nine infinite families of solutions (with s an integer polynomial in r), and also proves uniqueness (in the group-theoretic situation) for small values of the parameters.

Another generalization of MANNING's theorem can be obtained by relaxing the condition that G is doubly transitive on all non-trivial suborbits. Thus, it is proved in [11] that

<u>THEOREM 5.5</u>. *If* G *is primitive on* Ω *and* G_α *is doubly transitive on all non-diagonal suborbits except possibly one, with* $|G_\alpha| > 2$, *then* G *has rank at most* 4. *If the rank is* 4, *then the two doubly transitive suborbits of* G *are paired with each other, and the degrees, subdegrees, and intersection numbers are polynomials in a single integer parameter.*

The only known example of such a rank 4 group is $PSU(3,3^2)$ acting on 36 points, with subdegrees 1, 7, 7, 21. The stabiliser of a point is PSL(3,2), and the non-trivial suborbits can be identified with the points, lines, and flags of PG(2,2). Indeed, this is the only known primitive rank 4 group with non-trivial pairing of suborbits.

MANNING's theorem in fact implies a stronger statement than the one we have taken as a model for generalization: If G is primitive on Ω, and G_α is doubly transitive on its largest orbit, then $|G_\alpha| = 2$ or G is doubly transitive. Thus it would be desirable to bound the rank of a primitive group G under the assumption that G_α acts with prescribed rank, but not regularly, on its largest orbit. All that is known here is that if $\Delta(\alpha)$ is the largest G_α-orbit and $\Gamma(\alpha)$ any non-trivial G_α-orbit, then the permutation characters of G_α on $\Gamma(\alpha)$ and $\Delta(\alpha)$ must have a non-principal irreducible constituent in common, provided $|\Delta(\alpha)| > 1$. (For otherwise $G_{\alpha\gamma}$ is transitive on $\Delta(\alpha)$, for $\gamma \in \Gamma(\alpha)$; so $\Delta(\alpha)$ is a G_γ-orbit and hence is fixed by $\langle G_\alpha, G_\gamma \rangle$. But $\langle G_\alpha, G_\gamma \rangle = G$ unless $|G_\alpha| = 1$.)

Another general question, suggested by theorem 5.2, is this. Suppose G is primitive, Γ is a self-paired suborbit, and the actions of G_α on $\Gamma(\alpha)$ and $(\Gamma \circ \Gamma)(\alpha)$ are prescribed. When are these conditions consistent, and when do they imply a bound for the rank? (For example, if $G_\alpha = M_{22}$ acts on $\Gamma(\alpha)$ and $(\Gamma \circ \Gamma)(\alpha)$ as on the points and blocks of the Steiner system, then G has rank 3 and is isomorphic to HS. For $V_{16}Sp(4,2)$ on the points and blocks of the (16,4,3) design, there exists a rank 3 extension, Aut M_{22}; for $V_{64}G_2(2)$ on the points and blocks of the (64,4,3) design, neither answer is known.)

6. ALGEBRAIC RELATIONS AMONG SUBORBITS

We saw in section 1 that the rank of a transitive group is the sum of squares of multiplicities of irreducible constituents of the permutation character, while the number of self-paired suborbits is the sum of multiplicities of irreducibles of the first kind minus the sum for those of the second kind. If the multiplicities are $e_0=1,\ldots,e_s$, then the centraliser algebra (generated by the basis matrices) is a direct sum of matrix algebras of degrees $e_0,\ldots,e_s$ over $\mathbb{C}$. Thus all multiplicities are equal to 1 if and only if all pairs of basis matrices commute. This must be the case if the rank is at most 5, but need not be so for rank 6. (More generally, if $e > 1$, then $1_G + e\chi$ is not a permutation character. For in any transitive group there is an element g with no fixed points; if $\pi = 1_G + e\chi$, then $-1/e = \chi(g)$ is an algebraic integer.)

The case where the centraliser algebra is commutative is very important, and is discussed in section 29 of WIELANDT [45]. Certain group theoretic conditions (such as the existence of a regular abelian subgroup, or the existence of an element interchanging any pair of points, guarantee that this holds. Less trivially, GLAUBERMAN has remarked that if the Sylow 2-subgroups of G are cyclic or generalized quaternion, and G acts by conjugation on its set of involutions, then the centraliser algebra is commutative; he asks if this fact can be used to obtain an alternative proof of the theorem that such a group, if primitive, has a regular normal subgroup.

In other situations we may have only the weaker information that a particular pair C_i, C_j of basis matrices commute. This has the obvious consequence that $a_{ijk} = a_{jik}$ for all k; but sometimes it is possible to go further. I shall illustrate this with a discussion of a normal basis matrix (one which commutes with its transpose) corresponding to a doubly transitive subconstituent.

Suppose G_α is doubly transitive on $\Gamma(\alpha)$, with $\Gamma \neq \Gamma^*$ and $|\Gamma(\alpha)| = v$. Then $\Gamma^* \circ \Gamma$ is a single suborbit, with subdegree $v(v-1)/k$ for some k (section 5). Also, G_α is doubly transitive on $\Gamma^*(\alpha)$ (section 2), and so $\Gamma \circ \Gamma^*$ is a single suborbit, with subdegree $v(v-1)/k'$ for some k'. Counting in two ways quadrilaterals of the first kind in figure 6.1 gives

$$|\Omega| \frac{v(v-1)}{k} k(k-1) = |\Omega| \frac{v(v-1)}{k'} k'(k'-1),$$

so

$$k = k'.$$

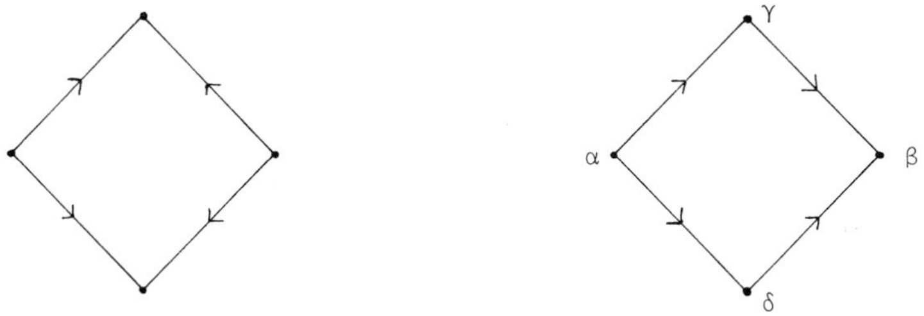

Fig. 6.1

Although the subdegrees are equal, the suborbits $\Gamma^* \circ \Gamma$ and $\Gamma \circ \Gamma^*$ may or may not be equal. An example where they are not equal is given in [8;I]; for one where they are equal, see theorem 5.5. If C, D, D' are the basis matrices of Γ, $\Gamma^* \circ \Gamma$, $\Gamma \circ \Gamma^*$ respectively, we have

$$C^T C = vI + kD, \quad C C^T = vI + kD';$$

so $\Gamma^* \circ \Gamma = \Gamma \circ \Gamma^*$ if and only if C is a normal matrix. We shall call Γ *normal* if this occurs.

THEOREM 6.1. *Γ is normal if and only if $|(\Gamma \circ \Gamma)(\alpha)| < v^2$. If this holds and G is primitive, then G has the same permutation character on $\Gamma(\alpha)$ and $\Gamma^*(\alpha)$.*

PROOF. The Γ-graph contains figures of the second kind in figure 6.1 if and only if Γ is normal (since $(\gamma,\delta) \in (\Gamma^* \circ \Gamma) \cap (\Gamma \circ \Gamma^*)$). Also, the graph contains such figures if and only if $|(\Gamma \circ \Gamma)(\alpha)| < v^2$ (for there are two paths of length 2 from α to β). Suppose Γ is normal and G has different permutation characters on $\Gamma(\alpha)$ and $\Gamma^*(\alpha)$. Then G is transitive on paths of length 2, so $\Gamma \circ \Gamma$ is a single suborbit, with subdegree v^2/l for some integer l. Counting these figures in two ways,

$$|\Omega| \frac{v(v-1)}{k} k^2 = |\Omega| \frac{v^2}{l} l(l-1),$$

so

$$(v-1)k = v(l-1).$$

Since Γ is normal, $l > 1$, and so v divides k; thus v = k, and G is imprimitive by theorem 5.1. (The structure of the graph is obvious.) □

If G is primitive, we can push the counting argument further. G_α has the same permutation character on $\Gamma(\alpha)$ and $\Gamma^*(\alpha)$, so these are the points and blocks of a symmetric design with parameters (v,K,Λ), possibly trivial. Then $\Gamma \circ \Gamma$ is the union of two suborbits with subdegrees vK/l and $v(v-K)/l'$ for some l, l'. Now the count gives

$$|\Omega| \frac{v(v-1)}{k} k^2 = |\Omega| \frac{vK}{l} l(l-1) + |\Omega| \frac{v(v-K)}{l'} l'(l'-1),$$

$$(v-1)k = K(l-1) + (v-K)(l'-1).$$

If the representations of G_α on $\Gamma(\alpha)$ and $\Gamma^*(\alpha)$ are equivalent, we can assume K = 1, whence it follows that $l = 1$, $l' = k+1$. If G_α is triply transitive on $\Gamma(\alpha)$ then this hypothesis holds, and in addition k = 1, $l' = 2$. (If, further, $G_\alpha^{\Gamma(\alpha)}$ is the symmetric or alternating group of sufficiently large degree, it can be shown that G has a regular normal subgroup.) Other solutions which actually occur are v = 7, k = 2, K = l = 4, l' = 1 ($PSU(3,3^2)$, degree 36, rank 4) and v = 11, k = 2, K = l = 5, l' = 1 (M_{12}, degree 144, rank 5).

The concept of normality has also been used in a rank-bounding theorem [11].

The formulae for the rank and number of self-paired suborbits show that, if the centraliser algebra is non-commutative, then there are non-self-paired suborbits. Is it even true that there will be non-normal suborbits? Does there exist an example of a basis matrix which is not even diagonalisable?

Sometimes it has been shown that there is an absolute bound, greater than 1, on the multiplicities. (For example, in some cases where the Sylow 2-subgroup has rank 2 or 3, and G acts by conjugation on its involutions, the multiplicities are at most 2 or 3.) It is tempting to conjecture that a general result of this kind holds, and it may be worth studying situations where such a bound exists. If the multiplicities are at most d, the centraliser algebra is a sum of matrix algebras with degrees at most d, and so it satisfies certain identities. These identities have combinatorial interpretations in the graphs, similar to that considered for normality; but I do not know how to exploit these conditions.

REFERENCES

[1] ATKINSON, M.D., *Doubly transitive but not doubly primitive permutation groups, I*, J. London Math. Soc. (2), 7 (1974) 632-634; *II*, ibid., to appear.

[2] BANNAI, E. & T. ITO, *On finite Moore graphs*, J. Fac. Sci. Univ. Tokyo, 20 (1973) 191-208.

[3] BENDER, H., *Transitive Gruppen gerader Ordnung, in denen jede Involution genau einen Punkt festlasst*, J. Algebra, 17 (1971) 527-554.

[4] BIGGS, N., *Finite groups of automorphisms*, London Math. Soc. Lecture Notes 6, C.U.P., 1971.

[5] BIGGS, N. & D.H. SMITH, *On trivalent graphs*, Bull. London Math. Soc., 3 (1971) 155-158.

[6] BRITTON, J.L., *The word problem*, Ann. of Math., 77 (1963) 16-32.

[7] BURNSIDE, W., *On some properties of groups of odd order*, Proc. London Math. Soc. (1), 33 (1901) 162-185.

[8] CAMERON, P.J., *Permutation groups with multiply transitive suborbits, I*, Proc. London Math. Soc. (3), 25 (1972) 427-440; *II*, Bull. London Math. Soc., to appear.

[9] CAMERON, P.J., *Extending symmetric designs*, J. Combinatorial Theory A, 14 (1973) 215-220.

[10] CAMERON, P.J., *Bounding the rank of certain permutation groups*, Math. Z., 124 (1972) 343-352.

[11] CAMERON, P.J., *Primitive groups with most suborbits doubly transitive*, Geometriae Dedicata, 1 (1973) 434-446.

[12] DAMERELL, R.M., *On Moore graphs*, Proc. Cambridge Philos. Soc., 74 (1973) 227-236.

[13] DELSARTE, P., *The association schemes of coding theory*, Mathematical Centre Tracts 55, 1974, pp. 139-157.

[14] DJOKOVIC, D.Z., *On regular graphs, I*, J. Combinatorial Theory, 10 (1971) 253-263; *II*, ibid., 12 (1972) 252-259.

[15] FEIT, W., *Characters of finite groups*, Benjamin, New York, 1967.

[16] FEIT, W. & G. HIGMAN, *The non-existence of certain generalized polygons*, J. Algebra, 1 (1964) 114-138.

[17] GARDINER, A.D., *Arc transitivity in graphs, I*, Quart. J. Math. Oxford Ser. (2), 24 (1973) 399-407; *II*, ibid., to appear.

[18] GARDINER, A.D., *Doubly primitive vertex stabilisers in graphs*, Math. Z., to appear.

[19] HERING, C., W.M. KANTOR & G. SEITZ, *Finite groups with a split* BN-*pair of rank 1*, J. Algebra, 20 (1972) 435-475.

[20] HIGMAN, D.G., *Finite permutation groups of rank 3*, Math. Z., 86 (1964) 145-156.

[21] HIGMAN, D.G., *Primitive rank 3 groups with a prime subdegree*, Math. Z., 91 (1966) 70-86.

[22] HIGMAN, D.G., *Intersection matrices for finite permutation groups*, J. Algebra, 6 (1967) 22-42.

[23] HIGMAN, D.G., *Combinatorial considerations about permutation groups*, Lecture Notes, Oxford, 1972.

[24] HIGMAN, G., B.H. NEUMANN & H. NEUMANN, *Embedding theorems for groups*, J. London Math. Soc. (1), 26 (1949) 247-254.

[25] KANTOR, W.M., *2-transitive designs*, this volume, pp. 365-418.

[26] KNAPP, W., *On the point stabilizer in a primitive permutation group*, Math. Z., 133 (1973) 137-168.

[27] MANNING, W.A., *Simply transitive primitive groups*, Trans. Amer. Math. Soc., 29 (1927) 815-825.

[28] MANNING, W.A., *A theorem concerning simply transitive primitive groups*, Bull. Amer. Math. Soc., 35 (1929) 330-332.

[29] MESNER, D.M., *A new family of partially balanced incomplete block designs with some Latin square design properties*, Ann. Math. Statist., 38 (1967) 571-581.

[30] NEUMANN, P.M., *Primitive permutation groups of degree 3p*, unpublished.

[31] QUIRIN, W.L., *Primitive permutation groups with small orbitals*, Math. Z., 122 (1971) 267-274.

[32] QUIRIN, W.L., *Extension of some results of Manning and Wielandt on primitive permutation groups*, Math. Z., 123 (1971) 223-230.

[33] SCOTT, L., *Unprimitive permutation groups*, in: *Theory of finite groups*, R. BRAUER & C-H. SAH (eds.), Benjamin, New York, 1969.

[34] SIMS, C.C., *Graphs and finite permutation groups*, *I*, Math. Z., 95 (1967) 76-86; *II*, ibid., 103 (1968) 276-281.

[35] SIMS, C.C., *Computational methods in the study of permutation groups*, in: *Computational problems in abstract algebra*, J. LEECH, (ed.), Pergamon Press, London, 1970, pp. 169-183.

[36] SMITH, D.H., *Primitive and imprimitive graphs*, Quart. J. Math. Oxford (2), 22 (1971) 551-557.

[37] SMITH, D.H., *Bounding the diameter of a distance-transitive graph*, J. Combinatorial Theory B, to appear.

[38] SMITH, M.S., *On rank 3 permutation groups*, to appear.

[39] THOMPSON, J.G., *Bounds for orders of maximal subgroups*, J. Algebra, 14 (1970) 135-138.

[40] TITS, J., *Sur la trialité et certains groupes qui s'en deduisent*, Publ. Math. IHES, 2 (1969) 14-60.

[41] TUTTE, W.T., *A family of cubical graphs*, Proc. Cambridge Philos. Soc., 43 (1947) 459-474.

[42] TUTTE, W.T., *On the symmetry of cubic graphs*, Canad. J. Math., 11 (1959) 621-624.

[43] WIELANDT, H., *Primitive Permutationsgruppen vom Grad 2p*, Math. Z., 63 (1956) 478-485.

[44] WIELANDT, H., *Subnormale Hülle in Permutationsgruppen*, Math. Z., 74 (1962) 297-298.

[45] WIELANDT, H., *Finite permutation groups*, Acad. Press, New York, 1964.

[46] WIELANDT, H., *Permutation groups through invariant relations and invariant functions*, Lecture Notes, Ohio State Univ., 1969.

[47] WIELANDT, H., *Subnormal subgroups and permutation groups*, Lecture Notes, Ohio State Univ., 1971.

[48] WONG, W.J., *Determination of a class of primitive permutation groups*, Math. Z., 99 (1967) 235-246.

GROUPS, POLAR SPACES AND RELATED STRUCTURES

E.E. SHULT

University of Florida, Gainesville, Fla. 32611, USA

INTRODUCTION

The purpose of this report is to give a simplified and more or less systematic account of certain developments at the interface between finite group theory and combinatorial theory. At the present time it is virtually possible to characterize the graphs associated with the so-called classical groups on the basis of an exceedingly crude graph-theoretic hypothesis. Such a theorem commands an obvious relevance to finite group theory since it may be used as a tool for diagnosing the presence of a "classical group" -that is, to tell whether a finite group G contains one of the groups $PGU(n,q^2)$ $PSO^{\varepsilon}(n,q)$, $(\varepsilon=\pm 1)$ or $PSp(2n,q)$. Of course in this case, the crudity (or simplicity, if one prefers) of the graph-theoretic hypothesis means that such a graph is more easily realized within a finite group, and this feature serves in making such a diagnostic tool more widely applicable. The phrase "virtually possible to characterize" appears above because certain extremal cases and certain rather tight open cases are also present. To be more specific, graphs which contain a vertex lying on an edge with every remaining vertex, and any generalized quadrangle may also appear along with the "classical-group-graphs" in the conclusion of the theorem. In practice, when the graph is realized in a finite group, say with vertices being a conjugacy class of subgroups and edges some convenient relation between these subgroups, the case that one vertex lies on an edge with each remaining vertex usually implies something very extreme for the group G, and so can usually be handled. Coming to grips with the case of generalized quadrangles is usually more difficult. Nonetheless nice applications of the theorem in finite groups exist (see section 10).

M. Hall, Jr. and J. H. van Lint (eds.), Combinatorics, 451-482.

In the hope that this theorem may prove useful in other problems, both inside and outside of finite group theory, I have tried to give a rather simplified presentation of this theorem. (Because some of the deeper results on polar spaces and buildings are necessarily deferred to Professor TITS' forthcoming book, the present treatment will bear a closer resemblance to a boyscout handbook on prepolar spaces and groups, rather than any sort of complete treatise.) The characterization theorem presented is really a linking together of a number of theorems. The chain of theorems begins with VELDKAMP's important work on finite polar spaces fifteen years ago, and this work was subsequently streamlined and extended to infinite polar spaces by TITS [32,29]. It is worth noting, therefore, that the presentation in sections 2 through 7 does not require the structures in question to be finite; but only that they have finite *rank*, a notion in this context roughly analogous to having finite dimension.

Possible variations of the characterization theorem to graphs of non-isotropic or non-singular projective points or to structures associated with other Chevalley groups is discussed in section 9. Section 10 concludes with a few historical notes concerning the origins of the problems and their applications to finite group theory.

1. THE MAIN THEOREMS

We begin with the basic characterization theorem for finite graphs. Throughout, the word "graph" means an undirected graph without loops or double edges. Thus a *graph* is a set V of vertices and a set E (called the *edge set*) which is a subset of $V^{(2)}$, the set of all 2-sets of elements of V. The graph is said to be *finite* if V is a finite set. A graph (V,E) is called *complete* if $E = V^{(2)}$. If X is a subset of V, the *subgraph* X means the graph $(X, E\cap X^{(2)})$. A *clique* means a complete subgraph of (V,E). (Note that in some quarters of this world, "clique" means "maximal complete subgraph"; it does not here.)

<u>THEOREM A.</u> *Let* $G = (V,E)$ *be a finite graph. Suppose, for each edge* (x,y) *in* E *there exists a clique* C(x,y) *containing* x *and* y *such that*

(i) $|C(x,y)| \geq 3$.

(ii) *If* $w \in V - C(x,y)$, *then* w *either lies on an edge with exactly one member of* C(x,y) *or on an edge with every member of* C(x,y).

Then one of the following conclusions holds:

(a) *There exists a vertex in* V *lying on an edge with every other vertex of* V.

(b) (V,E) $\simeq S_\pi$ *where* S_π *is the graph whose vertices are the points of a semiquadric* S_π *with edges being pairs of points of* S_π *which are perpendicular with respect to* π.

(c) *Each* C(x,y) *is uniquely determined by* x *and* y *and* (V, {C(x,y) | (x,y) $\in$ E}) *is a generalized quadrangle.*

(d) (V,E) *is totally disconnected; that is, the edge set* E *is empty.*

In case (b), π is either a non-degenerate polarity of a projective Desarguesian space P and S_π is the set of *absolute points* with respect to π (i.e. points p $\in$ P such that p $\in$ π(p)) or else π is a (proportionality class of) non-degenerate quadratic form(s) on P and S_π is the set of *singular points* of P with respect to π (i.e. points p $\in$ P such that π(p) = 0). In either case it makes sense to define "perpendicular with respect to π". Thus S_π denotes the absolute points under a non-degenerate unitary, orthogonal or symplectic polarity, or else the singular points of a non-degenerate quadratic form of a Desarguesian projective space over a field of characteristic 2. The point is that as a graph G is uniquely determined up to isomorphism.

The significance of case (c) is that the C(x,y) are actually maximal cliques, C(x,y) may be viewed as the "line" through x and y, and if L is a line and q a point in V not on L, then q is *collinear* (on a line) with exactly one member of L. It does not follow that all C(x,y)'s have the same cardinality as the following example shows: Let A and B be sets of size 3 and 4 respectively. Set V = A×B and let (p,q) $\in$ E if and only if p and q either agree in their A-coordinate or their B-coordinate, but not both. The graph becomes a 3×4 grid with rows and columns forming the seven "lines". If, however, there exists a vertex lying on at least three lines of a general graph in case (c), then it is not difficult to see that all "lines" -that is, the C(x,y)'s- have the same cardinality. Thus the "grids" are the only examples in which |C(x,y)| does not assume a constant value. For this reason (and for the reason that the "grids" are determined up to isomorphism) one frequently excludes these from the definition of generalized quadrangle (see PAYNE [16]).

Case (d) is a graph determined up to isomorphism.

As mentioned in the introduction, this theorem is pieced together from a number of other theorems. Moreover, this piecing together could have been done so that a version of the above theorem for infinite graphs could have been obtained. How this is done will become apparent (though somewhat clumsy to state) as we break up the above theorem into the pieces which make it work. For this purpose, we introduce an abstract concept designed to mimic the hypotheses of theorem A. This concept is due to BUEKENHOUT [8] (who unfortunately gave it a name rather awkward for me to use. I hope he will forgive me if I rename it for this report.):

DEFINITION 1.1. A *prepolar space* is a set of points P and a collection L of distinguished subsets of P called lines such that

(i) every line contains at least two points,

(ii) for each line L and point $p \in P-L$ either there is exactly one point of L lying on a line with p, or each point of L lies on some line with p.

If P is a prepolar space with the property that every line has cardinality at least 3, and if we let E be the set of collinear pairs of points of P then (P,E) is a graph satisfying the hypothesis of G in theorem A. Note that in the definition of a prepolar space, it is not assumed that different lines have the same cardinality, nor that two points lie in at most one line. This last avenue of generality even makes it appear inappropriate to even have called these blocks "lines". The following theorem shows that in all the interesting cases, we might just as well call them lines. We say a prepolar space is *linear* if every pair of points lies in at most one line.

THEOREM B. (BUEKENHOUT & SHULT [8]). *A prepolar space either contains a point collinear with all other points or is a linear prepolar space.*

In the case that a prepolar space P contains a point collinear with all other points, the space is said to be *degenerate*. If P contains no such point, P is said to be *non-degenerate*.

A *subspace* of a prepolar space (P,L) is a subset of mutually collinear points of P such that any line through two points of the subset lies entirely within the subset. A prepolar space has *rank* n if the greatest lower bound on the length of a tower of subspaces of (P,L) is n+1. (Note that the empty set and the subsets of P containing a single point are subspaces, so, for example, generalized quadrangles are simply prepolar spaces of rank 2.) One next defines a *polar space* S (following TITS' simplification of VELDKAMP's axioms) as a set S of points with a family of distinguished subsets closed under intersection called *subspaces* (of the polar space S) subject to certain axioms outlined more fully in section 5. One then proves

THEOREM C. (BUEKENHOUT & SHULT [8]). *Let* (P,L) *be a non-degenerate prepolar space of finite rank all of whose lines have cardinality* ≥ 3. *Then* (P,L) *together with its subspaces is a polar space.*

As we shall see shortly, every polar space together with its minimal proper subspaces as its collection of lines, is a prepolar space. We may therefore speak of the rank of a polar space as being its rank as a prepolar space. The next link in the chain is the fundamental theorem of TITS & VELDKAMP [29,32], classifying polar spaces of rank at least three.

THEOREM D. (TITS & VELDKAMP). *Let* S *be a polar space of finite rank* $n \geq 3$. *Then exactly one of the following situations is realized.*

(1) S *is a polar space* $S(\pi)$ *of a projective space with a polarity determined by a trace-valued* σ*-hermitian form.*

(2) S *is a polar space* $S(Q)$ *of a projective space with* Q *a non-degenerate pseudoquadratic form on a division ring* K *of characteristic* 2 *with respect to an antiautomorphism* σ *such that* $\sigma^2 = 1$ *and*

$$\{t \in K \mid t^\sigma = t\} \neq \{u + u^\sigma \mid u \in K\} .$$

(3) S *is a polar space* $S(\pi)$ *of a Desarguesian projective space coordinitized by a field of characteristic different from* 2, *equipped with a symplectic polarity* π.

(4) S *is a polar space of rank* 3 *whose maximal subspaces are Moufang planes (one polar space for each Cayley division algebra).*

(5) S *is a polar space of rank* 3 *corresponding to a* 3*-dimensional projective space* P *on a non-commutative division ring (*S *corresponds to the classical Klein quadric in the commutative case).*

Most of the technical terms, appearing in the statement of this theorem (for example σ-hermitian and pseudoquadratic) are defined in the next section. In case the polar space contains a finite number of points, cases (4) and (5) of the above theorem do not arise and one is left with a hermitian, symmetric or alternating (symplectic) polarity on a finite projective space, or the totally singular points with respect to a quadratic form on a Desarguesian projective space over a field of characteristic 2. A polar space of rank 2 is precisely a generalized quadrangle and it is now clear that theorems B, C and D, together, yield theorem A.

The next few sections are devoted to a description of sesquilinear forms, pseudoquadratic forms and their associated polar spaces.

2. SESQUILINEAR FORMS

Let V be a right vector space over a division ring K and let σ be an antiautomorphism of K. Then a *σ-sesquilinear form* is a biadditive mapping $f: V\times V \to K$ such that

$$f(xa,yb) = a^{\sigma} f(x,y)b \tag{2.1}$$

for all $x,y \in V$ and $a,b \in K$. We shall use the term "sesquilinear form" to mean "σ-sesquilinear for some (possibly unspecified) antiautomorphism σ of K". If c is a non-zero scalar in K, the mapping cf defined by $(cf)(x,y) = c\cdot f(x,y)$ is a σ'-sesquilinear form with σ' being the composition of σ with the inner automorphism of K induced by conjugation by c^{-1}. We say cf is *proportional* to f. Obviously proportionality is an equivalence relation on the set of sesquilinear forms.

A σ-sesquilinear form is called *reflexive* if $f(x,y) = 0$ implies $f(y,x) = 0$ for any pair of vectors x,y in V. The first observation is

PROPOSITION 2.1. *A σ-sesquilinear form f is reflexive if and only if there exists a non-zero scalar $\varepsilon \in K$ such that for every pair of vectors $x,y \in V$*

$$f(x,y) = f(y,x)^{\sigma}\cdot\varepsilon\ . \tag{2.2}$$

This is shown by proving that $h(x,y) = f(x,y)^{-1}f(y,x)^{\sigma}$ is a constant function on the subset of $V\times V$ consisting of pairs (u,v) such that $f(u,v) \neq 0$. Clearly if $f(x,v) = 0$, then $h(x,y) = h(x,y+v)$. Similarly if $f(u,y) = 0$, $h(x,y) = h(x+u,y)$. By passing through a chain of translations in each argument in this way, h can be shown to be a constant function on the support of f.

But if f satisfies (2.1), this relation can be iterated to yield the following relations between σ and ε:

$$\varepsilon^{\sigma} = \varepsilon^{-1}\ , \tag{2.3}$$

(2.4) σ^2 is the inner automorphism of K induced by conjugation by ε^{-1}.

A σ-sesquilinear form satisfying (2.2) (and hence (2.3) and (2.4)) is called (σ,ε)-*hermitian*. Several special cases are distinguished: If $\varepsilon = 1$, the form is simply called σ-*hermitian* and if $\varepsilon = -1$, it is called σ-*anti-hermitian*. If $\sigma = 1_K$, σ-hermitian forms are called *symmetric*, and σ-anti-hermitian forms are called *alternating*. Proposition 2.1 states that every reflexive sesquilinear form is a (σ,ε)-hermitian form. Starting with an arbitrary σ-sesquilinear form g we can always construct from it a reflexive form which is (σ,ε)-hermitian (for the same σ) by choosing ε so that (2.3) and (2.4) hold and setting

(2.5) $$f(x,y) = g(x,y) + g(y,x)^{\sigma}\cdot\varepsilon \ .$$

Are all of the reflexive sesquilinear forms obtained by such a semi-symmetrization process? The answer is "no". In case the reflexive (or (σ,ε)-hermitian) form f is obtained from a σ-sesquilinear form g via (2.5) we say that f is *trace-valued*.

PROPOSITION 2.2. *A* (σ,ε)-*hermitian form* f *is trace-valued if and only if for each vector* $x \in V$, $f(x,x)$ *lies in the subgroup* $\{t + t^{\sigma}\cdot\varepsilon \mid t \in K\}$ *of the additive group of* K.

For certain choices of (σ,ε,K), (σ,ε)-hermitian forms are always trace-valued. Indeed

PROPOSITION 2.3. *Assume* σ *is an antiautomorphism of a division ring* K *and* ε *is an element of* K *related to* σ *by* (2.3) *and* (2.4). *Then a necessary and sufficient condition that* f *be trace-valued is that*

(2.6) $$\{t + t^{\sigma}\cdot\varepsilon \mid t \in K\} = \{t \in K \mid t^{\sigma}\cdot\varepsilon = t\} \ .$$

This condition always holds if char $K \neq 2$, *or if* σ *acts non-trivially on the center of* K.

If f is a reflexive sesquilinear form, the symmetric relation on V defined by $R_f = \{(x,y) \mid x \in V,\ y \in V,\ f(x,y) = 0\}$ is called the *perpendicular relation* (with respect to f) and we write $x \perp_f y$ or $x \perp y$ if $(x,y) \in R_f$. Then as f is (σ,ε)-hermitian, $x^{\perp} = \{y \in V \mid x \perp y\}$ is a subspace of codimension at most one in V. If X is any subset of V, set $X^{\perp} = \{x \in V \mid f(y,x) = 0,\ \forall y \in X\}$, and write Rad V for $V^{\perp}$. We say f is *non-degenerate* if Rad V = (0). A subspace W of V is called *totally isotropic* (with respect to f) if $W \leq W^{\perp}$.

PROPOSITION 2.4. *If* X *and* Y *are two maximal totally isotropic subspaces,* $X \cap Y$ *has the same codimension in both* X *and* Y. *Consequently, any two maximal isotropic subspaces of* V *(*which exist by Zorn's lemma*) have the same dimension,* n.

This invariant cardinal number n is called the *Witt index* of f. To prove the first statement in proposition 2.4 suppose $\dim(Y/X\cap Y) > \dim(X/X\cap Y)$. Then if X_1 and Y_1 are complements of $X\cap Y$ in X and Y respectively, the space $X + (X_1^{\perp} \cap Y_1)$ is a totally isotropic space properly containing X, contrary to the maximality of X.

3. PSEUDOQUADRATIC FORMS

Again let K be a division ring, V a right vector space over K, σ an antiautomorphism of K and ε a non-zero element in K such that (2.3) and (2.4) hold. Following TITS [29], let $K_{\sigma,\varepsilon}$ denote the subgroup of the additive group of K, $\{t - t^{\sigma}\cdot\varepsilon \mid t \in K\}$. Set $K^{(\sigma,\varepsilon)} = K/K_{\sigma,\varepsilon}$, the quotient group.

A function $Q\colon V \to K^{(\sigma,\varepsilon)}$ is called a (σ,ε)-*quadratic form*, or a *pseudo-quadratic form relative to* σ *and* ε if there exists a σ-sesquilinear form $g\colon V\times V \to K$ such that

$$Q(x) = g(x,x) + K_{\sigma,\varepsilon}, \quad \forall x \in V. \tag{3.1}$$

PROPOSITION 3.1. $Q\colon V \to K^{(\sigma,\varepsilon)}$ *is a pseudoquadratic form with respect to* σ *and* ε *if and only if there exists a trace-valued* (σ,ε)*-hermitian form* $f\colon V\times V \to K$ *such that*

$$Q(x + y) = Q(x) + Q(y) + (f(x,y) + K_{\sigma,\varepsilon}), \quad \forall x,y \in V. \tag{3.2}$$

Given Q, the trace-valued form f of (3.2) is uniquely determined. If we write $f \doteq \beta Q$ we may think of β as a map from the set $Q_{\sigma,\varepsilon}$ of all (σ,ε)-quadratic forms into the set $S_{\sigma,\varepsilon}$ of all trace-valued (σ,ε)-hermitian forms.

PROPOSITION 3.2. *The mapping* $\beta\colon Q_{\sigma,\varepsilon} \to S_{\sigma,\varepsilon}$ *is onto. If* Q *is in* ker β, $Q(x)$ *lies in* $K^{\sigma,\varepsilon}/K_{\sigma,\varepsilon}$ *where* $K^{\sigma,\varepsilon}$ *is the subgroup* $\{t \in K \mid t + t^{\sigma}\cdot\varepsilon = 0\}$. β *is bijective if and only if* (2.6) *holds.*

A subspace X of V is *totally singular* if $Q(X) = 0$ in $K^{(\sigma,\varepsilon)}$. If X is a totally singular subspace for Q, it is clearly isotropic for the (σ,ε)-hermitian form βQ.

PROPOSITION 3.3. *The converse statement, that if* X *is a totally isotropic subspace with respect to* βQ, *then* X *is totally singular with respect to* Q, *is true if* (2.6) *holds.*

It can be shown, by redoing the proof of proposition 2.4, that

PROPOSITION 3.4. *If* X *and* Y *are two maximal totally singular subspaces of* V *then* $\dim(X/X\cap Y) = \dim(Y/X\cap Y)$, *whence* $\dim X = \dim Y$.

This uniform dimension of the maximal totally singular subspaces is called the *Witt index* of the pseudoquadratic form Q.

4. PROJECTIVE SPACES AND POLARITIES

A *projective space* P is a system of points P and a family L of distinguished subsets called lines such that

1) two points lie on exactly one line,
2) there exist four points, no three of which are collinear,
3) (Pasch's axiom) if L_1 and L_2 are two lines meeting at a point p, and if b and c are two points in $L_1-(p)$ and d and e are two points in $L_2-(p)$, then the line L_3 passing through b and d meets the line L_4 passing through c and e non-trivially.

A *subspace* X of a projective space P is a subset X of points such that every line L in L either lies in X or meets X in at most one point. Among the subspaces of P can be counted the empty set, called a *subspace of dimension* -1. If the subspace X consists of a single point we say that X is a 0-*dimensional subspace*; if X consists of a single line, X is 1-*dimensional*; and if some line is a maximal subspace of X, then all lines in X meet each other non-trivially, *each* line is a maximal subspace of X and X is called a 2-*dimensional subspace*, or a *projective plane*.

Let V be a right vector space of dimension at least 3 over a division ring K. If we let P denote the collection of 1-dimensional subspaces of V and let L be the set of 2-dimensional subspaces of V, each such subspace viewed as a collection of 1-dimensional subspaces (i.e. as subsets of P), then (P, L) is a projective space. Projective spaces constructed in this way are called *Desarguesian.* (Actually, "Desarguesian" is traditionally given an axiomatic definition equivalent to the one given here [31].) One of the most basic theorems of projective spaces is the following:

THEOREM 4.1. *Any projective space properly containing a projective plane as a subspace is Desarguesian.*

Suppose x is a point in the projective space P. Then it is easy to show that subspaces maximal with respect to not containing x (these exist by Zorn's lemma) are in fact maximal subspaces of P. Such subspaces are called *hyperplanes* of P. Let S denote the lattice of all subspaces of P. A mapping $\pi: S \to S$ is called a polarity if π reverses inclusion (that is, $X \leq Y$ implies $Y^{\pi} \leq X^{\pi}$) and for each point $x \in P$, x^{π} is either P or a hyperplane of P. (This definition follows TITS [32, p.128] and generalizes slightly the definition of a polarity given in DEMBOWSKI [10, p.42].) The *rank* of a polarity is the codimension of P^{π} in P. The polarity π is called *non-degenerate* if $P^{\pi} = \emptyset$.

If P is a Desarguesian projective space obtained from the right vector space V over a division ring K and f is a (σ,ε)-hermitian form on V, then the perpendicular map $X \to X^{\perp}$ (the "perpendicular" of X with respect to f) among the subspaces of V induces a polarity $\pi: S \to S$, where S is the set of subspaces of the projective space P. Clearly if f and π correspond in this way, f is non-degenerate if and only if π is. In this case we say that the polarity π is *represented by* f. If f is a trace-valued form we say that π is a polarity of *trace type*.

THEOREM 4.2. *Let* P *be the projective space of a vector space* V. *Every polarity of rank at least* 2 *in* P *is represented by a* (σ,ε)-*hermitian form* f *on* V.

Similarly, a pseudoquadratic form $Q: V \to K^{(\sigma,\varepsilon)}$ defines a perpendicular relation on the vectors of V which also induces a polarity of P. We are now in a position to discuss polar spaces.

5. ABSTRACT POLAR SPACES AND THE THEOREMS OF TITS AND VELDKAMP

The following definition is a simplification (due to TITS) of VELDKAMP's axioms.

DEFINITION 5.1. A *polar space* S is a set of points together with distinguished subsets called *subspaces* such that

(i) a subspace together with the subspaces it contains is a d-dimensional projective space with $-1 < d < n-1$ for some integer n called the *rank* of S;

(ii) given a subspace L of dimension n-1 and a point $p \in S-L$, there exists a unique subspace M containing p such that $\dim(M \cap L) = n-2$; it contains all points of L which lie together with p in some subspace of dimension one;

(iii) the intersection of any two subspaces is a subspace;

(iv) there exist disjoint subspaces of dimension n-1.

Let π be a polarity on a projective space P. Assume P is Desarguesian and π is represented by the trace-valued non-degenerate (σ,ε)-hermitian form f. Let $S(\pi)$ be the set of absolute points of P, so $S(\pi) = \{p \in P \mid p \in p^{\pi}\}$. Then calling the totally isotropic subspaces X of P (that is, those subspaces X such that $X < X$) "subspaces of $S(\pi)$", $S(\pi)$ becomes a polar space.

Similarly, let P be the projective space of a vector space V and let Q be a pseudoquadratic form on P. A singular point of P is a point corresponding to a 1-dimensional subspace of V whose vectors vanish under Q. Let S(Q) be the set of singular points of P and let the subspaces of S(Q) be those subspaces of P lying in S(Q). Then S(Q) becomes a polar space.

A complete proof of theorem D can be obtained from TITS' book [29]. The proof generally seems to proceed in two basic stages (though off in the margin there are exceptional cases for each stage): first one studies what occurs if the polar space can be embedded in a projective space (one reaches a virtual classification here); and second, one proves (with a few exceptions) that almost all polar spaces are embeddable. By an *embedding* of a polar space S into a projective space P we mean a triple (P,π,ϕ) where π is a polarity of P, ϕ is an injection of S into the set $S(\pi)$ of absolute points of P such that

(a) $\phi(S)$ spans P,

(b) for every subspace X of S, $\phi(X)$ is a subspace of P totally isotropic with respect to π.

It then turns out that when an embedding is possible the structure of $S(\pi)$ completely controls the structure of $\phi(S)$ (and hence S). This is because (assuming lines in S contain at least three points) if x and y are not collinear in S but $\phi(x)$ and $\phi(y)$ are perpendicular with respect to

π in P, then $\phi(x)$ is perpendicular to all of $\phi(S)$. As $\phi(S)$ spans P this places $\phi(x)$ in the radical, P^{π}. The classification begins with

THEOREM 5.1. (Part of Theorem 8.6 of TITS [29]). *Let* (P,π,ϕ) *be a projective embedding of a polar space* S *of rank* ≥ 2 *where* π *is represented by a* (σ,ε)*-hermitian form.*

(i) *If* σ *and* ε *satisfy condition* (2.6) (of proposition 2.3 above), π *is non-degenerate and* $\phi(S) = S(\pi)$.

(ii) *Suppose* σ *and* ε *do not satisfy* (2.6). *Then there exists an embedding* $(\overline{P},\overline{\pi},\overline{\phi})$ *of* S, *a morphism* $\mu\colon \overline{P}\to P$ *(as projective spaces) such that* $\overline{\phi}\cdot\mu = \phi$ *and* μ *and* $\overline{\pi}$ *induce* π *on* P, *and a pseudoquadratic form* $\overline{Q}$ *such that* $\overline{\pi} = \beta\overline{Q}$ (the "sesquilinearization" of Q) *and* $\overline{\phi}(S) = S(\overline{Q})$. *The morphism* μ *is unique up to isomorphism.*

Most of the rest of the proof emerges from

THEOREM 5.2. (Due originally to VELDKAMP). *A thick polar space of rank* ≥ 3 *whose maximal subspaces are Desarguesian is embeddable (that is, a* (P,π,ϕ) *exists).*

This gives only a vague idea, to be sure, but the full development will appear in TITS' monumental work [29].

We next describe the other links in the characterization theorem, those showing that non-degenerate prepolar spaces are polar spaces.

6. NON-DEGENERATE PREPOLAR SPACES ARE LINEAR

Let S be a prepolar space. If two points x and y lie together on at least one line of S we say that they are *collinear*. We may thus also regard S as a graph (undirected, without loops) with respect to the relation of being collinear. We write A(x) for the set of all points of S distinct from x but collinear with x, that is, the vertices adjacent to x in the graph. We define the radical of S by

(6.1) $$\mathrm{Rad}(S) = \{x \in S \mid \{x\} \cup A(x) = S\},$$

simply the set of points collinear with all remaining points. Recall that S is *non-degenerate* if and only if Rad(S) is empty.

We begin with a degeneracy criterion:

PROPOSITION 6.1. *Let* S *be a prepolar space and* L *a line of* S *and* b *some point of* L *such that* S = union A(u), *where* u *ranges over* L - {b}. *Then* Rad(S) *is non-empty.*

PROOF. If L has only 2 points, the result is obvious from the hypothesis on L. So we assume L has at least three points. Let Δ be the intersection of the A(u)'s as u ranges over L - {b}, and set $X_u = \{w \in S \mid A(w) \cap L = \{u\}\}$. Since S is a prepolar space this produces a partition of the points of S as

$$S = (L - \{b\}) + \Delta + \sum_u X_u \ , \tag{6.2}$$

where, in the last sum, u ranges over L - {b}.

Suppose an element $x \in X_u$ was adjacent to an element $y \in X_v$, where $u \neq v$. Then there exists a line M of S containing x and y. Then M meets Δ trivially since if $z \in M \cap \Delta$, then z is adjacent to u, so u, being adjacent to the two points z and x on M, is adjacent to y, contrary to the assumption that $y \in X_v$, $v \neq u$. By the basic property of a prepolar space, since b is not adjacent to x or y, b is adjacent to a unique point m on M. This point m does not lie in any X_w, $w \in L-\{b\}$. We have just seen that m cannot lie in Δ. So by (6.2) m lies in L - {b}. From the symmetry of u and v in the supposition at the beginning of this paragraph we may assume $m \neq u$. But then u is adjacent to the two points x and m on M and so is adjacent to y, contrary to the assumption that $y \in X_v$, once more. Thus the supposition at the beginning of this paragraph is false and so we may assume henceforward that if $u \neq v$, no element of X_u is adjacent to an element of X_v.

If all of the X_u's were empty then every point of L would lie in the radical of S. If exactly one of the X_u's, say X_v, were non-empty, then v would be an element of L - {b} lying in the radical of S. Thus we may assume that at least two of the X_u's, say X_u and X_v, are non-empty. Select $x \in X_u$, $y \in X_v$ and let M be a line passing through x and y. Then $x \in X_u \cap M$ implies $M \cap L = \{u\}$ and $M-\{u\} \subseteq X_u$. Since y is not adjacent to any member of X_u, by the previous paragraph, the basic axiom for prepolar spaces implies y is adjacent to u. Again, this contradicts $y \in X_v$, proving proposition 6.1. □

Notice that the assumption of proposition 6.1 differs only very slightly from the state of affairs forced by the axioms of a prepolar space, for always, if L is a line, the sets A(u) as u ranges over L, cover S. A glance at the points in the above proof at which Rad(S) is deduced to be non-empty shows that without loss of generality the conclusion of proposition 6.1

could have been sharpened to read "Rad(S) $\cap$ (L-{b}) is non-empty".

PROPOSITION 6.2. *Let* S *be a prepolar space and let* a,b *be two non-adjacent points of* S. *Then* $A^*(a) = A(a) \cup \{a\}$ *and* $A(a,b) = A(a) \cap A(b)$ *(equipped with those lines of* S *lying entirely inside these sets) are also prepolar spaces. Also*

$$\mathrm{Rad}(A(a,b)) \subseteq \mathrm{Rad}(A^*(a)).$$

PROOF. The first statement is easily verified. If z is a point in Rad(A(a,b)), and y is any point of A(a) distinct from z, either y lies in A(a,b), z lies on some line through a and y, or b is not adjacent to y and some line M through a and b does not contain z. In the first two cases z is adjacent to y, patently. In the third case, there is a unique point m in $M \cap A(b)$ and since z is adjacent to both m and a (both z and m differ from a since a does not lie in A(a,b)), z is adjacent to all members of M, in particular, to y. Thus in all cases z is adjacent to y and so, from the general choice of y, we have $z \in \mathrm{Rad}(A^*(a))$. □

In general if a is not adjacent to b we write A(a,b) for $A(a) \cap A(b)$.

PROPOSITION 6.3. *If* S *is a prepolar space and* a *and* b *are non-adjacent points of* S, *then*

$$\mathrm{Rad}(A(a,b)) \subseteq \mathrm{Rad}(S) .$$

PROOF. Assume $z \in \mathrm{Rad}(A(a,b))$. By proposition 6.2, z is adjacent to every member of A(a)-{z}. It remains therefore only to show that z is adjacent to any point w which is not adjacent to a. Since $z \in A(a)$, there is at least one line L through z and a and, moreover, w is adjacent to a unique point u on L. If u = z, z is adjacent to w. Suppose u ≠ z. Then b is not adjacent to u (since it is adjacent to z) and so $z \in A(u,b) \subseteq A^*(b)$ and $z \in \mathrm{Rad}(A(a,b)) \subseteq \mathrm{Rad}(A^*(b))$ (by (6.2)) imply

$$z \in \mathrm{Rad}(A(u,b)).$$

But then by proposition 6.2 once more,

$$z \in \mathrm{Rad}(A^*(u)).$$

Since w is an element of A(u) - {z}, this makes z adjacent to w also. Thus $z \in \mathrm{Rad}(S)$. □

PROPOSITION 6.4. *A non-degenerate prepolar space is linear.*

PROOF. Let S be a non-degenerate prepolar space and suppose by way of contradiction that S is not linear. Then there exist two distinct lines L_1 and L_2 such that $L_1 \cap L_2$ contains two points a and b. Since $L_1 \neq L_2$, at least one of the L_i's properly contains $L_1 \cap L_2$ and so if S were the union of the A(u)'s, u ranging over $L_1 \cap L_2$, then by proposition 6.1 Rad(S) would be non-empty, contrary to the assumption that S is non-degenerate. Thus we may assume that there exists a point c not adjacent to any member of $L_1 \cap L_2$. Then c is adjacent to a unique point x_i on $L_i - (L_1 \cap L_2)$, i=1,2. Since $L_1 \cap L_2$ contains two points adjacent to both x_1 and x_2, x_1 and x_2 are adjacent and so lie on some line M. Since c is adjacent to two members of M, c is adjacent to every member of M; thus $M \subseteq A(a,c) \cap A(b,c)$. We now have the configuration:

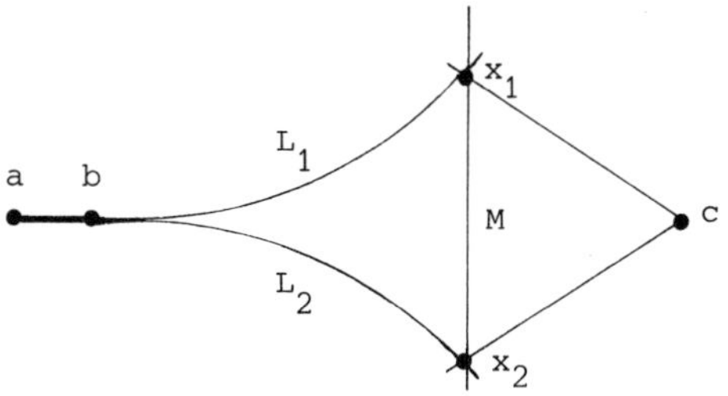

Choose u in A(a,c). We claim that either (i) u is adjacent to some member of $M - \{x_1\}$, or else (ii) $u = x_2$ and $M = \{x_1,x_2\}$.

If u is not adjacent to x_1 the claim is true.

If $u = x_2$, u is adjacent to some member of $M - \{x_1,x_2\}$ or else $M - \{x_1,x_2\}$ is empty. In either case the claim above is true.

Otherwise we may assume u to be distinct from x_1 and adjacent to both a and x_1. Then u is adjacent to all members of L_1, hence to at least two members of $L_1 \cap L_2$ and hence adjacent to all members of L_2, including x_2 on $M - \{x_1\}$. Thus in all cases the claim is justified.

If $M = \{x_1,x_2\}$ then by the claim x_2 is adjacent to every member of $A(a,c) - \{x_2\}$ so Rad(A(a,c)) contains x_1. Otherwise M contains three points, and the claim asserts that M is a line in the prepolar space A(a,c) and every point of A(a,c) is adjacent to some member of $M - \{x_1\}$. This is precisely the hypothesis of proposition 6.1 and so we may conclude again, that

Rad(A(a,c)) is non-empty. Thus in either case Rad(A(a,c)) is non-empty and so by proposition 6.3, Rad(S) is non-empty, contrary to the assumption that S is non-degenerate. □

REMARK. If one wished to investigate the relation of non-linearity of S with Rad(S) in more detail, one may use the observation following the proof of proposition 6.1. One can then see that M, in the above proof lies entirely inside Rad(S).

7. HOW THEOREM C WORKS

Throughout this section let S be a non-degenerate prepolar space.

PROPOSITION 7.1. *Choose a point* $a \in S$ *and assume* b *and* c *are two points in* $S - A^*(a)$. *Then* A(a,b) *and* A(a,c) *are isomorphic prepolar spaces.*

PROOF. For each $x \in A(a,b)$ let M_x be the unique (by proposition 6.4) line through x and a. Since c is not adjacent to a, $A(c) \cap M_x = \{x'\}$. We claim the mapping $x \to x'$ induces an isomorphism as prepolar spaces $A(a,b) \to A(a,c)$. By proposition 6.4, the "lines" of a non-degenerate prepolar space are the unique graph-theoretic cliques containing an edge and having the "one-or-all" adjacency property with points outside the cliques. Thus, since S is non-degenerate implies both A(a,b) and A(a,c) non-degenerate by proposition 6.3, it suffices merely to show that the mapping $x \to x'$ is a bijection preserving the relation of collinearity. It is easily seen that if the roles of b and c are reversed in the definition of x', the inverse mapping $x' \to x$ is produced. So bijectivity is obvious. Suppose x is adjacent to y in A(a,b). Then every point on M_x is adjacent to every on $M_y - \{a\}$ and so x' is adjacent to y', completing the proof. □

PROPOSITION 7.2. *Assume the lines of* S *contain at least three points. If* a,b,c,d *are points of* S *such that* (a,b) *and* (c,d) *are non-adjacent pairs, then* A(a,b) *and* A(c,d) *are isomorphic prepolar spaces.*

If a is not adjacent to c, we have from proposition 7.1,

$$A(a,b) \simeq A(a,c) = A(c,a) \simeq A(c,d) .$$

Otherwise, there is a line L through a and c containing at least three points and so, by proposition 6.1, there is a point f not in $A(a) \cup A(c)$. Then

$$A(a,b) \simeq A(a,f) \simeq A(c,f) \simeq A(c,d) \ ,$$

each isomorphism arising from an application of proposition 7.1.

One can see at this point that proposition 7.2 imposes a far-reaching uniformity across S. It is then not difficult to show

PROPOSITION 7.3. *If the lines of* S *contain at least three points, and* S *has finite rank* $n \geq 2$, *then*

(i) $A(a,b)$ *is a non-degenerate prepolar space of rank* $n-1$, *and*

(ii) *every maximal unrefinable tower of subspaces of* S *(beginning with the empty set) has the same length,* $n+1$.

One then proves

PROPOSITION 7.4. *If the lines of* S *contain at least three points and* S *has finite rank, then* S *contains two maximal subspaces which are disjoint.*

Actually something stronger than this can be proved without using proposition 7.3 and the assumption that all lines contain at least three points (see Proposition 8 of [8]).

Because of proposition 7.4, in proving that a non-degenerate prepolar space whose lines contain at least three elements is a polar space, it remains only to show that a maximal subspace M of S, together with the lines contained in it, form a projective space. We are assuming here that S has rank at least three. Then M contains a line as a proper subspace (proposition 7.3).

If rank S = 3, each line in M is a maximal subspace of M and in that case it is not too difficult to show that any two lines of M meet non-trivially. Since S is linear, this means any two lines in M meet at a unique point, so M with its lines is a (not necessarily Desarguesian) projective space, and so S is a polar space.

If rank S > 3, it suffices to show that Pasch's axiom holds for the lines in M. But if $L_1 \cap L_2 = \{p\}$ for two lines in M, it can be shown that a non-adjacent pair of points b and c exist such that L_1 and L_2 are two lines lying in a subspace of the prepolar space $A(b,c)$. Since $A(b,c)$ is a non-degenerate prepolar space of rank ≥ 3 (by propositions 6.4 and 7.3), induction on the rank shows that $M \cap A(b,c)$ with its lines is a projective space.

Since all lines of M connecting points in $L_1 \cup L_2$ lie in $M \cap A(b,c)$, Pasch's axiom holds for L_1 and L_2. From the general choice of L_1 and L_2, it follows that M is a projective space.

8. THE CASE OF THE GENERALIZED QUADRANGLES

The notion of a generalized quadrangle was first introduced by TITS [30], and precisely corresponds with the notion of a non-degenerate prepolar space of rank 2. Thus a generalized quadrangle is a non-empty collection L of subsets of a set P such that $p \in P$, $L \in L$ and $p \notin L$ imply the existence of a unique point $f(p,L)$ on L such that $\{p, f(p,L)\}$ is covered by a member of L, and any two members of L meet in at most one point, at least two of them being disjoint. We call the members of L lines, just as we have for all prepolar spaces. Note that f is a mapping from the set of non-incident pairs of $P \times L$ into P and observe that f is a surjection.

THEOREM 8.1. (BENSON [3]). *Let* (P,L) *be a generalized quadrangle with* P *finite. Assume*

(i) *each point of* P *lies on at least three lines, and*

(ii) *each line of* L *contains at least three points.*

Then there exist integers s *and* t *such that every point lies on* 1+t *lines and each line contains* 1+s *points.*

If (ii) fails, each line has 2 points, so (i) alone implies each line contains the same number of points. Similarly (ii) alone implies each point lies on the same number of lines. That (i) or (ii) may fail independently is shown by the examples in figure 8.1. The reader may verify for himself that if (i) or (ii) fails the quadrangle is either a "grid" as in figure 8.1(a) or the line-graph of a "grid", as in figure 8.1(b). If both (i) and (ii) fail, (P,L) is an ordinary quadrangle. We may therefore

(a) (b)

Figure 8.1. Generalized quadrangles in which hypothesis (i) or (ii) fails

assume for the remainder of this section that both (i) and (ii) hold, and (since this is the case of most relevance to finite groups and the casting of theorem A) that P is a finite set. Then the integers s and t of BENSON's theorem are defined. In that case, we shall say (P,L) is a generalized quadrangle of *order* (s,t).

It is easy to see that if (P,L) is a generalized quadrangle of order (s,t) and if we interchange the nomenclature of "point" and "line", we obtain a generalized quadrangle (P',L') (with $P' = L, L' = P$ and incidence matrix the transpose of that for (P,L)) of order (t,s) which we call the *dual* of (P,L).

Before going further, I would like to recommend to the reader the beautiful and timely surveys on generalized quadrangles by J.A. THAS [28] and by STANLEY PAYNE [16]. (As far as I know, PAYNE's notes are of more recent vintage and exist so far only in the form of mimeographed notes. It would be nice to see them published soon.)

The generalized quadrangles found among the polar spaces $S(\pi)$ and $S(Q)$ are as follows:

I. $S(\pi)$ when π is the polarity on projective 3-dimensional space obtained from a non-degenerate alternating form. (This quadrangle is associated with the groups $Sp(4,q)$ and has order (q,q).)

II. $S(Q)$ where Q is a non-degenerate quadratic form and P is projective 4-dimensional space. (This quadrangle is associated with the groups $O(5,q)$ and has order (q,q).)

III. $S(Q)$ where Q is a non-degenerate quadratic form of Witt index 2 and P is 5-dimensional projective space. (This quadrangle is associated with the groups $O^+(6,q)$ and has order (q,q^2).)

IV. $S(\pi)$ where π is the polarity on projective 3-dimensional space obtained from a σ-hermitian form over $GF(q^2)$ with σ the field automorphism of order 2. (This quadrangle is associated with the groups $U(4,q^2)$ and has order (q^2,q).)

V. $S(\pi)$ where π is the polarity on projective 4-dimensional space obtained from a σ-hermitian form over $GF(q^2)$ with σ the field automorphism of order 2. (This quadrangle is associated with the groups $U(5,q^2)$ and has order (q^2,q^3).)

Still other generalized quadrangles exist. TITS has constructed quadrangles of orders (q,q) and (q,q^2) which are generalizations of quadrangles of types II and III above. There are also generalized quadrangles of order (q-1,q+1) when q is a primepower. A special subclass of these, when q is even, was first given by AHRENS & SZEKERES [1] and independently by HALL [13]. It is a theorem of BENSON [4] that types I and II are dual to each other. In addition, types III and IV are also dual to each other. When Q is a non-degenerate quadratic form of Witt index 2 on a 4-dimensional vector space V, the associated totally singular point set S(Q) in $P \simeq PG(3,q)$ is a generalized quadrangle of type (q,1), that is (i) of theorem 8.1 does not hold here.

Now in using theorem A in the context of finite groups, one may frequently assume that the generalized quadrangles which may appear possess a fairly rich group of automorphisms. One then desires a theorem which may be used to characterize the generalized quadrangles. There are a number of theorems allowing one to characterize the quadrangles I through V listed above. Here is a sample:

THEOREM 8.2. (SINGLETON [20], BENSON [4]). *Suppose* (P,*L*) *is a finite generalized quadrangle such that for every non-collinear pair* x,y *of* P, *any point collinear with two of the set of points collinear with both* x *and* y *is in fact collinear with all* 1+t *points which are collinear with both* x *and* y. *Then* (P,*L*) *is a generalized quadrangle of type* I.

Dualization of this result yields a characterization of quadrangles of type II. Similarly TALLINI [25] and THAS [28a] have given a characterization of type III which dualizes to a characterization of type IV. There presently seems to be no known characterization of the quadrangles of type V on the basis of intrinsic local geometric properties.

However there is a recent marvelous theorem of BUEKENHOUT & LEFEVRE [7] that may prove useful in obtaining characterizations.

THEOREM 8.3. (BUEKENHOUT & LEFEVRE). *If the quadrangle* (P,*L*) *is finite and is embeddable in a projective space, then* $(P,L) \simeq S(\pi)$ *or* S(Q) *as polar spaces –i.e.* (P,*L*) *is one of the types* I-V *or has order* (q,1).

Finally, a recent theorem of TITS (in BUEKENHOUT [5, p.30]) shows that if a generalized quadrangle is exceedingly rich in automorphisms, then it is a quadrangle of "classical" type:

THEOREM 8.4. (TITS). *Let* (P,*L*) *be a finite generalized quadrangle, and assume the following two properties:*

(i) *If* A *and* B *are two lines through a point* p *and if* b *is any point on* B *distinct from* p, *then the subgroup of* Aut(P,L) *fixing* A *and* B *pointwise and stabilizing all lines through* p, *transitively permutes the lines passing through* $b \in B$ *which are distinct from* B.

(ii) *If* a *and* b *are two points lying on a line* L *and* M *is a line through* b *distinct from* L, *then the subgroup of* Aut(P,L) *fixing* L *pointwise and simultaneously stabilizing all lines through* a *and* b *is transitive on the points of* M - {b}.

Then either (P,L) *or its dual is an* S(π) *or an* S(Q).

Note that conditions (i) and (ii) of this theorem are dual to one another.

THEOREM 8.5. *Let* (P,L) *be a generalized quadrangle of order* (s,t). *Then*

(i) $|P| = (1+s)(1+st)$;

(ii) $|L| = (1+t)(1+st)$;

(iii) *each point is collinear with* s+st *points, and fails to be collinear with* s^2t *points.*

A point x and all the points collinear with it is denoted St(x) and is called the *star at* x (in the prepolar space terminology of section 6, this is just $A^*(x)$). We denote by C(St(x)), the subgroup of those automorphisms of (P,L) which leave St(x) fixed pointwise. We next observe

PROPOSITION 8.1. C(St(x)) *acts semiregularly on the set of points* P - St(x).

The following seems approachable,

CONJECTURE 8.1. *Assume* (P,L) *is a finite generalized quadrangle of order* (s,t), *and, for each* $x \in P$, C(St(x)) *has even order. Then* (P,L) *is type* I, IV *or* V, *with* q *a power of* 2.

Assume the hypothesis of the above conjecture, set G = Aut(P,L), let G_x be the subgroup of G fixing the point x, and write C(x) for C(St(x)). Then the following is easily proved:

(i) G is transitive on unordered pairs of collinear points. In particular G_x is transitive on St(x) - {x} and the stabilizer in G of a line is doubly transitive on the points in the line.

(ii) $G_x = N_G(C(x))$. C(x) is a 2-group, and is a trivial intersection group, that is it meets its G-conjugates which are distinct from itself trivially.

(iii) t is a power of 2, G_x induces a doubly transitive group (H,L_x) on the set of lines passing through x and (H,L_x) contains a normal subgroup (H_0,L_x) which is SL(2,q), Sz(q) or $U(3,q^2)$ acting in its natural representation on 1+t = 1+q, $1+q^2$, or $1+q^3$ letters, respectively.

(iv) If x and y are non-collinear points, there is a set C(x,y) of 1+w mutually non-collinear points containing x and y with the property that $A(u) \cap A(v) = A(x) \cap A(y)$ for all $u,v \in C(x,y)$. (Here, A(x) is the set of points collinear with x.) The system $L^* = \{C(u,v) \mid (u,v)$ a non-collinear pair in P$\}$ becomes in this way, a second system of lines for P, so that every pair of points in P either lies in a unique line from L or a unique line from L^* but not both. The two line sets are connected by the following interesting property: if $L \in L^*$ and $u \in P$ and u is L-collinear with two points of L, then u is L-collinear with every point of L.

(v) If x and y are not collinear, the subgroup <C(x),C(y)> stabilizes C(x,y) and induces on C(x,y) one of the permutation groups SL(2,q), Sz(q) or $U(3,q^2)$ acting doubly transitively on the 1+w points, with w = q, q^2 or q^3, respectively. If $w \neq t$ then SL(2,q) is obtained, $t = q^3$ and the subgroup H_0 of (iii) is $U(3,q^3)$.

It may be that the more general assumption that C(St(x)) contains an element of prime order p for each x in P, would also imply that (P,L) is type I, IV or V. In any event if the above conjecture could at least be proved, it would be useful in many instances of determining the 2-Sylow order of $G = \text{Aut}(P,L)$. For example, if we assume that $H = G_x$ and K = Stab(L), where L is a line containing x, then $H \cap K$ is the stabilizer of the flag (x,L) and G is an amalgam of H and K. A very pleasant pastime is this: One selects two transitive permutation groups whose one-point stabilizers are either isomorphic, or for which there is a subdirect product of each, which is not direct. These are the candidates for (H,L_x) and (K,L). If H is doubly primitive on the lines through x, then $K_0 = \ker[K \to \text{Sym}(L)]$ must coincide with C(St(x)). Similarly, if K is doubly primitive on the points in L, $H_0 = \ker[H \to \text{Sym}(L_x)]$ (where L_x is the set of lines through x) must also coincide with C(St(x)). The point is that if (s,t) is chosen so as not to coincide with cases I through V, the conjecture says C(St(x)) must have odd order and this means a 2-Sylow order of G is determined. Choices of H, $H \cap K$ and K giving low values of s and t are easily constructed; samples are:

	H	H∩K	K	(s,t)	|P|	|G|
1a.	E(16)Alt(5)	Alt(5)	Alt(6)	(5,15)	456	$2^9 3^2 5\cdot 19\cdot \|C\|$
1b.	E(16)Sym(5)	Sym(5)	Sym(6)	(5,15)	456	2 times case 1a
1c.	$E(16)D_{10}$	D_{10}	PSL(2,5)	(5,15)	456	$2^8 3\cdot 5\cdot 19\cdot \|C\|$
2a.	Alt(6)	D_{10}	PSL(2,5)	(5,35)	1056	$2^8 3^2 5\cdot 11\cdot \|C\|$
2b.	Sym(6)	$Z_5 Z_4$	PGL(2,5)	(5,35)	1056	2 times case 2a
3.	E(16)Sym(6)	Sym(6)	Sym(7)	(6,15)	567	$2^8 3^2 5\cdot 7^2 13\cdot \|C\|$
4.	Alt(8)	E(8)L(3,2)	$E(8^2)L(3,2)$	(7,14)	1485	$2^9 3^4 5\cdot 7\cdot 11\cdot \|C\|$
5.	$U(3,3^2)$	L(3,2)	E(8)L(3,2)	(7,35)	1944	$2^9 3^4 5\cdot 7\cdot 41\cdot \|C\|$
6.	E(16)Alt(6)	Alt(6)	Sym(7)	(13,15)	2744	$2^{10} 3^2 5\cdot 7^3\cdot \|C\|$
7.	Alt(8)	Sym(6)	E(16)Sym(6)	(15,27)	6496	$2^{11} 3^2 5\cdot 7^2 11\cdot \|C\|$

... and so on. If the conjecture is true, $|C| = |C(St(x))|$ is odd and the 2-Sylow order of G is known. In addition, part of the 2-fusion pattern is already prescribed in the two subgroups H and K, so one is presumably on his way to determining G.

9. VARIATIONS ON A THEME; OPEN QUESTIONS

A few open questions remain concerning prepolar spaces.

(1) *What is the exact relation between lines which meet at two or more points* (we call these "neighbor lines") *and the radical of* (P,L)*?*

One may consider the equivalence relation R defined on P by xRy if and only if $A^*(x) = A^*(y)$. Then one can show that $A^*(x) \subseteq A^*(y)$ implies either xRy or $y \in Rad(P)$. From this it follows that on the set $\bar{P}$ of equivalence classes under R, the structure of a prepolar space $\bar{L}$ can unambiguously be defined from L. The theorem is that $(\bar{P},\bar{L})$ is non-degenerate [8, Proposition 14]. Then from proposition 6.4, $(\bar{P},\bar{L})$ is linear. This means that if two lines L_1 and L_2 meet at two points, then at least one of the two points lies in Rad(P). From this one sees that lines meeting at three points must lie in Rad(P). Do lines meeting at exactly two points necessarily lie in Rad(P) also?

(2) *Can the structure of non-degenerate prepolar spaces containing lines of cardinality* 2 *be described?*

It is clear that if (P,L) is such a prepolar space then it contains many lines of cardinality 2. To see this, let $L = \{a,b\}$ be a line containing just two points. Then we have a decomposition $P = L+\Delta+X_a+X_b$ into disjoint sets, where $\Delta = A(a)\cap A(b)$, $X_a = A(a)-\Delta$, $X_b = A(b)-\Delta$. From non-degeneracy, both X_a and X_b are non-empty. Then the lines through point a either lie entirely inside $X_a \cup \{a\}$, or $\Delta \cup \{a\}$. Let M be a line through a lying in $X_a \cup \{a\}$. Then any point p in X_b is collinear with some point p' in M. Then if N is a line containing p and p' we see that $N\cap X_a = \{p'\}$, $N\cap X_b = \{p\}$ and $N \cap (\Delta\cup\{a,b\})$ is empty. Thus $N = \{p,p'\}$. There are thus $|X_b|$ such lines meeting M.

(3) *All of the graphs of theorem* A *corresponding to prepolar spaces of rank at least three involve structures associated with the groups* $Sp(2n,q)$, $O(2n+1,q)$, $O^{\varepsilon}(2n,q)$, $\varepsilon=\pm 1$, *and* $U(n,q^2)$. *These are the non-abelian simple sections of the finite Chevalley groups of types* B_n, C_n, D_n *and the twisted types* ${}^2D_n$ *and* ${}^2A_n$. *Does there exist a similar simple (and local) graph-theoretic hypothesis which could be used to characterize structures associated with other Chevalley and Steinberg groups?*

Those which come to mind are the groups of types A_n (the $PSL(n,q)$'s), ${}^3D_4$, G_2, F_4, E_6, E_7, E_8. Presumably ${}^2F_4$, ${}^2G_2$, ${}^2B_2$ are of such low rank as to be below the level of such a theorem. What does one use as a replacement for $S(\pi)$ or $S(Q)$ in these cases? One suspects that in the case of $PSL(n,q)$ one would use the line-graph of $PG(n-1,q)$, suggesting a hypothesis of the form:

<u>HYPOTHESIS H</u>. *If* (x,y) *is an edge in the graph G, there exists a clique* $C(x,y)$ *containing* $1+q+q^2$ *vertices such that every vertex* $z \in G-C(x,y)$ *is adjacent to* 0 *or* $1+q$ *members of* $C(x,y)$. *The sets* $A(z) \cap C(x,y)$ *which are non-empty as* z *ranges over* $G-C(x,y)$ *define a projective plane on* $C(x,y)$.

Possibly one can weaken hypothesis H. If $q = 1$, the socalled "triangular graphs" also have this property. But these correspond to the symmetric groups which may be thought of as what would be groups of type A_n if there were such a thing as a field containing one element.

Some work has already begun [6] on prepolar-like spaces that might be characteristic for the Chevalley groups of type E_6.

This point of view suggests still another question.

(4) *Is there an analogue of theorem* A *that could be used to characterize the graph of non-singular points in a projective space with a non-degenerate unitary polarity or the graph of non-singular points with square (or non-square) values under a non-degenerate quadratic form* Q?

That such a theorem may be possible is indicated by the following result:

THEOREM 9.1. *Let* G *be a finite graph satisfying the following hypothesis:*

(9.1) *(Cotriangle property) Given any non-adjacent pair of vertices* x *and* y, *there exists at least one third vertex* z *not adjacent to either* x *or* y, *such that any vertex in* $G-\{x,y,z\}$ *is adjacent to one or three members of* $\{x,y,z\}$.

Then the graphs G *are determined up to isomorphism.*

The graphs include the graphs $N(2n,2)$ of non-singular vectors with respect to a non-degenerate quadratic form in 2n-variables over GF(2).

Suppose G is a graph with the cotriangle property (9.1) and $G = X_1 + X_2 + \ldots + X_m$ is a partition of the vertices of G with each X_i non-empty and if $i \neq j$, every vertex of X_i being adjacent to every vertex of X_j. Then as subgraphs, each X_i has the cotriangle property. Clearly G is determined up to isomorphism by the isomorphism types of the X_i. We say G is *indecomposable* if no non-trivial partition of this type exists.

THEOREM 9.2. *If* G *has property* (9.1), *the following are equivalent:*

(i) $A(x) \cup \{x\} = A(y) \cup \{y\}$ *implies* $x = y$, *for all* $x, y \in G$.

(ii) *The vertex* z *of* (9.1) *is uniquely determined by* x *and* y.

The relation $A(x) \cup \{x\} = A(y) \cup \{y\}$ is clearly an equivalence relation on the vertices of G and if $\bar{G} = \{C_j\}$ is the family of equivalence classes of G with respect to this relation, and $i \neq j$, then either

(a) every vertex of C_i is adjacent to every vertex of C_j, or

(b) no vertex of C_i is adjacent to any vertex of C_j.

We can then make a graph on $\bar{G}$ by the relation (a). Then if G is indecomposable, so is $\bar{G}$. If G has the cotriangle property, so does $\bar{G}$, except now property (b) also holds. Finally G is determined up to isomorphism by the isomorphism type of $\bar{G}$ and the assignment of cardinalities $|C_i|$ to the vertices i of G.

Thus in proving theorem 9.1 we need only show

THEOREM 9.3. *If* G *is a finite indecomposable graph with the cotriangle property* (9.1) *for which* (b) *of theorem* 9.2 *holds, then either*

(i) $G \simeq N(2n,2)$,

(ii) $G \simeq Sp(2n,2)$,

(iii) $G \simeq T^*(n)$,

where $N(2n,2)$ *is the graph of non-singular vectors with respect to a non-degenerate quadratic form in* 2n *variables over* GF(2), $Sp(2n,2)$ *is the graph of non-zero vectors under the perpendicular relation of a non-degenerate symplectic form over* GF(2), *and* $T^*(n)$ *is the complement of the triangular graph on* n *letters.*

The proof is merely a modification of a theorem of J.J. SEIDEL [17]. Because of property (b) we may write $z = x+y$ unambiguously in the statement of the cotriangle property. (In SEIDEL's situation, the graph is already embedded in a symplectic space with "+" being ordinary vector addition, hence associative in his proof; also case (ii) is not possible in his situation.) SEIDEL's inspiration was in noticing that if X_a is the subgraph of vertices adjacent only to a in a cotriangle $\{a,b,c = a+b\}$, then X_a has the *triangle property* (that is, the hypothesis of theorem A with the $C(x,y)$'s all having just 3 points). The structure of $\Delta = A(a) \cap A(b) \cap A(c)$ is a little more difficult to see, but is determined up to isomorphism by X_a.

10. SOME GROUP-THEORETIC BACKGROUND AND SOME APPLICATIONS

Historically speaking theorem A is the coming together of two independent lines of development: on the one side, the development of the theory of polar spaces beginning with VELDKAMP's important work [32]; on the other side, the need to characterize graphs which arose in certain group-theoretic problems, eventually reaching the prepolar spaces. The comments in this section are confined to the group-theoretic side of the picture.

In one sense the dim beginnings are noticeable in the graph extension theorem. A well-known problem that arises in the theory of permutation groups is that of constructing a transitive extension. By saying that (G,X) is a permutation group, we mean that we have an injection $f\colon G \to \mathrm{Sym}(X)$. If x and y are two elements of X, the ability to reach y from x *via* a permutation in $f(G)$ is an equivalence relation on X, the equivalence classes being called G-*orbits*. We say G is *transitive* on X if G acts in one orbit on X.

Always, if Y is a subset of X, the permutations leaving Y fixed point-wise form a subgroup G_Y of G and by custom we write G_y for G_Y when Y = {y}. If G is transitive on X, all subgroups G_x, $x \in X$, form one conjugacy class of subgroups of G. We say G is *rank* n on X if G is transitive on X and G_x acts in n distinct orbits on X; a rank 2 group is called *doubly transitive* and is transitive on the set of ordered 2-sets from X. If (G,X) is doubly transitive, then G is called a *transitive extension* of $(G_x, X - \{x\})$. The problem is to reverse this process; find for which transitive groups $(G_x, X - \{x\})$, the doubly transitive group (G,X) exists.

As an example, there are 35 ways to partition 8 letters into two 4-sets and Sym(8) is a rank 3 permutation group (H,X) on the set X of partitions. The subgroup H_x fixing a partition x is $\mathrm{Sym}(4)\backslash Z_2$ acting with orbits of lengths 1, 18 and 16, so $H = H_x + H_x t_1 H_x + H_x t_2 H_x$. We can adjoin a new letter a to X and define a permutation z on $\{a\} \cup X$, transposing a and x, such that zt_1z and zt_2z lie in the set G = H+HzH. Then the set G is closed under composition of permutations and so G is a doubly transitive group on 36 letters and is a transitive extension of H. As it turns out H is also the full automorphism group of a graph which can be defined on X. The graph has valence 18 and is defined by the translates of the orbit of length 18 for H_x. Moreover, z can be chosen to centralize the subgroup H_x and induces automorphisms of the subgraphs defined by the H_x-orbits and A and B of lengths 18 and 16 respectively. However if $(x,y) \in AB$, then (x,y) is an edge if and only if (x^z,y^z) is not an edge.

This was the prototype of the so-called "graph extension theorem" [22] which gave a sufficient condition (involving graph-theoretic concepts) that a group (H,X) have a transitive extension $(G, \{a\} \cup X)$, namely

<u>HYPOTHESIS 10.1</u>. X *is a graph and* H = Aut(X) *is transitive on the vertices* X. *There exist automorphisms* h_1 *and* h_2 *of the subgraphs* $\Gamma = A(b)$ *and* $\Sigma = X - (\{b\} \cup A(b))$ *such that* $h_1 \times h_2$ *interchanges the set of adjacent pairs with the set of non-adjacent pairs in* $\Gamma \times \Sigma$.

Transitive extensions which arise in this way include PSL(2,q), $q \equiv 1 \pmod 4$, $U(3,q^2)$, q odd, groups of Ree type, the symmetric groups, the two 2-transitive representations of Sp(2n,2) and two doubly transitive sporadic groups, HS and (.3). Any doubly transitive group whose one-point-stabilizer contains a strongly closed subgroup of index two arises by virtue of hypothesis 10.1 [12,26]. The problem of classifying the doubly transitive groups which are transitive extensions coming from hypothesis 10.1 is still unsolved.

However, if one of the two automorphisms h_1 or h_2 of hypothesis 10.1 can be taken to be the identity (say, by composing them with the restriction of an automorphism of X fixing b) then the graph X is either a pentagon, or else has the "triangle property" -that is, the hypothesis of theorem A with C(x,y) assumed always to contain exactly three points. Regular graphs with this property were characterized in [23] (in fact SEIDEL proves a version of this result with the regularity of X relaxed [17]) before theorem A was ever proved. Because of this result the doubly transitive groups which arise from hypothesis 10.1 in this case must be PSL(2,5) on 6 letters, Sp(2n,2) on $2^{n-1}(2^n \pm 1)$ letters or the semidirect product V(2n,2)Sp(2n,2) on 2^{2n} letters. We thus have (logically, if not actually historically) an application of theorem A to doubly transitive groups.

A more general way of looking at the above construction is obtained by considering a new combinatorial object, the 2-graph, first introduced by GRAHAM HIGMAN in order to study the Conway group (.3) as a doubly transitive group on 276 letters. A *2-graph* (Ω,Δ) is a set of letters Ω and a family Δ of 3-sets from Ω such that any 4-set of Ω contains an even number of 3-sets belonging to Δ (the cases in which Δ contains all 3-sets or is empty, are regarded as *trivial* 2-graphs). A 2-graph is called *regular* if every pair of letters lies in the same number of sets in Δ. The transitive extensions obtained from the graph extension theorem are in 1-1-correspondence with the class of doubly transitive 2-graphs. Let X be the graph in hypothesis 10.1 and let a be the "new" point, and regard $\{a\} \cup X$ as a graph with $\{a\}$ as an isolated vertex and X as a subgraph. Then the 2-graph in question has $\{a\} \cup X$ as its set of letters, and all 3-sets of $\{a\} \cup X$ containing an odd number of edges as the family of 3-sets Δ. Because of hypothesis 10.1, the transitive extension constructed is in the automorphism group of the 2-graph $(\{a\} \cup X, \Delta)$. Because $G_b = H = \mathrm{Aut}(X)$, it is the *full* automorphism group. Those graphs G, for which the family Δ of 3-sets of vertices of G containing an odd number of edges defines a regular 2-graph (G,Δ) are characterized among all graphs by the fact that their (-1,0,1) adjacency matrices possess only two distinct eigenroots [17, Theorem 2.5]. If G is such a graph, and Y is a subgraph of G, we may switch with respect to Y; that is, obtain a new graph G' with Y and G-Y as subgraphs, by erasing all edges in $Y \times (G-Y)$ and declaring all non-adjacent pairs of $Y \times (G-Y)$ in G to be edges of G'. Then the 3-sets of G' possessing an odd number of edges is *still* Δ, so the same 2-graph is defined by *G'*. We may thus associate a 2-graph with a switching class of graphs, and SEIDEL [19] has made this association precise by showing that any two graphs which lead to the same 2-graph are actually switching-equivalent. It is interesting to note that SEIDEL first introduces the switching concept into graph theory in [18], well before its relevance to 2-graphs became known. The mathematical muse seems mysteriously to bring things forth at the right time!

I should mention that 2-graphs are interesting combinatorial objects to study in their own right. SEIDEL & GOETHALS [19] have shown that there is a unique regular 2-graph on 276 letters and in doing so have produced an elementary construction of this 2-graph from first principles. This gives us an elementary construction of Conway's group (.3) without the use of the Leech lattice, or even the existence of the larger Mathieu groups. Although there is an excellent development of 2-graphs in TAYLOR's thesis [27], I would recommend the reader to the more accessible and current survey of 2-graphs by SEIDEL [20].

Another application of theorem A stems originally from an earlier version [24] of the "triangle property" theorem [23] in which the vertices of the graph were actually involutions in a group, and edges were commuting pairs of involutions. As a corollary of that theorem (and GLAUBERMAN's Z^*-theorem [11]) one obtains the following non-simplicity criterion for a finite group.

THEOREM 10.1. *Let* K *be a non-empty union of classes of involutions in a finite group* G. *Suppose that if* t *and* s *are commuting members of* K, *then*

(i) ts ∈ K, *and*

(ii) *any element of* K *commutes with at least one involution in* ⟨t,s⟩.

Then either no two members of K *commute and* $Z^*(g)$ *is non-trivial, or else* G *has a normal elementary abelian* 2-*group. In either case* G *contains a non-trivial normal solvable subgroup.*

ALPERIN [2] generalized this theorem and gave a direct group-theoretic proof of it.

THEOREM 10.2. (ALPERIN) *Let* A *be a fours group in* G, *and set* $K = (A^{\#})^G$, *and suppose* $C_G(x) \cap A > 1$ *for all* $x \in K$. *Then* $A \cap O_2(G) > 1$.

By using theorem A one can prove an "odd p" version of theorem 10.1, namely:

THEOREM 10.4. (SHULT [24a]). *Let* K *be a union of classes of cyclic groups of prime order* p *in* G. *Suppose that whenever* X *and* Y *are distinct members of* K *which commute with one another, then*

(i) *at least three* Z_p*'s in* <X,Y> *lie in* K, *and*

(ii) $C_G(P) \cap \langle X,Y \rangle \cap K$ *is non-trivial for each* P *in* K.

Suppose, further, that at least two members of K *commute with one another. Then* G *contains a non-trivial normal elementary* p*-group* N, *central in* <K>.

Does there exist a similar "odd p" version of ALPERIN's generalization of hypothesis 10.1? Such a theorem would have a hypothesis referring to a fixed subgroup A of type $Z_p \times Z_p$, with $C_G(X) \cap A > 1$ for each X in K, the set of G-conjugates of Z_p's in A. Indeed, this suggests a variation on the theorems of BRODKEY [9] and LAFFEY [14]: Suppose A is a subgroup of G of type $Z_p \times Z_p$ and suppose A meets all of its G-conjugates non-trivially. Then show $O_p(G)$ is non-trivial. So far these generalizations remain to be proved.

These are meagre beginnings, but it is the author's belief that the applications of theorem A (and similar theorems) to group-theoretic problems has just begun.

REFERENCES

[1] AHRENS, B.W. & G. SZEKERES, *On a combinatorial generalization of 27 lines associated with a cubic surface*, J. Austral. Math. Soc., 10 (1969) 485-492.

[2] ALPERIN, J.L., *On fours groups*, Illinois J. Math., 16 (1972) 349-351.

[3] BENSON, C., *On the structure of generalized quadrangles*, J. Algebra, 15 (1970) 443-454.

[4] BENSON, C., *Generalized quadrangles and (B,N)-pairs*, thesis, Cornell University, 1965.

[5] BUEKENHOUT, F., *Characterizations of semi-quadrics; a survey*, preprint.

[6] BUEKENHOUT, F., Personal communication.

[7] BUEKENHOUT, F. & C. LEFEVRE, *Generalized quadrangles in projective spaces*, to appear.

[8] BUEKENHOUT, F. & E.E. SHULT, *On the foundations of polar geometry*, to appear in Geomtriae Dedicata.

[9] BRODKEY, J.S., *A note on finite groups with an abelian Sylow subgroup*, Proc. Amer. Math. Soc., 14 (1963) 132-133.

[10] DEMBOWSKI, P., *Finite geometries*, Ergebnisse der Mathematik 44, Springer-Verlag, Berlin etc., 1968.

[11] GLAUBERMAN, G., *Central elements in core-free groups*, J. Algebra, 4 (1966) 403-420.

[12] HALL Jr., M. & E.E. SHULT, *Equiangular lines, the graph extension theorem and transfer in triply transitive groups*, to appear in Math. Z.

[13] HALL Jr., M., *Affine generalized quadrilaterals*, in: *Studies in pure mathematics*, L. MIRSKY (ed.), Academic Press, London, 1971, pp. 113-116.

[14] LAFFEY, T.J., *A problem on cyclic subgroups of finite groups*, Proc. Edinburgh Math. Soc., 18 (Series II) (1973) 247-250.

[15] LINT, J.H. VAN & J.J. SEIDEL, *Equilateral point sets in elliptic geometry*, Kon. Nederl. Akad. Wetensch. Proc. A, 69 (= Indag. Math. 28) (1966) 335-348.

[16] PAYNE, S., *Finite generalized quadrangles; a survey*, mimeographed notes.

[17] SEIDEL, J.J., *On 2-graphs and Shult's characterization of symplectic and orthogonal geometries over* GF(2), Tech. University Eindhoven, T.H. Report 73-WSK-02, 1973.

[18] SEIDEL, J.J., *Strongly regular graphs of* L_2*-type and of triangular type*, Kon. Nederl. Akad. Wetensch. Proc. A, 70 (= Indag. Math. 29) (1967) 188-196.

[19] SEIDEL, J.J., Personal communication.

[20] SEIDEL, J.J., *Survey of 2-graphs*, preprint for an address before the Boca Raton Conference, 1974.

[21] SINGLETON, R.R., *Minimal regular graphs of maximal even girth*, J. Combinatorial Theory, 1 (1966) 306-332.

[22] SHULT, E.E., *The graph extension theorem*, Proc. Amer. Math. Soc., 33 (1972) 278-284.

[23] SHULT, E.E., *Characterizations of certain classes of graphs*, J. Combinatorial Theory B, 13 (1972) 1-26.

[24] SHULT, E.E., *A characterization of the groups* Sp(2n,2), J. Algebra, 15 (1970) 543-553.

[24a] SHULT, E.E., *On subgroups of type* $Z_p \times Z_p$, to appear in J. Algebra.

[25] TALLINI, G., *Ruled graphic systems*, in: Atti Convegno Geometria e sue Applicazioni, Perugia, 1971, pp. 403-411.

[26] TAYLOR, D.E., *Monomial representations and strong graphs*, in: Proc. First Autralian Conf. Combinatorial Math., Univ. of Newcastle, 1972, pp. 197-201.

[27] TAYLOR, D.E., *Some topics in the theory of finite groups*, thesis, Oxford Univ., 1971.

[28] THAS, J.A., *4-gonal configurations*, to appear in Geometriae Dedicata.

[28a] THAS, J.A., *4-gonal configurations with parameters* $r = q^2+1$ *and* $k = q+1$. To appear.

[29] TITS, J., *Buildings and* B,N-*pairs of spherical type*, Lecture Notes in Mathematics, Springer-Verlag, Berlin etc., 1974.

[30] TITS, J., *Sur la trialité et certains groups que s'en deduisent*, Publ. Math. I.H.E.S., Paris, 2 (1959) 14-60.

[31] VEBLEN, O. & J.W. YOUNG, *Projective geometry I*, Ginn, Boston, 1916.

[32] VELDKAMP, F.D., *Polar geometry I-V*, Kon. Nederl. Akad. Wet. Proc. A, 62 (1959) 512-551; A 63 (1960) 207-212 (= Indag. Math. 21 resp. 22).